CHAPMAN & HALL/CRC APPLIED MATHEMATICS
AND NONLINEAR SCIENCE SERIES

Mathematics of
Quantum Computation
and Quantum Technology

T0203805

Published Titles

Computing with hp-ADAPTIVE FINITE ELEMENTS, Volume 1, One and Two Dimensional Elliptic and Maxwell Problems, Leszek Demkowicz

Computing with hp-ADAPTIVE FINITE ELEMENTS, Volume 2, Frontiers: Three Dimensional Elliptic and Maxwell Problems with Applications, Leszek Demkowicz, Jason Kurtz, David Pardo, Maciej Paszyński, Waldemar Rachowicz, and Adam Zdunek

CRC Standard Curves and Surfaces with Mathematica®: *Second Edition,* David H. von Seggern

Exact Solutions and Invariant Subspaces of Nonlinear Partial Differential Equations in Mechanics and Physics, Victor A. Galaktionov and Sergey R. Svirshchevskii

Geometric Sturmian Theory of Nonlinear Parabolic Equations and Applications, Victor A. Galaktionov

Introduction to Fuzzy Systems, Guanrong Chen and Trung Tat Pham

Introduction to non-Kerr Law Optical Solitons, Anjan Biswas and Swapan Konar

Introduction to Partial Differential Equations with MATLAB®, Matthew P. Coleman

Introduction to Quantum Control and Dynamics, Domenico D'Alessandro

Mathematical Methods in Physics and Engineering with Mathematica, Ferdinand F. Cap

Mathematics of Quantum Computation and Quantum Technology, Goong Chen, Louis Kauffman, and Samuel J. Lomonaco

Optimal Estimation of Dynamic Systems, John L. Crassidis and John L. Junkins

Quantum Computing Devices: Principles, Designs, and Analysis, Goong Chen, David A. Church, Berthold-Georg Englert, Carsten Henkel, Bernd Rohwedder, Marlan O. Scully, and M. Suhail Zubairy

Stochastic Partial Differential Equations, Pao-Liu Chow

Forthcoming Titles

Mathematical Theory of Quantum Computation, Goong Chen and Zijian Diao

Mixed Boundary Value Problems, Dean G. Duffy

Multi-Resolution Methods for Modeling and Control of Dynamical Systems, John L. Junkins and Puneet Singla

CHAPMAN & HALL/CRC APPLIED MATHEMATICS
AND NONLINEAR SCIENCE SERIES

Mathematics of Quantum Computation and Quantum Technology

Edited by

Goong Chen
Louis Kauffman
Samuel J. Lomonaco

CRC Press
Taylor & Francis Group
Boca Raton London New York

CRC Press is an imprint of the
Taylor & Francis Group, an **informa** business
A CHAPMAN & HALL BOOK

CRC Press
Taylor & Francis Group
6000 Broken Sound Parkway NW, Suite 300
Boca Raton, FL 33487-2742

First issued in paperback 2019

ISBN-13: 978-1-58488-899-4 (hbk)
ISBN-13: 978-0-367-38861-4 (pbk)

Library of Congress Cataloging-in-Publication Data

Mathematics of quantum computation and quantum technology / editors, Goong
    Chen, Louis Kauffman, Samuel J. Lomonaco.
        p. cm. -- (Chapman & Hall/CRC applied mathematics and nonlinear
        science series ; no. 14)
    Includes bibliographical references and index.
    ISBN 978-1-58488-899-4 (hardback : alk. paper)
        1. Quantum computers--Mathematics. 2. Quantum theory--Mathematics. I.
    Chen, Goong, 1950- II. Kauffman, Louis H., 1945- III. Lomonaco, Samuel J. IV.
    Title. V. Series.

QA76.889.M383 2007
004.1--dc22   2007024955

Visit the Taylor & Francis Web site at
http://www.taylorandfrancis.com

and the CRC Press Web site at
http://www.crcpress.com

Preface

Quantum computing is a vast and fascinating interdisciplinary project of the 21st century. Research and development in this monumental enterprise involve just about every field of science and engineering. In this volume, we focus on two important disciplines–mathematics and physics. The choice is made with good reasons:

(1) Mathematics and mathematicians have played major roles in the development of quantum computation. In the middle 1990's Peter Shor's quantum factoring algorithm generated tremendous enthusiasm for the push to build the quantum computer. Shor's algorithm was the first example showing that, in principle, quantum computers could out-perform classical digital computers on problems of significance. This opened wide the field for discovery of new quantum algorithms. This search has gone on in tandem with the search for new principles and techniques that will make the new computers practical and actual. Mathematics participates in this process at all levels.

(2) Physicists and engineers are the principal players in the design and evolution of the quantum computer. The technological aspect of building a scalable quantum computer within a reasonable time horizon is crucial for the field to remain viable. Recent advances in hardware development are extremely encouraging. Such quantum technology is the material cornerstone of quantum computing. The impact of this research goes well beyond quantum computing, reaching nanotechnology, chemical physics, condensed matter physics and the fundamental nature of matter at the quantum mechanical level.

Items (1) and (2), mathematics and physics, are inextricably linked. In this field of research we see mathematicians and physicists not only working together, but building common language and techniques that move back and forth across the disciplines. It has been remarkable to watch, over the last few years, the extraordinary clarity of articulation of basic quantum physical principles that is now available both for physicists and mathematicians. This is far beyond a matter of simple exposition. New points of view on quantum theory are emerging from these studies, and aspects of quantum theory (such as non-locality and teleportation), previously thought to be matters of philosophy, are now understood to be at the very basis of quantum information theory and the practice of quantum computation.

Based on the firm understanding that mathematics and physics are equal partners in the continuing discovery of quantum computing, the three editors of this book organized an NSF conference entitled "Mathematics of Quantum Computation and Quantum Technology", held at Texas A&M University in November 2005. During the 3-day conference, many central topics were reported and examined, and vivid discussions ensued. The funding organizations were NSF, IMA (Institute of Mathematics and Its Applications, Minneapolis, U.S.A.) and the Texas A&M University. We are especially grateful to Dr. Henry Warchall of NSF for providing the largest share of participant support via NSF Grant DMS 0531131.

The present volume contains materials much broader and deeper than what were presented at the conference, due to the generous time frame for the authors to prepare their manuscripts. This is evidenced in the large number of chapters, sixteen of them in all, as well as over six hundred of pages of papers. More specifically, this volume consists of four parts:

Part I: Quantum Computing—quantum algorithms and hidden subgroups, quantum search, algorithmic complexity and quantum simulation;

Part II: Quantum Technology—math tools, quantum wave functions, SQUIDs, optical quantum computing;

Part III: Quantum Information—quantum error correction, quantum cryptography, quantum entanglement and communication; and

Part IV: Quantum Topology, Categorical Algebra and Logic—knot theory, category, algebra and logic.

A Panel Report to NSF containing recommendations for federal funding on the mathematical research on quantum computing is also attached at the very end as an Appendix.

This book was written collectively by the authors of its many and diverse chapters. We are indebted to them for their invaluable contributions. We also wish to thank the reviewers (several of them are not coauthors of any book chapters) for their helpful reports and comments.

Ms. Robin Campbell has done the high quality editorial work in processing and compiling the book chapters. Mr. Bob Stern of the Taylor and Francis

Group has expedited the book publication in every way. Working together with them on this book project was indeed a great pleasure.

Goong Chen
Louis Kauffman
Samuel J. Lomonaco

Contributors

Samson Abramsky
Oxford University Computing
Laboratory
Wolfson Building
Parks, Road
Oxford, OX1 3QD
England

Graeme Ahokas
Department of Computer Science
University of Calgary
Calgary, Alberta
Canada
and
Institute for Quantum Information
Science
University of Calgary
Calgary, Alberta
Canada

Salah A. Aly
Department of Computer Science
Texas A&M University
College Station, TX 77843-3112

Syed M. Assad
Department of Physics
National University of Singapore
2 Science Drive 3
Singapore 117542
Singapore

Dominic W. Berry
Centre for Quantum Computer
Technology
Macquarie University

Sydney, New South Wales
Australia
and
Department of Physics
The University of Queensland
Brisbane, Queensland
Australia

C. Bracher
Physics Department
Bryn Mawr College
Bryn Mawr, PA 19010

Howard E. Brandt
U.S. Army Research Laboratory
Adelphi, MD 20783-1197

Goong Chen
Department of Mathematics
and Institute for Quantum Studies
Texas A&M University
College Station, TX 77843-3368

Richard Cleve
David R. Cheriton School of
Computer Science and
Institute for Quantum Computing
University of Waterloo
Waterloo, Ontario, Canada
and
Perimeter Institute for Theoretical
Physics
Waterloo, Ontario, Canada

Bob Coecke
Oxford University Computing
Laboratory
Wolfson Building
Parks Road
Oxford OX1 3QD
UK

Leon Cohen
City University of New York
695 Park Avenue
New York, NY 10021

Zijian Diao
Department of Mathematics
Ohio University-Eastern
St. Clairsville, OH 43950

Jonathan P. Dowling
Hearne Institute for Theoretical
Physics
Department of Physics
and Astronomy
Louisiana State University
Baton Rouge, LA 70803-4001

Berthold-Georg Englert
Department of Physics
National University of Singapore
2 Science Drive 3
Singapore 117542
Singapore

Louis H. Kauffman
Department of Mathematics,
Statistics and Computer Science
University of Illinois at Chicago
(UIC)
Chicago, IL 60607

Andreas Klappenecker
Department of Computer Science
Texas A&M University
College Station, TX 77843-3112

M. Kleber
Physik-Department T30
Technische Universität München
James-Franck-Straße
85747 Garching
Germany

T. Kramer
Physics Department
Harvard University
One Oxford Street
Cambridge, MA 02138

Hwang Lee
Hearne Institute for Theoretical
Physics
Department of Physics and
Astronomy
Louisiana State University
Baton Rouge, LA 70803-4001

Fu-li Li
Department of Applied Physics
Xi'an Jiaotong University
Xi'an 710049
China

Samuel J. Lomonaco
Department of Computer Science
and Computer Engineering
University of Maryland
Baltimore County (UMBC)
Baltimore, MD 21250

Dusko Pavlovic
Kestrel Institute
3260 Hillview Avenue
Palo Alto, CA 94304

V. Ramakrishna
Department of Mathematics
University of Texas at Dallas
Box 830688
Richardson, TX 75083-0688

J. Maurice Rojas
Department of Mathematics
Texas A&M University
College Station, TX 77843-3368

Barry C. Sanders
Institute for Quantum Information
Science
University of Calgary
Calgary, Alberta
Canada

Pradeep Kiran Sarvepalli
Department of Computer Science
Texas A&M University
College Station, TX 77843-3112

Peter Shiue
Department of Mathematical
Science
University of Nevada-Las Vegas
Las Vegas, NV 89154

Federico M. Spedalieri
Department of Electrical Engineering
University of California, Los Angeles
Los Angeles, CA 90095-1594

Jun Suzuki
Department of Physics
National University of Singapore
2 Science Drive 3
Singapore 117542
Singapore

M. Tseng
Department of Mathematics
University of Texas at Dallas
Box 830688
Richardson, TX 75083-0688

Zhigang Zhang
Department of Mathematics
Texas A&M University
College Station, TX 77843-3368

H. Zhou
Mathematics Department
University of Texas-Dallas
Box 830688
Richardson, TX 75083-0688

Suhail Zubairy
Department of Physics and
Institute for Quantum Studies
Texas A&M University
College Station, TX 77843-4242

Contents

Quantum Computation

Chapter 1

Quantum Hidden Subgroup Algorithms: An Algorithmic Toolkit

Samuel J. Lomonaco and Louis H. Kauffman

Abstract   One of the most promising and versatile approaches to creating new quantum algorithms is based on the quantum hidden subgroup (QHS) paradigm, originally suggested by Alexei Kitaev. This class of quantum algorithms encompasses the Deutsch–Jozsa, Simon, Shor algorithms, and many more.

In this paper, our strategy for finding new quantum algorithms is to decompose Shor's quantum factoring algorithm into its basic primitives, then to generalize these primitives, and finally to show how to reassemble them into new QHS algorithms. Taking an alphabetic building blocks approach, we use these primitives to form an algorithmic toolkit for the creation of new quantum algorithms, such as wandering Shor algorithms, continuous Shor algorithms, the quantum circle algorithm, the dual Shor algorithm, a QHS algorithm for Feynman integrals, free QHS algorithms, and more.

Toward the end of this paper, we show how Grover's algorithm is most surprisingly almost a QHS algorithm, and how this result suggests the possibility of an even more complete "algorithmic toolkit" beyond QHS algorithms.

1.1   Introduction

One major obstacle to the fulfillment of the promise of quantum computing is the current scarcity of quantum algorithms. Quantum computing researchers simply have not yet found enough quantum algorithms to determine whether or not future quantum computers will be general purpose or special purpose computing devices. As a result, much more research is crucially needed to determine the algorithmic limits of quantum computing.

One of the most promising and versatile approaches to creating new quantum algorithms is based on the quantum hidden subgroup (QHS) paradigm, originally suggested by Alexei Kitaev [20]. This class of quantum algorithms encompasses the Deutsch–Jozsa, Simon, Shor algorithms, and many more.

In this paper, our strategy for finding new quantum algorithms is to decompose Shor's quantum factoring algorithm into its basic primitives, then to generalize these primitives, and finally to show how to reassemble them into new QHS algorithms. Taking an alphabetic building blocks approach, we will use these primitives to form an algorithmic toolkit for the creation of new quantum algorithms, such as wandering Shor algorithms, continuous Shor algorithms, the quantum circle algorithm, the dual Shor algorithm, a QHS algorithm for Feynman integrals, free QHS algorithms, and more.

Toward the end of this paper, we show how Grover's algorithm is most surprisingly almost a QHS algorithm, and how this suggests the possibility of an even more complete algorithmic toolkit beyond QHS algorithms.

1.2   An example of Shor's quantum factoring algorithm

Before discussing how Shor's algorithm can be decomposed into its primitive components, let's take a quick look at an example of the execution of Shor's factoring algorithm. As we discuss this example, we suggest that the reader, as an exercise, try to find the basic QHS primitives that make up this algorithm. Can you see them?

Shor's quantum factoring algorithm reduces the task of factoring a positive integer N to first finding a random integer a relatively prime to N, and then next to determining the period P of the following function

$$\mathbb{Z} \xrightarrow{\varphi} \mathbb{Z} \bmod N$$

$$x \longmapsto a^x \bmod N\,,$$

where $\mathbb{Z}$ denotes the additive group of integers, and where $\mathbb{Z} \bmod N$ denotes the integers $\bmod N$ under multiplication.[1]

Since $\mathbb{Z}$ is an infinite group, Shor chooses to work instead with the finite additive cyclic group $\mathbb{Z}_Q$ of order $Q = 2^m$, where $N^2 \le Q < 2N^2$, and with the "approximating" map

$$\mathbb{Z}_Q \xrightarrow{\widetilde{\varphi}} \mathbb{Z} \bmod N$$

$$x \longmapsto a^x \bmod N\,,\, 0 \le x < Q.$$

We begin by constructing a quantum system with two quantum registers

$$|\text{LEFT\_REGISTER}\rangle\, |\text{RIGHT\_REGISTER}\rangle\,,$$

the left intended for holding the arguments x of $\widetilde{\varphi}$, the right for holding the corresponding values of $\widetilde{\varphi}$. This quantum system has been constructed with a unitary transformation

$$U_{\widetilde{\varphi}} : |x\rangle\, |1\rangle \longmapsto |x\rangle\, |\widetilde{\varphi}(x)\rangle$$

implementing the "approximating" map $\widetilde{\varphi}$.

As an example, let us use Shor's algorithm to factor the integer $N = 21$, assuming that $a = 2$ has been randomly chosen. Thus, $Q = 2^9 = 512$.

Unknown to us, the period is $P = 6$, and hence, $Q = 6 \cdot 85 + 2$.

We proceed by executing the following steps:

Shor Algorithm Example

STEP 0 Initialize

$$|\psi_0\rangle = |0\rangle\, |1\rangle$$

---

[1]A random integer a with $\gcd(a, N) = 1$ is found by selecting a random integer, and then applying the Euclidean algorithm to determine whether or not it is relatively prime to N. If not, then the gcd is a non-trivial factor of N, and there is no need to proceed further. However, this possibility is highly unlikely if N is large.

STEP 1 Apply the inverse Fourier transform[2]

$$\mathscr{F}^{-1} : |u\rangle \longmapsto \frac{1}{\sqrt{512}} \sum_{x=0}^{511} \omega^{-ux} |x\rangle$$

to the left register, where $\omega = \exp(2\pi i/512)$ is a primitive 512-th root of unity, to obtain

$$|\psi_1\rangle = \frac{1}{\sqrt{512}} \sum_{x=0}^{511} |x\rangle |1\rangle \ .$$

STEP 2 Apply the unitary transformation

$$U_{\widetilde{\varphi}} : |x\rangle |1\rangle \longmapsto |x\rangle |2^x \bmod 21\rangle$$

to obtain

$$|\psi_2\rangle = \frac{1}{\sqrt{512}} \sum_{x=0}^{511} |x\rangle |2^x \bmod 21\rangle \ .$$

STEP 3 Apply the Fourier transform

$$\mathscr{F} : |x\rangle \longmapsto \frac{1}{\sqrt{512}} \sum_{y=0}^{511} \omega^{xy} |y\rangle$$

to the left register to obtain

$$\begin{aligned}
|\psi_3\rangle &= \frac{1}{512} \sum_{x=0}^{511} \sum_{y=0}^{511} \omega^{xy} |y\rangle |2^x \bmod 21\rangle \\
&= \frac{1}{512} \sum_{y=0}^{511} |y\rangle \left(\sum_{x=0}^{511} \omega^{xy} |2^x \bmod 21\rangle \right) \\
&= \frac{1}{512} \sum_{y=0}^{511} |y\rangle |\Upsilon(y)\rangle
\end{aligned}$$

where

$$|\Upsilon(y)\rangle = \sum_{x=0}^{511} \omega^{xy} |2^x \bmod 21\rangle \ .$$

---

[2]Actually, for this step, the original Shor algorithm uses instead the Hadamard transform, which for step 1, has the same effect as the 512-point Fourier transform.

STEP 4 Measure the left register. Then with Probability

$$Prob_{\widetilde{\varphi}}(y) = \frac{\langle \Upsilon(y) \mid \Upsilon(y) \rangle}{(512)^2}$$

the state will "collapse" to $|y\rangle$ with the value measured being the integer y, where $0 \le y < Q$.

A plot of $Prob_{\widetilde{\varphi}}(y)$ is shown in Fig. 1.1. (See [21] and [25] for details.)

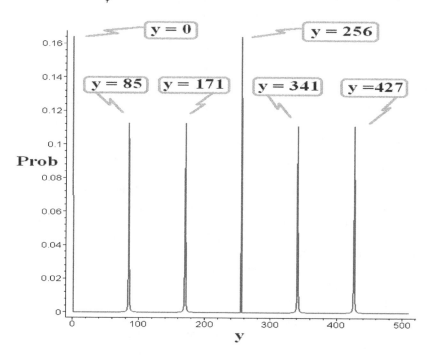

FIGURE 1.1: A plot of **Prob$_{\widetilde{\varphi}}$(y)**.

The peaks in the above plot of $Prob_{\widetilde{\varphi}}(y)$ occur at the integers

$$y = 0, \; 85, \; 171, \; 256, \; 341, \; 427.$$

The probability that at least one of these six integers will occur is quite high. It is actually 0.78^+. Indeed, the probability distribution has been intentionally engineered to make the probability of these particular integers as high as possible. And there is a good reason for doing so.

The above six integers are those for which the corresponding rational y/Q is "closest" to a rational of the form d/P. By "closest" we mean that

$$\left| \frac{y}{Q} - \frac{d}{P} \right| < \frac{1}{2Q} < \frac{1}{2P^2} .$$

In particular,

$$\frac{0}{512}, \frac{85}{512}, \frac{171}{512}, \frac{256}{512}, \frac{341}{512}, \frac{427}{512}$$

are rationals respectively "closest" to the rationals

$$\frac{0}{6}, \frac{1}{6}, \frac{2}{6}, \frac{3}{6}, \frac{4}{6}, \frac{5}{6} .$$

The six rational numbers $0/6$, $1/6$, ... , $5/6$ are "closest" in the sense that they are convergents of the continued fraction expansions of $0/512$, $85/512$, ... , $427/512$, respectively. Hence, each of the six rationals $0/6$, $1/6$, ... , $5/6$ can be found using the standard continued fraction recursion formulas.

But, we are not searching for rationals of the form d/P. Instead, we seek only the denominator $P = 6$.

Unfortunately, the denominator $P = 6$ can only be obtained from the continued fraction recursion when the numerator and denominator of d/P are relatively prime. Given that the algorithm has selected one of the random integers 0, 85, ... , 427, the probability that the corresponding rational d/P has relatively prime numerator and denominator is $\phi(6)/6 = 1/3$, where $\phi(-)$ denotes the Euler phi (totient) function. So the probability of finding $P = 6$ is actually not 0.78^+, but is instead 0.23^-.

As it turns out, if we repeat the algorithm $O(\lg\lg N)$ times,[3] we will obtain the desired period P with probability bounded below by approximately $4/\pi^2$. However, this is not the end of the story. Once we have in our possession a candidate P' for the actual period $P = 6$, the only way we can be sure we have the correct period P is to test P' by computing $2^{P'} \bmod 21$. If the result is 1, we are certain we have found the correct period P. This last part of the computation is done by the repeated squaring algorithm.[4]

---

[3]We use lg to denote log_2, i.e., the log to the base 2.

[4]By the repeated squaring algorithm, we mean the algorithm which computes $a^{P'} \bmod N$ via the expression

$$a^{P'} = \prod_j \left(a^{2^j} \right)^{P'_j},$$

where $P' = \sum_j P'_j 2^j$ is the radix 2 expansion of P'.

1.3 Definition of quantum hidden subgroup (QHS) algorithms

Now that we have taken a quick look at Shor's algorithm, let's see how it can be decomposed into its primitive algorithmic components. We will first need to answer the following question:

What is a quantum hidden subgroup algorithm?

But before we can answer this question, we need to provide an answer to an even more fundamental question:

What is a hidden subgroup problem?

DEFINITION 1.1   *A map $\varphi : G \longrightarrow S$ from a group G into a set S is said to have* **hidden subgroup structure** *if there exists a subgroup K_φ of G, called a* **hidden subgroup**, *and an injection $\iota_\varphi : G/K_\varphi \longrightarrow S$, called a* **hidden injection**, *such that the diagram*

$$
\begin{array}{ccc}
G & \overset{\varphi}{\longrightarrow} & S \\
{\scriptstyle v} \searrow & & \nearrow {\scriptstyle \iota_\varphi} \\
& G/K_\varphi &
\end{array}
$$

is commutative,[5] where G/K_φ denotes the collection of right cosets of K_φ in G, and where $v : G \longrightarrow G/K_\varphi$ is the natural surjection of G onto G/K_φ. We refer to the group G as the **ambient group** *and to the set S as the* **target set**. *If K_φ is a normal subgroup of G, then $H_\varphi = G/K_\varphi$ is a group, called the* **hidden quotient group**, *and $v : G \longrightarrow G/K_\varphi$ is an epimorphism, called the* **hidden epimorphism**. *We will call the above diagram the* **hidden subgroup structure** *of the map $\varphi : G \longrightarrow S$. (See [25],[20].)*

REMARK 1.1   The underlying intuition motivating this formal definition is as follows: Given a natural surjection (or epimorphism)

---

[5]By saying that this diagram is commutative, we mean $\varphi = \iota_\varphi \circ v$. The notion generalizes in an obvious way to more complicated diagrams.

$v : G \longrightarrow G/K_\varphi$, an "archvillain with malice aforethought" hides the algebraic structure of v by intentionally renaming all the elements of G/K_φ, and "maliciously tossing in for good measure" some extra elements to form a set S and a map $\varphi : G \longrightarrow S$. ∎

The hidden subgroup problem can be stated as follows:

Hidden Subgroup Problem (HSP). *Let* $\varphi : G \longrightarrow S$ *be a map with hidden subgroup structure. The problem of determining a hidden subgroup K_φ of G is called a **hidden subgroup problem (HSP)**. An algorithm solving this problem is called a **hidden subgroup algorithm**.*

The corresponding quantum form of this HSP is stated as follows:

Hidden Subgroup Problem (Quantum Version). *Let* $\varphi : G \longrightarrow S$ *be a map with hidden subgroup structure. Construct a quantum implementation of the map φ as follows:*

Let $\mathcal{H}_G$ and $\mathcal{H}_S$ be Hilbert spaces defined respectively by the orthonormal bases $\{|g\rangle : g \in G\}$ and $\{|s\rangle : s \in S\}$ and let $s_0 = \varphi(1)$, where 1 denotes the identity[6] of the ambient group A. Finally, let U_φ be a unitary transformation such that

$$\mathcal{H}_G \otimes \mathcal{H}_S \longrightarrow \mathcal{H}_G \otimes \mathcal{H}_S$$
$$|g\rangle |s_0\rangle \longmapsto |g\rangle |\varphi(g)\rangle \ .$$

*Determine the hidden subgroup K_φ with bounded probability of error by making as few queries as possible to the blackbox U_φ. A quantum algorithm solving this problem is called a **quantum hidden subgroup (QHS) algorithm**.*

1.4   The generic QHS algorithm

We are now in a position to construct one of the fundamental algorithmic primitives found in Shor's algorithm.

Let $\varphi : G \longrightarrow S$ be a map from a group G to a set S with hidden subgroup structure. We assume that all representations of G are equivalent to unitary representations.[7] Let $\widehat{G}$ denote a **complete set of distinct irreducible unitary representations** of G. Using multiplicative notation for G, we let 1 denote the

---

[6] We are using multiplicative notation for the group G.

[7] This is true for all finite groups as well as for a large class of infinite groups.

identity of G, and let s_0 denote its image in S. Finally, let $\widehat{1}$ denote the **trivial representation** of G.

REMARK 1.2   If G is abelian, then $\widehat{G}$ becomes the **dual group** of characters. ∎

The generic QHS algorithm is given below:

Generic Quantum Subroutine QRAND(φ)

Step 0 │ Initialization

$$|\psi_0\rangle = \left|\widehat{1}\right\rangle |s_0\rangle \in \mathcal{H}_{\widehat{G}} \otimes \mathcal{H}_S .$$

Step 1 │ Application of the inverse Fourier transform $\mathcal{F}_G^{-1}$ of G to the left register

$$|\psi_1\rangle = \frac{1}{\sqrt{|G|}} \sum_{g \in G} |g\rangle |s_0\rangle \in \mathcal{H}_G \otimes \mathcal{H}_S ,$$

where $|G|$ denotes the cardinality of the group G.

Step 2 │ Application of the unitary transformation U_φ

$$|\psi_2\rangle = \frac{1}{\sqrt{|G|}} \sum_{g \in G} |g\rangle |\varphi(g)\rangle \in \mathcal{H}_G \otimes \mathcal{H}_S .$$

Step 3 │ Application of the Fourier transform $\mathcal{F}_G$ of G to the left register

$$|\psi_3\rangle = \frac{1}{|G|} \sum_{\gamma \in \widehat{G}} |\gamma| \, Trace\left(\sum_{g \in G} \gamma^\dagger(g) |\gamma\rangle\right) |\varphi(g)\rangle$$

$$= \frac{1}{|G|} \sum_{\gamma \in \widehat{G}} |\gamma| \, Trace\left(|\gamma\rangle |\Phi(\gamma^\dagger)\rangle\right) \in \mathcal{H}_{\widehat{G}} \otimes \mathcal{H}_S ,$$

where $|\gamma|$ denotes the degree of the representation γ, where $\gamma^\dagger$ denotes the contragradient representation (i.e., $\gamma^\dagger(g) = \gamma\left(g^{-1}\right)^T = \overline{\gamma(g)}^T$), where $Trace\left(\gamma^\dagger |\gamma\rangle\right) = \sum_{i=1}^{|\gamma|} \sum_{j=1}^{|\gamma|} \overline{\gamma_{ji}(g)} |\gamma_{ij}\rangle$, and where $\left|\Phi\left(\gamma_{ij}^\dagger\right)\right\rangle = \sum_{g \in G} \overline{\gamma_{ji}(g)} |\varphi(g)\rangle$.

Step 4 Measurement of the left quantum register with respect to the orthonormal basis

$$\left\{ \left| \gamma_{ij} \right\rangle : \gamma \in \widehat{G}, 1 \leq i, j \leq |\gamma| \right\} \ .$$

Thus, with probability

$$Prob_{\varphi}\left(\gamma_{ij} \right) = \frac{|\gamma|^2 \left\langle \Phi\left(\gamma_{ij}^{\dagger} \right) | \Phi\left(\gamma_{ij}^{\dagger} \right) \right\rangle}{|G|^2} \ ,$$

the resulting measured value is the entry $\gamma_{\mathbf{ij}}$, and the quantum system "collapses" to the state

$$\left| \psi_4 \right\rangle = \frac{\left| \gamma_{ij} \right\rangle \left| \Phi\left(\gamma_{ij}^{\dagger} \right) \right\rangle}{\sqrt{\left\langle \Phi\left(\gamma_{ij}^{\dagger} \right) | \Phi\left(\gamma_{ij}^{\dagger} \right) \right\rangle}} \in \mathcal{H}_{\widehat{G}} \otimes \mathcal{H}_S$$

Step 5 Step 5. Output γ_{ij}, and stop.

1.5 Pushing and lifting hidden subgroup problems (HSPs)

But Shor's algorithm consists of more than the primitive QRAND.

For many (but not all) hidden subgroup problems (HSPs) $\varphi : G \longrightarrow S$, the corresponding generic QHS algorithm QRAND either is not physically implementable or is too expensive to implement physically. For example, the HSP φ is usually not physically implementable if the ambient group is infinite (e.g., G is the infinite cyclic group $\mathbb{Z}$), and is too expensive to implement if the ambient group is too large (e.g., G is the symmetric group $\mathbb{S}_{10^{100}}$). In this case, there is a standard generic way of "tweaking" the HSP to get around this problem, which we will call **pushing**.

DEFINITION 1.2   *Let $\varphi : G \longrightarrow S$ be a map from a group G to a set S. A map $\widetilde{\varphi} : \widetilde{G} \longrightarrow S$ from a group $\widetilde{G}$ to the set S is said to be a* **push** *of φ, written*

$$\widetilde{\varphi} = Push\left(\varphi \right) \ ,$$

provided there exists an epimorphism $\nu : G \longrightarrow \widetilde{G}$ *from G onto $\widetilde{G}$, and a transversal*[8] $\tau : \widetilde{G} \longrightarrow G$ *of ν such that $\widetilde{\varphi} = \varphi \circ \tau$, i.e., such that the following diagram is commutative*

$$ G \overset{\varphi}{\longrightarrow} S $$
$$ \uparrow \tau \nearrow \widetilde{\varphi} $$
$$ \widetilde{G} $$

If the epimorphism μ and the transversal τ are chosen in an appropriate way, then execution of the generic QHS subroutine with input $\widetilde{\varphi} = Push(\varphi)$, i.e., execution of

$$ \text{QRAND}(\widetilde{\varphi}) , $$

will with high probability produce an irreducible representation $\widetilde{\gamma}$ of the group $\widetilde{G}$ which is sufficiently close to an irreducible representation γ of the group G. If this is the case, then there is a polynomial time classical algorithm which upon input $\widetilde{\gamma}$ produces the representation γ.

Obviously, much more can be said about pushing. But unfortunately that would take us far afield from the objectives of this paper. For more information on pushing, we refer the reader to [27].

It would be remiss not to mention that the above algorithmic primitive of pushing suggests the definition of a second primitive which we will call **lifting**.

DEFINITION 1.3   *Let $\varphi : G \longrightarrow S$ be a map from a group G to a set S. A map $\underline{\varphi} : \underline{G} \longrightarrow S$ from a group $\underline{G}$ to the set S is said to be a* **lift** *of φ, written*

$$ \underline{\varphi} = Lift(\varphi) , $$

provided there exists a morphism $\eta : \underline{G} \longrightarrow G$ from $\underline{G}$ to G such that

---

[8]Let $\nu : A \longrightarrow B$ be an epimorphism from a group A to a group B. Then a transversal τ of ν is a map $\tau : B \longrightarrow A$ such that $\nu \circ \tau : B \longrightarrow A$ is the identity map $b \longmapsto b$. (It immediately follows that τ is an injection.) In other words, a transversal τ of an epimorphism ν is a map which maps each element b of B to an element of A contained in the coset b, i.e., to a coset representative of b.

$\underline{\varphi} = \varphi \circ \eta$, *i.e., such that the following diagram is commutative*

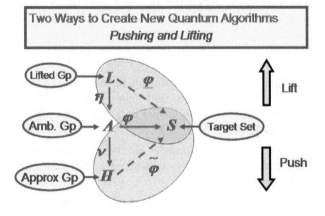

FIGURE 1.2: Pushing and Lifting HSPs.

1.6 Shor's algorithm revisited

We are now in position to describe Shor's algorithm in terms of its primitive components. In particular, we can now see that Shor's factoring algorithm is a classic example of a QHS algorithm created from the push of an HSP.

Let N be the integer to be factored. Let $\mathbb{Z}$ denote the additive group of integers, and $\mathbb{Z}_N^{\times}$ denote the integers $\bmod N$ under multiplication.

Shor's algorithm is a QHS algorithm that solves the following HSP

$$\varphi : \mathbb{Z} \longrightarrow \mathbb{Z}_N^{\times}$$

$$m \longmapsto a^m \bmod N$$

with unknown hidden subgroup structure given by the following commutative diagram

$$
\begin{array}{ccc}
\mathbb{Z} & \xrightarrow{\ \varphi\ } & \mathbb{Z}_N^{\times} \\
{\scriptstyle v}\searrow & & \nearrow{\scriptstyle \iota} \\
& \mathbb{Z}/P\mathbb{Z} &
\end{array} \ ,
$$

where a is an integer relatively prime to N, where P is the hidden integer period of the map $\varphi : \mathbb{Z} \longrightarrow \mathbb{Z}_N^{\times}$, where $P\mathbb{Z}$ is the additive subgroup of all integer multiples of P (i.e., the hidden subgroup), where $v : \mathbb{Z} \longrightarrow \mathbb{Z}/P\mathbb{Z}$ is the natural epimorphism of the integers onto the quotient group $\mathbb{Z}/P\mathbb{Z}$ (i.e., the hidden epimorphism), and where $\iota : \mathbb{Z}/P\mathbb{Z} \longrightarrow \mathbb{Z}_N^{\times}$ is the hidden monomorphism.

An obstacle to creating a physically implementable algorithm for this HSP is that the domain $\mathbb{Z}$ of φ is infinite. As observed by Shor, a way to work around this difficulty is to push the HSP.

In particular, as illustrated by the following commutative diagram

$$
\begin{array}{ccc}
\mathbb{Z} & \xrightarrow{\ \varphi\ } & \mathbb{Z}_N^{\times} \\
{\scriptstyle \mu}\searrow{\scriptstyle \tau} & & \nearrow \\
& \mathbb{Z}_Q &
\end{array} \quad \widetilde{\varphi} = Push\,(\varphi) = \varphi \circ \tau \ ,
$$

a push $\widetilde{\varphi} = Push\,(\varphi)$ is constructed by selecting the epimorphism $\mu : \mathbb{Z} \longrightarrow \mathbb{Z}_Q$ of $\mathbb{Z}$ onto the finite cyclic group $\mathbb{Z}_Q$ of order Q, where the integer Q is the unique power of 2 such that $N^2 \le Q < 2N^2$, and then choosing the transversal[9]

$$\tau : \mathbb{Z}_Q \longrightarrow \mathbb{Z} \ ,$$

$$m \bmod Q \longmapsto m$$

where $0 \le m < Q$. *This push* $\widetilde{\varphi} = Push\,(\varphi)$ *is called* **Shor's oracle.**

---

[9]A **transversal** for an epimorphism $\alpha_\varphi : \mathbb{Z} \longrightarrow \mathbb{Z}_Q$ is an injection $\tau_\varphi : \mathbb{Z}_Q \longrightarrow \mathbb{Z}$ such that $\alpha_\varphi \circ \tau_\varphi$ is the identity map on $\mathbb{Z}_Q$, i.e., a map that takes each element of $\mathbb{Z}_Q$ onto a coset representative of the element in $\mathbb{Z}$.

Shor's algorithm consists in first executing the quantum subroutine $\mathrm{QRAND}(\widetilde{\varphi})$, thereby producing a random character

$$\gamma_{y/Q} : m \bmod Q \longmapsto \frac{my}{Q} \bmod 1$$

of the finite cyclic group $\mathbb{Z}_Q$. The transversal τ used in pushing has been engineered in such a way as to assure that the character $\gamma_{y/Q}$ is sufficiently close to a character

$$\gamma_{d/P} : k \bmod P \longmapsto \frac{kd}{P} \bmod 1$$

of the hidden quotient group $\mathbb{Z}/P\mathbb{Z} = \mathbb{Z}_P$. In this case, sufficiently close means that

$$\left| \frac{y}{Q} - \frac{d}{P} \right| \le \frac{1}{2P^2} ,$$

which means that d/P is a continued fraction convergent of y/Q, and thus can be found by the classical polynomial time continued fraction algorithm.[10]

1.7   Wandering Shor algorithms, a.k.a. vintage Shor algorithms

Now let's use the primitives described in Sections 1.3, 1.4, and 1.5 to create other new QHS algorithms, called wandering Shor algorithms.

Wandering Shor algorithms are essentially QHS algorithms on free abelian finite rank n groups A which, with each iteration, first select a random cyclic direct summand $\mathbb{Z}$ of the group A, and then apply one iteration of the standard Shor algorithm to produce a random character of the "approximating" finite group $\widetilde{A} = \mathbb{Z}_Q$, called a **group probe**.[11] In this way, three different wandering Shor algorithms are created in [25]. The first two wandering Shor algorithms given in [25] are quantum algorithms which find the order P of a maximal cyclic subgroup of the hidden quotient group H_φ. The third computes the entire hidden quotient group H_φ.

---

[10]The characters $\gamma_{y/Q}$ and $\gamma_{d/P}$ can in the obvious way be identified with points of the unit circle in the complex plane. With this identification, we can see that this inequality is equivalent to saying the the chordal distance between these two rational points on the unit circle is less than or equal to $1/2P^2$. Hence, Shor's algorithm is using the topology of the unit circle.

[11]By a group probe $\widetilde{A}$, we mean an epimorphic image of the ambient group A.

The first step in creating a wandering Shor algorithm is to find the right generalization of one of the primitives found in Shor's algorithm, namely, the transversal $\iota : \mathbb{Z}_Q \longrightarrow \mathbb{Z}$ of Shor's factoring algorithm. In other words, we need to construct the "correct" generalization of the transversal from $\mathbb{Z}_Q$ to a free abelian group A of rank n. For this reason, we have created the following definition:

DEFINITION 1.4   *Let A be the free abelian group of rank n, let $v : A \longrightarrow \mathbb{Z}_Q$ be an epimorphism onto the cyclic group $\mathbb{Z}_Q$ of order Q with selected generator $\tilde{a}$. A transversal[12] $\iota : \mathbb{Z}_Q \longrightarrow A$ of v is said to be a* **Shor transversal** *provided that:*

  1) $\iota(n\tilde{a}) = n\iota(\tilde{a})$ *for all $0 \leq n < Q$, and*

  2) *For each (free abelian) basis $a_1', a_2', \ldots, a_n'$ of A, the coefficients $\lambda_1', \lambda_2', \ldots, \lambda_n'$ of $\iota(\tilde{a}) = \sum_j \lambda_j' a_j'$ satisfy $\gcd(\lambda_1', \lambda_2', \ldots, \lambda_n') = 1$.*

REMARK 1.3   Later, when we construct a generalization of Shor transversals to free groups of finite rank n, we will see that the first condition simply states that a Shor transversal is nothing more than a 2-sided Schreier transversal. The second condition of the above definition simply says that ι maps the generator $\tilde{a}$ of $\mathbb{Z}_Q$ onto a generator of a free direct summand $\mathbb{Z}$ of A. (For more details, please refer to Section 1.12 of this paper.)   ∎

REMARK 1.4   In [25], we show how to use the extended Euclidean algorithm to construct the epimorphism $v : A \longrightarrow \mathbb{Z}_Q$ and the transversal $\iota : \mathbb{Z}_Q \longrightarrow A$.   ∎

Flow charts for the three wandering Shor algorithms created in [25] are given in Figs. 1.3 through 1.5. In [25], these were also called **vintage Shor algorithms**.

The algorithmic complexities of the above wandering Shor algorithms are given in [25]. For example, the first wandering Shor algorithm is of time com-

---

[12]For a definition of the transversal of an epimorphism, please refer to footnote 8.

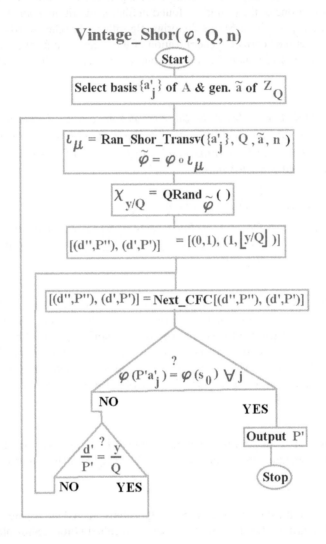

FIGURE 1.3: Flowchart for the first wandering Shor algorithm (a.k.a. a vintage Shor algorithm). This algorithm finds the order P of a maximal cyclic subgroup of the hidden quotient group H_φ.

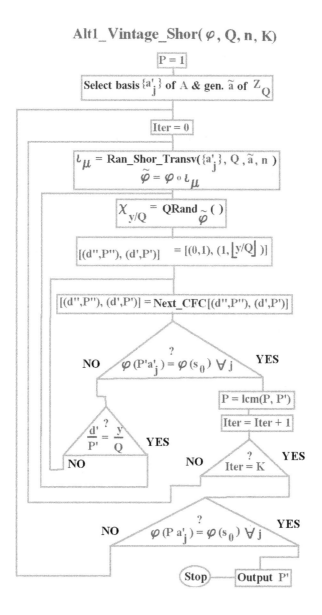

FIGURE 1.4: Flowchart for the second wandering Shor algorithm (a.k.a. a v-intage Shor algorithm). This algorithm finds the order P of a maximal cyclic subgroup of the hidden quotient group H_φ.

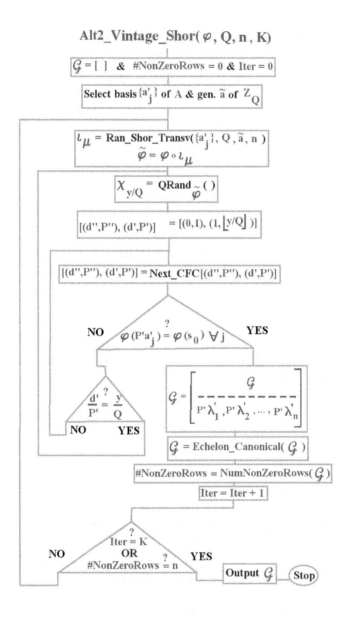

FIGURE 1.5: Flowchart for the third wandering Shor algorithm, a.k.a. a v-intage Shor algorithm. This algorithm finds the entire hidden quotient group H_φ.

plexity[13]

$$O\left(n^2 (\lg N)^3 (\lg \lg N)^{n+1}\right) ,$$

where n is the rank of the free abelian group A. This can be readily deduced from the abbreviated flowchart given in Fig. 1.6.

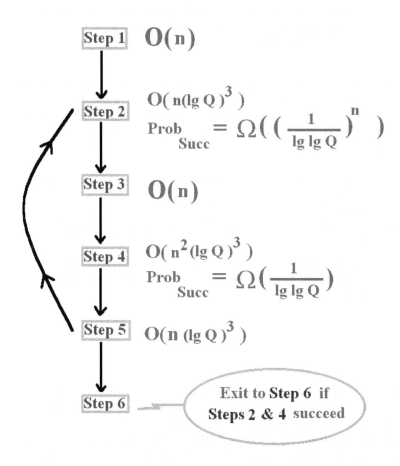

FIGURE 1.6: Abbreviated flowchart for the first wandering Shor algorithm.

---

[13] We use lg to denote the *log* to the base 2, i.e., log_2.

1.8   Continuous (variable) Shor algorithms

In [27] and in [29], the algorithmic primitives found in above sections of
this paper were used to create a class of algorithms called continuous Shor
algorithms. By a **continuous variable Shor algorithm**, we mean a quantum
hidden subgroup algorithm that finds the hidden period P of an admissible
function $\varphi : \mathbb{R} \longrightarrow \mathbb{R}$ from the reals $\mathbb{R}$ to itself.

REMARK 1.5   By an admissible function, we mean a function belonging to any sufficiently well behaved class of functions. For example,
the class of functions which are Lebesgue integrable on every closed interval of $\mathbb{R}$. There are many other classes of functions that work equally
as well. ∎

Actually, the papers [27], [29] give in succession three such continuous Shor
algorithms, each successively more general than the previous.

For the first algorithm, we assume that the unknown hidden period P is
an integer. The algorithm is then constructed by using rigged Hilbert spaces
[4], [10], linear combinations of Dirac delta functions, and a subtle extension
of the Fourier transform found in the generic QHS subroutine $\text{QRAND}(\varphi)$,
which has been described previously in Section 1.4 of this paper. In Step 5 of
$\text{QRAND}(\varphi)$, the observable

$$A = \int\limits_{-\infty}^{\infty} dy \frac{\lfloor Qy \rfloor}{Q} |y\rangle \langle y|$$

is measured, where Q is an integer chosen so that $Q \geq 2P^2$. It then follows
that the output of this algorithm is a rational m/Q which is a convergent of the
continued fraction expansion of a rational of the form n/P.

The above quantum algorithm is then extended to a second quantum al-
gorithm that finds the hidden period P of functions $\varphi : \mathbb{R} \longrightarrow \mathbb{R}$, where the
unknown period P is a rational.

Finally, the second algorithm is extended to a third algorithm which finds the
hidden period P of functions $\varphi : \mathbb{R} \longrightarrow \mathbb{R}$, when P is an *arbitrary real number*.
We point out that for the third and last algorithm to work, we must impose a
very restrictive condition on the map $\varphi : \mathbb{R} \longrightarrow \mathbb{R}$, i.e., the condition that the
map φ is continuous.

1.9 The quantum circle and the dual Shor algorithms

We have shown in previous sections how the mathematical primitives of pushing and lifting can be used to create new quantum algorithms. In particular, we have described how pushing and lifting can be used to derive new HSPs from an HSP $\varphi : G \longrightarrow S$ on an arbitrary group G. We now see how group duality can be exploited by these two primitives to create even more quantum algorithms.

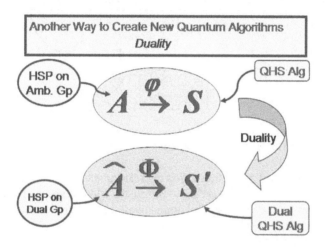

FIGURE 1.7: Using duality to create new QHS algorithms.

To this end, we assume that G is an *abelian* group. Hence, its dual group of characters $\widehat{G}$ exists.[14] It now follows that pushing and lifting can also be used to derive new HSPs $\Phi : \widehat{G} \longrightarrow S'$ on the dual group $\widehat{G}$. In [27], this method is used to create a number of new quantum algorithms derived from Shor-like HSPs $\varphi : \mathbb{Z} \longrightarrow S$.

A roadmap is shown in Fig. 1.8 of the developmental steps taken to find and to create a new QHS algorithm on $\mathbb{Z}_Q$, which is (in the sense described below)

---

[14] If G is non-abelian, then its dual is not a group, but instead the representation algebra $\mathscr{A}$ over the group ring $\mathbb{C}G$. The methods described in this section can also be used to create new quantum algorithms for HSPs $\Phi : \mathscr{A} \longrightarrow S$ on the representation algebra $\mathscr{A}$.

dual to Shor's original algorithm. We call the algorithm developed in the final
step of Fig. 1.8 the **dual Shor algorithm**.

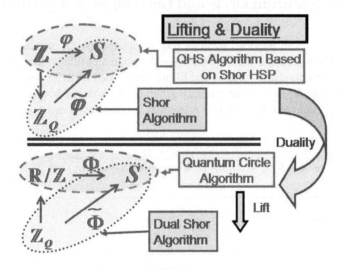

FIGURE 1.8: Roadmap for creating the dual Shor algorithm.

As indicated in Fig. 1.5, our first step is to create an intermediate QHS algo-
rithm based on a Shor-like HSP $\varphi : \mathbb{Z} \longrightarrow S$ from the additive group of integers
$\mathbb{Z}$ to a target set S. The resulting algorithm "lives" in the infinite dimensional
space $\mathcal{H}_{\mathbb{Z}}$ defined by the orthonormal basis $\{\langle n| : n \in \mathbb{Z}\}$. This is a physically
unimplementable quantum algorithm created as a first stepping stone in our
algorithmic development sequence. Intuitively, this algorithm can be viewed
as a "distillation" or a "purification" of Shor's original algorithm.

As a next step, **duality** is used to create the **quantum circle algorithm**. This
is accomplished by devising a QHS algorithm for an HSP $\Phi : \mathbb{R}/\mathbb{Z} \longrightarrow S$ on
the dual group $\mathbb{R}/\mathbb{Z}$ of the additive group of integers $\mathbb{Z}$. (By $\mathbb{R}/\mathbb{Z}$, we mean
the **additive group of reals mod 1**, which is isomorphic to the multiplicative
group $\{e^{2\pi i\theta} : 0 \leq \theta < 1\}$, i.e., the **unit circle** in the complex plane.) Once a-
gain, this is probably a physically unimplementable quantum algorithm.[15] But
its utility lies in the fact that it leads to the physically implementable quantum
algorithm created in the last and final developmental step, as indicated in Fig.
1.8. For in the final step, a physically implementable QHS algorithm is created

---

[15]There is a possibility that the quantum circle algorithm may have a physical implementation in
terms of quantum optics.

by **lifting** the HSP $\Phi : \mathbb{R}/\mathbb{Z} \longrightarrow S$ to an HSP $\widetilde{\Phi} : \mathbb{Z}_Q \longrightarrow S$. For the obvious reason, we call the resulting algorithm a **dual Shor algorithm**.

For detailed descriptions of each of these quantum algorithms, i.e., the "distilled" Shor, the quantum circle, and the dual Shor algorithms, the reader is referred to [27] and [29].

We give below brief descriptions of the quantum circle and the dual Shor algorithms.

For the **quantum circle algorithm**, we make use of the following spaces (each of which is used in quantum optics):

- The rigged Hilbert space $H_{\mathbb{R}/\mathbb{Z}}$ with orthonormal basis $\{|x\rangle : x \in \mathbb{R}/\mathbb{Z}\}$. By "orthonormal" we mean that $\langle x|y\rangle = \delta(x-y)$, where "$\delta$" denotes the Dirac delta function. The elements of $H_{\mathbb{R}/\mathbb{Z}}$ are **formal integrals** of the form $\oint dx\, f(x)|x\rangle$. (The physicist Dirac in his classic book [6] on quantum mechanics refers to these integrals as infinite sums. See also [4] and [10].)

- The complex vector space $H_{\mathbb{Z}}$ of formal sums

$$\left\{ \sum_{n=-\infty}^{\infty} a_n |\mathbf{n}\rangle : a_n \in C \ \forall n \in \mathbb{Z} \right\}$$

with orthonormal basis $\{|n\rangle : n \in Z\}$. By "orthonormal" we mean that $\langle n|m\rangle = \delta_{nm}$, where δ_{nm} denotes the Kronecker delta.

We can now design an algorithm which solves the following hidden subgroup problem:

Hidden Subgroup Problem for the Circle. *Let $\Phi : \mathbb{R}/\mathbb{Z} \longrightarrow \mathbb{C}$ be an admissible function from the circle group $\mathbb{R}/\mathbb{Z}$ to the complex numbers $\mathbb{C}$ with hidden rational period $\alpha \in \mathbb{Q}/\mathbb{Z}$, where $\mathbb{Q}/\mathbb{Z}$ denotes the rational circle, i.e., the rationals* mod 1.

REMARK 1.6 By an admissible function, we mean a function belonging to any sufficiently well behaved class of functions. For example, the class of functions which are Lebesgue integrable on $\mathbb{R}/\mathbb{Z}$. There, are many other classes of functions that work equally as well. ∎

PROPOSITION 1.1

If $\alpha = a_1/a_2$ (with $\gcd(a_1, a_2) = 1$) is a rational period of a function $\Phi : \mathbb{R}/\mathbb{Z} \longrightarrow C$, then $1/a_2$ is also a period of Φ. Hence, the minimal rational period of Φ is always a reciprocal integer mod 1.

The following quantum algorithm finds the reciprocal integer period of the function Φ.

$$\boxed{\text{CIRCLE-ALGORITHM}(\Phi)}$$

$\boxed{\text{Step 0}}$ Initialization

$$|\psi_0\rangle = |0\rangle |0\rangle \in \mathcal{H}_{\mathbb{Z}} \otimes \mathcal{H}_{\mathbb{C}} \, .$$

$\boxed{\text{Step 1}}$ Application of the inverse Fourier transform $\mathcal{F}^{-1} \otimes 1$

$$|\psi_1\rangle = \int dx \, e^{2\pi i \cdot 0} |x\rangle |0\rangle = \int dx \, |x\rangle |0\rangle \in \mathcal{H}_{\mathbb{R}/\mathbb{Z}} \otimes \mathcal{H}_{\mathbb{C}} \, .$$

$\boxed{\text{Step 2}}$ Step 2. Application of the unitary transformation $U_\varphi : |x\rangle |u\rangle \mapsto |x\rangle |u + \Phi(x)\rangle$

$$|\psi_2\rangle = \int dx \, |x\rangle |\Phi(x)\rangle \in \mathcal{H}_{\mathbb{R}/\mathbb{Z}} \oplus \mathcal{H}_{\mathbb{C}} \, .$$

$\boxed{\text{Step 3}}$ Application of the Fourier transform $\mathcal{F} \otimes 1$.

REMARK 1.7   Letting $x_m = x - \frac{m}{a}$, we have

$$\int dx \, e^{2\pi i n x} |\Phi(x)\rangle = \sum_{m=0}^{a-1} \int_{m/a}^{(m+1)/a} dx \, e^{-2\pi i n x} |\Phi(x)\rangle$$

$$= \sum_{m=0}^{a-1} \int_0^{1/a} dx_m \, e^{-2\pi i n \left(x_m + \frac{m}{a}\right)} \left| \Phi\left(x_m + \frac{m}{a}\right)\right\rangle$$

$$= \left(\sum_{m=0}^{a-1} e^{-2\pi i n m/a} \right) \int_0^{1/a} dx \, e^{-2\pi i n x} |\Phi(x)\rangle \, ,$$

where $1/a$ is the unknown reciprocal period. But

$$\sum_{m=0}^{a-1} e^{-2\pi i n m/a} = a\delta_{n=0 \bmod a} = \begin{cases} a \text{ if } n = 0 \bmod a \\ 0 \text{ otherwise} \end{cases}$$

Hence,

$$|\psi_3\rangle = \sum_{n\in\mathbb{Z}} |n\rangle \int dx\, e^{-2\pi inx} |\Phi(x)\rangle$$

$$= \left(\sum_{n\in\mathbb{Z}} |n\rangle\, \delta_{n=0 \bmod a}\right) \int_0^{1/a} dx\, e^{-2\pi inx} |\Phi(x)\rangle$$

$$= \left(\sum_{\ell\in\mathbb{Z}} |\ell a\rangle\right) \left(\int_0^{1/a} dx\, e^{-2\pi inx} |\Phi(x)\rangle\right) = \sum_{\ell\in\mathbb{Z}} |\ell a\rangle\, |\Omega(\ell a)\rangle\,.$$

∎

Step 4 Measurement of

$$|\psi_3\rangle = \sum_{\ell\in\mathbb{Z}} |\ell a\rangle\, |\Omega(\ell \mathbf{a})\rangle \in \mathscr{H}_{\mathbb{Z}} \otimes \mathscr{H}_{\mathbb{C}}$$

with respect to the observable

$$\sum_{n\in\mathbb{Z}} n\, |n\rangle\, \langle n|$$

to produce a random eigenvalue ℓa.

REMARK 1.8   The above quantum circle algorithm can be extended to a quantum algorithm which finds the hidden period α of a function Φ: $\mathbb{R}/\mathbb{Z} \longrightarrow \mathbb{C}$, when α is an arbitrary real number $\bmod 1$. But in creating this extended quantum algorithm, a very restrictive condition must be imposed on the map $\Phi : \mathbb{R}/\mathbb{Z} \longrightarrow \mathbb{C}$, namely, the condition that Φ be continuous. ∎

We now give a brief description of the **dual Shor algorithm**.

The dual Shor algorithm is a QHS algorithm created by making a discrete approximation of the quantum circle algorithm. More specifically, it is created by lifting the QHS circle algorithm for $\varphi : \mathbb{R}/\mathbb{Z} \longrightarrow \mathbb{C}$ to the finite cyclic group $\mathbb{Z}_Q$, as illustrated in the commutative diagram given below:

$$
\begin{array}{l}
\mathbb{Z}_Q \\
\mu \downarrow \quad \searrow \widetilde{\varphi} = Push(\varphi) = \varphi \circ \mu \\
\mathbb{R}/\mathbb{Z} \longrightarrow S
\end{array}
$$

Intuitively, just as in Shor's algorithm, the circle group $\mathbb{R}/\mathbb{Z}$ is approximated with the finite cyclic group $\mathbb{Z}_Q$, where the group $\mathbb{Z}_Q$ is identified with the additive group

$$\left\{\frac{0}{Q},\frac{1}{Q},\ldots,\frac{Q-1}{Q}\right\} \bmod 1 \,,$$

and where the hidden subgroup $\mathbb{Z}_P$ is identified with the additive group

$$\left\{\frac{0}{P},\frac{1}{P},\ldots,\frac{P-1}{P}\right\} \bmod 1 \,,$$

with $P = a_2$.

This is a physically implementable quantum algorithm. In a certain sense, it is actually faster than Shor's algorithm because the last step of Shor's algorithm uses the standard continued fraction algorithm to determine the unknown period. On the other hand, the last step of the dual Shor algorithm uses the much faster Euclidean algorithm to compute the greatest common divisor of the integers $\ell_1 a, \ell_2 a, \ell_3 a, \ldots$, thereby determining the desired reciprocal integer period $1/a$. For more details, please refer to [27] and [29].

1.10   A QHS algorithm for Feynman integrals

We now discuss a QHS algorithm based on Feynman path integrals. This quantum algorithm was developed at the Mathematical Sciences Research Institute (MSRI) in Berkeley, California when the first author of this paper was challenged with an invitation to give a talk on the relation between Feynman path integrals and quantum computing at an MSRI conference on Feynman path integrals.

Until recently, both authors of this paper thought that the quantum algorithm to be described below was a highly speculative quantum algorithm because the existence of Feynman path integrals is very difficult (if not impossible) to determine in a mathematically rigorous fashion. But surprisingly, Jeremy Becnel in his doctoral dissertation [1] actually succeeded in creating a firm mathematical foundation for this algorithm.

We should mention, however, that the physical implementability of this algorithm is still to be determined.

DEFINITION 1.5   *Let* PATHS *be the real vector space of all contin-*

uous paths $x : [0, 1] \longrightarrow \mathbb{R}^n$ *which are* L^2 *with respect to the inner product*

$$x \cdot y = \int_0^1 ds \; x(s) y(s)$$

with scalar multiplication and vector sum defined as

- $(\lambda x)(s) = \lambda x(s)$

- $(x + y)(s) = x(s) + y(s)$

We wish to create a QHS algorithm for the following hidden subgroup problem:

Hidden Subgroup Problem for PATHS. *Let* $\varphi :$ PATHS $\longrightarrow C$ *be a functional with a hidden subspace* V *of* PATHS *such that*

$$\varphi(x + v) = \varphi(x) \quad \forall v \in V$$

Our objective is to create a QHS algorithm which solves the above problem, i.e., which finds the hidden subspace V.

DEFINITION 1.6 *Let* $\mathcal{H}_{\text{PATHS}}$ *be the rigged Hilbert space with orthonormal basis* $\{|x\rangle : x \in$ PATHS$\}$, *and with bracket product* $\langle x | y \rangle = \delta(x - y)$.

We will use the following observation to create the QHS algorithm:

Observation. PATHS $= \bigcup_{v \in V} (v + V^{\perp})$, *where* $V^{\perp}$ *denotes the orthogonal complement of the hidden vector subspace* V.

The QHS algorithm for Feynman path integral is given below:

$$\boxed{\text{FEYNMAN}(\varphi)}$$

$\boxed{\text{Step 0}}$ Initialize

$$|\psi_0\rangle = |0\rangle |0\rangle \in \mathcal{H}_{\text{PATHS}} \otimes \mathcal{H}_{\mathbb{C}} \, .$$

$\boxed{\text{Step 1}}$ Apply $\mathscr{F}^{-1} \otimes 1$

$$|\psi_1\rangle = \int_{\text{PATHS}} \mathscr{D}x \, e^{2\pi i x \cdot 0} |x\rangle |0\rangle = \int_{\text{PATHS}} \mathscr{D}x \, |x\rangle |0\rangle \, .$$

Step 2 Apply $U_\varphi : |x\rangle |u\rangle \mapsto |x\rangle |u + \varphi(x)\rangle$

$$|\psi_2\rangle = \int\limits_{\text{PATHS}} \mathscr{D}x \, |x\rangle |\varphi(x)\rangle \ .$$

Step 3 Apply $\mathscr{F} \otimes 1$

$$
\begin{aligned}
|\psi_3\rangle &= \int\limits_{\text{PATHS}} \mathscr{D}y \int\limits_{\text{PATHS}} \mathscr{D}x \, e^{-2\pi i x \cdot y} |y\rangle |\varphi(x)\rangle \\
&= \int\limits_{\text{PATHS}} \mathscr{D}y \, |y\rangle \int\limits_{\text{PATHS}} \mathscr{D}x \, e^{-2\pi i x \cdot y} |\varphi(x)\rangle \ .
\end{aligned}
$$

But

$$
\begin{aligned}
&\int\limits_{\text{PATHS}} \mathscr{D}x \, e^{-2\pi i x \cdot y} |\varphi(x)\rangle \\
&= \int\limits_{V} \mathscr{D}v \int\limits_{v+V^\perp} \mathscr{D}x \, e^{-2\pi i x \cdot y} |\varphi(x)\rangle \\
&= \int\limits_{V} \mathscr{D}v \int\limits_{V^\perp} \mathscr{D}x \, e^{-2\pi i (v+x) \cdot y} |\varphi(v+x)\rangle \\
&= \int\limits_{V} \mathscr{D}v \, e^{-2\pi i v \cdot y} \int\limits_{V^\perp} \mathscr{D}x \, e^{-2\pi i x \cdot y} |\varphi(x)\rangle \ .
\end{aligned}
$$

However,

$$\int\limits_{V} \mathscr{D}v \, e^{-2\pi i v \cdot y} = \int\limits_{V^\perp} \mathscr{D}u \, \delta(y-u) \ .$$

So,

$$
\begin{aligned}
|\psi_3\rangle &= \int\limits_{\text{PATHS}_n} \mathscr{D}y \, |y\rangle \int\limits_{V} \mathscr{D}v \, e^{-2\pi i v \cdot y} \int\limits_{V^\perp} \mathscr{D}x \, e^{-2\pi i x \cdot y} |\varphi(x)\rangle \\
&= \int\limits_{\text{PATHS}_n} \mathscr{D}y \, |y\rangle \int\limits_{V^\perp} \mathscr{D}u \, \delta(y-u) \int\limits_{V^\perp} \mathscr{D}x \, e^{-2\pi i x \cdot y} |\varphi(x)\rangle \\
&= \int\limits_{V^\perp} \mathscr{D}u \, |u\rangle \int\limits_{V^\perp} \mathscr{D}x \, e^{-2\pi i x \cdot u} |\varphi(x)\rangle \\
&= \int\limits_{V^\perp} \mathscr{D}u \, |u\rangle |\Omega(u)\rangle \ .
\end{aligned}
$$

Step 4 | Measure

$$|\psi_3\rangle = \int_{V^\perp} \mathscr{D}u \, |u\rangle \, |\Omega(u)\rangle$$

with respect to the observable

$$A = \int_{\text{PATHS}} \mathscr{D}w \, |w\rangle \langle w|$$

to produce a random element of $V^\perp$.

The above algorithm suggests an intriguing question.

Question. *Can the above QHS Feynman integral algorithm be modified in such a way as to create a quantum algorithm for the Jones polynomial? In other words, can it be modified by replacing Paths with the space of gauge connections, and making suitable modifications?*

This question is motivated by the fact that the integral over gauge transformations

$$\widehat{\psi}(K) = \int \mathscr{D}A \, \psi(A) \, \mathscr{W}_K(A)$$

looks very much like a Fourier transform, where

$$\mathscr{W}_K(A) = tr\left(P\exp\left(\oint_K A\right)\right)$$

denotes the **Wilson loop** over the knot K.

1.11 QHS algorithms on free groups

In this and the following section of this paper, our objective is to show that a free group is the most natural domain for QHS algorithms. In retrospect, this is not so surprising if one takes a discerning look at Shor's factoring algorithm, for in Section 1.6, we have seen that Shor's algorithm is essentially a QHS algorithm on the free group $\mathbb{Z}$ which has been pushed onto the finite group $\mathbb{Z}_Q$.

In particular, let $\varphi : G \longrightarrow S$ be a map with hidden subgroup structure from a finitely generated (f.g.) group G to a set S. We assume that the hidden subgroup K is a normal subgroup of G of finite index. Then the objectives of this section are to demonstrate the following:

- Every hidden subgroup problem (HSP) $\varphi : G \longrightarrow S$ on an arbitrary f.g. group G can be lifted to an HSP $\widetilde{\varphi} : F \longrightarrow S$ on a free group F of finite rank.

- Moreover, a solution for the lifted HSP $\widetilde{\varphi} : F \longrightarrow S$ is for all practical purposes the same as the solution for the original HSP $\varphi : G \longrightarrow S$.

Thus, one need only investigate QHS algorithms for free groups of finite rank!

Before we can describe the above results, we need to review a number of definitions. We begin with the definition of a free group:

DEFINITION 1.7   *[Universal Definition] A group F is said to be **free** of finite rank n if there exists a finite set of n generators $X = \{x_1, x_2, \ldots, x_n\}$ such that, for every group G and for every map $f : X \longrightarrow G$ of the set X into the group G, the map f extends to a morphism $\widetilde{f} : F \longrightarrow G$. We call the set X a **free basis** of the group F, and frequently denote the group F by $F\left(x_1, x_2, \ldots, x_n\right)$, .*

REMARK 1.9   It follows from this definition that the morphism $\widetilde{f}$ is unique.   ∎

The intuitive idea encapsulated by this definition is that a free group is an unconstrained group (very much analogous to a physical system without boundary conditions.) In other words, a group is free provided it has a set of generators such that the only relations among those generators are those required for F to be a group. For example,

- $x_i x_i^{-1} = 1$ is an allowed relation.

- $x_i x_j = x_j x_i$ is not an allowed relation for $i \neq j$.

- $x_i^3 = 1$ is not an allowed relation.

As an immediate consequence of the above definition, we have the following proposition:

PROPOSITION 1.2
Let G be an arbitrary f.g. group with a finite set of n generators $\{g_1, g_2, \ldots, g_n\}$, and let $F = F\left(x_1, x_2, \ldots, x_n\right)$ be the free group of rank n with free basis $\{x_1, x_2, \ldots, x_n\}$.

Then by the above definition, the map $x_j \longmapsto g_j$ $(j = 1, 2, \ldots, n)$ induces a unique epimorphism $v : F \longrightarrow G$ from F onto G. With this epimorphism, every HSP $\varphi : G \longrightarrow S$ on the group G uniquely lifts to the HSP $\widetilde{\varphi} = \varphi \circ v : F \longrightarrow S$ on the free group F.

Moreover, if K and $\widetilde{K}$ are the hidden subgroups of the HSPs φ and $\widetilde{\varphi}$, respectively, the corresponding hidden quotient groups G/K and $F/\widetilde{K}$ of these two HSPs are isomorphic. Hence, every solution of the HSP $\widetilde{\varphi} : F \longrightarrow S$ immediately produces a solution of the original HSP $\varphi : G \longrightarrow S$.

We close this section with the definition of a group presentation, a concept that will be needed in the next section for generalizing Shor's algorithm to free groups.

DEFINITION 1.8   *Let G be a group. A* **group presentation**

$$\left(x_1, x_2, \ldots, x_n : r_1, r_2, \ldots, r_m \right)$$

for G is a set of free generators $x_1, x_2, \ldots, x_n$ of a free group F and a set of words $r_1, r_2, \ldots, r_n$ in $F\left(x_1, x_2, \ldots, x_n\right)$, called **relators**, *such that the group G is isomorphic to the quotient group $F\left(x_1, x_2, \ldots, x_n\right)/Cons$ $\left(r_1, r_2, \ldots, r_n\right)$, where $Cons\left(r_1, r_2, \ldots, r_n\right)$, called the* **consequence** *of $r_1, r_2,$ $\ldots, r_n$, is the smallest normal subgroup of $F\left(x_1, x_2, \ldots, x_n\right)$ containing the relators $r_1, r_2, \ldots, r_n$.*

The intuition captured by the above definition is that $x_1, x_2, \ldots, x_n$ are the generators of G, and $r_1 = 1, r_2 = 1, \ldots, r_n = 1$ is a complete set of relations among these generators, i.e., every relation among the generators of G is a **consequence** of (i.e., derivable from) the relations $r_1 = 1, r_2 = 1, \ldots, r_n = 1$. For example,

- $\left(x_1, x_2, \ldots, x_n :\right)$ and $\left(x_1, x_2, \ldots, x_n : x_1 x_1^{-1}, x_2^5 x_2^{-5}, x_3 x_4 x_4^{-1} x_3^{-1}\right)$ are both presentations of the free group $F\left(x_1, x_2, \ldots, x_n\right)$.

- $\left(x : x^Q\right)$ and $\left(x : x^a, x^b\right)$ are both presentations of the cyclic group $\mathbb{Z}_Q$ of order Q, where a and b are integers such that $\gcd(a, b) = Q$.

- $\left(x_1, x_2 : x_1^3, x_2^2, \left(x_1 x_2\right)^2\right)$ is a presentation of the symmetric group S_3 on three symbols.

1.12   Generalizing Shor's algorithm to free groups

The objective of this section is to generalize Shor's algorithm to free groups of finite rank.[16] The chief obstacle to accomplishing this goal is finding a correct generalization of the Shor transversal

$$\mathbb{Z}_Q \xrightarrow{\ \tau\ } \mathbb{Z}$$

$$n \bmod Q \longmapsto n \ (\, 0 \leq n < Q)$$

Unfortunately, there appear to be few mathematical clues indicating how to go about making such a generalization. However, as we shall see, the generalization of the Shor transversal to the transversal found in the wandering Shor algorithm does provide a crucial clue, suggesting that a generalized Shor transversal must be a 2-sided Schreier transversal. (See Section 1.7.)

We begin by formulating a constructive approach to free groups:

DEFINITION 1.9   *Let $F\left(x_1, x_2, \ldots, x_n\right)$ be a free group with free basis $x_1, x_2, \ldots, x_n$. Then a **word** is a finite string of the symbols $x_1, x_1^{-1}, x_2, x_2^{-1}, \ldots, x_n, x_n^{-1}$. A **reduced word** is a word in which there is no substring of the form $x_j x_j^{-1}$ or $x_j^{-1} x_j$. Two words are said to be **equivalent** if one can be transformed into the other by applying a finite number of substring insertions or deletions of the form $x_j x_j^{-1}$ or $x_j^{-1} x_j$. We denote an **arbitrary word** w by $w = a_1 a_2 \cdots a_\ell$, where each $a_j = x_{k_j}^{\pm 1}$. The **length** $|w|$ of a word $w = a_1 a_2 \cdots a_\ell$ is number of symbols $x_{k_j}^{\pm 1}$ that appear in w, i.e., $|w| = \ell$.*

For example, $x_2 x_1^{-1} x_1 x_1^{-1} x_5^{-1} x_5^{-1} x_5^{-1} x_5$ is a word of length 8 which is equivalent to the reduced word $x_2 x_1^{-1} x_5^{-1} x_5^{-1}$ of length 4.

It easily follows that:

PROPOSITION 1.3
A free group $F\left(x_1, x_2, \ldots, x_n\right)$ is simply the set of reduced words together

---

[16]We remind the reader that, in Section 1.6, we showed that Shor's algorithm is essentially a QHS algorithm on the free group $\mathbb{Z}$ of rank 1 constructed by a push onto the cyclic group $\mathbb{Z}_Q$. In light of this and of the results outlined in the previous section, it is a natural objective to generalize Shor's algorithm to free groups of finite rank.

with the obvious definition of product, i.e., concatenation followed by full reduction.

We can now use this constructive approach to create a special kind of transversal $\tau : G \longrightarrow F$ of an epimorphism $\nu : F \longrightarrow G$, called a 2-sided Schreier transversal [14]:

DEFINITION 1.10 *A set $\mathcal{W}$ of reduced words in a free group $F = F(x_1, x_2, \ldots, x_n)$ is said to be a **2-sided Schreier system** provided*

- *The empty word 1 lies in $\mathcal{W}$.*

- $w = a_1 a_2 \cdots a_{\ell-1} a_\ell \in \mathcal{W} \Rightarrow w_{Left} = a_1 a_2 \cdots a_{\ell-1} \in \mathcal{W}$, *and*

- $w = a_1 a_2 \cdots a_{\ell-1} a_\ell \in \mathcal{W} \Rightarrow w_{Right} = a_2 \cdots a_{\ell-1} a_\ell \in \mathcal{W}$.

*Given an epimorphism $\nu : F \longrightarrow G$ of the free group F onto a group G, a **2-sided Schreier transversal** $\tau : G \longrightarrow F$ for ν is a transversal of ν for which there exists a 2-sided Schreier system such that $\tau(G) = \mathcal{W}$. A 2-sided Schreier transversal is said to be **minimal** provided the length of each word w is less than or equal to the length of each reduced word in the coset $w Ker(\nu) = Ker(\nu) w$, where $Ker(\nu)$ denotes the kernel of the epimorphism ν.*

The wandering Shor algorithm found in Section 1.7 suggests that a correct generalization of the Shor transversal $n \bmod N \longmapsto n$ $(0 \le n < Q)$ must at least have the property that it is a minimal 2-sided Schreier transversal. Whatever other additional properties this generalization must have is simply not clear.

In [31], we construct and investigate a number of different QHS algorithms on free groups that arise from the application of various additional conditions imposed upon the minimal 2-sided Schreier transversal requirement. In this section, we only give a descriptive sketch of the simplest of these algorithms, i.e., a QHS algorithm on free groups with only the minimal 2-sided Schreier transversal requirement imposed.

Let $F = F(x_1, x_2, \ldots, x_n)$ be the free group of finite rank n with free basis $X = \{x_1, x_2, \ldots, x_n\}$, and let $\varphi : F \longrightarrow S$ be an HSP on the free group F. We assume that the hidden subgroup K is normal and of finite index in F. (Please note that $K = Ker(\varphi) = \varphi^{-1}\varphi(1)$.)

- Choose a finite group probe G with presentation $(x_1, x_2, \ldots, x_n : r_1, r_2, \ldots, r_m)_\nu$, where the subscript ν denotes the epimorphism $\nu : F \longrightarrow G$ induced by the map $x_j \longmapsto x_j Cons(r_2, \ldots, r_m)$.

- Choose a minimal 2-sided Schreier transversal $\tau : G \longrightarrow F$ of the epimorphism $v : F \longrightarrow G$.

- Finally, construct the push

$$\widetilde{\varphi} = Push\,(\varphi) = \varphi \circ \tau : G \longrightarrow S.$$

Our generalized Shor algorithm for the free group F consists of the following steps:

Shor's Algorithm Generalized to Free Groups

Step 1 Call QRAND($\widetilde{\varphi}$) to produce a word s'_j in F close to a word s_j lying in $\varphi^{-1}\varphi(1)$.

Step 2 With input s'_j, use a polytime classical algorithm to determine s_j. (See [31].)

Step 3 Repeat Steps 1 and 2 until enough relators s_j's are found to produce a presentation

$$\left(x_1, x_2, \ldots, x_n : s_1, s_2, \ldots, s_\ell\right)$$

of the hidden subgroup F/K, then output the presentation $\left(x_1, x_2, \ldots, x_n : s_1, s_2, \ldots, s_\ell\right)$, and STOP.

Obviously, much more needs to be said. For example, we have not explained how one chooses the relators r_j so that $G = (x_1, x_2, \ldots, x_n : r_1, r_2, \ldots, r_m)$ is a good group probe. Moreover, we have not explained what classical algorithm is used to transform the words s'_j into the relators s_j. For more details, we refer the reader to [31].

1.13   Is Grover's algorithm a QHS algorithm?

In this section, our objective is to factor Grover's algorithm into the QHS primitives developed in the previous sections of this paper. As a result, we will show that Grover's algorithm is more closely related to Shor's algorithm than one might at first expect. In particular, we will show that Grover's algorithm is a QHS algorithm in the sense that it solves an HSP $\varphi : S_N \longrightarrow S$, which we will refer to as the **Grover HSP**. However, we will then show that the standard QHS algorithm for this HSP cannot possibly find a solution.

We begin with a question:

Does Grover's algorithm have symmetries that we can exploit?

The problem solved by Grover's algorithm [11], [12], [13], [24] is that of finding an unknown integer label j_0 in an unstructured database with items labeled by the integers:

$$0, 1, 2, \ldots, j_0, \ldots, N-1 = 2^n - 1 \;,$$

given the oracle

$$f(j) = \begin{cases} 1 \text{ if } j = j_0 \;, \\ 0 \text{ otherwise} \;. \end{cases}$$

Let $\mathscr{H}$ be the Hilbert space with orthonormal basis $|0\rangle, |1\rangle, |2\rangle, \ldots,$ $|N-1\rangle$. Grover's oracle is essentially given by the unitary transformation

$$I_{|j_0\rangle} : \mathscr{H} \longrightarrow \mathscr{H}$$
$$|j\rangle \longmapsto (-1)^{f(j)} |j\rangle \;,$$

where $I_{|j_0\rangle} = I - 2 |j_0\rangle \langle j_0|$ is inversion in the hyperplane orthogonal to $|j\rangle$. Let W denote the Hadamard transformation on the Hilbert space H. Then Grover's algorithm is as follows:

Grover's Algorithm

Step 0 (Initialization)

$$|\psi\rangle \longleftarrow W|0\rangle = \frac{1}{\sqrt{N}} \sum_{j=0}^{N-1} |j\rangle$$
$$k \longleftarrow 0 \;.$$

Step 1 Loop until $k \approx \pi \sqrt{N}/4$

$$|\psi\rangle \longleftarrow Q|\psi\rangle = -W I_{|0\rangle} W I_{|j_0\rangle} |\psi\rangle$$
$$k \longleftarrow k+1 \;.$$

Step 2 Measure $|\psi\rangle$ with respect to the standard basis

$$|0\rangle,|1\rangle,|2\rangle,\ldots,|N-1\rangle$$

to obtain the unknown state $\left|j_0\right\rangle$ with

$$Prob \geq 1 - \frac{1}{N}.$$

But where is the hidden symmetry in Grover's algorithm?

Let S_N be the symmetric group on the symbols $0,1,2,\ldots,N-1$. Then Grover's algorithm is invariant under the **hidden subgroup** $Stab_{j_0} = \{g \in S_N : g(j_0) = j_0\} \subset S_N$, called the **stabilizer subgroup** for j_0, i.e., Grover's algorithm is invariant under the group action

$$Stab_{j_0} \times \mathcal{H} \quad\longrightarrow\quad \mathcal{H}$$

$$\left(g, \sum_{j=0}^{N-1} a_j |j\rangle\right) \quad\longmapsto\quad \sum_{j=0}^{N-1} a_j |g(j)\rangle$$

Moreover, if we know the hidden subgroup $Stab_{j_0}$, then we know j_0, and vice versa. In other words, the problem of finding the unknown label j_0 is informationally the same as the problem of finding the hidden subgroup $Stab_{j_0}$.

Let $(ij) \in S_N$ denote the permutation that interchanges integers i and j, and leaves all other integers fixed. Thus, (ij) is a transposition if $i \neq j$, and the identity permutation 1 if $i = j$.

PROPOSITION 1.4

The set $\left\{(0j_0),(1j_0),(2j_0),\ldots,((N-1)j_0)\right\}$ is a complete set of distinct coset representatives for the hidden subgroup $Stab_{j_0}$ of S_N, i.e., the coset space $S_N/Stab_{j_0}$ is given by the following complete set of distinct cosets:

$$S_N/Stab_{j_0} = \left\{ \begin{array}{l} (0j_0)\, Stab_{j_0},\ (1j_0)\, Stab_{j_0},\ (2j_0)\, Stab_{j_0},\ \ldots, \\ ((N-1)j_0)\, Stab_{j_0} \end{array} \right\}$$

We can now see that Grover's algorithm is a hidden subgroup algorithm in the sense that it is a quantum algorithm which solves the following hidden subgroup problem:

Grover's Hidden Subgroup Problem. *Let* $\varphi : S_N \longrightarrow S$ *be a map from the symmetric group* S_N *to a set* $S = \{0,1,2,\ldots,N-1\}$ *with hidden subgroup* *structure given by the commutative diagram*

$$
\begin{array}{ccc}
S_N & \longrightarrow & S \\
{\scriptstyle v_{j_0}}\searrow & & \nearrow {\scriptstyle \iota} \\
& S_N/Stab_{j_0} &
\end{array}\ ,
$$

where $v_{j_0} : S_N \longrightarrow S_N/Stab_{j_0}$ *is the natural surjection of* S_N *on to the coset space* $S_N/Stab_{j_0}$, *and where*

$$
\iota : S_N/Stab_{j_0} \longrightarrow S
$$
$$
(jj_0)\, Stab_{j_0} \longmapsto j
$$

is the unknown relabeling (bijection) of the coset space $S_N/Stab_{j_0}$ *onto the set* S. *Find the hidden subgroup* $Stab_{j_0}$ *with bounded probability of error.*

Now let us compare Shor's algorithm with Grover's.

From Section 1.6, we know that Shor's algorithm [21], [25], [35], [36] solves the hidden subgroup problem $\varphi : \mathbb{Z} \longrightarrow \mathbb{Z}_N$ with hidden subgroup structure

$$
\begin{array}{ccc}
\mathbb{Z} & \longrightarrow & \mathbb{Z}_N \\
{\scriptstyle v}\searrow & & \nearrow {\scriptstyle \iota} \\
& \mathbb{Z}/P\mathbb{Z} &
\end{array}
$$

Moreover, as stated in Section 1.6, Shor has created his algorithm by pushing[17] the above hidden subgroup problem $\varphi : \mathbb{Z} \longrightarrow \mathbb{Z}_N$ to the hidden subgroup problem $\widetilde{\varphi} : \mathbb{Z}_Q \longrightarrow \mathbb{Z}_N$ (called Shor's oracle), where the hidden subgroup structure of $\widetilde{\varphi}$ is given by the commutative diagram

$$
\begin{array}{ccc}
\mathbb{Z} & \longrightarrow & \mathbb{Z}_N \\
{\scriptstyle \alpha}\searrow\!\!\!\searrow {\scriptstyle \tau} & & \nearrow {\scriptstyle \widetilde{\varphi}=\varphi\circ\tau} \\
& \mathbb{Z}_Q &
\end{array}\ ,
$$

where α is the natural epimorphism of $\mathbb{Z}$ onto $\mathbb{Z}_Q$, and where τ is Shor's chosen transversal for the epimorphism α.

Surprisingly, Grover's algorithm, viewed as an algorithm that solves the Grover hidden subgroup problem, is very similar to Shor's algorithm.

---

[17]See Section 1.5 for a definition of pushing.

Like Shor's algorithm, Grover's algorithm solves a hidden subgroup problem, i.e., the Grover hidden subgroup problem $\varphi : S_N \longrightarrow S$ with hidden subgroup structure

$$
\begin{array}{ccc}
S_N & \longrightarrow & S \\
\scriptstyle\nu \searrow & & \nearrow \scriptstyle\iota \\
& S_N/Stab_{j_0} &
\end{array} \quad ,
$$

where $S = \{0,1,2,\ldots,N-1\}$ denotes the set resulting from an unknown relabeling (bijection)

$$
(jj_0)\, Stab_{j_0} \longmapsto j
$$

of the coset space

$$
S_N/Stab_{j_0} = \Big\{ (0j_0)\, Stab_{j_0}, (1j_0)\, Stab_{j_0}, (2j_0)\, Stab_{j_0}, \ldots,
$$
$$
((N-1)\, j_0)\, Stab_{j_0} \Big\} .
$$

Also, like Shor's algorithm, we can think of Grover's algorithm as one created by pushing the Grover hidden subgroup problem $\varphi : S_N \longrightarrow S$ to the hidden subgroup problem $\widetilde{\varphi} : S_N/Stab_{j_0} \longrightarrow S$, where the pushing is defined by the following commutative diagram

$$
\begin{array}{ccc}
S_N & \longrightarrow & S = S_N/Stab_{j_0} \\
\scriptstyle\alpha \searrow \scriptstyle\tau & & \nearrow \scriptstyle\widetilde{\varphi} = \varphi \circ \tau \\
& S_N/Stab_0 &
\end{array} \quad ,
$$

where $\alpha : S_N \longrightarrow S_N/Stab_0$ denotes the natural surjection of S_N onto the coset space $S_N/Stab_0$, and where $\tau : S_N/Stab_0 \longrightarrow S_N$ denotes the transversal of α given by

$$
\begin{array}{c}
S_N/Stab_0 \longrightarrow S_N \\
(j0)\, Stab_0 \longmapsto (j0)
\end{array} \quad .
$$

Again, also like Shor's algorithm, the map $\widetilde{\varphi}$ given by

$$
\begin{array}{c}
S_N/Stab_0 \longrightarrow S_N/Stab_{j_0} = S \\
(j0)\, Stab_0 \longmapsto (jj_0)\, Stab_{j_0} = j
\end{array}
$$

is (if $j_0 \neq 0$) actually a disguised Grover's oracle. For the map $\widetilde{\varphi}$ can easily be shown to simply to be

$$\widetilde{\varphi}\left((j0)Stab_0\right) = \begin{cases} (j0)Stab_{j_0} & \text{if } j = j_0 , \\ Stab_{j_0} & \text{otherwise} , \end{cases}$$

which is informationally the same as Grover's oracle

$$f(j) = \begin{cases} j \text{ if } j = j_0 , \\ 1 \text{ otherwise} . \end{cases}$$

Hence, we can conclude that Grover's algorithm is a quantum algorithm very much like Shor's algorithm, in that it is a quantum algorithm that solves the Grover hidden subgroup problem.

However, this appears to be where the similarity between Grover's and Shor's algorithms ends. The standard non-abelian QHS algorithm for S_N cannot find the hidden subgroup $Stab_{j_0}$ for each of following two reasons:

- Since the subgroups $Stab_j$ are not normal subgroups of S_N, it follows from the work of Hallgren et al. [16], [17] that the standard non-abelian hidden subgroup algorithm will find the largest normal subgroup of S_N lying in $Stab_{j_0}$. But unfortunately, the largest normal subgroup of S_N lying in $Stab_j$ is the trivial subgroup of S_N.

- The subgroups $Stab_0, Stab_1, \ldots, Stab_{N-1}$ are mutually conjugate subgroups of S_N. Moreover, one can not hope to use this QHS approach to Grover's algorithm to find a faster quantum algorithm. For Zalka [40] has shown that Grover's algorithm is optimal.

As stated previously, the arguments given in this section suggest that Grover's and Shor's algorithms are more closely related than one might at first expect. Although we have shown that the standard non-abelian QHS algorithm on S_N can not solve the Grover hidden subgroup problem, there still remains an intriguing question:

Question. *Is there some modification (or extension) of the standard QHS algorithm on the symmetric group S_N that actually solves Grover's hidden subgroup problem?*

For a more in-depth discussion of the results found in this section, we refer the reader to [30].

1.14 Beyond QHS algorithms: A suggestion of a meta-scheme for creating new quantum algorithms

In this paper, we have decomposed Shor's quantum factoring algorithm into primitives, generalized these primitives, and then reassembled them into a wealth of new QHS algorithms. But as the results found in the previous section suggest, this list of quantum algorithmic primitives is far from complete. This is expressed by the following question:

Question. *Where can we find more algorithmic primitives to create a more well rounded toolkit for quantum algorithmic development?*

The previous section suggests that indeed all quantum algorithms may well be hidden subgroup algorithms in the sense that they all find hidden symmetries, i.e., hidden subgroups. This is suggestive of the following meta-procedure for quantum algorithm development:

Meta-Procedure for Quantum Algorithm Development

| Meta-Step 1 | Explicitly state the problem to be solved. |

| Meta-Step 2 | Rephrase the problem as a hidden symmetry problem. |

| Meta-Step 3 | Create a quantum algorithm to find the hidden symmetry. |

Question. *Can this meta-procedure be made more explicit?*
Perhaps some reader of this paper will be able to answer this question.

Acknowledgement

This work is partially supported by the Defense Advanced Research Projects Agency (DARPA) and Air Force Research Laboratory, Air Force Materiel Command, USAF, under agreement number F30602-01-2-0522. The U.S. Government is authorized to reproduce and distribute reprints for Governmental purposes notwithstanding any copyright annotation thereon. This work is also partially supported by the Institute for Scientific Interchange (ISI), Torino, the National Institute of Standards and Technology (NIST), the Mathematical Sciences Research Institute (MSRI), the Isaac Newton Institute for Mathematical Sciences, and the L-O-O-P fund.

References

[1] Becnel, Jeremy James, Doctoral dissertation, (March, 2006). (http://etd.lsu.edu/docs/available/etd-06222006-133421/)

[2] Bernstein, Ethan, and Umesh Vazirani, **Quantum Complexity Theory**, SIAM Journal of Computing, Vol. 26, No. 5 (1997), 1411–1473.

[3] Biham, Eli, Ofer Biham, David Biron, Markus Grassl, and Daniel A. Lidar, **Grover's quantum search algorithm for an arbitrary ininitial amplitude distribution**, Phys Rev A 60 (1999), 2742–2745.

[4] Bohm, A., **"The Rigged Hilbert Space and Quantum Mechanics,"** Springer–Verlag, (1978).

[5] Cleve, Richard, Artur Ekert, Chiara Macchiavello, and Michele Mosca, **Quantum Algorithms Revisited,** Phil. Trans. Roy. Soc. Lond., A, (1997). http://xxx.lanl.gov/abs/quant-ph/9708016

[6] Dirac, P.A.M., **"The Principles of Quantum Mechanics,"** (Fourth edition), Oxford Science Publications, (1991).

[7] Ekert, Artur K. and Richard Jozsa, **Quantum computation and Shor's factoring algorithm,** Rev. Mod. Phys., 68 (1996), 733–753.

[8] Ettinger, Mark, and Peter Hoyer, **On Quantum Algorithms for Noncommutative Hidden Subgroups,** (1998). http://xxx.lanl.gov/abs/quant-ph/9807029

[9] Fulton, William, and Joe Harris, **"Representation Theory,"** Springer–Verlag, (1991).

[10] Gadella, M., and F. Gomez, A unified mathematical formalism for the Dirac formulation of quantum mechanics, Plenum Publishing, (2002), 815–869.

[11] Grover, Lov K., in Proc. 28th Annual ACM Symposium on the Theory of Computation, ACM Press, New York, (1996), 212–219.

[12] Grover, Lov K., **Quantum mechanics helps in searching for a needle in a haystack,** Phys. Rev. Lett., 79(2) (1997). (http://xxx.lanl.gov/abs/quant-ph/9706033)

[13] Grover, Lov K., **A framework for fast quantum mechanical algorithms,** http://xxx.lanl.gov/abs/quant-ph/9711043

[14] Hall, Marshall, **"The Theory of Groups," Macmillan Publishing,** (1967).

[15] Hales, Lisa R., The quantum Fourier transform and extensions of the abelian hidden subgroup problem, (UC Berkeley thesis), http://xxx.lanl.gov/abs/quant-ph/0212002.

[16] Hallgren, Sean, Alexander Russell, Amnon Ta-Shma, **The Hidden subgroup problem and quantum computation using group representations,** Proceedings of the Thirty-Second Annual ACM Symposium on Theory of Computing, Portland, Oregon, May 2000, 627–635.

[17] Hallgren, Sean, Alexander Russell, Amnon Ta-Shma, **The Hidden subgroup problem and quantum computation using group representations,** SIAM J. Comput., Vol. 32, No. 4 (2003), 916–934.

[18] Ivanyos, Gabor, Frederic Magniez, and Miklos Santha, **Efficient quantum algorithms for some instances of the non-Abelian hidden subgroup problem,** (2001). http://xxx.lanl.gov/abs/quant-ph/0102014

[19] Jozsa, Richard, **Quantum factoring, discrete logarithms and the hidden subgroup problem,** IEEE Computing in Science and Engineering, (to appear). http://xxx.lanl.gov/abs/quant-ph/0012084

[20] Kitaev, A., **Quantum measurement and the abelian stabilizer problem,** (1995), quant-ph preprint archive 9511026.

[21] Lomonaco, Samuel J., Jr., **Shor's Quantum Factoring Algorithm,** AMS PSAPM/58, (2002), 161–179. (http://arxiv.org/abs/quant-ph/0010034)

[22] Lomonaco, Samuel J., Jr., (ed.), **"Quantum Computation: A Grand Mathematical Challenge for the Twenty-First Century and the Millennium,"** Proceedings of the Symposia of Applied Mathematics, vol. 58, American Mathematical Society, Providence, Rhode Island, (2002). (358 pages)

(http://www.ams.org/bookstore?fn=20&arg1=whatsnew&item=
PSAPM-58)
(http://www.csee.umbc.edu/~lomonaco/ams/Lecture_Notes.html)

[23] Lomonaco, Samuel J., Jr., and Howard E. Brandt, (eds.), **"Quantum Computation and Information,"** AMS Contemporary Mathematics, vol. 305, American Mathematical Society, Providence, RI, (2002). (310 pages)
(http://www.csee.umbc.edu/~lomonaco/ams/Special.html)

[24] Lomonaco, Samuel J., Jr., **Grover's quantum search algorithm,** AMS PSAPM/58, (2002), 181–192.
(http://arxiv.org/abs/quant-ph/0010040)

[25] Lomonaco, Samuel J., Jr., and Louis H. Kauffman, **Quantum Hidden Subgroup Algorithms: A Mathematical Perspective,** AMS CONM/ 305 (2002), 139–202.
(http://arxiv.org/abs/quant-ph/0201095)

[26] Lomonaco, Samuel L., Jr., **A Rosetta stone for quantum mechanics with an introduction to quantum computation,** AMS PSAPM/58, (2002), 3–65. (http://arxiv.org/abs/quant-ph/0007045)

[27] Lomonaco, Samuel J., Jr., and Louis H. Kauffman, **Continuous quantum hidden subgroup algorithms**, SPIE Proceedings on Quantum Information and Computation, Vol. 5105, 11 (2003), 80–89. (http://arxiv.org/abs/quant-ph/0304084)

[28] Lomonaco, Samuel J., Jr., and, Louis H. Kauffman, **Quantum Hidden Subgroup Algorithms: The Devil Is in the Details,** 2004 Proceedings of SPIE Proceedings on Quantum Information and Computation, (2004), 137–141. http://arxiv.org/abs/quant-ph/0403229

[29] Lomonaco, Samuel J., Jr., and Louis H. Kauffman, **A Continuous Variable Shor Algorithm,** AMS CONM/381, (2005), 97–108. (http://arxiv.org/abs/quant-ph/0210141)

[30] Lomonaco, Samuel J., and Louis H. Kauffman, **Is Grover's algorithm a quantum hidden subgroup algorithm?** (http://arxiv.org/abs/quant-ph/0603140).

[31] Lomonaco, Samuel J., Jr., and Louis H. Kauffman, **Quantum hidden subgroup algorithms on free groups,** (in preparation).

[32] Lomonaco, Samuel J., Jr., **The non-abelian Fourier transform and quantum computation,** MSRI Streaming Video, (2000),

http://www.msri.org/publications/ln/msri/2000/qcomputing/
lomonaco/1/index.html

[33] Mosca, Michelle, and Artur Ekert, **The Hidden Subgroup Problem and Eigenvalue Estimation on a Quantum Computer,** Proceedings of the 1st NASA International Conference on Quantum Computing and Quantum Communication, Springer–Verlag, (2001). (http://xxx.lanl.gov/abs/quant-ph/9903071)

[34] Russell, Alexander, and Amnon Ta-Shma, **Normal Subgroup Reconstruction and Quantum Computation Using Group Representations,** STOC, (2000).

[35] Shor, Peter W., **Polynomial time algorithms for prime factorization and discrete logarithms on a quantum computer,** SIAM J. on Computing, 26(5) (1997), 1484–1509. (http://xxx.lanl.gov/abs/quant-ph/9508027)

[36] Shor, Peter W., **Introduction to quantum algorithms,** AMS PSAPM/58, (2002), 143–159. (http://xxx.lanl.gov/abs/quant-ph/0005003)

[37] Vazirani, Umesh, **On the power of quantum computation,** Philosophical Transactions of the Royal Society of London, Series A, 354:1759–1768, August 1998.

[38] Vazirani, Umesh, **A survey of quantum complexity theory,** AMS P-SAPM/58, (2002), 193–217.

[39] van Dam, Wim, and Lawrence, Ip, **Quantum Algorithms for Hidden Coset Problems,**
http://www.cs.caltech.edu/~hallgren/hcp.pdf

[40] Zalka, Christof, **Grover's quantum searching algorithm is optimal,** Phys. Rev. A, Vol. 60, No. 4 (1999), 2746–2751. (http://xxx.lanl.gov/abs/quant-ph/9711070)

Chapter 2

A Realization Scheme for Quantum Multi-Object Search

Zijian Diao, Goong Chen, and Peter Shiue

Abstract   We study the quantum circuit design using 1-bit and 2-bit unitary gates for the iterations of the multi-object quantum search algorithm. The oracle block is designed in order to efficiently implement any sign-flipping operations. A chief ingredient in the design is the permutation operator which maps a set of search targets to another set on which the sign-flipping operation can be easily done. Such a proposed algorithmic approach implicates a minimal symmetric group generation problem: how to generate elements of a symmetric group using the smallest number of concatenations with a set of given generators. For the general case, this is an open problem. We indicate how the complexity issues depend on the solution of this problem through simple examples.

2.1   Introduction

The quantum search algorithm due to L.K. Grover [12] has the advantage of a quadratic speedup over the classical serial search on an unsorted database. Grover's algorithm deals with single-object search. Its quantum circuit design is given in [11]. When there is more than one search target, as is prevalent in

most search problems, algorithms for multi-object search have been studied in
[2]–[8].

For multi-object search problems, the number of *items satisfying the search
criterion* (i..e, *search targets*) is not known a priori in general. This results in
a *quantum counting* problem for which eigenvalue estimates must be made in
order to determine the cardinality (see k in (2.1) below) of the search target
set; see [5, 6]. No quantum circuit design for the general multi-object search
algorithm is yet available, even though some *block diagram* has been suggested
in [6].

Let $D = \{w_i \,|\, i = 1, 2, \ldots, N\}$, where $N = 2^n$, be an unsorted database which
is encoded as basis quantum states $\widehat{D} = \{|w_i\rangle \,|\, i = 1, 2, \ldots, N\}$. Without loss
of generality, we assume that the set of search targets is

$$W = \{|w_1\rangle, |w_2\rangle, \ldots, |w_k\rangle\}. \tag{2.1}$$

Elements in W are identified through queries with the (block box) oracle function f:

$$f(w_i) = \begin{cases} 1 \text{ if } 1 \leq i \leq k, \\ 0 \text{ if } k+1 \leq i \leq N. \end{cases} \tag{2.2}$$

Recall from [7] that the unitary operator corresponding to the generalized
Grover search engine is given by

$$U = -\mathbf{I}_s \mathbf{I}_f, \tag{2.3}$$

where

$$\mathbf{I}_s = \mathbf{1} - 2|s\rangle\langle s|, \quad |s\rangle \equiv \frac{1}{\sqrt{N}} \sum_{i=1}^{N} |w_i\rangle, \tag{2.4}$$

is the "inversion about the average" operator, while

$$\mathbf{I}_f = \mathbf{1} - 2 \sum_{i=1}^{k} |w_i\rangle\langle w_i| \tag{2.5}$$

is the "selective sign-flipping" operator, since

$$\mathbf{I}_f |w_i\rangle = \begin{cases} -|w_i\rangle, & \text{if } 1 \leq i \leq k, \\ |w_i\rangle, & \text{if } k+1 \leq i \leq N. \end{cases} \tag{2.6}$$

The iterations

$$U^j |s\rangle \tag{2.7}$$

are performed and stopped at $j \approx \frac{\pi}{4}\sqrt{\frac{N}{k}}$. A measurement on the quantum system will yield a state in W with large probability.

Note that the oracle function f in (2.2) is in a black box and is not known explicitly. Without a priori knowledge of the search targets, the realization of (2.5) on the quantum computer is utterly non-trivial. For complexity theorists, the use of an oracle function f is a standard practice where f is readily available as a separate computing unit and the complexity involved for the construction and operation of f is entirely ignored. However, in the context of quantum computers, in order to have a complete design which does not depend on any other stand-alone units, and to exploit the entanglement between quantum subsystems, the quantum oracle has to be integrated with other components of the system. In theory, the "standard" way to implement $\mathbf{I}_f$ is by the well-known Deutsch's f-c-n "gate"

$$U_f : |w\rangle|y\rangle \longrightarrow |w\rangle|y \oplus f(w)\rangle \qquad (2.8)$$

where $|w\rangle \in \widehat{D}$ and $|y\rangle$, the auxiliary register, is chosen to be $|y\rangle = \frac{1}{\sqrt{2}}(|0\rangle - |1\rangle)$, leading to

$$U_f \left(|w\rangle \otimes \left[\frac{1}{\sqrt{2}}(|0\rangle - |1\rangle) \right] \right) = (-1)^{f(w)}|w\rangle \otimes \left[\frac{1}{\sqrt{2}}(|0\rangle - |1\rangle) \right]. \qquad (2.9)$$

However, Deutsch's gate (2.8) is not an elementary gate. The action of U_f, a linear operator, is determined by the implicitly *nonlinear* oracle function f. This approach still treats the quantum oracle as a separate module working independently, instead of an integral part of the whole quantum system. Furthermore, unless the computational structure of f is given explicitly, it is highly puzzling to us whether and how it will indeed be possible in the future to realize (2.8) quantum mechanically without the need of using elementary 1-bit and 2-bit unitary gates. As a matter of fact, all current physical implementations of quantum algorithms construct the quantum oracles via *"hard wiring"*, i.e., adapting the layout of the circuit according to the (known) distribution of the function values of f. The main thrust of this paper is to propose a "hard wiring" design to realize (2.8) with elementary gates.

In the quantum circuit design for (2.5) (or, equivalently, for (2.8)), it is totally reasonable to expect that the complexity of the "hard wiring" circuit depends on k in certain way. Therefore, for a single oracle call, there is a clear distinction between its complexity in theoretical discussion, where it is considered to be carried out in one step, and that in the practical implementation, where the *hidden complexity* associated with its construction via elementary gates must be accounted for. At present, our approach proposed here is mostly a *viability study.* The optimal design and its corresponding complexity analysis merit a separate paper, which we hope to present in the sequel.

Return to the multi-object search equations (2.3) and (2.7). In comparison with the quantum circuit design for the single-object search and in view of the commentary in the preceding two paragraphs, we understand that the main difference is in the oracle block $\mathcal{O}$ (cf. [11, Theorem 8]). In the next few sections, we ready ourselves in the redesign of this portion.

2.2 Circuit design for the multi-object sign-flipping operator

The task of $\mathbf{I}_f$ is to selectively flip the signs of the target states. For the single object case, we can construct $\mathbf{I}_f$ with polynomial complexity using basic 1-qubit and 2-qubit quantum gates [11]. For multi-object case, we may directly concatenate k selective sign-flipping operators of each of the k target states. However, the complexity of this construction is proportional to the number of search targets, which becomes very inefficient when k is large. A better design is to divide the targets into groups and flip the signs of states in each group together.

Example 2.1
Let $n = 4$ and assume the search targets be $E = \{|1100\rangle, |1101\rangle\, |1110\rangle,$ $|1111\rangle\}$. We can flip the signs of all the states in E together, without resorting to four sign-flipping operators tailored to the four targets individually. See Fig. 2.1 for details.   □

We summarize the strategy of our design of I_f first.

1. Partition the set W of search objects into subsets W_i with proper cardinality.

2. Via permutation p_i, map each W_i onto a set E_i of states whose signs are easy to flip together, e.g., E in Example 2.1.

3. Flip the signs of states in W_i through the operations on E_i.

We start by partitioning the set W of target states into $m + 1$ sets of W_i's, according to the binary expansion of k, $k = (k_m k_{m-1} \ldots k_2 k_1 k_0)_2$, i.e., $k =$

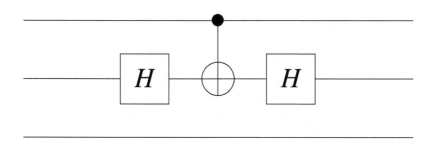

FIGURE 2.1: Circuit design of $\mathbf{I}_f$ for Example 2.1. H denotes the usual Hadamard transform. The concatenation of the two Hadamard gates and the CNOT gate on the first two qubits maps $|11\rangle$ to $-|11\rangle$, hence the signs of all four states in E with leading qubits $|11\rangle$ are flipped together.

$k_m 2^m + k_{m-1} 2^{m-1} + \ldots + k_1 2^1 + k_0 2^0$. Note that $m < n$, unless all states are search targets. Each set W_i contains $k_i 2^i$ states, for $i = 0, 1, \ldots, m$. W_i might be empty.

Example 2.2

(i) Let $k = 7$ and $W = \{|w_1\rangle, |w_2\rangle, \ldots, |w_7\rangle\}$. Then W can be partitioned to the following:

$$W = W_2 \dot\cup W_1 \dot\cup W_0$$

with $W_2 = \{|w_1\rangle, |w_2\rangle, |w_3\rangle, |w_4\rangle\}$, $W_1 = \{|w_5\rangle, |w_6\rangle\}$, and $W_0 = \{|w_7\rangle\}$, where $\dot\cup$ denotes disjoint union.

(ii) If $k = 10$ and $W = \{|w_1\rangle, |w_2\rangle, \ldots, |w_{10}\rangle\}$. Then

$$W = W_3 \dot\cup W_1,$$

where $W_3 = \{|w_1\rangle, |w_3\rangle, |w_4\rangle, |w_5\rangle, |w_7\rangle, |w_8\rangle, |w_9\rangle, |w_{10}\rangle\}$, $W_1 = \{|w_2\rangle, |w_6\rangle\}$, and $W_2 = W_0 = \emptyset$.

Note that the partition is non-unique. The only thing that matters for now is the cardinality of each set W_i, $i = 0, 1, \ldots, m$. □

We flip the signs of basis states in W by flipping the signs of the states in W_i for $i = 0, 1, 2, \ldots, m$. For each W_i, we construct a circuit block B_i. If W_i is empty, no action is needed. We now delineate the circuit design for a generic block B_i in three steps.

Step 1. Let us denote the basis states in W_i as $W_i = \{|w_{i,1}\rangle, |w_{i,2}\rangle, \ldots, |w_{i,2^i}\rangle\}$. Each $w_{i,j}$ is an n-bit string of 0 and 1's. Define E_i to be the set consisting of all the states whose first $n - i$ bits are all 1's. Clearly, $|E_i| = 2^i = |W_i|$. We construct the quantum circuit P_i which implements the permutation p_i mapping W_i onto E_i. One feasible, albeit inefficient, implementation is to pair up each $w_{i,j}$ with a state in E_i and do 2^i transpositions, as described in Table 2.1.

$$
\begin{array}{ll}
& \overbrace{\quad}^{(n-i)\text{ bits}} \overbrace{\quad}^{i\text{ bits}} \\
w_{i,1} & \leftrightarrow 11\cdots1\,00\cdots00; \quad (w_{i,1} \quad 11\cdots100\cdots00) \\
w_{i,2} & \leftrightarrow 11\cdots1\,00\cdots01; \quad (w_{i,2} \quad 11\cdots100\cdots01) \\
w_{i,3} & \leftrightarrow 11\cdots1\,00\cdots10; \quad (w_{i,3} \quad 11\cdots100\cdots10) \\
\vdots & \qquad\qquad\qquad\qquad\quad \vdots \\
w_{i,2^i-1} & \leftrightarrow 11\cdots1\,11\cdots10; \quad (w_{i,2^i-1}11\cdots111\cdots10) \\
w_{i,2^i} & \leftrightarrow 11\cdots1\,11\cdots11; \quad (w_{i,2^i} \quad 11\cdots111\cdots11).
\end{array}
$$

Table 2.1 The transpositions of the states in W_i with those in E_i. The left column of the table signifies that the two sides of the double arrow "$\leftrightarrow$" are mutually transposed. We use the 2-cycles on the right column to denote the corresponding transpositions on the left column.

Example 2.3
  Assume that $n = 7$ and $i = 4$. For $j = 4$, say we have

$$w_{i,j} = w_{4,4} = 0001111.$$

We want to perform the permutation

$$0001111 \leftrightarrow 1110011. \tag{2.10}$$

For ease of discussion, we make the following list:

$$
\begin{array}{lll}
s_1\colon 0001111; & s_2\colon 0011111; & s_3\colon 0111111; \\
s_4\colon 1111111; & s_5\colon 1110111; & s_6\colon 1110011.
\end{array}
$$

Note that each successive pair of symbols s_i and s_{i+1} differs by only one bit.

Then the transposition (2.10) can be achieved through the following sequence of transpositions (cf. the notation used in Table 2.1):

$$(s_1\ s_2)(s_2\ s_3)(s_3\ s_4)(s_4\ s_5)(s_5\ s_6)(s_4\ s_5)(s_3\ s_4)(s_2\ s_3)(s_1\ s_2). \qquad (2.11)$$

Note that through the above permutations, s_1 becomes s_6 and s_6 becomes s_1, achieving (2.10), while $s_2, s_3, \ldots, s_5$ remain unchanged. Several permutations in (2.11) are duplicated. Thus we only need to construct $(s_1\ \ s_2)$, $(s_2\ \ s_3)$, $(s_3\ \ s_4)$, $(s_4\ \ s_5)$ and $(s_5\ \ s_6)$ in order to achieve (2.10).

The circuit design in Fig. 2.2 realizes the permutation $(s_1\ s_2) = (0001111\ 0011111)$. The circuit diagrams for any other $(s_i\ s_{i+1})$ in (2.11) are similar. $\Box$

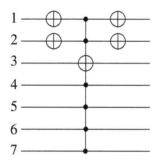

FIGURE 2.2: Circuit diagram for the permutation $(s_1\ \ s_2) = (0001111\ 0011111)$. Note that the third bit is flipped when and only when the remaining bits are, respectively, 0,0,1,1,1,1, in sequential order.

Step 2. Construct $\mathcal{O}_i$, which flips the signs of any states whose first $n - i$ leading bits are all 1's, i.e., states in E_i. The circuit block is given in Fig. 2.3.

Recall from [11] that K_{n-i} is the key transformation on (the first) $n - i$ bits, defined by

$$K_{n-i} = \mathbf{1}_{n-i} - 2|\overbrace{11\cdots 1}^{n-i}\rangle\langle 11\cdots 1|,$$

where $\mathbf{1}_{n-i}$ is the identity operator on the first $n - i$ bits. Its construction in terms of elementary gates is given in [11, Fig. 9]. The $\mathcal{O}_i$ block in Fig. 2.3 thus represents the unitary transformation

$$K_{n-i} \otimes \mathbf{1}_i, \text{ where } \mathbf{1}_i \text{ is the identity operator on the last } i \text{ bits.}$$

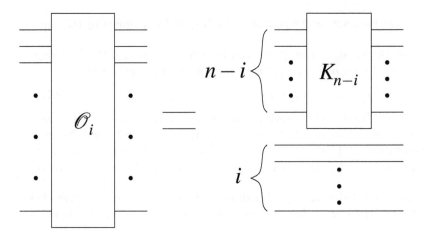

FIGURE 2.3: The $\mathcal{O}_i$ block, which flips the signs of any states whose first $n-i$ bits are all 1's. K_{n-i} is the key transformation for the first $n-i$ bits.

Step 3. Piece together $\mathcal{O}_i$, P_i, and P_i^{-1} (the circuit implementing p_i^{-1}, the inverse of p_i), to obtain B_i, for $i=0,1,\ldots,m$. See Fig. 2.4.

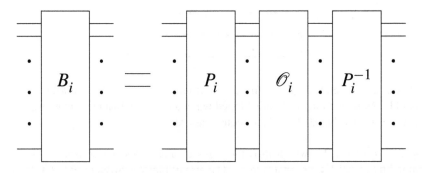

FIGURE 2.4: The block $B_i, i=0,1,\ldots,m$.

Further, concatenate all the B_i blocks for $i=0,1,\ldots,m$ to form the $\mathcal{O}$ (oracle) block. See Fig. 2.5.

Example 2.4

Let $n=2$ and assume the search targets be $W=\{|00\rangle,|01\rangle\}$. Then $k=2$ and only one block B_1 for $W_1=\{|00\rangle,|01\rangle\}$ is needed. See Fig. 2.6 for

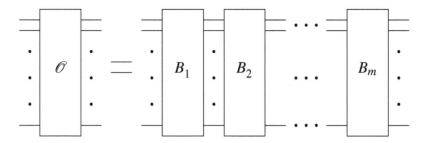

FIGURE 2.5: The oracle block $\mathcal{O}$, formed by concatenating $B_0, B_1, \ldots, B_m$.

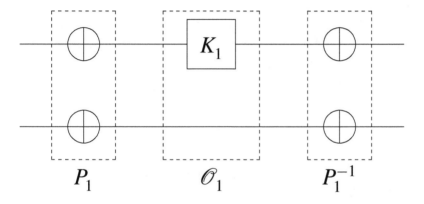

FIGURE 2.6: The circuit design of B_1 for Example 2.4.

the circuit design of B_1.

If, as in [11], what we have available are the following elementary gates:

- 1-bit unitary gates

$$
U_{\theta, \phi} = \begin{bmatrix} \cos\theta & -ie^{-i\phi}\sin\theta \\ -ie^{-i\phi}\sin\theta & \cos\theta \end{bmatrix}, \quad 0 \le \theta, \phi \le 2\pi. \quad (2.12)
$$

- 2-bit quantum phase gates

$$Q_\eta = \begin{bmatrix} 1 & 0 & 0 & 0 \\ 0 & 1 & 0 & 0 \\ 0 & 0 & 1 & 0 \\ 0 & 0 & 0 & e^{i\eta} \end{bmatrix}, \quad 0 \le \eta \le 2\pi. \tag{2.13}$$

The circuit design for B_1 and, consequently, $\mathcal{O}$, is given in Fig. 2.7.

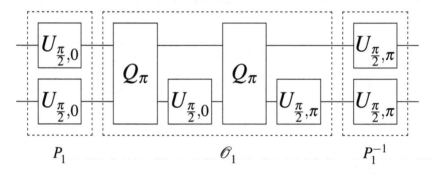

FIGURE 2.7: The circuit design of B_1 and $\mathcal{O}$ for Example 2.4, using $U_{\theta,\phi}$ and Q_η as elementary gates.

Note that P_1 here performs the permutations $(00 \quad 11)(01 \quad 10)$.   □

THEOREM 2.1
Let $U_\mathcal{O}$ denote the unitary operator corresponding to the operation performed by $\mathcal{O}$. Then

$$U_\mathcal{O}|w_j\rangle = \begin{cases} (-1)|w_j\rangle, & \text{if } |w_j\rangle \in W, \\ |w_j\rangle, & \text{if } |w_j\rangle \notin W. \end{cases}$$

PROOF   If $|w_j\rangle \in W$, then $|w_j\rangle \in W_{i_0}$ for some unique i_0, $i_0 \in \{0,1,\ldots,m\}$. Therefore

$$B_i|w_j\rangle = \begin{cases} -|w_j\rangle, & \text{if } i = i_0, \\ |w_j\rangle, & \text{if } i \ne i_0. \end{cases}$$

Thus $U_{\mathcal{O}}|w_j\rangle = B_m B_{m-1} \cdots B_1 B_0 |w_j\rangle = -|w_j\rangle$.

If $|w_j\rangle \notin W$, then $B_i|w_j\rangle = |w_j\rangle$ for all $i \in \{0, 1, \ldots, m\}$. Therefore $U_{\mathcal{O}}|w_j\rangle = B_m B_{m-1} \cdots B_1 B_0 |w_j\rangle = |w_j\rangle$. ∎

Hence $U_{\mathcal{O}}$ indeed corresponds to the sign-flipping operator $\mathbf{I}_f$ in (2.8). Finally, the overall circuit blocks are given in Fig. 2.8.

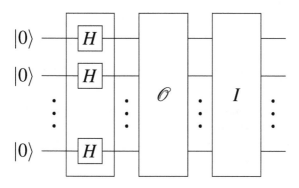

FIGURE 2.8: The block diagram for the multi-object search iteration (2.7). The block I performs the inversion about average operation $\mathbf{I}_s$, whose circuit design is the same as in [11].

2.3   Additional discussion

Any universal set of quantum gates [1] can be used to construct the component circuitries required in Section 2.2. In particular, the 1-bit and 2-bit gates $U_{\theta,\phi}$ and Q_η in (2.12) and (2.13) are universal. Therefore, they can be used for the purpose of this paper. See some relevant results in [11].

But there is a special case we need to address. That is, when $n - i = 1$. We need to construct K_1, which is simply the transformation $|0\rangle \rightarrow |0\rangle$ and $|1\rangle \rightarrow -|1\rangle$. However, this is not directly constructible with the 1-bit gates $U_{\theta,\phi}$, as they are *special unitary*, i.e., all $U_{\theta,\phi}$ have determinant equal to 1.

Two solutions are possible:

1. Use an auxiliary qubit which is set to $|1\rangle$. Bind this auxiliary qubit with the first work qubit with a phase shift gate Q_π. If the leading work qubit is $|1\rangle$, then Q_π maps $|11\rangle$ to $-|11\rangle$. Otherwise, Q_π leaves $|10\rangle$ unchanged. Ignore the auxiliary qubit, the sign of the leading work qubit is flipped. In other words, we have the equivalent network as shown in Fig. 2.9.

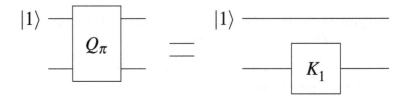

FIGURE 2.9: Construction of K_1 using Q_π and an auxiliary qubit.

2. This one is slightly more complicated than the previous one, but no auxiliary qubit is needed. The idea is to use the first qubit to flip the sign of the second qubit, no matter what it is, so that the sign of the overall state is flipped. See the captions and circuits in Fig. 2.10.

The second solution, in particular, points out one possible realization of the 1-bit phase shift operator

$$
\begin{bmatrix} e^{i\phi} & 0 \\ 0 & e^{i\phi} \end{bmatrix},
\tag{2.14}
$$

which was not possible using the $U_{\theta,\phi}$ gates alone. The circuit is given in Fig. 2.11.

2.4   Complexity issues

Following the analysis of the quantum circuit design for the single-object search ([11]), we know that we have linear circuit complexity to construct the I block using elementary 1-bit and 2-bit gates. To be exact, using $U_{\theta,\phi}$ and Q_η gates, the total number of gates needed is $24n - 74$, where n is the number of qubits involved. However, the construction of the $\mathcal{O}$ block for the multi-object case is more complicated than that of the single-object case. Since we have broken up the $\mathcal{O}$ block into $m+1$ blocks B_0, B_1, ..., B_m, where m can

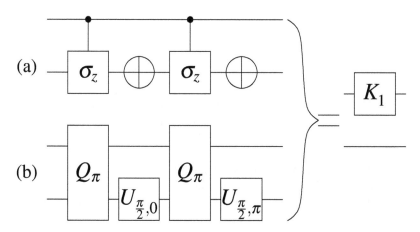

FIGURE 2.10: (a) Construction of K_1 without auxiliary qubits, where σ_z is the standard Pauli matrix $\begin{bmatrix} 1 & 0 \\ 0 & -1 \end{bmatrix}$. If the first qubit is $|0\rangle$, then the NOT gate will be applied on the second qubit twice. Hence nothing is changed. If the first qubit is $|1\rangle$, then no matter what the second qubit is, its sign is going to be flipped exactly once. So the function of this circuit is exactly what we expected. (b) The components circuits in (a) are rewritten in terms of U and Q gates.

be as large as $n-1$, and each B_i utilizes a K_{n-i} block, which requires linear complexity itself, we would not expect our design to have linear complexity as in the single object case. Even so, it is still highly desirable if we can achieve the design with as much simplicity as possible. As suggested by the summary of our design in Section 2.2, there are several flexibilities that we can exploit in order to achieve optimal complexity.

1. The partition of W into W_i's is not unique. We only enforce the cardinality requirement.

2. The choice of permutation $p_i : W_i \rightarrow E_i$ is not unique. In fact, there are $2^i!$ of them.

3. The implementation of each permutation p_i or p_i^{-1} in terms of elementary 1-bit and 2-bit gates is not unique.

Taking all these factors into account, we can formulate the optimal design of the block $\mathcal{O}$ as a minimization problem:

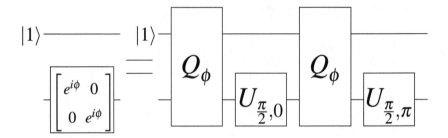

FIGURE 2.11: The 1-bit phase gate in equation (2.14) can be realized using the elementary gates $U_{\theta,\phi}$ and Q_η. The first auxiliary qubit is always set to $|1\rangle$. The net operation done on the second qubit is exactly (2.14) by ignoring the auxiliary qubit.

$$\min_{\mathbb{P}} \sum_i \min_{p_i : W_i \to E_i} \{c(p_i) + c(p_i^{-1})\}, \qquad (2.15)$$

where $\mathbb{P}$ denotes all possible partitions of W into W_i's; $c(p_i)$ and $c(p_i^{-1})$ denote the complexities of p_i and p_i^{-1}, respectively. We have omitted the complexity of $\mathcal{O}_i$, since it stays the same in our design. Because the complexity of the permutations p_i and p_i^{-1} constitutes the basic elements of our complexity analysis, we elaborate on this issue in the following.

The block P_i implements the permutation which maps the 2^i states in W_i onto those in E_i. A "brute-force" way of constructing P_i is to pair up the states in W_i and E_i, implement 2^i transpositions separately following the approach given in Example 2.2, and then concatenate them together. In total, the number of transposition blocks to construct for the overall search circuit will be linear in k, the number of target states. When k is much smaller than N, the total number of items in the database, as in the cases when quantum search algorithm is most powerful, the complexity of the circuit is still quite satisfactory, since each transposition requires only $O(\log^2 N)$ elementary gates ([11]). Nevertheless, unfortunately, when k is large, the complexity of this kind of construction becomes unacceptable.

We should note that, in general, the "brute-force" approach is far from being optimal and there is much room for improvement. For example, in Example 2.4, we did not use this approach to implement the permutation $(00 \quad 11)(01 \quad 10)$. Instead, we use two NOT gates to realize the product of those two transpositions in one step. The resultant circuit is much simpler than the one by concatenation of the two transpositions constructed separately. Let us look at another example.

Example 2.5

Let $n = 4$ and assume the search targets be $W = \{|0000\rangle, |0001\rangle, |0010\rangle,$ $|0011\rangle, |0100\rangle, |0101\rangle, |0110\rangle, |0111\rangle\}$, i.e., all states with leading qubit being 0. Clearly, $W_0 = W_1 = W_2 = \emptyset$ and $W_3 = W$. If we had followed the approach as given above, we would have to construct 8 transpositions, namely, (0000 1000), (0001 1001), (0010 1010), 0011 1011), (0100 1100), (0101 1101), (0110 1110), and (0111 1111). However, there exists a much more elegant way to implement P_3. All we need to do here is to negate the first qubit and the correctness of this approach is trivial to verify. This cuts down the circuit complexity to a constant for this example. ☐

We can rephrase our discussion of implementing permutation p_i via elementary gates under the framework of group theory. Now this task is reduced to a special case of the optimal generation of finite symmetric groups using a set of generators, e.g., but not limited to, the set of transpositions:

Problem: Let S_{2^n} be the symmetric group on 2^n elements, and let $G = \{g_1, g_2, \ldots, g_L\}$ be a set of generators for S_{2^n}. Define $c(p)$, the complexity of a permutation $p \in S_{2^n}$, by

$$c(p) = \min_{p = g_{i_1} g_{i_2} \cdots g_{i_l}} l. \tag{2.16}$$

Given p, what is $c(p)$, the minimum number of g_i's (repetition counted) needed to generate p, and what is the best (shortest) generation? ☐

With our problem in mind, we can encode the 2^n elements by their binary representation, and formulate the operations of the elementary gates by permutations on these elements. We may take the generating set G to be the permutations resulted from any sets of universal gates, in particular, the following fundamental gates:

NOT-gate: it flips the value of one qubit from 0 to 1, and 1 to 0.

Controlled-NOT-gate: it flips the value of a designated qubit depending on other control qubits.

It is well-known that these gates form a universal generating set of S_{2^n}. However, it is not clear what is the most efficient way to generate any given permutation $P \in S_{2^n}$ using the permutations induced by them.

Example 2.6

Let $n = 2$. Consider the minimum generation of S_4 via the following 4 permutations

NOT-gate on bit 1: $N(1) = (00 \quad 01)(10 \quad 11)$,
NOT-gate on bit 2: $N(2) = (00 \quad 10)(01 \quad 11)$,

Controlled-NOT-gate, bit 1 controlling bit2:
$\Lambda_1(2) = (01 \quad 11)$,
Controlled-NOT-gate, bit 2 controlling bit1:
$\Lambda_2(1) = (10 \quad 11)$.

We may generate all the $4! = 24$ permutations with these four permutations. Fig. 2.12 gives a minimum generation using breadth first search [9, p. 469].

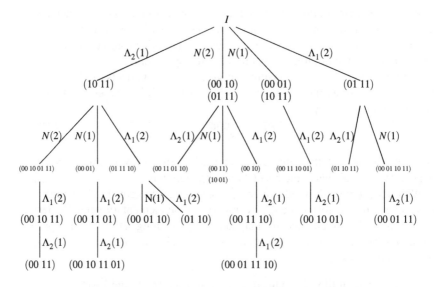

FIGURE 2.12: Generation tree of the symmetric group S_4.

We can see that the depth of this tree is 4. That is, we need at most 4 permutations (elementary gates) to implement any permutation in S_{2^2}. And we can read out the optimal generation by following the branches of the tree.

The analogous group generation problem has been studied in [15]. It is well known that the set $B = \{(1 \quad i) | 2 \le i \le n\}$ generates S_n. Given $\alpha \in S_n$, let $l_B(\alpha)$ be the complexity of α using generators from B. Let l_B be the largest $l_B(\alpha)$ of all elements in S_n. One can show that $l_B = 4$

in S_4. See Table 2.2, where the left-hand-side is written in terms of the list notation of a permutation, e.g., $(3,1,4,2)$ stands for the permutation $1 \to 3$, $2 \to 1$, $3 \to 4$, and $4 \to 2$. [15] has shown that l_B is $3m$ in S_{2m+1}. It seems that a formula of l_B in S_{2m} is an interesting open question.   ☐

Since any computable function can be embedded into a reversible function, which can be viewed as a permutation, we can reformulate the computation of functions by generating permutations. For the complexity issues, we also consider the basic bit operations as in arithmetic complexity theory [13]. The difference is, we have translated all the basic bit operations into permutations in a certain symmetric group, and the complexity is considered in the context of group generation. This kind of symmetric group framework has also been used in the enumeration problems in combinatorics [10, 14]. The difficulty of our problem lies in the fact that symmetric groups are non-abelian. More research is needed in order to understand better ways to do multi-object quantum search.

References

[1] A. Barenco, C.H. Bennett, R. Cleve, D. DiVincenzo, N. Margolus, P. Shor, T. Sleator and H. Weinfurter, Elementary gates for quantum computation, *Phys. Rev.* **A52**, 3457 (1995).

[2] E. Biham, O. Biham, D. Biron, M. Grassl and D.A. Lidar, Grover's quantum search algorithm for an arbitrary initial amplitude distribution, *Phys. Rev.* **A60** (1999), 2742–2745.

[3] O. Biham, E. Biham, D. Biron, M. Grassl and D.A. Lidar, Generalized Grover search algorithm for arbitrary initial amplitude distribution, in *Quantum Computing and Quantum Communications*, Lecture Notes in Comp. Sci., vol. 1509, Springer–Verlag, New York, 1998, pp. 140–147.

[4] M. Boyer, G. Brassard, P. Høyer and A. Tapp, Tight bounds on quantum searching, *Fortsch, Phys.* **46** (1998), 493–506.

[5] G. Brassard, P. Høyer and A. Tapp, Quantum counting, quant-ph/9805082, May 1998.

[6] G. Chen and Z. Diao, Quantum multi-object search algorithm with the

$(2, 1, 3, 4)$ $=$ $(1 \quad 2)$

$(3, 2, 1, 4)$ $=$ $(1 \quad 3)$

$(4, 2, 3, 1)$ $=$ $(1 \quad 4)$

$(1, 2, 3, 4)$ $=$ $(1 \quad 2)(1 \quad 2)$

$(3, 1, 2, 4)$ $=$ $(1 \quad 3)(1 \quad 2)$

$(4, 1, 3, 2)$ $=$ $(1 \quad 4)(1 \quad 2)$

$(2, 3, 1, 4)$ $=$ $(1 \quad 2)(1 \quad 3)$

$(4, 2, 1, 3)$ $=$ $(1 \quad 4)(1 \quad 3)$

$(2, 4, 3, 1)$ $=$ $(1 \quad 2)(1 \quad 4)$

$(3, 2, 4, 1)$ $=$ $(1 \quad 3)(1 \quad 4)$

$(1, 3, 2, 4)$ $=$ $(1 \quad 2)(1 \quad 3)(1 \quad 2)$

$(4, 1, 2, 3)$ $=$ $(1 \quad 4)(1 \quad 3)(1 \quad 2)$

$(1, 4, 3, 2)$ $=$ $(1 \quad 2)(1 \quad 4)(1 \quad 2)$

$(3, 1, 4, 2)$ $=$ $(1 \quad 3)(1 \quad 4)(1 \quad 2)$

$(4, 3, 1, 2)$ $=$ $(1 \quad 4)(1 \quad 2)(1 \quad 3)$

$(2, 4, 1, 3)$ $=$ $(1 \quad 2)(1 \quad 4)(1 \quad 3)$

$(1, 2, 4, 3)$ $=$ $(1 \quad 3)(1 \quad 4)(1 \quad 3)$

$(3, 4, 2, 1)$ $=$ $(1 \quad 3)(1 \quad 2)(1 \quad 4)$

$(2, 3, 4, 1)$ $=$ $(1 \quad 2)(1 \quad 3)(1 \quad 4)$

$(4, 3, 2, 1)$ $=$ $(1 \quad 4)(1 \quad 2)(1 \quad 3)(1 \quad 2)$

$(1, 4, 2, 3)$ $=$ $(1 \quad 2)(1 \quad 4)(1 \quad 3)(1 \quad 2)$

$(2, 1, 4, 3)$ $=$ $(1 \quad 3)(1 \quad 4)(1 \quad 3)(1 \quad 2)$

$(3, 4, 1, 2)$ $=$ $(1 \quad 3)(1 \quad 2)(1 \quad 4)(1 \quad 2)$

$(1, 3, 4, 2)$ $=$ $(1 \quad 2)(1 \quad 3)(1 \quad 4)(1 \quad 2)$

Table 2.2 Generation table of S_4 with 2-cycles $(1 \quad 2)$, $(1 \quad 3)$, and $(1 \quad 4)$.

availability of partial information, *Z. Naturforsch.* **A56a** (2001), 879–888.

[7] G. Chen, S.A. Fulling and J. Chen, Generalization of Grover's algorithm to multi-object search in quantum computing, Part I: continuous time and discrete time, in Chap. 6 of *"Mathematics of Quantum Computation"*, edited by R.K. Brylinski and G. Chen, CRC Press, Boca Raton, Florida, 2002, 135–160.

[8] G. Chen and S. Sun, ibid, Part II: general unitary transformations, in Chap. 7 of the same book, 161–168.

[9] T. Cormen, C. Leiserson, and R. Rivest, *Introduction to algorithms*, MIT Press, 1989.

[10] N.B. De Bruijn, Polya's theory of counting, in *Applied Combinatorial Mathematics*, E.F. Beckenbach ed., John Wiley and Sons, 1964.

[11] Z. Diao, M.S. Zubairy and G. Chen, A quantum circuit design for Grover's algorithm, *Z. Naturforsch.* **A57a** (2002), 701–708.

[12] L.K. Grover, Quantum mechanics helps in searching for a needle in a haystack, *Phys. Rev. Lett* **78** (1997), 325–328.

[13] D. Knuth, *The art of computer programming*, v.2, Addison-Wesley, 1997.

[14] J.H. Kwak and J. Lee, Enumeration of graph covering, surface branched coverings and related group theory, *Combinatorial and Computational Mathematics, present and future*, World Scientific, 2001, 97–161.

[15] I. Pak, Reduced decompositions of permutations in term of star transposition, generalized Catalan numbers, and k-ary trees, *Discrete Math.* **204** (1999), 329–335.

Chapter 3

On Interpolating between Quantum and Classical Complexity Classes

J. Maurice Rojas

Abstract   We reveal a natural algebraic problem whose complexity appears to interpolate between the well-known complexity classes **BQP** and **NP**: $\star$ Decide whether a univariate polynomial with exactly m monomial terms has a p-adic rational root. In particular, we show that while $(\star)$ is doable in quantum randomized polynomial time when $m = 2$, $(\star)$ is nearly **NP**-complete for general m. In particular, $(\star)$ is in **NP** for most inputs and, under a plausible hypothesis involving primes in arithmetic progression (implied by the generalized Riemann hypothesis for certain cyclotomic fields), a randomized polynomial time algorithm for $(\star)$ would imply the widely disbelieved inclusion **NP** $\subseteq$ **BPP**. This type of quantum/classical interpolation phenomenon appears to be new. As a consequence we can also address recent questions on the complexity of polynomial factorization posed by Cox, and by Karpinski and Shparlinski.

3.1 Introduction and main results

Thanks to quantum computation, we now have exponential speed-ups for important practical problems such as integer factoring (IF) and discrete logarithm (DL) [49]. However, a fundamental open question that remains is whether there are any **NP-complete** problems admitting exponential speed-ups via quantum computation. (We briefly review the complexity classes **NP** and **BQP**, as well as a few more, in Section 3.2 below.) Succinctly, this is the $\mathbf{NP}[.5] \overset{?}{\subseteq} \mathbf{BQP}$ question [6], and a positive answer would imply that quantum computation can also provide efficient algorithms for myriad problems (all at least as hard IF or DL) that have occupied practitioners in optimization and computer science for decades. The truth of the inclusion $\mathbf{NP} \subseteq \mathbf{BQP}$ is currently unknown as of early 2007. However, in light of important derandomization results [22], there is reason to believe the opposite (and also unknown) inclusion $\mathbf{BQP} \subseteq \mathbf{NP}$.

We propose an algebraic approach to these questions by illustrating a decision problem, involving sparse polynomials over $\mathbb{Q}_p$ (the p-adic rationals), whose complexity appears to interpolate between the complexity classes **BQP** and **NP**. Roughly speaking, "interpolation" here means that we have a decision problem, with computational complexity an increasing function of a parameter m, such that our problem...

(a) ...can be solved by a quantum computer in polynomial time, with error probability $< \frac{1}{3}$, for small values of m,
(b) ...can be used to simulate any computation in **BQP**, for small values of m,
(c) ...is **NP**-hard for large values of m, and
(d) ...can be solved in **NP** for large values of m.

Given a problem satisfying properties (a)–(d), one could then obtain the inclusion $\mathbf{BQP} \subseteq \mathbf{NP}$. Furthermore, one could then in principle study the transition from **BQP** to **NP** by analyzing the complexity of our interpolating problem for "mid-range" values of m. Our p-adic problem stated in the main theorem below satisfies Properties (a), (d) (for most inputs) and, under a plausible number-theoretic assumption clarified below, Property (c) as well. We will discuss the difficulty behind attaining all 4 properties shortly.

First, let us review some necessary terminology: For any ring R containing the integers $\mathbb{Z}$, let $\mathbf{FEAS}_R$—the R-**feasibility problem**—denote the problem of deciding whether a given system of polynomials $f_1, \ldots, f_k$ chosen from $\mathbb{Z}[x_1, \ldots, x_n]$ has a root in R^n. Observe then that $\mathbf{FEAS}_{\mathbb{R}}$ and $\mathbf{FEAS}_{\mathbb{Q}}$ are respectively the central problems of algorithmic real algebraic geometry and algorithmic arithmetic geometry (see Section 3.1.1 below for further details).

To measure the "size" of an input polynomial in our complexity estimates, we will essentially just count the number of bits needed to write down the coefficients and exponents in its monomial term expansion. This is the **sparse** input size, as opposed to the "dense" input size used frequently in computational algebra.

DEFINITION 3.1 *Let* $f(x):=\sum_{i=1}^{m} c_i x^{a_i} \in \mathbb{Z}[x_1,\ldots,x_n]$ *where* $x^{a_i}:=$ $x_1^{a_{1i}}\cdots x_n^{a_{ni}}$, $c_i \neq 0$ *for all* i, *and the* a_i *are distinct. We call such an* f *an* **n-variate m-nomial** *and define*

$$\text{size}(f) := \sum_{i=1}^{m} \left(1 + \lceil \log_2(2 + |c_i|) \rceil + \lceil \log_2(2 + |a_{1,i}|) \rceil \right.$$
$$\left. + \cdots + \lceil \log_2(2 + |a_{n,i}|) \rceil \right),$$

and $\text{size}_p(f) := \text{size}(f) + \log(2 + p)$. *(We also extend* size, *and thereby* size_p, *additively to polynomial systems.) Finally, for any collection* $\mathscr{F}$ *of polynomial systems with integer coefficients, let* $\textbf{FEAS}_R(\mathscr{F})$ *denote the natural restriction of* $\textbf{FEAS}_R$ *to inputs in* $\mathscr{F}$. $\diamond$

Observe that $\text{size}(a + bx^{99} + cx^d) = O(\log d)$ if we fix a, b, c. The degree of a polynomial can thus sometimes be exponential in its sparse size. Since it is not hard to show that $\textbf{FEAS}_{\mathbb{Q}_p}(\mathbb{Z}[x_1]) \in \textbf{P}$ when p is fixed (cf. Section 3.3 below), it will be more natural to take the size of an input prime p into account as well, and we do so as follows.

DEFINITION 3.2 *Let* $\textbf{FEAS}_{\mathbb{Q}_{\text{primes}}}$ *(resp.* $\textbf{FEAS}_{\mathbb{Q}_{\text{primes}}}(\mathscr{F})$*) denote the union of problems* $\bigcup_{p \text{ prime}} \textbf{FEAS}_{\mathbb{Q}_p}$ *(resp.* $\bigcup_{p \text{ prime}} \textbf{FEAS}_{\mathbb{Q}_p}(\mathscr{F})$*), so that a prime* p *is also part of the input, and the underlying input size is* size_p. *Also let* Q_n *denote the product of the first* n *primes and define* $\mathscr{U}_m := \{f \in \mathbb{Z}[x_1] \mid f \text{ has} \leq m \text{ monomial terms}\}$. $\diamond$

Observe that $\mathbb{Z}[x_1]$ is thus the union $\bigcup_{m \geq 0} \mathscr{U}_m$. Our results will make use of the following plausible number-theoretic hypothesis.

FLAT PRIMES HYPOTHESIS (FPH)

Following the notation above, there is an absolute constant $C \geq 1$ such that for any $n \in \mathbb{N}$, the set $\{1 + kQ_n \mid k \in \{1,\ldots,2^{n^C}\}\}$ contains at least $\frac{2^{n^C}}{n}$ primes.

Assumptions at least as strong as FPH are routinely used, and widely believed, in the cryptology and algorithmic number theory communities (see, e.g., [37, 36, 26, 45, 20]). In particular, we will see in Section 3.2.1 below how FPH is implied by the generalized Riemann hypothesis (GRH) for the number fields $\{\mathbb{Q}(\omega_{Q_n})\}_{n\in\mathbb{N}}$, where ω_M denotes a primitive $M^{\underline{th}}$ root of unity,[1] but can still hold under certain failures of the latter hypotheses.

MAIN THEOREM

Following the notation above, $\mathbf{FEAS}_{\mathbb{Q}_{\text{primes}}}(\mathcal{U}_2)\in\mathbf{BQP}.$ *Also,* $\mathbf{FEAS}_{\mathbb{Q}_{\text{primes}}}$ $(\mathbb{Z}[x_1])\in\mathbf{NP}$ *for "most" inputs in the following sense: For any* $f\in\mathbb{Z}[x_1]$ *and* $\varepsilon>0,$ *a fraction of at least* $1-\varepsilon$ *of the primes* p *with* $O\left(\log\left(\frac{1}{\varepsilon}\right)+\text{size}(f)\right)$ *digits are such that the solvability of* f *over* $\mathbb{Q}_p$ *admits a succinct certificate. Finally, assuming the truth of FPH, if* $\mathbf{FEAS}_{\mathbb{Q}_{\text{primes}}}(\mathbb{Z}[x_1])\in\mathcal{C}$ *for some complexity class* $\mathcal{C},$ *then* $\mathbf{NP}\subseteq\mathbf{BPP}\cup\mathcal{C}.$ *In particular, assuming the truth of FPH,* $\mathbf{FEAS}_{\mathbb{Q}_{\text{primes}}}(\mathbb{Z}[x_1])\in\mathbf{BQP}\implies\mathbf{NP}\subseteq\mathbf{BQP}.$

Our main result thus suggests that sparse polynomials can provide a tool to shed light on the difference between **BQP** and **NP**. Indeed, one consequence of our results is a new family of problems which admit (or are likely to admit) **BQP** algorithms: even the complexity of $\mathbf{FEAS}_{\mathbb{Q}_{\text{primes}}}(\mathcal{U}_3)$ is currently unknown, so the problems $\{\mathbf{FEAS}_{\mathbb{Q}_{\text{primes}}}(\mathcal{U}_m)\}_{m\geq3}$ provide a new context— distinct from integer factoring or discrete logarithm—to study quantum speed-up over classical methods.

REMARK 3.1   While it has been known since the late 1990's that $\mathbf{FEAS}_{\mathbb{Q}_{\text{primes}}}\in\mathbf{EXPTIME}$ [32, 33] (relative to our notion of input size), we are unaware of any earlier algorithms yielding $\mathbf{FEAS}_{\mathbb{Q}_{\text{primes}}}(\mathcal{F})\in\mathbf{BQP}$, for any non-trivial family of polynomial systems $\mathcal{F}$. Also, while it is not hard to show that $\mathbf{FEAS}_{\mathbb{Q}_{\text{primes}}}$ is **NP**-hard from scratch, there appear to be no earlier results indicating the smallest n such that $\mathbf{FEAS}_{\mathbb{Q}_{\text{primes}}}$ $(\mathbb{Z}[x_1,\ldots,x_n])$ is **NP**-hard.   ∎

The author is unaware of any other natural algebraic problem that at least partially interpolates between **BQP** and **NP** in the sense above. The only other problem known to interpolate between **BQP** and **some** classical complexity class arises from very recent results on the complexity of approximating a certain braid invariant—the famous **Jones polynomial**, for certain classes of braids, evaluated at an $m^{\underline{th}}$ root of unity—and involves a complexity class apparently higher than **NP**. In brief: (a') [2] gives a **BQP** algorithm that computes an additive approximation for arbitrary m, (b') seminal work of Freedman,

---
[1] i.e., a complex number ω_M with $\omega_M^M=1$; and $\omega_M^d=1\implies M|d.$

Kitaev, Larsen, and Wang shows that such approximations can simulate any **BQP** computation, already for $m=5$ [15, 16], and (c') [53] shows that for arbitrary m, computing the most significant bit of the absolute value of the Jones polynomial is **PP**-hard. In particular, the very notion of **BQP**-completeness is subtle: the Jones polynomial provides the **only** known non-trivial **BQP**-complete problem [3] (as opposed to the hundreds of **NP**-complete problems now known [17]), and the definition is technically rather different from that of **NP**-completeness [27, 3].

Thus, while we do not know whether $\mathbf{FEAS}_{\mathbb{Q}_{\text{primes}}}(\mathscr{U}_2)$ is **BQP**-complete in any rigorous sense, our results nevertheless provide a new potential source for quantum/classical complexity interpolation. Note also that the **BQP**-completeness of the integer factoring and discrete logarithm are open questions as well.

Recall that a univariate polynomial has a root in a field K iff it possesses a degree 1 factor with coefficients in K. Independent of its connection to quantum computing, our main theorem also provides a new complexity limit for polynomial factorization over $\mathbb{Q}_p[x_1]$. In particular, the main theorem shows that finding even just the low degree factors for **sparse** polynomials (with $\log p$ a summand in the sparse input size) is likely **not** doable in randomized polynomial time. This complements Chistov's earlier deterministic polynomial time algorithm for dense polynomials and fixed p [9]. Our main theorem also provides an interesting contrast to earlier work of Lenstra [30], who showed that—over the ring $\mathbb{Q}[x_1]$ instead—one can find all **low** degree factors of a sparse polynomial in polynomial time (thus improving the famous Lenstra–Lenstra–Lovasz algorithm [31]).

One can also naturally ask if detecting a **degenerate** root in $\mathbb{Q}_p$ for f (i.e., a degree 1 factor over $\mathbb{Q}_p$ whose square also divides f) is as hard as detecting arbitary roots in $\mathbb{Q}_p$. Via our techniques, we can easily prove essentially the same complexity lower-bound as above for the latter problem.

COROLLARY 3.1

Using $\text{size}_p(f)$ *as our notion of input size, suppose we can decide for any input prime p and $f \in \mathbb{Z}[x_1]$ whether f is divisible by the square of a degree 1 polynomial in $\mathbb{Q}_p[x_1]$, within some complexity class $\mathscr{C}$. Then, assuming the truth of FPH,* $\mathbf{NP} \subseteq \mathscr{C} \cup \mathbf{BPP}$.

Let $\mathbb{F}_p$ denote the finite field with p elements. Corollary 3.1 then complements an analogous earlier result of Karpinski and Shparlinski (independent of the truth of FPH) for detecting degenerate roots in $\mathbb{C}$ and the algebraic closure of $\mathbb{F}_p$.

Note also that while the truth of GRH usually implies algorithmic speed-ups (in contexts such as primality testing [37], complex dimension computation [26], detection of rational points [45], and class group computation [20]), the main theorem and Corollary 3.1 instead reveal complexity **speed-limits** implied by GRH.

3.1.1   Open questions and the relevance of ultrametric complexity

Complexity results over one ring sometimes inspire and motivate analogous results over other rings. An important early instance of such a transfer was the work of Paul Cohen on quantifier elimination over $\mathbb{R}$ and $\mathbb{Q}_p$ [10]. To close this introduction, let us briefly review how results over $\mathbb{Q}_p$ can be useful over $\mathbb{Q}$, and then raise some natural questions arising from our main results.

First, recall that the decidability of **FEAS**$_{\mathbb{Q}}$ is a major open problem: decidability for the special case of cubic polynomials in two variables would already be enough to yield significant new results in the direction of the Birch–Swinnerton–Dyer conjecture (see, e.g., [50, Ch. 8]), and the latter conjecture is central in modern number theory (see, e.g., [21]). The fact that **FEAS**$_{\mathbb{Z}}$ is undecidable is the famous negative solution of Hilbert's Tenth Problem, due to Matiyasevitch and Davis, Putnam, and Robinson [34, 13], and is sometimes taken as evidence that **FEAS**$_{\mathbb{Q}}$ may be undecidable as well (see also [42]).

From a more positive direction, much work has gone into using p-adic methods to find an algorithm for **FEAS**$_{\mathbb{Q}}(\mathbb{Z}[x,y])$ (i.e., deciding the existence of rational points on algebraic curves), via extensions of the **Hasse Principle**[2] (see, e.g., [11, 41, 43]). Algorithmic results over the p-adics are also central in many other computational results: polynomial time factoring algorithms over $\mathbb{Q}[x_1]$ [31], computational complexity [46], and elliptic curve cryptography [29].

Our results thus provide another step toward understanding the complexity of solving polynomial equations over $\mathbb{Q}_p$, and reveal yet another connection between quantum complexity and number theory. Let us now consider some possible extensions of our results. First, let **FEAS**$_{\mathbb{F}_{\text{primes}}}$ denote the obvious finite field analogue of **FEAS**$_{\mathbb{Q}_{\text{primes}}}$.

Question 1 *Is* **FEAS**$_{\mathbb{F}_{\text{primes}}}(\mathbb{Z}[x_1])$ **NP**-*hard?*

---

[2]If $F(x_1,\ldots,x_n) = 0$ is any polynomial equation and Z_K is its zero set in K^n, then the Hasse Principle is the assumption that $[Z_{\mathbb{C}}$ smooth, $Z_{\mathbb{R}} \neq \emptyset$, and $Z_{\mathbb{Q}_p} \neq \emptyset$ for all primes $p] \Longrightarrow Z_{\mathbb{Q}}$ is nonempty as well. The Hasse Principle is a theorem when $Z_{\mathbb{C}}$ is a smooth quadratic hypersurface or a smooth curve of genus zero, but fails in subtle ways already for curves of genus one (see, e.g., [40]).

Question 2 *Given a prime p and an* $f \in \mathbb{F}_p[x_1]$, *is it* **NP**-*hard to decide whether f is divisible by the square of a degree* 1 *polynomial in* $\mathbb{F}_p[x_1]$ *(relative to* $\mathrm{size}_p(f)$*)?*

David A. Cox asked the author whether $\mathbf{FEAS}_{\mathbb{F}_{\mathrm{primes}}} (\mathbb{Z}[x_1]) \overset{?}{\in} \mathbf{P}$ around August 2004 [12], and Erich Kaltofen posed a variant of Question 1—$\mathbf{FEAS}_{\mathbb{F}_{\mathrm{primes}}}$ $(\mathscr{U}_3) \overset{?}{\in} \mathbf{P}$—a bit earlier in [23]. Karpinski and Shparlinski raised Question 2 toward the end of [24]. Since Hensel's Lemma (cf. Section 3.2 below) allows one to find roots in $\mathbb{Q}_p$ via computations in the rings $\mathbb{Z}/p^\ell\mathbb{Z}$, the main theorem thus provides some evidence toward positive answers for Questions 1 and 2. Note in particular that a positive answer to Question 1 would provide a definitive complexity lower bound for polynomial factorization over $\mathbb{F}_p[x_1]$, since randomized polynomial time algorithms (relative to the **dense** encoding) are already known (e.g., Berlekamp's algorithm [4, Sec. 7.4]).

On a more speculative note, one may wonder if quantum computation can produce new speed-ups by circumventing the dependence of certain algorithms on GRH. This is motivated by Hallgren's recent discovery of a **BQP** algorithm for deciding whether the class number of a number field of constant degree is equal to a given integer [20]: The best classical complexity upper bound for the latter problem is **NP** $\cap$ **coNP**, obtainable so far only under the assumption of GRH [8, 35]. Unfortunately, the precise relation between **BQP** and **NP** $\cap$ **coNP** is not clear. However, could it be that quantum computation can eliminate the need for GRH in an even more direct way? For instance:

Question 3 *Is there a quantum algorithm which generates, within a number of qubit operations polynomial in n, a prime of the form* $kQ_n + 1$ *with probability* $> \frac{2}{3}$?

Our main results are proved in Section 3.3, after the development of some necessary theory in Section 3.2 below. For the convenience of the reader, we will recall the definitions of all relevant complexity classes and review certain types of generalized Riemann hypotheses.

3.2 Background and ancillary results

Recall the containments of complexity classes $\mathbf{P} \subseteq \mathbf{BPP} \subseteq \mathbf{BQP} \subseteq \mathbf{PP} \subseteq$ **PSPACE** and $\mathbf{P} \subseteq \mathbf{NP} \cap \mathbf{coNP} \subseteq \mathbf{NP} \cup \mathbf{coNP} \subseteq \mathbf{PP}$, and the fact that the properness of **every** preceding containment is a major open problem [38, 6]. (Indeed, as of early 2007, it is still not known whether even the containment **P** ⊆ **PSPACE** is proper!) We briefly review the definitions of the aforementioned complexity classes below (see [38, 6] for a full and rigourous treatment):

P The family of decision problems which can be done within (classical) polynomial-time.

BPP The family of decision problems admitting (classical) randomized polynomial-time algorithms that terminate with an answer that is correct with probability at least[3] $\frac{2}{3}$.

BQP The family of decision problems admitting **quantum** randomized polynomial-time algorithms that terminate with an answer that is correct with probability at least[3] $\frac{2}{3}$ [6].

NP The family of decision problems where a ``Yes'' answer can be **certified** within (classical) polynomial-time.

coNP The family of decision problems where a ``No'' answer can be **certified** within (classical) polynomial-time.

PP The family of decision problems admitting (classical) randomized polynomial-time algorithms that terminate with an answer that is correct with probability strictly greater than $\frac{1}{2}$.

PSPACE The family of decision problems solvable within polynomial-time, provided a number of processors exponential in the input size is allowed.

Now recall that **3CNFSAT** is the famous seminal **NP**-complete problem [17] which consists of deciding whether a Boolean sentence of the form $B(X) = C_1(X) \wedge \cdots \wedge C_k(X)$ has a satisfying assignment, where C_i is of one of the following forms:
$$X_i \vee X_j \vee X_k, \quad \neg X_i \vee X_j \vee X_k, \quad \neg X_i \vee \neg X_j \vee X_k, \quad \neg X_i \vee \neg X_j \vee \neg X_k,$$
$i,j,k \in [3n]$, and a satisfying assigment consists of an assigment of values from $\{0,1\}$ to the variables $X_1, \ldots, X_{3n}$ which makes the equality $B(X) = 1$ true.[4] Each C_i is called a **clause**.

We will need a clever reduction from feasibility testing for univariate polynomial systems over certain fields to **3CNFSAT**. First, note that the nonzero

---

[3]It is easily shown that we can replace $\frac{2}{3}$ by any constant strictly greater than $\frac{1}{2}$ and still obtain the same family of problems [38].

[4]Throughout this paper, for Boolean expressions, we will always identify 0 with ``False'' and 1 with ``True''.

polynomials in $\mathbb{Z}[x_1]$ form a **lattice** [51] with respective to the operations of least common multiple and greatest common divisor.

DEFINITION 3.3   *Letting Q_n denote the product of the first n primes, let us inductively define a homomorphism $\mathscr{P}_n$—the ($n^{\underline{th}}$) **Plaisted morphism**—from certain Boolean polynomials in the variables $X_1, \ldots, X_n$ to $\mathbb{Z}[x_1]$, as follows: (1) $\mathscr{P}_n(0) := 1$, (2) $\mathscr{P}_n(X_i) := x_1^{Q_n/p_i} - 1$, (3) $\mathscr{P}_n(\neg B)$ $:= \frac{x^{Q_n} - 1}{\mathscr{P}_n(B)}$, for any Boolean polynomial B for which $\mathscr{P}_n(B)$ has already been defined, (4) $\mathscr{P}_n(B_1 \vee B_2) := \mathrm{lcm}(\mathscr{P}_n(B_1), \mathscr{P}_n(B_2))$, for any Boolean polynomials B_1 and B_2 for which $\mathscr{P}_n(B_1)$ and $\mathscr{P}_n(B_2)$ have already been defined.* $\diamond$

LEMMA 3.1
For all $n \in \mathbb{N}$ and all clauses $C(X_i, X_j, X_k)$ with $i, j, k \leq n$, we have that $\mathscr{P}_n(C)$ can be computed within time polynomial in n, and $\mathrm{size}(\mathscr{P}_n(C)) = O(n^2)$. Furthermore, if K is any field possessing Q_n distinct $Q_n^{\underline{th}}$ roots of unity, then a **3CNFSAT** *instance $B(X) := C_1(X) \wedge \cdots \wedge C_k(X)$ has a satisfying assignment iff the zero set in K of the polynomial system $F_B := (\mathscr{P}_n(C_1), \ldots, \mathscr{P}_n(C_k))$ has a root ζ satisfying $\zeta^{Q_n} - 1$.*

David Alan Plaisted proved the special case $K = \mathbb{C}$ of the above lemma in [39]. His proof extends with no difficulty whatsoever to the more general family of fields detailed above. Other than an earlier independent observation of Kaltofen and Koiran [25], we are unaware of any other variant of Plaisted's reduction involving a field other than $\mathbb{C}$.

Let us now recall a version of Hensel's Lemma sufficiently general for our proof of our main theorem, along with a useful characterization of certain finite rings. Recall that $\mathbb{Z}_p$ denotes the p-adic integers, which can be identified with base-p digit sequences extending infinitely to the left. For any ring R, we also let R^* denote the group of multiplicatively invertible elements of R.

HENSEL'S LEMMA 1
(See, e.g., [44, Pg. 48].) Suppose $f \in \mathbb{Z}_p[x_1]$ and $x \in \mathbb{Z}_p$ satisfies $f(x) \equiv 0$ (mod p^ℓ) and $\mathrm{ord}_p f'(x) < \frac{\ell}{2}$. Then there is a root $\zeta \in \mathbb{Z}_p$ of f with $\zeta \equiv x$ (mod $p^{\ell - \mathrm{ord}_p f'(x)}$) and $\mathrm{ord}_p f'(\zeta) = \mathrm{ord}_p f'(x)$.

LEMMA 3.2
Given any cyclic group G, $a \in G$, and an integer d, the equation $x^d =$

a has a solution iff the order of a divides $\frac{\#G}{\gcd(d,\#G)}$. In particular, F_q^* is cyclic for any prime power q, and $(\mathbb{Z}/p^\ell\mathbb{Z})^*$ is cyclic for any (p,ℓ) with p an odd prime or $\ell \leq 2$. Finally, for $\ell \geq 3$, $(\mathbb{Z}/2^\ell\mathbb{Z})^* \cong \{-1,1\} \times \{1,5,5^2,5^3,\ldots,5^{2^{\ell-2}-1}\} \mod 2^\ell \}$.

The last lemma is standard (see, e.g., [4, Ch. 5]).

We will also need the following result on an efficient randomized reduction of $\mathbf{FEAS}_K(\mathbb{Z}[x_1]^k)$ to $\mathbf{FEAS}_K(\mathbb{Z}[x_1]^2)$. Recall that $\mathbb{C}_p$—the p-adic complex numbers—is the metric closure of the algebraic closure of $\mathbb{Q}_p$, and that $\mathbb{C}_p$ is algebraically closed.

LEMMA 3.3

Suppose $f_1,\ldots,f_k \in \mathbb{Z}[x_1] \setminus \{0\}$ are polynomials of degree $\leq d$, with $k \geq 3$. Also let $Z_K(f_1,\ldots,f_k)$ denote the set of common zeroes of $f_1,\ldots,f_k$ in some field K. Then, if $a = (a_1,\ldots,a_k)$ and $b = (b_1,\ldots,b_k)$ are chosen uniformly randomly from $\{1,\ldots,18dk^2\}^{2k}$, we have
$$\mathrm{Prob}\left(Z_K\left(\textstyle\sum_{i=1}^k a_i f_i, \sum_{i=1}^k b_i f_i\right) = Z_K(f_1,\ldots,f_k)\right) \geq \tfrac{8}{9}$$
for any $K \in \{\mathbb{C}, \mathbb{C}_p\}$.

While there are certainly earlier results that are more general than Lemma 3.3 (see, e.g., [19, Sec. 3.4.1] or [26, Thm. 5.6]), Lemma 3.3 is more direct and self-contained for our purposes. For the convenience of the reader, we provide its proof.

PROOF OF LEMMA 3.3

Assume $f_i(x) := \sum_{j=0}^d c_{i,j} x^i$ for all $i \in \{1,\ldots,k\}$. Let $W := \left(\bigcup_{i=1}^\ell Z_K(f_i)\right) \setminus Z_K(f_1,\ldots,f_k)$ and $\varphi(u,\zeta) := \sum_{i=1}^k u_i f_i(\zeta)$ for any $\zeta \in W$. Note that $\#W \leq kd$ and that for any fixed $\zeta \in W$, the polynomial $\varphi(u,\zeta)$ is linear in u and not identically zero. By Schwartz's Lemma [48], for any fixed $\zeta \in W$, there are at most kN^{k-1} points $u \in \{1,\ldots,N\}^k$ with $\varphi(u,\zeta) = 0$. So then, there at most dk^2N^{k-1} points $u \in \{1,\ldots,N\}^k$ with $\varphi(u,\zeta) = 0$ for some $\zeta \in W$.

Clearly then, the probability that a uniformly randomly chosen pair $(a,b) \in \{1,\ldots,N\}^{2k}$ satisfies $\varphi(a,\zeta) = \varphi(b,\zeta) = 0$ for some $\zeta \in W$ is bounded above by $\frac{2dk^2}{N}$. So taking $N = 18dk^2$ we are done.

Let us also recall the *p*-adic **Newton polygon**, which allows us to easily read off the norms of p-adic roots of polynomials. In particular, recall that the convex hull of any subset $S \subseteq \mathbb{R}^2$ is the smallest convex set containing S. Also, for any prime p and $x \in \mathbb{Z}_p$, recall that the *p*-adic **valuation**, $\mathrm{ord}_p x$, is the greatest k such that $p^k | x$. We then extend $\mathrm{ord}_p(\cdot)$ to $\mathbb{Q}_p$ by $\mathrm{ord}_p\left(\frac{a}{b}\right) := \mathrm{ord}_p(a) - \mathrm{ord}_p(b)$

for any $a,b \in \mathbb{Z}_p$, and let $|x|_p := p^{-\text{ord}_p x}$ denote the **p-adic norm**. The norm $|\cdot|_p$ defines a natural metric satisfying the ultrametric inequality and, along with $\text{ord}_p(\cdot)$, extends naturally to the **p-adic complex numbers** $\mathbb{C}_p$ (the metric completion of the algebraic closure of $\mathbb{Q}_p$).

LEMMA 3.4

(See, e.g., [44].) Given any polynomial $f(x_1) := \sum_{i=1}^{m} c_i x^{a_i} \in \mathbb{Z}[x_1]$, we define its **p-adic Newton polygon**, $\text{Newt}_p(f)$, as the convex hull of the points $\{(a_i, \text{ord}_p c_i) \mid i \in \{1,\ldots,m\}\}$. Then the number of roots of f in $\mathbb{C}_p$ with p-adic valuation v is **exactly** the horizontal length of the face of $\text{Newt}_p(f)$ with normal $(v,1)$.

Example 3.1

For the polynomial $f(x_1) := 243x^6 - 3646x^5 + 18240x^4 - 35310x^3 + 29305x^2 - 8868x + 36$, the polygon $\text{Newt}_3(f)$ can easily be verified to resemble the following illustration:

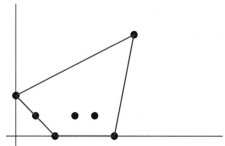

Note in particular that there are exactly 3 "lower" edges, and their respective horizontal lengths and inner normals are 2, 3, 1, and $(1,1)$, $(0,1)$, and $(-5,1)$. Lemma 3.4 then tells us that f has exactly 6 roots in $\mathbb{C}_3$: 2 with 3-adic valuation 1, 3 with 3-adic valuation 0, and 1 with 3-adic valuation -5. Indeed, one can check that the roots of f are exactly 6, 1, and $\frac{1}{243}$, with respective multiplicities 2, 3, and 1. ◇ ⬚

We now move on to some final background from analytic number theory that we will need.

3.2.1 Review of Riemann hypotheses

Primordial versions of the connection between analysis and number theory are not hard to derive from scratch and have been known at least since the

19th century. For example, letting $\zeta(s) := \sum_{n=1}^{\infty} \frac{1}{n^s}$ denote the usual **Riemann zeta function** (for any real number $s > 1$), one can easily derive with a bit of calculus (see, e.g., [52, pp. 30–32]) that

$$\zeta(s) = \prod_{p \text{ prime}} \frac{1}{1 - \frac{1}{p^s}}, \text{ and thus } -\frac{\zeta'(s)}{\zeta(s)} = \sum_{n=1}^{\infty} \frac{\Lambda(n)}{n^s},$$

where Λ is the classical Mangoldt function which sends n to $\log p$ or 0, according as $n = p^m$ for some prime p (and some positive integer m) or not. For a deeper connection, recall that $\pi(x)$ denotes the number of primes (in $\mathbb{N}$) $\leq x$ and that the **prime number theorem (PNT)** is the asymptotic formula $\pi(x) \sim \frac{x}{\log x}$ for $x \longrightarrow \infty$. Remarkably then, the first proofs of PNT, by Hadamard and de la Vallée–Poussin (independently, in 1896), were based essentially on the fact that $\zeta(\beta + i\gamma)$ has **no** zeroes on the vertical line $\beta = 1$.[5]

More precisely, writing $\rho = \beta + i\gamma$ for real β and γ, recall that ζ admits an analytic continuation to the complex plane sans the point 1 [52, Sec. 2].[6] In particular, the only zeroes of ζ outside the **critical strip** $\{\rho = \beta + i\gamma \mid 0 < \beta < 1\}$ are the so-called **trivial** zeroes $\{-2, -4, -6, \ldots, \}$. Furthermore the zeroes of ζ in the critical strip are symmetric about the **critical line** $\beta = \frac{1}{2}$ and the real axis. The **Riemann hypothesis (RH)**, from 1859, is then the following assertion:

> **(RH)** All zeroes $\rho = \beta + i\gamma$ of ζ with $\beta > 0$ lie on the critical line $\beta = \frac{1}{2}$.

Among myriad hitherto unprovably sharp statements in algorithmic number theory, it is known that RH is true $\Longleftrightarrow \left| \pi(x) - \int_2^x \frac{dt}{\log t} \right| = O(\sqrt{x} \log x)$ [52]. In particular, RH is widely agreed to be the most important problem in modern mathematics. Since May 24, 2000, RH even enjoys a bounty of one million US dollars thanks to the Clay Mathematics Foundation.

Let us now consider the extension of RH to primes in arithmetic progressions: For any primitive $M^{\underline{th}}$ root of unity ω_M, define the **(cyclotomic) Dedekind zeta function** via the formula $\zeta_{\mathbb{Q}(\omega_M)}(s) := \sum_{\mathfrak{a}} \frac{1}{(\mathcal{N}\mathfrak{a})^s}$, where $\mathfrak{a}$ ranges over all nonzero ideals of $\mathbb{Z}[\omega_M]$ (the ring of algebraic integers in $\mathbb{Q}(\omega_M)$), $\mathcal{N}$ denotes the norm function, and $s > 1$ [4]. Then, like ζ, the function $\zeta_{\mathbb{Q}(\omega_M)}$ also admits an analytic continuation to $\mathbb{C} \setminus \{1\}$ (which we'll also call $\zeta_{\mathbb{Q}(\omega_M)}$), $\zeta_{\mathbb{Q}(\omega_M)}$ has trivial zeroes $\{-2, -4, -6, \ldots, \}$, and all other zeroes of $\zeta_{\mathbb{Q}(\omega_M)}$ lie in the

---

[5]Shikau Ikehara later showed in 1931 that PNT is in fact **equivalent** to the fact that ζ has no zeroes on the vertical line $\beta = 1$ (the proof is reproduced in [14]).

[6]We'll abuse notation henceforth by letting ζ denote the analytic continuation of ζ to $\mathbb{C} \setminus \{1\}$.

critical strip $(0,1) \times \mathbb{R}$ [28]. (The zeroes of $\zeta_{\mathbb{Q}(\omega_M)}$ in the critical strip are also symmetric about the critical line $\frac{1}{2} \times \mathbb{R}$ and the real axis.) We then define the following statement:

(**GRH**$_{\mathbb{Q}(\omega_M)}$)[7] For any primitive $M^{\underline{th}}$ root of unity ω_M, all the zeroes $\rho = \beta + i\gamma$ of $\zeta_{\mathbb{Q}(\omega_M)}$ with $\beta > 0$ lie on the critical line $\beta = \frac{1}{2}$.

In particular, letting $\pi(x, M)$ denote the number of primes p congruent to 1 mod M satisfying $p \leq x$, it is known that GRH$_{\mathbb{Q}(\omega_M)}$ is true

$$\iff \left| \pi(x, M) - \frac{1}{\varphi(M)} \int_2^x \frac{dt}{\log t} \right| = O\left(\sqrt{x}(\log x + \log M) \right),$$

where $\varphi(M)$ is the number of $k \in \{1, \ldots, M-1\}$ relatively prime to M. (This follows routinely from the conditional effective Chebotarev theorem of [28, Thm. 1.1], taking $K = \mathbb{Q}$ and $L = \mathbb{Q}(\omega_M)$ in the notation there. One also needs to recall that the discriminant of $\mathbb{Q}(\omega_M)$ is bounded from above by $M^{\varphi(M)}$ [4, Ch. 8, pg. 260].)

From the very last estimate, an elementary calculation shows that FPH is implied by the truth of the hypotheses $\{GRH_{\mathbb{Q}(\omega_{Q_n})}\}_{n \in \mathbb{N}}$. However, we point out that FPH can **still** hold even in the presence of infinitely many non-trivial zeta zeroes off the critical line. For instance, if we instead make the weaker assumption that there is an $\varepsilon > 0$ such that all the non-trivial zeroes of $\{\zeta_{\mathbb{Q}(\omega_{Q_n})}\}_{n \in \mathbb{N}}$ have real part $\leq \frac{1}{2} + \varepsilon$, then one can still prove the weaker inequality

$$\left| \pi(x, M) - \frac{1}{\varphi(M)} \int_2^x \frac{dt}{\log t} \right| = O\left(x^{\frac{1}{2}+\varepsilon}(\log x + \log M) \right)$$

(see, e.g., [5]). Another elementary calculation then shows that this looser deviation bound **still** suffices to yield FPH. In fact, one can even have non-trivial zeroes of $\zeta_{\mathbb{Q}(\omega_{Q_n})}$ approach the line $\{\beta = 1\}$ arbitrarily closely, provided they do not approach too quickly as a function of n. (See [47] for further details.)

---

[7]There is definitely conflicting notation in the literature as to what the "extended" Riemann hypothesis or "generalized" Riemann hypothesis are. We thus hope to dissipate any possible confusion via subscripts clearly declaring the field we are working with.

3.3   The proofs of our main results

3.3.1   The univariate threshold over $\mathbb{Q}_p$: proving the main theorem

The first assertion—that $\textbf{FEAS}_{\mathbb{Q}_{\text{primes}}}(\mathcal{U}_2) \in \textbf{BQP}$—rests upon a quantum algorithm for finding the multiplicative order of an element of $(\mathbb{Z}/p^\ell\mathbb{Z})^*$ (see [49, 7]), once we make a suitable reduction from $\textbf{FEAS}_{\mathbb{Q}_{\text{primes}}}$. The second assertion—that the larger problem $\textbf{FEAS}_{\mathbb{Q}_{\text{primes}}}(\mathbb{Z}[x_1])$ lies in $\textbf{NP}$—follows from an application of the Newton polygon. The final assertion—that $\textbf{FEAS}_{\mathbb{Q}_{\text{primes}}}(\mathbb{Z}[x_1])$ is $\textbf{NP}$-hard under randomized reductions—relies on properties of primes in specially chosen arithmetic progressions, via our generalization (cf. Section 3.2) of an earlier trick of Plaisted [39].

Before proceeding, we will need a final (elementary) quantitative bound on p-adic roots and Newton polygons, and the computation of sizes/logarithms.

PROPOSITION 3.1

For any $f \in \mathbb{Z}[x_1]$ and $\zeta \in \mathbb{Q}_p$, $|\text{ord}_p\zeta| \leq \text{size}(f)$ and $\text{size}(p^{\text{size}(f)}) \leq \text{size}(f)\frac{\log p}{\log 2}$. Also, within time quadratic in $\text{size}(f)$ (resp. $\text{size}_p(f)$), we can compute an integer in the interval $[\text{size}(f), 2\text{size}(f)]$ (resp. $[\text{size}_p(f), 2\text{size}_p(f)])$. Finally, the number of primes for which $\text{ord}_p f'(\zeta) > 0$ for some root $\zeta \in \mathbb{C}_p$ of f is $O(v_f + \deg f \log \deg f)$ where v_f is maximum of the base 2 logarithms of the coefficients of f.

In particular, the only non-trivial portion is the final assertion, which follows easily once one applies the classical Hadamard estimates to the product formula for the discriminant of f [18, Ch. 12].

Proof that $\textbf{FEAS}_{\mathbb{Q}_{\text{primes}}}(\mathcal{U}_2) \in \textbf{BQP}$: First note that it clearly suffices to show that we can decide (with error probability $< \frac{1}{3}$, say) whether the polynomial $f(x):=x^d - \alpha$ has a root in $\mathbb{Q}_p$, using a number of qubit operations polynomial in $\text{size}(\alpha) + \log d$. (This is because we can divide by a suitable constant, and arithmetic over $\mathbb{Q}$ is doable in polynomial time.) The case $\alpha = 0$ always results in the root 0, so let us assume $\alpha \neq 0$. Clearly then, any p-adic root ζ of $x^d - \alpha$ satisfies $d\,\text{ord}_p\zeta = \text{ord}_p\alpha$. Since we can compute $\text{ord}_p\alpha$ and reductions of integers mod d in $\textbf{P}$ [4, Ch. 5], we can then clearly assume that $d|\text{ord}_p\alpha$ (for otherwise, there can be no root over $\mathbb{Q}_p$). Moreover, by rescaling x by an

appropriate power of p (thanks to Proposition 3.1) we can assume further that
$\mathrm{ord}_p \alpha = 0$.

Now note that $f'(\zeta) = d\zeta^{d-1}$ and thus $\mathrm{ord}_p f'(\zeta) = \mathrm{ord}_p(d)$. So by Hensel's
Lemma, it suffices to decide whether the mod p^ℓ reduction of f has a root in
$\mathbb{Z}/p^\ell \mathbb{Z}$, for $\ell = 1 + 2\mathrm{ord}_p d$. (Note in particular that $\mathrm{size}(p^\ell) = O(\log(p)\log(d))$
which is polynomial in our notion of input size.) By Lemma 3.2, we can
easily decide the latter feasibility problem, given the multiplicative order of α
in $(\mathbb{Z}/p^\ell \mathbb{Z})^*$; and we can do the latter in **BQP** by Shor's seminal algorithm
for computing order in a cyclic group [49, pp. 1498–1501], provided $p^\ell \notin$
$\{8, 16, 32, \ldots\}$. So the first assertion is proved for $p^\ell \notin \{8, 16, 32, \ldots\}$.

To dispose of the remaining cases $p^\ell \in \{8, 16, 32, \ldots\}$, write $\alpha = (-1)^a 5^b$
and observe that such an expression is unique, by the last part of Lemma 3.2.
The first part of Lemma 3.2 then easily yields that $x^d - \alpha$ has a root iff
$$(a \text{ odd} \implies d \text{ is odd}) \wedge (\text{the order of } 5^b \text{ divides } \tfrac{2^{\ell-2}}{\gcd(d, 2^{\ell-2})}).$$
In particular, we see that $x^d - \alpha$ **always** has a root when d is odd, so we can
assume henceforth that d is even.

Letting $\flat$ be the order of 5^b, it is then easy to check that the order of α
is either $\flat$ or $2\flat$, according to whether a is even or odd. Moreover, since d
is even, we see that $x^d - \alpha$ can have no roots in $(\mathbb{Z}/2^\ell \mathbb{Z})^*$ when a is odd.
So we can now reduce the feasibility of $x^d - \alpha$ to **two** order computations as
follows: Compute, now via Boneh and Lipton's quantum algorithm for order
computation in Abelian groups [7, Thm. 2], the order of α and $-\alpha$. Observe
then that a is odd iff the order of α is larger (and then $x^d - \alpha$ has no roots
in $(\mathbb{Z}/2^\ell \mathbb{Z})^*$), so we can assume henceforth that α has the smaller order. To
conclude, we then declare that $x^d - \alpha$ has a root in $\mathbb{Q}_2$ iff the order of α divides
$\tfrac{2^{\ell-2}}{d}$. This last step is correct, thanks to the first part of Lemma 3.2, so we at
last obtain $\mathbf{FEAS}_{\mathbb{Q}_{\mathrm{primes}}}(\mathscr{U}_2) \in \mathbf{BQP}$.

Proof that $\mathbf{FEAS}_{\mathbb{Q}_{\mathrm{primes}}}(\mathbb{Z}[x_1]) \in \mathbf{NP}$ **for "most" inputs:** First note that de-

---

tecting the existence of 0 as a root of f is easily done in linear time, simply by
checking whether all exponents are positive. Furthermore, by Proposition 3.1,
we can rescale x_1 (inducing at worst quadratic growth in $\mathrm{size}_p(f)$) so that all
the roots of f in $\mathbb{Q}_p$ lie in $\mathbb{Z}_p$.

So let us assume without loss of generality that $x_1 \nmid f$ and that all the roots
of $\mathbb{Q}_p$ lie in $\mathbb{Z}_p$, and proceed to define the following succinct certificate for
$\mathbf{FEAS}_{\mathbb{Q}_{\mathrm{primes}}}(\mathbb{Z}[x_1])$: any pair of the form $(v, \zeta) \in \mathbb{Z} \times (\mathbb{Z}/p^k \mathbb{Z})$ with k a fixed
positive integer, ζ a root of f over $\mathbb{Z}/p^k \mathbb{Z}$ and $(v, 1)$ an inner edge normal of
$\mathrm{Newt}_p(f)$. Since verifying the desired properties for (v, ζ) is clearly doable
within time polynomial in $\mathrm{size}_p(f)$, thanks to our earlier algorithmic observa-

tions on $\mathbf{FEAS}_{\mathbb{Q}_{\text{primes}}}$ $(\mathscr{U}_2)$, we now need only show that f has a root in $\mathbb{Q}_p$ iff such a certificate exists.

If f indeed has a root $\zeta \in \mathbb{Q}_p$ then $\text{ord}_p\zeta$ is a positive integer and f has a root $\bar{\zeta} \in \mathbb{Z}/p^k\mathbb{Z}$ with $\zeta \equiv \bar{\zeta} \pmod{p^k}$ for **all** $k \geq 1$. So our stated certificate exists when f has a root in $\mathbb{Q}_p$.

To see the converse, assume momentarily that $\text{ord}_p f'(\zeta) \leq (k-1)/2$ for all roots $\zeta \in \mathbb{C}_p$ of f. Then, given a certificate as stated above, Hensel's Lemma immediately implies that f has a root in $\mathbb{Q}_p$. The case where $\text{ord}_p f'(\zeta)$ is large for all roots of $\mathbb{Q}_p$ thus presents a difficulty, but Proposition 3.1 immediately implies that if simply consider primes with $\log O(v_f + \deg f \log(\deg f))$ digits, we can in fact assume that $\text{ord}_p f'(\zeta) = 0$ for all but a vanishingly small fraction of p. Since $\log(v_f + \deg f) = O(\text{size}(f))$, we immediately obtain our desired assertion.

Proof that FEAS$_{\mathbb{Q}_{\text{primes}}}$ $(\mathbb{Z}[x_1])$ **is NP-hard under Randomized Reductions:**

First note that $\text{size}(Q_n) = O(n \log n)$, via the prime number theorem. Observe then that the truth of FPH implies that we can efficiently find a prime p of the form $kQ_n + 1$, with $k \in \{1, \ldots, 2^{n^C}\}$, via random sampling, as follows: Pick a uniformly randomly integer from $\{1, \ldots, 2^{n^C}\}$ and using, say, the famous polynomial-time AKS primality testing algorithm [1], verify whether $kQ_n + 1$ is prime. We repeat this until we either find a prime, or fail $9n$ consecutive times.

Via the elementary estimate $(1 - \frac{1}{B})^{Bt} < \frac{1}{t}$, valid for all $B, t > 1$, we then easily obtain that our method results in a prime with probability at least $\frac{8}{9}$. Since $\text{size}(1 + 2^{n^C}Q_n) = O(\log(2^{n^C}Q_n)) = O(n^C + n \log n)$, it is clear that our simple algorithm requires a number of bit operations just polynomial in n. Moreover, the number of random bits needed is clearly $O(n^C)$.

Having now probabilistically generated a prime $p = 1 + kQ_n$, Lemma 3.1 then immediately yields the implication "$\mathbf{FEAS}_{\mathbb{Q}_{\text{primes}}}$ $(\mathscr{U}\mathscr{S}) \in \mathscr{C} \Longrightarrow \mathbf{NP} \in \mathscr{C} \cup \mathbf{BPP}$," where $\mathscr{U}\mathscr{S} := \{(f_1, \ldots, f_k) \mid f_i \in \mathbb{Z}[x_1], \ k \in \mathbb{N}\}$: Indeed, if $\mathbf{FEAS}_{\mathbb{Q}_{\text{primes}}}$ $(\mathscr{U}\mathscr{S}) \in \mathscr{C}$ for some complexity class $\mathscr{C}$, then we could combine our hypothetical $\mathscr{C}$ algorithm for $\mathbf{FEAS}_{\mathbb{Q}_{\text{primes}}}$ $(\mathscr{U}\mathscr{S})$ with our randomized prime generation routine (and the Plaisted morphism for $K = \mathbb{Q}_p$) to obtain an algorithm with complexity in $\mathscr{C} \cup \mathbf{BPP}$ for any **3CNFSAT** instance.

So now we need only show that this hardness persists if we reduce $\mathscr{U}\mathscr{S}$ to systems consisting of just one univariate sparse polynomial. Clearly, we can at least reduce to pairs of polynomials via Lemma 3.3, so now we need only reduce from pairs to singletons.

Toward this end, suppose $a \in \mathbb{Z}$ is a non-square mod p and p is odd. Clearly then, the only root in $\mathbb{F}_p$ of (the mod p reduction of) the quadratic form $q(x,y):=x^2-ay^2$ is $(0,0)$. Furthermore, by considering the valuations of x and y, it is also easily checked that the only root of q in $\mathbb{Q}_p$ is $(0,0)$. Thus, given any $(f,g) \in \mathbb{Z}[x_1]^2$, we can form $q(f,g)$ (which has size $O(\text{size}(f)+\text{size}(g)+\text{size}(p)))$ to obtain a polynomial time reduction of $\textbf{FEAS}_{\mathbb{Q}_\text{primes}}(\mathbb{Z}[x_1]^2)$ to $\textbf{FEAS}_{\mathbb{Q}_\text{primes}}(\mathbb{Z}[x_1])$, assuming we can find a quadratic non-residue efficiently. (If $p=2$ then we can simply use $q(x,y):=x^2+xy+y^2$ and then there is no need at all for a quadratic non-residue.) However, this can easily be done by picking two random $a \in \mathbb{F}_p$: With probability at least $\frac{3}{4}$, at least one of these numbers will be a quadratic non-residue (and this can be checked in **P** by computing $a^{(p-1)/2}$ via recursive squaring). So we are done.

3.3.2   Detecting square-freeness: proving corollary 3.1

Given any $f \in \mathbb{Z}[x_1]$, observe that f has a root in $\mathbb{Q}_p$ iff f^2 is divisible by the square of a degree 1 polynomial in $\mathbb{Q}_p[x_1]$. Moreover, since $\text{size}(f^2)=O(\text{size}(f)^2)$, we thus obtain a polynomial-time reduction of the problem considered by Corollary 3.1 to $\textbf{FEAS}_{\mathbb{Q}_\text{primes}}(\mathbb{Z}[x_1])$. So we are done.

In recent joint work with Sean Hallgren and Bjorn Poonen, the author has extended the main theorem to finite fields with a prime number of elements. Also, it appears that the assumption of FPH can be removed over p-adic fields, but not yet over finite fields. ◇

Acknowledgements

The author thanks Leonid Gurvits, Erich Kaltofen, and David Alan Plaisted for their kind encouragement. In particular, Leonid Gurvits, Sean Hallgren, and Erich Kaltofen respectively pointed out the references [15], [27], and [25]. I also thank Dan J. Bernstein, Jan Denef, Sidney W. Graham, Sean Hallgren, and Igor Shparlinski for some useful conversations and e-mails during the conception of this work.

Special thanks to Mark Danny Rintoul III, Sean Hallgren, Bernie Shiffman, and Steve Zelditch for their warm hospitality during visits to Sandia National Laboratories, NEC Laboratories, and Johns Hopkins University, where this work was completed.

References

[1] Agrawal, Manindra; Kayal, Neeraj; and Saxena, Nitin, *"PRIMES is in P,"* submitted for publication, downloadable from http://www.cse.iitk.ac.in/news/primality.html

[2] Aharonov, Dorit; Jones, Vaughan; and Landau, Zeph, *"A Polynomial Quantum Algorithm for Approximating the Jones Polynomial,"* Math ArXiV preprint quant-ph/0511096.

[3] Aharonov, Dorit and Arad, Itai, *"The **BQP**-Hardness of Approximating the Jones Polynomial,"* Math ArXiV preprint quant-ph/0605181.

[4] Bach, Eric and Shallit, Jeff, *Algorithmic Number Theory, Vol. I: Efficient Algorithms,* MIT Press, Cambridge, MA, 1996.

[5] Bach, Eric; Giesbrecht, Mark; and McInnes, *"The complexity of number theoretic problems,"* Technical Report No. 247/91, Dept. Computer Science, Univ. Toronto, January 1991.

[6] Bernstein, Ethan and Vazirani, Umesh, *"Quantum Complexity Theory,"* SIAM Journal of Computation **26**, no. 5, pp. 1411–1473, October, 1997.

[7] Boneh, Dan and Lipton, Richard J., *"Quantum Cryptanalysis of Hidden Linear Functions,"* Advances in cryptology—CRYPTO '95 (Santa Barbara, CA, 1995), pp. 424–437, Lecture Notes in Comput. Sci., 963, Springer, Berlin, 1995.

[8] Buchmann, J. and Williams, H. C., *"On the existence of a short proof for the value of the class number and regulator of a real quadratic field,"* NATO Advanced Science Institutes Series C, Vol. 256, Kluwer, Dordrecht (1989), pp. 327–345.

[9] Chistov, Alexander L., *"Efficient Factoring [of] Polynomials over Local Fields and its Applications,"* in I. Satake, editor, Proc. 1990 International Congress of Mathematicians, pp. 1509–1519, Springer–Verlag, 1991.

[10] Cohen, Paul J., *"Decision procedures for real and p-adic fields,"* Comm. Pure Appl. Math. 22 (1969), pp. 131–151.

[11] Colliot-Thelene, Jean-Louis, *"The Hasse principle in a pencil of algebraic varieties,"* Number theory (Tiruchirapalli, 1996), pp. 19–39, Contemp. Math., 210, Amer. Math. Soc., Providence, RI, 1998.

[12] Cox, David Alan, *personal communication via e-mail,* August 2004.

[13] *Hilbert's Tenth Problem: Relations with Arithmetic and Algebraic Geometry,* Papers from a workshop held at Ghent University, Ghent, November 2–5, 1999. Edited by Jan Denef, Leonard Lipshitz, Thanases Pheidas and Jan Van Geel. Contemporary Mathematics, 270, American Mathematical Society, Providence, RI, 2000.

[14] Dym, H. and McKean, H. P., *Fourier Series and Integrals,* Probability and Mathematical Statistics, vol. 14, Academic Press, 1972.

[15] Freedman, Michael; Kitaev, Alexander; and Wang, Z., *"Simulation of Topological Field Theories by Quantum Computers,"* Commun. Math. Phys. **227** (2002), pp. 587–603.

[16] Freedman, Michael; Larsen, Michael; and Wang, Z., *"A Modular Functor which is Universal for Quantum Computation,"* Commun. Math. Phys. **227** (2002), no. 3, pp. 605–622.

[17] Garey, Michael R. and Johnson, David S. *Computers and Intractability: A Guide to the Theory of NP-Completeness,* A Series of Books in the Mathematical Sciences, W. H. Freeman and Co., San Francisco, Calif., 1979, x+338 pp.

[18] Gel'fand, Israel Moseyevitch; Kapranov, Misha M.; and Zelevinsky, Andrei V.; *Discriminants, Resultants and Multidimensional Determinants,* Birkhäuser, Boston, 1994.

[19] Giusti, Marc and Heintz, Joos, *"La détermination des points isolés et la dimension d'une variété algébrique peut se faire en temps polynomial,"* Computational Algebraic Geometry and Commutative Algebra (Cortona, 1991), Sympos. Math. XXXIV, pp. 216–256, Cambridge University Press, 1993.

[20] Hallgren, Sean, *"Fast quantum algorithms for computing the unit group and class group of a number field,"* STOC'05: Proceedings of the 37th Annual ACM Symposium on Theory of Computing, pp. 468–474, ACM, New York, 2005.

[21] Hindry, Marc and Silverman, Joseph H., *Introduction to Diophantine Geometry,* Graduate Texts in Mathematics, vol. 201, Springer–Verlag, 2000.

[22] Impagliazzo, Russell and Wigderson, Avi, "**P** = **BPP** if **EXPTIME** Requires Exponential Circuits: Derandomizing the XOR Lemma," STOC '97 (El Paso, TX), pp. 220–229, ACM, New York, 1999.

[23] Kaltofen, Erich, *"Polynomial factorization: a success story,"* In ISSAC 2003 Proc. 2003 Internat. Symp. Symbolic Algebraic Comput. (New York, N.Y., 2003), J. R. Sendra, Ed., ACM Press, pp. 3–4.

[24] Karpinski, Marek and Shparlinski, Igor, *"On the computational hardness of testing square-freeness of sparse polynomials,"* Applied algebra, algebraic algorithms and error-correcting codes (Honolulu, HI, 1999), pp. 492–497, Lecture Notes in Comput. Sci., 1719, Springer, Berlin, 1999.

[25] Kaltofen, Erich and Koiran, Pascal, *"Finding small degree factors of multivariate supersparse (Lacunary) Polynomials over algebraic number fields,"* in ISSAC '06, Proc. 2006 Internat. Symp. Symbolic Algebraic Comput., to appear, ACM Press.

[26] Koiran, Pascal, *"Randomized and Deterministic Algorithms for the Dimension of Algebraic Varieties,"* Proceedings of the 38th Annual IEEE Computer Society Conference on Foundations of Computer Science (FOCS), Oct. 20–22, 1997, ACM Press.

[27] Knill, Emanuel and Laflamme, Raymond, *"Quantum Computation and Quadratically Signed Weight Enumerators,"* Math ArXiV preprint `quant-ph/9909094`.

[28] Lagarias, Jeff and Odlyzko, Andrew, *"Effective Versions of the Chebotarev Density Theorem,"* Algebraic Number Fields: L-functions and Galois Properties (Proc. Sympos. Univ. Durham, Durham, 1975), 409–464, Academic Press, London, 1977.

[29] Lauder, Alan G. B., *"Counting solutions to equations in many variables over finite fields,"* Found. Comput. Math. 4 (2004), no. 3, pp. 221–267.

[30] Lenstra (Jr.), Hendrik W., *"Finding Small Degree Factors of Lacunary Polynomials,"* Number Theory in Progress, Vol. 1 (Zakopane-Kóscielisko, 1997), pp. 267–276, de Gruyter, Berlin, 1999.

[31] Lenstra, Arjen K.; Lenstra, Hendrik W., Jr.; Lovász, L., *"Factoring polynomials with rational coefficients,"* Math. Ann. 261 (1982), no. 4, pp. 515–534.

[32] Maller, Michael and Whitehead, Jennifer, *"Computational complexity over the 2-adic numbers,"* The mathematics of numerical analysis (Park

City, UT, 1995), pp. 513–521, Lectures in Appl. Math., 32, Amer. Math. Soc., Providence, RI, 1996.

[33] Maller, Michael and Whitehead, Jennifer, *"Computational complexity over the p-adic numbers,"* J. Complexity 13 (1997), no. 2, pp. 195–207.

[34] Matiyasevich, Yuri V., *"On Recursive Unsolvability of Hilbert's Tenth Problem,"* Logic, Methodology and Philosophy of Science, IV (Proc. Fourth Internat. Congr., Bucharest, 1971), pp. 89–110, Studies in Logic and Foundations of Math., Vol. 74, North-Holland, Amsterdam, 1973.

[35] McCurley, Kevin S., *"Short Cryptographic key distribution and computation in class groups,"* NATO Advanced Science Institutes Series C, Vol. 256, Kluwer, Dordrecht (1989), pp. 459–479.

[36] Mihailescu, Preda, *"Fast generation of provable primes using search in arithmetic progressions,"* Advances in cryptology—CRYPTO '94 (Santa Barbara, CA, 1994), pp. 282–293, Lecture Notes in Comput. Sci., 839, Springer, Berlin, 1994.

[37] Miller, Gary L., *"Riemann's Hypothesis and Tests for Primality,"* J. Comput. System Sci. **13** (1976), no. 3, 300–317.

[38] Papadimitriou, Christos H., *Computational Complexity,* Addison-Wesley, 1995.

[39] Plaisted, David A., *"New NP-Hard and NP-Complete Polynomial and Integer Divisibility Problems,"* Theoret. Comput. Sci. 31 (1984), no. 1–2, 125–138.

[40] Poonen, Bjorn, *"An explicit algebraic family of genus-one curves violating the Hasse principle,"* 21st Journées Arithmétiques (Rome, 2001), J. Théor. Nombres Bordeaux 13 (2001), no. 1, pp. 263–274.

[41] _____, *"The Hasse principle for complete intersections in projective space,"* Rational points on algebraic varieties, pp. 307–311, Progr. Math., 199, Birkhuser, Basel, 2001.

[42] _____, *"Hilbert's tenth problem and Mazur's conjecture for large subrings of* $\mathbb{Q}$*,"* J. Amer. Math. Soc. 16 (2003), no. 4, pp. 981–990.

[43] _____, *"Heuristics for the Brauer-Manin Obstruction for Curves,"* Experimental Mathematics, to appear. Also available as Math ArXiV preprint `math.NT/0507329` .

[44] Robert, Alain M., *A course in p-adic analysis,* Graduate Texts in Mathematics, 198, Springer–Verlag, New York, 2000.

[45] Rojas, J. Maurice, *"Computational Arithmetic Geometry I: Sentences Nearly in the Polynomial Hierarchy,"* J. Comput. System Sci., STOC '99 special issue, vol. 62, no. 2, march 2001, pp. 216–235.

[46] _____, *"Additive Complexity and the Roots of Polynomials Over Number Fields and p-adic Fields,"* Proceedings of ANTS-V (5th Annual Algorithmic Number Theory Symposium, University of Sydney, July 7–12, 2002), Lecture Notes in Computer Science #2369, Springer–Verlag (2002), pp. 506–515.

[47] _____, *"Dedekind Zeta Functions and the Complexity of Computing Complex Dimension,"* preprint.

[48] Schwartz, Jacob T., *"Fast Probabilistic Algorithms for Verification of Polynomial Identities,"* J. of the ACM 27, 701–717, 1980.

[49] Shor, Peter W., *"Polynomial-time algorithms for prime factorization and discrete logarithms on a quantum computer,"* SIAM J. Comput. 26 (1997), no. 5, pp. 1484–1509.

[50] Silverman, Joseph H., *The Arithmetic of Elliptic Curves,* Graduate Texts in Mathematics, vol. 106, Springer–Verlag, 1996.

[51] Stanley, Richard, *Enumerative combinatorics,* Vol. 1. with a foreword by Gian-Carlo Rota, corrected reprint of the 1986 original, Cambridge Studies in Advanced Mathematics, 49, Cambridge University Press, Cambridge, 1997.

[52] Tenenbaum, Gérald and Mendès France, Michel, *The Prime Numbers and Their Distribution,* Student Mathematical Library, vol. 6, AMS Press, Rhode Island, 2000.

[53] Yard, Jon and Wocjan, Pawel, *"The Jones Polynomial: Quantum Algorithms and Applications in Quantum Complexity Theory,"* Math ArXiV preprint quant-ph/0603069.

Chapter 4

Quantum Algorithms for Hamiltonian Simulation

Dominic W. Berry, Graeme Ahokas, Richard Cleve, and Barry
C. Sanders

Abstract   Arguably one of the most important applications of quantum
computers is the simulation of quantum systems. In the case where the Hamiltonian consists of a sum of interaction terms between small subsystems, the
simulation is thought to be exponentially more efficient than classical simulation. More generally, evolution under suitably specified sparse Hamiltonians
may be efficiently simulated. In recent work we have shown that the complexity of simulating evolution under a Hamiltonian is very close to linear in the
evolution time. In addition, we have shown that in the general case of a sparse
Hamiltonian the complexity grows slowly with respect to the number of qubits.
In this chapter we review these results.

4.1   Introduction

An intriguing feature of quantum systems is that, in general, they are inefficient to simulate on classical computers. This prompted Feynman's 1982
conjecture that quantum systems could be used to efficiently simulate other
quantum systems [1]. Later work showed that a quantum computer, if built,
could efficiently simulate general quantum systems [2, 3, 4, 5].

There are a range of other algorithms which have been developed for quantum computers. Shor's algorithm [6] allows efficient factorization of numbers, and could be used for breaking the most commonly used encryption. More recently, algorithms have been found for other mathematical problems, some of which could be used for codebreaking [7]. A more general algorithm is Grover's search algorithm [8]. This algorithm is designed to search for inputs to a function that produce a desired output. The speedup is only quadratic, so this algorithm does not give the dramatic speedup of more specialized algorithms.

Of the known quantum algorithms that give exponential speedup,[1] simulation of physical systems has the widest applicability. It could be used, for example, in chemistry for predicting the properties of molecules. In Refs. [2, 3, 4, 5] the problem considered is the evolution of a system under a Hamiltonian. (It should be noted that our notion of simulation is distinct from the problem of finding the ground state of Hamiltonians. In the latter case, it does not appear to be possible to achieve an exponential speedup. The ground state of a Hamiltonian can be used to encode a search problem [9], and Ref. [10] shows that, in the black-box setting, it is not possible to achieve an exponential speedup for search problems.)

In the work of Lloyd [3] it is required that the quantum system is composed of small subsystems, and the Hamiltonian consists of a sum of interactions which only involve a small number of subsystems. A more general situation was considered by Aharonov and Ta-Shma (ATS) [4]. They do not require a tensor product structure to the Hamiltonian, but require that it is sparse and there is an efficient method of calculating the nonzero entries in a given column of the Hamiltonian. The Hamiltonians considered by Lloyd are sparse, and are therefore included in this generalization. There are also a range of other problems which produce such Hamiltonians. These Hamiltonians can also arise as encodings of computational problems, such as simulations of quantum walks [11, 12, 13, 14, 15].

In our recent work [5] we improved upon the efficiency of the schemes of ATS and Lloyd by applying the higher order integrators of Suzuki [16, 17]. Our work contains a number of results:

1. In order to simulate evolution over time t, our scheme requires a number of steps which scales as $t^{1+1/2k}$, where k is the order of the integrator and may be chosen to be an arbitrarily large integer.

2. We found upper bounds on the error, which enable us to estimate the optimal order k for a given evolution time t.

---

[1]Exponential speedup over the best known classical algorithms.

3. We showed that general sublinear scaling in t is not possible. This means that the simulation scheme using the integrators of Suzuki is close to optimal in the evolution time.

4. We provided a superior method for decomposing the Hamiltonian into a sum for the problem considered by ATS. This dramatically reduces the scaling from polynomial in the number of qubits to close to constant.

In Sections 4.2 and 4.3 we review the results of Refs. [3, 4], and we review our results from Ref. [5] in Sections 4.4 to 4.6.

4.2 Simulation method of Lloyd

The problem considered by Lloyd [3] is as follows. The quantum system is composed of N "variables", or subsystems, and the total Hamiltonian consists of a sum of interaction terms

$$H = \sum_{j=1}^{m} H_j. \tag{4.1}$$

Each interaction term H_j acts on at most k_{L} of the subsystems, with maximum dimension of d_j. Lloyd also allows the Hamiltonian to depend on time. The vast majority of quantum systems have Hamiltonians of this type, because interactions only occur between a small number of subsystems, not jointly over all subsystems.

Because each H_j acts on a Hilbert space of dimension d_j, the number of operations required to simulate evolution under H_j scales as d_j^2. In order to approximate evolution under the Hamiltonian H, it is therefore desired to simulate evolution under a sequence of the individual Hamiltonians H_j. Lloyd uses the approximation

$$e^{iHt} \approx (e^{iH_1 t/r} \dots e^{iH_m t/r})^r. \tag{4.2}$$

Because the number of steps required for simulation of each Hamiltonian H_j scales as d_j^2, the total number of steps scales as $r\sum_{j=1}^{m} d_j^2 \le rmd^2$, where $d = \max\{d_j\}$.

Lloyd gives the bound on the error using this approximation as

$$\|r(e^{iHt/r} - 1 - iHt/r)\|_{\sup}. \tag{4.3}$$

Here the norm gives the maximum expectation value of the operator. Lloyd also gives the alternative expression

$$e^{iHt} = (e^{iH_1t/r} \ldots e^{iH_mt/r})^r + \sum_{i>j}[H_i, H_j]t^2/2r + \sum_{l \geq 3}E(l), \qquad (4.4)$$

where the $E(l)$ are bounded by $\|E(l)\|_{\sup} \leq r\|Ht/r\|_{\sup}^l/l!$. This implies that, in order to obtain accuracy ε for simulation over time t, the number of time-slices r needs to scale as t^2m^2/ε. This implies that the total number of steps required scales as $t^2m^3d^2/\varepsilon$.

Overall, the simulation is tractable provided the number of steps is a polynomial function of the number of subsystems N. If the interactions do not involve more than k_L subsystems, then the number of interactions m is bounded by N^{k_L}. If k_L and d do not increase with N, then the number of steps is of order N^{3k_L}, which is polynomial in N.

4.3   Simulation method of ATS

A more general form of simulation was considered by Aharonov and Ta-Shma [4]. They consider simulation of an arbitrary row-sparse Hamiltonian. That is, the Hamiltonian may be represented by a matrix with only a moderate number of nonzero elements in each row. It is always possible to represent the Hamiltonian by a matrix which is diagonal simply by choosing the appropriate basis. However, in practice the Hamiltonian is provided in a certain basis and it is not efficient to determine the diagonal representation. In the following we will regard the Hamiltonian as a matrix, without specifying that the matrix is just a representation of the Hamiltonian.

It is easily seen that the Hamiltonians considered by Lloyd are row-sparse. As each interaction Hamiltonian H_j only acts on a subsystem of dimension d_j, it has no more than d_j elements in each row. The overall Hamiltonian then has no more than md elements in each row. As this scales polynomially with the number of subsystems, the overall Hamiltonian is row-sparse.

The main difference between the problem considered by Lloyd and that considered by ATS is that for Lloyd's problem the decomposition of the Hamiltonian is given, whereas all ATS assume is that there is some method of calculating the nonzero terms in the columns of the Hamiltonian.

In the case where the Hamiltonian H has at most D nonzero elements in each row and acts upon a system of dimension no larger than 2^n (so it may be represented on n qubits), ATS show a method for decomposing H into a sum of no more than $(D+1)^2 n^6$ terms

$$H = \sum_{j=1}^{(D+1)^2 n^6} H_j. \qquad (4.5)$$

Each H_j is 1-sparse; that is, there is no more than one nonzero element in each row/column. This also implies that it is 2×2 combinatorially block diagonal (it is equivalent to a 2×2 block diagonal matrix under an appropriate permutation of the basis states).

The decomposition is essentially equivalent to the coloring problem for an undirected graph. The nodes of the graph correspond to the basis states, and the edges correspond to the nonzero elements of the Hamiltonian. The problem is to color the graph such that no two edges with the same color connect the same vertex. Each color then corresponds to a different Hamiltonian H_j.

The method ATS use for the coloring is to use the color

$$\vec{j} = (v, x \bmod v, y \bmod v, \mathrm{rind}_H(x,y), \mathrm{cind}_H(x,y)) \qquad (4.6)$$

(where rind_H and cind_H are defined below). Here the convention is taken that $x \leq y$. We are taking the color to be a vector of integers, and will use subscripts for the different components. If $x = y$, v is set as 1, otherwise it is set as the first integer in the range $[2 \ldots n^2]$ such that $x \neq y \bmod v$.

For convenience we define the function f which gives the nonzero elements in each column. If the nonzero elements in column x are $y_1, \ldots, y_{D'}$, where $D' \leq D$, then $f(x,i) = (y_i, H_{x,y_i})$ for $i \leq D'$, and $f(x,i) = (x,0)$ for $i > D'$. We use subscript y for the first component of f, and subscript H for the second component of f.

We may give the definitions of cind_H and rind_H succinctly using this function. If $H_{x,y} \neq 0$, then

$$f_y(y, \mathrm{cind}_H(x,y)) = x, \qquad f_y(x, \mathrm{rind}_H(x,y)) = y. \qquad (4.7)$$

That is, cind_H gives the column index of x and rind_H gives the row index of y. If $H_{x,y} = 0$, then $\mathrm{rind}_H(x,y)$ and $\mathrm{cind}_H(x,y)$ are both taken to be zero.

Given this coloring, one wishes to determine a function which outputs the nonzero element row number and value for each $H_{\vec{j}}$. We may give this function as $g(x, \vec{j}) = (y, (H_{\vec{j}})_{x,y})$, where x is the column number, $\vec{j}$ is the color, y is the row number and $(H_{\vec{j}})_{x,y}$ is the required elements. This function can be determined in the following way. There are three cases where a nonzero result is given:

1. The color corresponds to the diagonal elements. We require $j_1 = 1$ (corresponding to $v = 1$). In addition, for consistency we require $j_2 = j_3 = 0$, $j_4 = j_5$ and $f_y(x, j_4) = x$. Then we output $y = x$ and $(H_j)_{x,y} = H_{x,x}$.

2. The color corresponds to off-diagonal elements and the nonzero element is in the lower triangle (so $x < y$). We require $j_1 > 1$ (corresponding to $v > 1$) to give the off-diagonal elements. If $x < y$ then we also require x mod $j_1 = j_2$. For consistency, we also require $f_y(x, j_4)$ mod $j_1 = j_3$, $f_y(f_y(x, j_4), j_5) = x$, $x < f_y(x, j_4)$ and j_1 to be the first integer such that $x \neq f_y(x, j_4)$ mod j_1. Then $y = f_y(x, j_4)$ and $(H_{\vec{j}})_{x,y} = H_{x,y}$.

3. The color corresponds to off-diagonal elements and the nonzero element is in the upper triangle (so $y < x$). We require $j_1 > 1$ (corresponding to $v > 1$) to give the off-diagonal elements. If $y < x$ then we also require x mod $j_1 = j_3$, and for consistency with the coloring scheme we require $f_y(x, j_5)$ mod $j_1 = j_2$, $f_y(f_y(x, j_5), j_4) = x$, $f_y(x, j_5) < x$ and j_1 is the first integer such that $x \neq f_y(x, j_5)$ mod j_1. Then $y = f_y(x, j_5)$ and $(H_{\vec{j}})_{x,y} = H_{y,x}$.

In all other cases the output is simply $(H_{\vec{j}})_{x,y} = 0$ and $y = x$. ATS do not explicitly give this function, though it is implicit from their coloring method. It is easily seen that the coloring gives at most one nonzero element in each column. It is not possible for both cases 2 and 3 to hold, because some of the conditions for these cases would imply that x mod $j_1 = j_2 = j_3$, but $j_2 = j_3$ violates the conditions that $f_y(x, j_4)$ mod $j_1 = j_3$ and j_1 is the first integer such that $x \neq f_y(x, j_4)$ mod j_1.

After giving this coloring scheme, ATS show that it is possible to efficiently simulate the individual $H_{\vec{j}}$. Here we summarize their method, with some minor differences. Let the row number be y for the nonzero element in column x; then the nonzero element in column y is in row x. We let $\tilde{U}_x$ be the approximation of the unitary on those basis states, and $m_x = \min\{x, y\}$ and $M_x = \max\{x, y\}$.

Given that the black-box function f may be represented by a unitary U_f, it is possible to obtain a unitary U_g for the black-box function $g(x, \vec{j})$ such that

$$U_g |x, \vec{j}, 0\rangle = |x, \vec{j}, y, (H_{\vec{j}})_{x,y}\rangle |\phi_{x,\vec{j}}\rangle, \qquad (4.8)$$

where $|\phi_{x,\vec{j}}\rangle$ represents additional ancilla states produced by the calculation. From this it is possible to derive a unitary $T_{\vec{j}}$ such that

$$T_{\vec{j}} |x, 0\rangle = |x, m_x, M_x, \tilde{U}_x\rangle \qquad (4.9)$$

In order to remove the additional ancilla states, it is necessary to apply U_g, copy the output, and apply $U_g^\dagger$. Another unitary T is then defined such that

$$T|v\rangle|m_x, M_x, \tilde{U}_x\rangle = (\tilde{U}_x|v\rangle)|m_x, M_x, \tilde{U}_x\rangle. \qquad (4.10)$$

To simulate the unitary, one first applies $T_{\bar{j}}$ to produce a state with the approximation of the unitary, applies T to implement the unitary, then applies $T_{\bar{j}}^\dagger$ to remove the extra ancilla states. Overall, U_g is applied twice and $U_g^\dagger$ is applied twice.

To see the action of this on basis state $|x\rangle$, let us take the action of $\tilde{U}_x$ on $|x\rangle$ to give the state $\tilde{\alpha}|m_x\rangle + \tilde{\beta}|M_x\rangle$. Then the sequence of transformations gives

$$\begin{aligned}
T_{\bar{j}}^\dagger T T_{\bar{j}}|x, 0\rangle &= T_{\bar{j}}^\dagger T|x, m_x, M_x, \tilde{U}_x\rangle \\
&= T_{\bar{j}}^\dagger \left(\tilde{\alpha}|m_x, m_x, M_x, \tilde{U}_x\rangle + \tilde{\beta}|M_x, m_x, M_x, \tilde{U}_x\rangle \right) \\
&= \tilde{\alpha}|m_x, 0\rangle + \tilde{\beta}|M_x, 0\rangle.
\end{aligned} \qquad (4.11)$$

From the definition, $m_{m_x} = m_{M_x} = m_x$ and $M_{m_x} = M_{M_x} = M_x$; hence performing $T_{\bar{j}}^\dagger$ correctly removes the additional states. It is essential that $T_{\bar{j}}$ does not add additional states which depend on x, because then it would not be possible to perform this step.

4.4   Higher order integrators

In our work [5] we improve upon the work of Lloyd and ATS in two main ways. We apply higher order integrators to improve the scaling of the complexity with time, and we apply an improved coloring method. The higher-order integrators of Suzuki are defined in the following way [16, 17]. The first order integrator is

$$S_2(\lambda) = \prod_{j=1}^{m} e^{H_j \lambda/2} \prod_{j'=m}^{1} e^{H_{j'} \lambda/2}, \qquad (4.12)$$

which is the basic Lie–Trotter product formula. The higher order integrators are obtained via the recursion relation

$$S_{2k}(\lambda) = [S_{2k-2}(p_k\lambda)]^2 S_{2k-2}((1 - 4p_k)\lambda)[S_{2k-2}(p_k\lambda)]^2 \qquad (4.13)$$

with $p_k = (4 - 4^{1/(2k-1)})^{-1}$ for $k > 1$. Suzuki proves that [16]

$$\left\| \exp\left(\sum_{j=1}^{m} H_j \lambda \right) - S_{2k}(\lambda) \right\| \in O(|\lambda|^{2k+1}) \tag{4.14}$$

for $|\lambda| \to 0$. The parameter λ corresponds to $-it$ for Hamiltonian evolution.

We can deduce from Eq. (4.14) another bound that is more quantitatively precise. Our result is

LEMMA 4.1

Using integrators of order $k \geq 2$ and dividing the time into r intervals, we have the bound

$$\left\| \exp\left(-it \sum_{j=1}^{m} H_j \right) - [S_{2k}(-it/r)]^r \right\| \leq \frac{\mu_k (2m5^{k-1} q_k \tau)^{2k+1}}{(2k+1)! r^{2k}}, \tag{4.15}$$

where $\tau = t \times \max \|H_j\|$, $q_k = \prod_{k'=2}^{k} |1 - 4p_{k'}|$, $\kappa_k = (2q_k 5^{k-1})^{-(2k+1)}$,

$$\mu_k = (1 + \kappa_k) e^{\delta_1} [(e^{\delta_2} - 1)/\delta_2], \tag{4.16}$$

and we have the restrictions

$$2m5^{k-1} q_k \tau/r \leq \delta_1,$$
$$(1 + \kappa_k) e^{\delta_1} (2m5^{k-1} q_k \tau)^{2k+1} / [(2k+1)! r^{2k}] \leq \delta_2. \tag{4.17}$$

Before proceeding to the proof, we note that $\kappa_2 \approx 8.2 \times 10^{-5}$, and κ_k rapidly approaches zero for large k. The δ_1 and δ_2 may be made small to obtain tighter bounds, though this requires more stringent requirements in Eqs. (4.17). For large k and small $\delta_{1,2}$, $\mu_k \approx 1$, and the upper bound is approximately $(2m5^{k-1} q_k \tau)^{2k+1} / [(2k+1)! r^{2k}]$. We now proceed to the proof.

PROOF   If we take a Taylor expansion of both terms in the left-hand side (LHS) of Eq. (4.14), then the terms containing λ to powers less than $2k+1$ must cancel because the expression is of order $|\lambda|^{2k+1}$. Terms in the Taylor expansion with λ^l for $l \geq 2k+1$ contain a product of l of the H_j terms, so

$$\exp\left(\sum_{j=1}^{m} H_j \lambda \right) = S_{2k}(\lambda) + \sum_{l=2k+1}^{\infty} \lambda^l$$

$$\times \left[\sum_{p=1}^{L_l} C_p^l \prod_{q=1}^{l} H_{j_{pq}} + \sum_{p=1}^{\bar{L}_l} \bar{C}_p^l \prod_{q=1}^{l} H_{j_{pq}} \right]. \qquad (4.18)$$

Here L_l is the number of terms in the Taylor expansion of the exponential at order l, and C_p^l are the constants in that expansion. The quantities $\bar{L}_l$ and $\bar{C}_p^l$ are the corresponding quantities for the Taylor expansion of the integrator $S_{2k}(\lambda)$.

To determine a bound on the correction term, we first determine bounds on the quantities L_l and C_p^l. Expanding $(H_1 + \cdots + H_m)^l$ yields m^l terms, so $L_l = m^l$. In addition, we have $C_p^l = 1/l!$ from the multiplying factor in the Taylor expansion of the exponential.

It is somewhat more complicated for the integrator. The integrator $S_2(\lambda)$ consists of a product of $2m-1$ exponentials. The minus 1 comes about because $e^{H_m\lambda/2} \times e^{H_m\lambda/2}$ may be simplified to $e^{H_m\lambda}$. Each of the powers in the exponentials contains multiplying factors of $1/2$, except for this central exponential where the multiplying factor is 1.

Then in using the recursion relation (4.13), the number of exponentials changes according to the map $x \mapsto 5x - 4$. The minus 4 is because the exponentials at the ends are combined. This gives the total number of exponentials as $2(m-1)5^{k-1} + 1$. In keeping track of the multiplying factors in the exponentials, it is convenient to keep track of the exponentials at the ends and the exponentials in the center separately.

Using x and y for the maximum magnitudes of the multiplying factors for the inner elements and outer elements, respectively, the recursion relation gives the map $x \mapsto \max\{p_k x, |1 - 4p_k|x, 2p_k y, |1 - 3p_k|y\}$ and $y \mapsto p_k y$. It turns out that the element which gives the maximum in the map for x is always $|1 - 4p_k|x$, and x always exceeds y. We therefore have the maximum multiplying factor in the exponentials as $q_k = \prod_{k'=2}^{k} |1 - 4p_{k'}|$. We take $k \geq 2$, because we are not concerned with the error for the low order integrators.

The Taylor expansion for $S_{2k}(t)$ may be determined by expanding each of the exponentials individually and performing the multiplication. To place a bound on the contribution to the error from terms containing λ^l, we can replace each of the terms in this expansion with the upper bounds on their norms. Thus the bounds may be obtained from the expansion of

$$(1 + |q_k \Lambda \lambda| + |q_k \Lambda \lambda|^2/2! + \ldots)^{2(m-1)5^{k-1}+1}, \qquad (4.19)$$

where $\Lambda \equiv \max \|H_j\|$. This is just the expansion of $\exp\{|q_k \Lambda \lambda|[2(m-$

$1)5^{k-1} + 1]\}$, so the terms containing λ^l may be bounded by

$$\frac{\{|q_k \Lambda \lambda|[2(m-1)5^{k-1} + 1]\}^l}{l!}. \tag{4.20}$$

Combining this with the earlier results and using standard inequalities gives

$$\left\| \sum_{l=2k+1}^{\infty} \lambda^l \left[\sum_{p=1}^{L_l} C_p^l \prod_{q=1}^{l} H_{j_{pq}} + \sum_{p=1}^{\bar{L}_l} \bar{C}_p^l \prod_{q=1}^{l} H_{j_{pq}} \right] \right\|$$

$$\leq \sum_{l=2k+1}^{\infty} \frac{|\lambda \Lambda|^l}{l!} \left[m^l + q_k^l [2(m-1)5^{k-1} + 1]^l \right]$$

$$\leq \frac{\{|\lambda \Lambda q_k|[2(m-1)5^{k-1} + 1]\}^{2k+1}}{(2k+1)!} \exp\{|\lambda \Lambda q_k|[2(m-1)5^{k-1} + 1]\}$$

$$+ \frac{|\lambda \Lambda m|^{2k+1}}{(2k+1)!} \exp|\lambda \Lambda m|. \tag{4.21}$$

From this point onward the derivation differs from that in Ref. [5], which gives a slightly weaker bound. Here we make fewer simplifications than in [5], giving a more complicated but tighter bound.

Simplifying Eq. (4.21) gives the inequality

$$\left\| \exp\left(\lambda \sum_{j=1}^{m} H_j \right) - S_{2k}(\lambda) \right\| \leq \frac{(1 + \kappa_k) e^{\delta_1} |2m5^{k-1} q_k \Lambda \lambda|^{2k+1}}{(2k+1)!}, \tag{4.22}$$

where $\kappa_k = (2q_k 5^{k-1})^{-(2k+1)}$, and we have the restriction $|2m5^{k-1} q_k \Lambda \lambda| \leq \delta_1$. Substituting $\lambda = -it/r$ where r is an integer, and taking the power of r, gives the error bound

$$\left\| \exp\left(-it \sum_{j=1}^{m} H_j \right) - [S_{2k}(-it/r)]^r \right\|$$

$$\leq \left[1 + \frac{(1 + \kappa_k) e^{\delta_1} (2m5^{k-1} q_k \tau/r)^{2k+1}}{(2k+1)!} \right]^r - 1$$

$$\leq \exp\left[\frac{(1 + \kappa_k) e^{\delta_1} (2m5^{k-1} q_k \tau)^{2k+1}}{(2k+1)! r^{2k}} \right] - 1$$

$$\leq \frac{e^{\delta_2} - 1}{\delta_2} \frac{(1 + \kappa_k) e^{\delta_1} (2m5^{k-1} q_k \tau)^{2k+1}}{(2k+1)! r^{2k}}, \tag{4.23}$$

for $2m5^{k-1}q_k\tau/r \le \delta_1$, and in the last line we have used the restriction $(1+\kappa_k)e^{\delta_1}(2m5^{k-1}q_k\tau)^{2k+1}/[(2k+1)!r^{2k}] \le \delta_2$. We therefore obtain the bound given in Eq. (4.15). ∎

We can also give a bound without requiring the extra conditions. Directly using Eq. (4.21) gives the bound

$$
\left\| \exp\left(-it \sum_{j=1}^{m} H_j \right) - [S_{2k}(-it/r)]^r \right\| \le \left(1 + \frac{(\tau m/r)^{2k+1}}{(2k+1)!} \exp(\tau m/r) \right.
$$
$$
\left. + \frac{\{(\tau q_k/r)[2(m-1)5^{k-1}+1]\}^{2k+1}}{(2k+1)!} \exp\{(\tau q_k/r)[2(m-1)5^{k-1}+1]\} \right)^r
$$
$$
- 1. \tag{4.24}
$$

The scaling is somewhat less obvious for this expression than for Lemma 4.1. However, this expression provides a slightly tighter bound, and does not require additional conditions.

To obtain an understanding of how tight the bounds are, consider the example of the Hamiltonian consisting of the spin operator J_x for a spin 50 system. In the basis of J_z eigenstates this operator is tridiagonal. It is straightforward to decompose this Hamiltonian into a sum of two Hamiltonians which are 1-sparse, so $m = 2$.

We take the example of evolution over the time period $t = \pi/4$, $k = 2$ and a range of values of r. For each value of r, the actual error using the integrator was determined, as well as the limit given by Eq. (4.24). In Ref. [5], the alternative bound of $2(2m\tau5^{k-1})^{2k+1}/r^{2k}$ was given. The bound using this expression was also determined for each value of r.

The three values are plotted in Fig. 4.1. Both upper bounds are above the actual error (as would be expected). The upper bound given by Eq. (4.24) is only about three orders of magnitude above the actual error. In contrast, the upper bound from Ref. [5] is many orders of magnitude larger. Thus we find that the upper bound given here is a far tighter bound.

Next we consider the number of exponentials, $N_{\exp}$, required to achieve a certain level of accuracy. The result is as given in the following theorem:

THEOREM 4.1
When the permissible error, as measured by the trace distance between states, is bounded by ε, $N_{\exp}$ is bounded by

$$
N_{\exp} \le \frac{m5^{2k}(mq_k\tau)^{1+1/2k}}{[(2k+1)!\varepsilon]^{1/2k}}, \tag{4.25}
$$

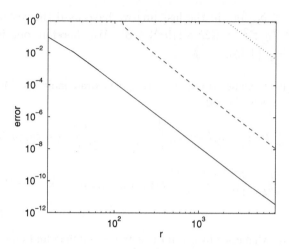

FIGURE 4.1: The error in approximating the evolution under the Hamiltonian $H = J_x$ for spin 50 using different numbers of subdivisions r. The actual error is shown as the solid line, the bound of Eq. (4.24) is shown as the dashed line and the bound from Ref. [5] is shown as the dotted line.

where $\tau = t \times \max \|H_j\|$, $k \geq 2$ is an integer, and we have the restriction $(2k+1)!\varepsilon \leq 1 \leq 2m5^{k-1}q_k\tau$.

PROOF   First note that placing limits on the norm of the difference of the unitaries is equivalent to placing a limit on the trace distance between the output states. This is because

$$\|U_1 - U_2\| \geq \|U_1|\psi\rangle - U_2|\psi\rangle\|$$
$$\geq \frac{1}{2}\mathrm{Tr}\left|U_1|\psi\rangle\langle\psi|U_1^\dagger - U_2|\psi\rangle\langle\psi|U_2^\dagger\right|$$
$$= D\left(U_1|\psi\rangle\langle\psi|U_1^\dagger, U_2|\psi\rangle\langle\psi|U_2^\dagger\right), \qquad (4.26)$$

where the function D is the trace distance.

Now let us take

$$r = \left\lceil (2m5^{k-1}q_k\tau)^{1+1/2k}\left[\frac{\mu_k}{(2k+1)!\varepsilon}\right]^{1/2k}\right\rceil, \qquad (4.27)$$

and take $\delta_1 = \delta_2 = 1$. With this choice of r, the restrictions $(2k+1)!\varepsilon \leq 1 \leq 2m5^{k-1}q_k\tau$ imply that Eqs. (4.17) hold. In addition, these restrictions mean that the magnitude of the expression in the ceiling function in Eq. (4.27) is at least 1, so the error in approximating r by this expression is no more than a factor of 2. In addition, the right-hand side of Eq. (4.15) does not exceed ε, so the error is no more than ε.

Because the number of exponentials in $S_{2k}(\lambda)$ does not exceed $2m5^{k-1}$, we have $N_{exp} \leq 2m5^{k-1}r$. Taking r as in Eq. (4.27), we obtain the upper bound on N_{exp} given in Eq. (4.25). ∎

By taking k to be sufficiently large, it is possible to obtain scaling that is arbitrarily close to linear in τ. However, for a given value of τ, taking k to be too large will increase N_{exp}. We can obtain an estimate of the optimum value of k in the following way. First replace q_k with 1 in Eq. (4.25) and omit $[(2k+1)!]^{1/2k}$ from the denominator. These simplifications only increase the bound. Now re-express Eq. (4.25) as

$$N_{exp} \leq m^2 \tau e^{2k\ln 5 + \ln(m\tau/\varepsilon)/2k}. \qquad (4.28)$$

The value of k that minimizes this expression is

$$k = \text{round}\left[\frac{1}{2}\sqrt{\log_5(m\tau/\varepsilon) + 1}\right]. \qquad (4.29)$$

Adding 1 and rounding takes account of the fact that k must take integer values. Adopting this value of k provides the upper bound

$$N_{exp} \leq 2m^2 \tau e^{2\sqrt{\ln 5 \ln(m\tau/\varepsilon)}}. \qquad (4.30)$$

It can be shown that this result holds with the conditions $\varepsilon \leq 1 \leq m\tau/25$.

If we include q_k we obtain a slightly different estimate for k. Numerically it is found that q_k scales approximately as 3^{-k}. Therefore, including q_k gives the approximate optimal value for k as

$$k \approx \frac{1}{2}\sqrt{\log_{5/\sqrt{3}}(m\tau/\varepsilon)}. \qquad (4.31)$$

This result is not qualitatively different.

4.5 Linear limit on simulation time

Our results show that the simulation of any Hamiltonian may be performed arbitrarily close to linearly in the scaled time τ. This is essentially optimal, because it is not possible to perform the simulation sublinear in τ. The result is

THEOREM 4.2

For all positive integers N there exists a 2-sparse Hamiltonian H such that simulating the evolution of H for scaled time $\tau = t\|H\| = \pi N/2$ within precision $1/4$ requires at least $\tau/2\pi$ black-box queries to H.

Here the situation is similar to that for the problem of ATS. There is a black-box function which gives the nonzero elements in each column of the Hamiltonian, and we quantify the difficulty of the calculation by the number of black-box queries. Note that we have not specified any limit on the dimension of H. In fact, we will require that the number of qubits can grow at least logarithmically with respect to τ. It can be seen that this is essential, because if the dimension was limited it would be possible to classically simulate the evolution by diagonalizing the Hamiltonian, and the complexity of the calculation would not increase indefinitely with τ.

PROOF   The proof is based upon simulating a Hamiltonian which determines the parity of N bits. It has been shown that the parity of N bits requires $N/2$ queries to compute within error $1/4$ [18]; therefore it is not possible to simulate a Hamiltonian which determines the parity any more efficiently.

The Hamiltonian which we consider is based upon the J_x operator with J_z basis states. For spin $J = N/2$, the matrix elements of J_x are

$$\langle j+1|J_x|j\rangle = \langle j|J_x|j+1\rangle = \sqrt{(N-j)(j+1)}/2, \qquad (4.32)$$

where state $|j\rangle$ is an eigenstate of J_z with eigenvalue $j - N/2$. From standard properties of rotation operators, $e^{-i\pi J_x}|0\rangle = |N\rangle$ and $\|J_x\| = J = N/2$.

In order to produce a Hamiltonian which calculates the parity of the bits $X_1, \ldots, X_N$, we add a qubit to the basis states and define the Hamiltonian such that

$$\langle l', j+1|H|l, j\rangle = \langle l, j|H|l', j+1\rangle = \sqrt{(N-j)(j+1)}/2 \qquad (4.33)$$

for values of l and l' such that $l \oplus l' = X_{j+1}$ (where $\oplus$ is XOR). This Hamiltonian corresponds to a graph with two disjoint lines which "cross over" at the positions where bits X_j are 1.

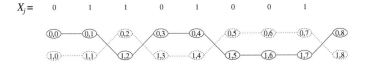

FIGURE 4.2: Graph representing an example of the Hamiltonian in the proof of Theorem 4.2. States are represented by ellipses, and nonzero elements of the Hamiltonian are indicated by lines. The sequence of states $\left| l_j, j \right\rangle$ with $l_0 = 0$ is indicated by the solid line.

The graph corresponding to a Hamiltonian of this type is shown in Fig. 4.2. The system separates into two distinct sets of states which are not connected. If the system starts in one of the states on the path indicated by the solid line, it cannot evolve under the Hamiltonian to a state on the dotted line.

We may determine a sequence of bits $l_0, \ldots, l_N$ such that $l_j \oplus l_{j+1} = X_{j+1}$. The Hamiltonian acting on the set of states $\left| l_j, j \right\rangle$ will then be identical to J_x acting on the states $\left| j \right\rangle$. It is therefore clear that $e^{-i\pi H} \left| l_0, 0 \right\rangle = \left| l_N, N \right\rangle$. If $l_0 = 0$, then l_j is the parity of bits X_1 to X_j, and l_N gives the parity of all N bits. Thus, starting with the state $\left| 0, 0 \right\rangle$ and simulating the evolution $e^{-i\pi H}$, we obtain the state $\left| l_N, N \right\rangle$. Measuring the state of the ancilla qubit then gives the parity, l_N. If the error in the simulation is less than $1/4$, then the probability of error in determining the parity will be less than $1/4$.

To determine the nonzero elements in a column of H we make two queries to X_j. In particular, for the column corresponding to basis state $\left| l, j \right\rangle$, we make a query to X_j and X_{j+1}. It is then straightforward to determine the nonzero elements in the column from Eq. (4.33). If we make fewer than $N/4$ queries to H, then there are fewer than $N/2$ queries to the X_j. This means that the probability for error in determining the parity must be at least $1/4$, and therefore the simulation cannot be achieved with error less than $1/4$.

Since $\|H\| = N/2$ and the time of the simulation is π, the scaled time for this theorem is $\tau = \|H\| t = \pi N/2$. Thus the simulation of H requires at least $N/4 = \tau/2\pi$ queries to obtain error less than $1/4$. ∎

We used this result to show a general result for the number of exponentials required in integrators. This result holds for general integrators which apply to all Hamiltonians.

COROLLARY 4.1

There is no general integrator for Hamiltonians of the form $H = H_1 + H_2$ such that (trace distance) error $< 1/4$ may be achieved with the number of exponentials $N_{\mathrm{exp}} < t\|H\|/2\pi$.

PROOF   We take H as in the preceding proof. This Hamiltonian may be expressed in the form $H = H_1 + H_2$ by taking H_1 to be the Hamiltonian with $\langle l', j+1|H_1|l,j\rangle$ nonzero only for even j, and H_2 to be the Hamiltonian with $\langle l', j+1|H_2|l,j\rangle$ nonzero only for odd j. To determine the nonzero elements in a column of H_1 or H_2 we require only one query to the X_j. For example, for H_1, if j is odd, then we perform a query to X_j; otherwise we perform a query to X_{j+1}.

Both H_1 and H_2 are 1-sparse, and may be efficiently simulated with only two queries to X_j. This result may be shown in the following way. Via one call to X_j, one may calculate m_x, M_x and $\tilde{U}_x$ [where the column index x represents (l, j)]. Therefore, by standard methods one may derive a unitary $\tilde{T}_p$ such that

$$\tilde{T}_p|x,0\rangle = |x, m_x, M_x, \tilde{U}_x\rangle|\phi_x\rangle, \qquad (4.34)$$

where $p = 1$ or 2 for H_1 or H_2, respectively. This is the equivalent of $T_{\vec{j}}$ in Eq. (4.9), except that it produces the additional ancilla $|\phi_x\rangle$.

For the theorem we assume that the parity X_j is given by a unitary that does not produce additional ancilla states

$$X|j,0\rangle = |j, X_j\rangle. \qquad (4.35)$$

Whether we perform a query to X_j or X_{j+1} will depend on whether j is odd or even and whether $p = 1$ or 2. For convenience we denote the result by X_x. The unitary $\tilde{T}_p$ may be expressed as the product of an initial unitary, X, and a final unitary

$$\tilde{T}_p = \tilde{T}_{p,2} X \tilde{T}_{p,1}. \qquad (4.36)$$

The unitary $\tilde{T}_{p,2}$ applies to the output subsystem which contains X_x, whereas $\tilde{T}_{p,1}$ does not act on this subsystem. Let us represent by T_p the sequence of operations $\tilde{T}_{p,2} X \tilde{T}_{p,1}$, followed by copying the outputs m_x, M_x and $\tilde{U}_x$, and applying $\tilde{T}_{p,1}^\dagger \tilde{T}_{p,2}^\dagger$. As $\tilde{T}_{p,1}$ does not act on the subsystem which contains X_x, this sequence of operations gives the map

$$T_p|x,0\rangle = |x, m_x, M_x, \tilde{U}_x, X_x\rangle. \qquad (4.37)$$

The definition of X_x implies that it depends on m_x and M_x, but the same value is obtained for $x = m_x$ or $x = M_x$. It is therefore possible to apply the sequence $T_p^\dagger T T_p$ to correctly apply $\tilde{U}_x$, as in Eq. (4.10). Overall X is applied once and $X^\dagger$ is applied once.

Hence the simulation may be performed with the number of calls to X_j no more than twice the number of exponentials $N_{\exp}$. If $N_{\exp} < t\|H\|/2\pi$, then the total number of queries to the X_j is less than $t\|H\|/\pi$. Taking $t = \pi$ and $\|H\| = N/2$, the number of queries is less than $N/2$. However, from the proof of Theorem 4.2 this Hamiltonian cannot be simulated over time $t = \pi$ with error less than $1/4$ if the number of queries is less than $N/2$. Hence error rate $< 1/4$ cannot be achieved with $N_{\exp} < t\|H\|/2\pi$.

∎

4.6   Efficient decomposition of Hamiltonian

In Section 4.3 we explained the ATS method for decomposing the Hamiltonian. Their method employs an efficient decomposition of a general sparse Hamiltonian into a sum of $m = (D+1)^2 n^6$ 1-sparse Hamiltonians H_j: $H = \sum_{j=1}^m H_j$. Using the standard Lie–Trotter formula the number of time-slices r scales as $m^{1.5}$. The total number of exponentials therefore scales as $mr \propto m^{2.5} \propto n^{15}$ for the ATS method. This is also the scaling of the number of black-box calls for the method of ATS. Here we show that the decomposition can be performed much better—with a reduction to $m = 6D^2$, so the number of exponentials is independent of n—and at $\log^* n$ cost as quantified by the number of black-box calls.

The function $\log^* n \equiv \min\{r|\log_2^{(r)} n < 2\}$ is the iterated logarithm function and may be regarded as being "nearly constant". It is convenient to think of the $\log^*$ of a number as being the smallest height of a tower of powers of 2 that exceeds the number. For example $65536 = 2^{2^{2^2}}$ so $\log^* 65536 = 4$, which is the height of the tower of powers of 2, and a tower of height 5 yields (approximately) 2×10^{19728}, so we can see that $\log^* n$ is very small for all reasonable values of n.

In Ref. [5] we showed that

LEMMA 4.2

There exists a decomposition $H = \sum_{j=1}^m H_j$, where each H_j is 1-sparse, *such that $m = 6D^2$ and each query to any H_j can be simulated by making $O(\log^* n)$ queries to H.*

Here we summarize the proof; the complete proof is given in Ref. [5]. In order to perform this decomposition, we use a more efficient graph coloring than that used by ATS. We use the vector for the color j given by

$$\vec{j} = (v, \mathrm{rind}_H(x,y), \mathrm{cind}_H(x,y)), \tag{4.38}$$

for pairs where $H_{x,y} = 0$ and $x \le y$. Here rind_H and cind_H are defined as for ATS, except we do not require the additional value of zero (which ATS require for cases where $H_{x,y} = 0$). This is because we do not need colors for pairs where there is no edge.

Just using the pair $(\mathrm{rind}_H(x,y), \mathrm{cind}_H(x,y))$ would not be sufficient for a coloring. This is because it would be possible to have three row numbers w, x, and y, such that $w < x < y$, y and x are the number $\mathrm{rind}_H(x,y)$ neighbors of x and w, respectively, and x and w are the number $\mathrm{cind}_H(x,y)$ neighbors of y and x, respectively. Therefore it is necessary to add the additional parameter v. We only require 6 alternative values for v, so the total number of alternative values of $\vec{j}$ is only $6D^2$. In comparison ATS require a total of $(D+1)^2 n^6$ values of $\vec{j}$.

The values of v are assigned in a way which uses ideas from deterministic coin tossing [19, 20]. First one determines a sequence of values x_l^0 such that $x_0^0 = x$, $x_1^0 = y$, and the following pairs x_l^0, x_{l+1}^0 satisfy

$$(\mathrm{rind}_H(x,y), \mathrm{cind}_H(x,y)) = (\mathrm{rind}_H(x_l^0, x_{l+1}^0), \mathrm{cind}_H(x_l^0, x_{l+1}^0)). \tag{4.39}$$

This sequence usually terminates very quickly. If it does not, these indices are only determined up to $x_{z_n+1}^{(0)}$, where z_n is the number of times we must iterate $l \mapsto 2\lceil \log_2 l \rceil$ (starting at 2^n) to obtain 6 or less. It can be shown that z_n is approximately $\log^* n$.

Next, values of $x_l^{(1)}$ are determined in the following way. The first bit where $x_l^{(0)}$ differs from $x_{l+1}^{(0)}$ is determined, and the value (for $x_l^{(0)}$) and position of this bit are recorded as $x_l^{(1)}$. At the end of the chain, $x_l^{(1)}$ is the first bit of $x_l^{(0)}$ followed by zeros. This procedure is repeated up to $x_l^{(z_n)}$, and we take $v = x_0^{(z_n)}$. It can be shown that there are only 6 possible values for $x_0^{(z_n)}$, and the value obtained for the pair (w,x) differs from that for (x,y).

To illustrate this procedure, let us consider the Hamiltonian for which a portion of the graph is shown in Fig. 4.3. The calculation for v for the edge between x and y is illustrated in Table 4.1, and the corresponding calculation for the edge between w and x is illustrated in Table 4.2. In the tables $n = 18$,

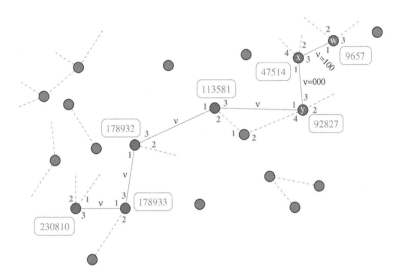

FIGURE 4.3: A portion of the graph for the example given in Tables 4.1 and 4.2. The bold labels indicate the ordering of the edges around each vertex. Each edge is labeled by the labels at each end of the edge, and the additional parameter v. The values of v for the edges (w,x) and (x,y) determined in Tables 4.1 and 4.2 are shown. In the sequence of solid edges, each edge has the same labels, so it is necessary for the v to differ to ensure that adjoining edges have distinct labels. The numbers in the first columns of Tables 4.1 and 4.2 are the binary representations of the vertex numbers given here.

so the numbers of different possible values in columns 1 to 5 are 2^{18}, 36, 12, 8, and 6. In this case z_n is equal to 4, and we have therefore determined the sequence of $x_i^{(0)}$ up to $x_5^{(0)}$.

As an example of the calculation of $x_i^{(1)}$, the second bit of $x_0^{(0)}$ differs from the corresponding bit of $x_1^{(0)}$. The second bit of $x_1^{(0)}$ is 0, so this is the first bit of $x_0^{(1)}$. We subtract 1 from the bit position to obtain 1, and take the remaining bits of $x_0^{(1)}$ to be the binary representation of 1. For the case of $x_5^{(0)}$, this is the end of the chain, so we simply take $x_5^{(1)}$ to be the first bit of $x_5^{(0)}$, which is 1, followed by zeros.

$l\backslash p$	0	1	2	3	4
0	001011100110011010	000001	0100	000	**000**
1	010110101010011011	000010	1100	100	*100*
2	011011101110101101	000000	0001	*000*	*000*
3	101011101011110100	010001	*1001*	*100*	*100*
4	101011101011110101	*000001*	*0000*	000	000
5	111000010110011010	100000	1000	100	100

Table 4.1 Example values of $x_l^{(p)}$ under our scheme for calculating v. The value of v obtained is in the upper right, and is shown in bold. For this example $n = 18$ and $z_n = 4$. The values in italics are those that may differ from $w_{l+1}^{(p)}$ (there are no corresponding values for the bottom row).

$l\backslash p$	0	1	2	3	4
0	000010010110111001	000010	1100	100	**100**
1	001011100110011010	000001	0100	000	000
2	010110101010011011	000010	1100	100	*001*
3	011011101110101101	000000	0001	*111*	*100*
4	101011101011110100	010001	*0000*	*000*	*000*
5	101011101011110101	*100000*	*1000*	100	100

Table 4.2 Example values of $w_l^{(p)}$ under our scheme for calculating v. The value of v obtained is in the upper right, and is shown in bold. For this example $n = 18$ and $z_n = 4$. The values in italics are those which may differ from $x_{l-1}^{(p)}$.

We use the notation $w_l^{(p)}$ for the values given in Table 4.2. This example illustrates the case where the sequence of $w_l^{(0)}$ (with $w_l^{(0)} = x_{l-1}^{(0)}$) ends before the sequence of $x_l^{(0)}$. This means that $w_5^{(1)} \neq x_4^{(1)}$, and the differences propagate so that $w_2^{(4)} \neq x_1^{(4)}$. However, $w_1^{(4)}$ is still equal to $x_0^{(4)}$.

Now $w_0^{(4)}$ gives the value of v for the edge between w and x. Because $w_1^{(4)} = x_0^{(4)}$, $w_1^{(4)}$ is equal to the value of v for the edge between x and y. The method for calculation ensures that $w_1^{(4)}$ differs from $w_0^{(4)}$, so we obtain different values of v for these two edges, as required.

Using this lemma, we have shown the following general theorem on the number of black-box calls required for Hamiltonian simulation.

THEOREM 4.3

The Hamiltonian H may be simulated within error ε for time t with the number of black-box calls

$$N_{\text{bb}} \in O\left((\log^* n)d^2 5^{2k}(d^2 q_k \tau)^{1+1/2k}/[(2k+1)!\varepsilon]^{1/2k}\right) \qquad (4.40)$$

with $\tau = t\|H\|$ and k an integer ≥ 2.

PROOF   Overall the number of Hamiltonians $H_{\vec{j}}$ in the decomposition is $m = 6D^2$. To calculate $g(x, \vec{j})$, it is necessary to call the black-box $2(z_n + 2)$ times.

Given a unitary U_f representing the black-box function f, one may obtain a unitary operator U_g satisfying

$$U_g|x, \vec{j}, 0\rangle = |x, \vec{j}, y, (H_{\vec{j}})_{x,y}\rangle|\phi_{x,\vec{j}}\rangle, \qquad (4.41)$$

where the additional ancilla states $|\phi_{x,\vec{j}}\rangle$ are produced by the calculation. As discussed in Section 4.3, the Hamiltonian $H_{\vec{j}}$ may be simulated via two applications of U_g and two applications of $U_g^{\dagger}$. As z_n is of order $\log^* n$, the number of black-box calls to f for the simulation of each $H_{\vec{j}}$ is $O(\log^* n)$. Using these values, along with Eq. (4.25), we obtain the number of black-box queries as in Eq. (4.40). ∎

In this theorem we have quantified the complexity of the calculation simply by the number of black-box calls. It is also necessary to apply a number of auxiliary operations in addition to each black-box call. In determining g from

f, we must make a calculation for v. It is necessary to perform bit comparisons between a maximum of $z_n + 2$ numbers in the first step, and each has n bits. This requires $O(n \log^* n)$ operations. In the next steps the number of bits is $O(\log_2 n)$ bits or less, which is negligible. Hence the number of auxiliary operations is

$$O\left(n(\log^* n)d^2 5^{2k}(d^2 q_k \tau)^{1+1/2k}/[(2k+1)!\varepsilon]^{1/2k}\right). \qquad (4.42)$$

In comparison the implicit scaling in Ref. [4] was n^{16}.

4.7 Conclusions

It is possible to efficiently simulate physical systems provided the Hamiltonian for the system is sparse. Lloyd showed this in the case where the system is composed of many small subsystems, and the Hamiltonian is a sum of interaction terms between these subsystems [3]. In this case the Hamiltonian is sparse. Aharonov and Ta-Shma [4] showed this for the case of general sparse Hamiltonians where the Hamiltonian is not given as a sum of simple terms. They show how to decompose the Hamiltonian into a sum of 1-sparse Hamiltonians.

The schemes given by Lloyd and ATS are still somewhat inefficient; the complexity scales as t^2 (for Lloyd's method) or $t^{1.5}$ (for the method of ATS). In addition, the number of black-box calls in the method of ATS scales as the 15th power of n (the number of qubits), and uses $(D+1)^2 n^6$ Hamiltonians in the sum. This was improved somewhat by Childs [21], who found scaling of n^2 for the number of black-box calls.

In our work we improved significantly upon these results. We applied the higher-order integrators of Suzuki [16, 17] to reduce the scaling to $t^{1+1/2k}$ for arbitrary integer k. We placed an upper limit on the error, and used this to estimate the optimum value of k to use. In addition, we showed that this scaling is close to optimal, because it is not possible to achieve sublinear scaling.

We also provided a superior method for decomposing the Hamiltonian into a sum. The scaling of the number of black-box calls is effectively independent of the number of qubits, and the total number in the sum is just $6D^2$, rather than $(D+1)^2 n^6$ as for the method of ATS. The problem is analogous to deterministic coin tossing [19, 20], and the scaling is the same. In the case of deterministic coin tossing this scaling was proven to be optimal. This suggests that the scaling is optimal, though the proof can not be directly applied to this case.

Acknowledgments

This project has been supported by the Australian Research Council, Canada's NSERC, iCORE, CIAR and MITACS. R.C. thanks Andrew Childs for helpful discussions.

References

[1] R. P. Feynman, *Int. J. Theoret. Phys.* **21**, 467 (1982).

[2] D. Deutsch, *Proc. Roy. Soc. Lon. A* **400**, 97 (1985).

[3] S. Lloyd, *Science* **273**, 1073 (1996).

[4] D. Aharonov and A. Ta-Shma, *Proc. 35th Annual ACM Symp. on Theory of Computing*, 20 (2003).

[5] D. W. Berry, G. Ahokas, R. Cleve, and B. C. Sanders, Comm. Math. Phys. **270**, 359 (2007).

[6] P. W. Shor, *Proc. 35th IEEE Symp. on Foundations of Computer Science*, 124 (1994).

[7] S. Hallgren, *Proc. 34th Annual ACM Symp. on Theory of Computing*, 653 (2002).

[8] L. Grover, *Phys. Rev. Lett.* **79**, 325 (1997).

[9] E. Farhi, J. Goldstone, S. Gutmann, and M. Sipser, http://arxiv.org/abs/quant-ph/0001106 (2000).

[10] C. H. Bennett, E. Bernstein, G. Brassard, and U. Vazirani, *SIAM J. Comput.*, **26**, 1510 (1997).

[11] A. Childs, E. Farhi, and S. Gutmann, *J. Quant. Inf. Proc.* **1**, 35 (2002).

[12] N. Shenvi, J. Kempe, and K. B. Whaley, *Phys. Rev. A* **67**, 052307 (2003).

[13] A. Childs and J. Goldstone, http://arxiv.org/abs/quant-ph/0306054 (2003).

[14] A. Ambainis, *Proc. 45th IEEE Symp. on Foundations of Computer Science*, 22 (2004).

[15] A. Ambainis, J. Kempe, and A. Rivosh, *Proc. of ACM-SIAM Symp. on Discrete Algorithms*, 1099 (2005).

[16] M. Suzuki, *Phys. Lett. A* **146**, 319 (1990).

[17] M. Suzuki, *J. Math. Phys.* **32**, 400 (1991).

[18] R. Beals, H. Buhrman, R. Cleve, M. Mosca, and R. de Wolf, *J. ACM* **48**, 778 (2001).

[19] R. Cole and U. Vishkin, *Inform. and Control* **70**, 32 (1986).

[20] N. Linial, *SIAM J. Comput.* **21**, 193 (1992).

[21] A. M. Childs, *Ph.D. Thesis*, Massachusetts Institute of Technology, (2004).

Quantum Technology

Chapter 5

New Mathematical Tools for Quantum Technology

C. Bracher, M. Kleber, and T. Kramer

Abstract   Progress in manufacturing technology has allowed us to probe the behavior of devices on a smaller and faster scale than ever before. With increasing miniaturization, quantum effects come to dominate the transport properties of these devices; between collisions, carriers undergo *ballistic motion* under the influence of local electric and magnetic fields. The often surprising properties of quantum ballistic transport are currently elucidated in "clean" atomic physics experiments. From a theoretical viewpoint, the electron dynamics is governed by *ballistic propagators* and *Green functions*, intriguing quantities at the crossroads of classical and quantum mechanics. Here, we briefly describe the propagator method, some ballistic Green functions, and their application in a diverse range of problems in atomic and solid state physics, such as photodetachment, atom lasers, scanning tunneling microscopy, and the quantum Hall effect.

5.1   Physics in small dimensions

The laws of quantum mechanics provide a means for a successful interpretation of measurements and experiments on a microscopically small scale. It goes without saying that we can never "observe" directly what's going on in an atom or molecule. To understand what nature is telling us we must learn its

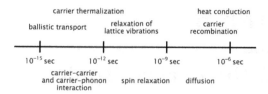

FIGURE 5.1: Timescales in semiconductors.

language. Its grammar follows the mathematical rules of quantum mechanics. Without mathematical tools we would not be able to describe intriguing processes such as, for example the mapping of quantum states of light to intrinsic atomic states. Fortunately, the necessary formalism is often powerful, elegant and quite easy to comprehend. In our contribution we will demonstrate the revival of an established mathematical tool in the important field of quantum technology. This tool is known under the name of Green function or Green's function in honor of George Green.[1]

Many problems in electrodynamics, hydrodynamics, heat conduction, a-coustics, etc., require the solutions of inhomogeneous linear differential equations. It is there where Green functions come to full power. The corresponding mathematical approach is the same in all branches of physics—as long as we are dealing with linear, ordinary or partial differential equations. In quantum mechanics, Green functions enjoy the advantage of having a physical meaning: The single-particle Green function $G(\mathbf{r}, \mathbf{r}'; E)$ is the relative probability amplitude for a particle to move with energy E from an arbitrary point $\mathbf{r}'$ to another point $\mathbf{r}$. Probability amplitudes are known to be essential in all kinds of quantum problems. In his book on *'The Character of Physical Law'* Feynman [1] notes that "... *everything that can be deduced from the ideas of the existence of quantum mechanical probability amplitudes, strange though they are, will work, ... one hundred percent....*"

In this tutorial we present some basic features of electron and atom motion in external fields. External electric, magnetic and even gravitational fields are well suited to control the motion of particles in quantum devices. It is not our purpose to dive into the technical depths of ultrasmall electronics research and technology. We only want to illustrate how useful *single-particle* Green

---

[1]Green was an almost entirely self-taught English mathematician and physicist who in 1828 published an essay entitled: *"On the Applications of Mathematical Analysis to the Theories of Electricity and Magnetism."* In this essay he obtained integral representations for the solutions of problems connected with the Laplace operator.

functions and propagators can be for basic problems in quantum technology whenever a microscopic description of quantum transport is necessary. A single particle description is appropriate for devices with low particle densities where interaction processes can be neglected. But they are also useful in more general cases. Indeed, microscopically small particles travel freely on length scales of the order of the free mean path ℓ, which is the distance that an electron travels before its initial phase is destroyed for whatever reasons [2, 3]. The mean free paths depend strongly on the material under consideration and they are much affected by temperature. Particles that travel freely are called *ballistic* particles.

In Fig. 5.1 we show timescales for the motion of electrons in a typical semiconductor. We invoke the uncertainty principle $\Delta p \Delta x \sim \hbar$ to get a feeling for the time domain of ballistic motion. Motion with well defined momentum requires $\Delta p \ll p$. For a quasi-classical description, the electrons are required to be well localized compared to the mean free path, i. e., $\Delta x \ll \ell = p\tau/m$ with τ being the time for ballistic motion (see Fig. 5.1). It follows that $\Delta p \Delta x \ll p\ell = p^2\tau/m$. Suppose we have thermal electrons; then we can use the equipartition law of classical statistics, $p^2\tau/m \sim k_B T \tau$ with k_B being the Boltzmann constant and T the temperature. Comparing with the uncertainty principle we obtain $\tau \gg \hbar/(k_B T)$ which for room temperature is of the order 10^{-13} to 10^{-14} seconds. For thermal electrons this last inequality is frequently met. However, for electrons moving with high energies of the order eV the inequality is no longer fulfilled. In this case one has to treat ballistic transport fully quantum mechanically. In the following we present examples where quantum transport is essential. First, however, we review some useful mathematical tools in a nut shell.

Clearly, we cannot cite all relevant literature in this field. This would be an impossibly difficult and lengthy job. But the interested reader will find a wealth of literature in the research articles cited in this tutorial.

5.2 Propagators and Green functions

A time-dependent treatment of the flow of charge carriers is based on the time-dependent Schrödinger equation. In this context it is useful to summarize a few aspects of the initial-value problem for a wave function known at $t = t_0$,

$$\psi(\mathbf{r}, t = t_0) = \psi_0(\mathbf{r}) = \langle \mathbf{r} | \psi(t = t_0) \rangle. \tag{5.1}$$

The corresponding ket vector $|\psi(t)\rangle$ evolves according to the basic law of quantum mechanics,

$$(i\hbar\partial_t - H)|\psi(t)\rangle = 0. \tag{5.2}$$

The formal solution of (5.2) is conveniently written in terms of the time evolution operator $U(t,t_0)$:

$$|\psi(t)\rangle = U(t,t_0)|\psi(t_0)\rangle. \tag{5.3}$$

In coordinate space (5.3) reads

$$\langle\mathbf{r}|\psi(t)\rangle = \int d^3\mathbf{r}'\,\langle\mathbf{r}|U(t,t_0)|\mathbf{r}'\rangle\langle\mathbf{r}'|\psi(t_0)\rangle. \tag{5.4}$$

The integral kernel of (5.4) is called propagator K:

$$K(\mathbf{r},t|\mathbf{r}',t_0) \equiv \langle\mathbf{r}|U(t,t_0)|\mathbf{r}'\rangle. \tag{5.5}$$

Obviously, $K(\mathbf{r},t|\mathbf{r}',t_0)$ is the time evolution matrix $U(t,t_0)$ in coordinate space representation. From the last two equations, we have

$$\lim_{t\to t_0} K(\mathbf{r},t|\mathbf{r}',t_0) = \delta^3(\mathbf{r}-\mathbf{r}'). \tag{5.6}$$

The (time-) retarded Green function must vanish for $t < t_0$. It is usually defined by

$$G(\mathbf{r},t;\mathbf{r}',t_0) = \frac{1}{i\hbar}\Theta(t-t_0)K(\mathbf{r},t|\mathbf{r}',t_0). \tag{5.7}$$

From this definition, and the fact that the propagator is a solution of the time-dependent Schrödinger equation, the retarded Green function is seen to satisfy the differential equation

$$(i\hbar\partial_t - H)\,G(\mathbf{r},t;\mathbf{r}',t_0) = \delta^3(\mathbf{r}-\mathbf{r}')\delta(t-t_0). \tag{5.8}$$

The delta function $\delta(t-t_0)$ on the right hand side of (5.6) originates from the step function $\Theta(t-t_0)$ in the definition of G.

For time-dependent Hamiltonians, the propagator will depend separately on t and t_0. For time-independent Hamiltonians, the propagator depends only on the time difference $t - t_0$. In the latter case the Laplace transform of the propagator

$$G(\mathbf{r},\mathbf{r}';E) = \frac{1}{i\hbar}\int_0^\infty dt\,\exp(iEt/\hbar)\,K(\mathbf{r},t|\mathbf{r}',0) \tag{5.9}$$

generates the energy (-dependent) Green function,

$$G(\mathbf{r},\mathbf{r}';E) = \lim_{\eta\to 0^+}\left\langle\mathbf{r}\left|\frac{1}{E-H+i\eta}\right|\mathbf{r}'\right\rangle. \tag{5.10}$$

$G(\mathbf{r},\mathbf{r}';E)$ is the amplitude for travel of a particle from $\mathbf{r}$ to $\mathbf{r}'$ out of a point source and, as a function of energy. This feature will emerge if we evaluate (5.10) explicitly. We should also mention that the appearance of the infinitesimally small, positive imaginary term $i\eta$ in (5.10) has a simple reason: To enforce convergence of the integral in (5.9) one has to replace E by $E+i\eta$. The physical meaning of such a small shift into the complex energy plane becomes evident if one evaluates (5.10) for a free particle. The result [4],

$$G_{\text{free}}(\mathbf{r},\mathbf{r}';E) = -\frac{m}{2\pi\hbar^2}\frac{\exp(ik|\mathbf{r}-\mathbf{r}'|)}{|\mathbf{r}-\mathbf{r}'|}, \tag{5.11}$$

is well known from scattering theory: For $\mathbf{r}'$ fixed and $\mathbf{r}$ variable, G_{free} describes an *outgoing* spherical wave that originates from a point source at $\mathbf{r}=\mathbf{r}'$. Had we Fourier transformed the time-advanced Green function instead of the time-retarded Green function, we would, of course, have ended up with an incoming spherical wave instead of an outgoing wave.

Propagators contain all necessary information about the motion of a particle. Unfortunately it is not always possible to find a closed-form solution for K or G. For potentials which are *at most quadratic in the coordinates*, the propagator assumes the canonical form [5]

$$K(\mathbf{r},t|\mathbf{r}',0) = A(t)\,\exp\left[iS_{cl}(\mathbf{r},\mathbf{r}';t)/\hbar\right], \tag{5.12}$$

where S_{cl} is the corresponding *classical action*, and where $A(t)$ is a time-dependent factor independent of the particle's position. However, nonquadratic potentials such as the Coulomb potential generally do not have the canonical form (5.12). Explicit expressions for propagators can be found, for example, in [5, 6, 7, 8, 9, 10, 11].

5.2.0.0.1 The Moshinsky shutter: This example illustrates how the free propagator

$$K_{\text{free}}(\mathbf{r},t|\mathbf{r}',t_0) = \left[\frac{m}{2\pi i\hbar(t-t_0)}\right]^{1/2}\exp\left[\frac{im(\mathbf{r}-\mathbf{r}')^2}{2\hbar(t-t_0)}\right], \tag{5.13}$$

is used to solve an initial value problem that describes the flow of quantum particles. It is of interest in the context of the quantum mechanical propagation of a signal. Moshinsky [12] has analyzed the spreading of such a signal. He considered a monochromatic beam of noninteracting particles of mass m and energy E_k. The particles are supposed to move parallel to the x–axis from left

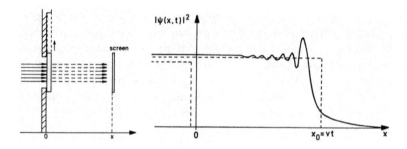

FIGURE 5.2: The Moshinsky shutter: The shutter is removed at $t_0 = 0$. The quantum particles start propagating towards the screen. When will the particles arrive at the screen? The distribution of the traveling particles is shown on the right hand side where the full line corresponds to the quantum result (17) and, where the classical propagation is represented by dotted lines.

to right. The beam is stopped (and absorbed) by a shutter at $x = 0$ (see Fig. 5.2). The signal is given at $t = t_0 = 0$ when the shutter is opened.

The sudden removal of the shutter marks the beginning of a "quantum race" where the particles run along the positive x-axis. In order to elucidate the spreading of the signal all one has to do is to calculate $\psi(x,t)$ starting with

$$\psi_0(x) = \psi(x,t=0) = \Theta(-x)e^{ikx}. \tag{5.14}$$

Using (5.4) one finds

$$\langle x|\psi(t \geq 0)\rangle = M(x;k;\hbar t/m), \tag{5.15}$$

where the Moshinsky function M is defined in terms of the complementary error function [13],

$$M(x;k;\tau) = \frac{1}{2}\exp\left(ikx - ik^2\tau/2\right)\, erfc\left[\frac{x-k\tau}{(2i\tau)^{1/2}}\right] \tag{5.16}$$

with $i^{1/2} = \exp(i\pi/4)$.

An interesting property of (5.15) is revealed when we evaluate the particle number probability. Introducing $u = (\hbar kt/m - x)/(\pi\hbar t/m)^{1/2}$, we obtain

$$|\langle x|\psi(t)\rangle|^2 = \frac{1}{2}\left\{\left[\frac{1}{2}+C(u)\right]^2 + \left[\frac{1}{2}+S(u)\right]^2\right\}. \tag{5.17}$$

The functions $C(u)$ and $S(u)$ are the well-known Fresnel integrals [13]. The corresponding probability pattern is called diffraction in time because it arises when the shutter is opened for a finite time t. Although transient effects are important by themselves [14] we won't discuss them in more detail here. In what follows, we will discuss stationary quantum transport.

5.3 Quantum sources

In real-space representation, propagators and Green functions describe the motion of quantum particles from some initial point $\mathbf{r}'$ to a final point $\mathbf{r}$. But where do the particles come from? One may think of two different situations: i) the particles have been around all the time like electrons in an atom, or, ii) the particles are generated by a source, a situation which is quite familiar from scattering theory where a beam of particles is generated by an accelerator in a region far away from the target. In mesoscopic physics and nanotechnology, however, there is usually no such large spatial separation. Let us motivate the introduction of coherent quantum sources of particles and illustrate their properties by means of an example.

5.3.1 Photoelectrons emitted from a quantum source

We may consider the photoeffect as a two-step process as illustrated in Fig. 5.3. The time evolution of the emitted electron is of course governed by the rules of quantum mechanics. In the absence of any interaction between photon and electron, the electron under consideration is attached to the atom and is described by the bound-state wave function $\psi_{\text{atom}}(\mathbf{r})$. Let us consider a dilute gas of independent atoms where the interaction of the photoelectron with neighboring atoms can be neglected. In the presence of a photon field this wave function will obtain a small scattering component $\psi_{\text{sc}}(\mathbf{r})$ that allows the electron to leave the atom. For a dipole-allowed transition, the dipole operator $\hat{D}(\mathbf{r}) \propto \hat{\boldsymbol{\varepsilon}} \cdot \mathbf{r}$ is responsible for transferring the electron from its initial bound state $|\psi_{\text{atom}}\rangle$ to a continuum state $|\psi_{\text{sc}}\rangle$. Under steady-state conditions with many atoms (each having the Hamiltonian $\hat{H}_{\text{atom}}$) and weak monochromatic light we must solve the problem:

$$\left[E - \hat{H}_{\text{atom}} - \hat{H}_{\text{rad}} - \hat{D} \cdot (\hat{a} + \hat{a}^{\dagger})\right]\left(|\psi_{\text{atom}}\rangle |1\rangle + |\psi_{\text{sc}}\rangle |0\rangle\right) = 0 \qquad (5.18)$$

FIGURE 5.3: Two steps to create a photoelectron: Step 1 (left panel): The photon transfers its energy to the initially bound electron. Step 2 (right panel): The photoelectron escapes from the absorption region.

where $E = E_{atom} + h\nu$ is the energy sum of electron and photon. The unperturbed Hamilton operator of the radiation field with field operators $\hat{a}$ and $\hat{a}^\dagger$ is denoted by $\hat{H}_{rad}$, with the zero-point energy being subtracted. As usual, $|1\rangle$ characterizes the presence of the photon and $|0\rangle$ its absence after absorption. Projection onto the zero-photon state $\langle 0|$ yields the desired equation for the scattering solution,

$$\left[E - \hat{H}_{atom}(\mathbf{r})\right]\psi_{sc}(\mathbf{r}) = \hat{D}(\mathbf{r})\psi_{atom}(\mathbf{r}) \equiv \sigma(\mathbf{r}). \qquad (5.19)$$

We can interpret $\hat{D}|\psi\rangle$ as a *source function* $|\sigma\rangle$ for the photoelectrons: The dipole operator *prepares* the electron in a continuum state but can be neglected once the electron has left the atom.

5.3.2 Currents generated by quantum sources

The last two equations can be generalized to a situation where the scattered particle experiences some final-state interaction. For example, the presence of a final-state Coulomb interaction or of an external field can be readily taken into account in Eq. (5.19) by writing

$$\left[E - H_0 - W(\mathbf{r})\right]\psi_{sc}(\mathbf{r}) = \sigma(\mathbf{r}), \qquad (5.20)$$

where H_0 is the Hamiltonian of a free particle and where $W(\mathbf{r})$ represents the interaction of the emitted particle (for example the photoelectron) with its environment. Here and in the following we omit the hat symbol for the operator ($H_0 \equiv \hat{H}_0$). In analogy to other inhomogeneous field equations, e.g., Maxwell's equations, the right-hand term $\sigma(\mathbf{r})$ in (5.20) is again identified as a source for the scattered wave $\psi_{sc}(\mathbf{r})$.

We now turn to the mathematical aspects of (5.20). Introducing the energy Green function $G(\mathbf{r}, \mathbf{r}'; E)$ for the Hamiltonian H defined via [15]:

$$\left[E - H_0 - W(\mathbf{r}) \right] G(\mathbf{r}, \mathbf{r}'; E) = \delta^3(\mathbf{r} - \mathbf{r}') , \qquad (5.21)$$

a solution to (5.20) in terms of a convolution integral reads:

$$\psi_{sc}(\mathbf{r}) = \int d^3 r' \, G(\mathbf{r}, \mathbf{r}'; E) \sigma(\mathbf{r}') . \qquad (5.22)$$

In general, this result is not unique. However, any two solutions $\psi_{sc}^1(\mathbf{r})$ and $\psi_{sc}^2(\mathbf{r})$ differ only by an eigenfunction $\psi_{hom}(\mathbf{r})$ of the homogeneous Schrödinger equation, with $H = H_0 + W$ and $H \psi_{hom}(\mathbf{r}) = E \psi_{hom}(\mathbf{r})$. The ambiguity in $\psi_{sc}(\mathbf{r})$ is resolved by the demand that $G(\mathbf{r}, \mathbf{r}'; E)$ presents a retarded solution characterized by outgoing-wave behavior as $r \to \infty$. Formally, this enforces the same choice as in Eq. (5.10). It is then easy to decompose the Green function (5.10) into real and imaginary parts

$$G(\mathbf{r}, \mathbf{r}'; E) = \left\langle \mathbf{r} \left| PP \left(\frac{1}{E - H} \right) - i\pi \delta(E - H) \right| \mathbf{r}' \right\rangle , \qquad (5.23)$$

where $PP(\ldots)$ denotes the Cauchy principal value of the energy integration.

Defining the current density in the scattered wave in the usual fashion by $\mathbf{j}(\mathbf{r}) = \hbar \Im[\psi_{sc}(\mathbf{r})^* \nabla \psi_{sc}(\mathbf{r})]/M$ (where for simplicity we omitted the vector potential $\mathbf{A}(\mathbf{r})$, see [16]), the inhomogeneous Schrödinger equation (5.20) gives rise to a modified equation of continuity [17, 18]:

$$\nabla \cdot \mathbf{j}(\mathbf{r}) = -\frac{2}{\hbar} \Im[\sigma(\mathbf{r})^* \psi_{sc}(\mathbf{r})] , \qquad (5.24)$$

where $\Im[x]$ stands for the imaginary part of x. Thus, the inhomogeneity $\sigma(\mathbf{r})$ acts as a source for the particle current $\mathbf{j}(\mathbf{r})$. By integration over the source volume, and inserting (5.22), we obtain a bilinear expression for the total particle current $J(E)$, i. e., the total scattering rate:

$$J(E) = -\frac{2}{\hbar} \Im \left[\int d^3 r \int d^3 r' \, \sigma(\mathbf{r})^* G(\mathbf{r}, \mathbf{r}'; E) \sigma(\mathbf{r}') \right] . \qquad (5.25)$$

Some important identities concerning the total current $J(E)$ are most easily recognized in a formal Dirac bra-ket representation. In view of (5.23), we may express $J(E)$ by

$$J(E) = -\frac{2}{\hbar} \Im[\langle \sigma | G | \sigma \rangle] = \frac{2\pi}{\hbar} \langle \sigma | \delta(E - H) | \sigma \rangle , \qquad (5.26)$$

from which the sum rule immediately follows [18]:

$$\int_{-\infty}^{\infty} dE\, J(E) = \frac{2\pi}{\hbar} \langle \sigma | \sigma \rangle = \frac{2\pi}{\hbar} \int d^3 r |\sigma(\mathbf{r})|^2 , \qquad (5.27)$$

(provided this integral exists).

5.3.3 Recovering Fermi's golden rule

In order to connect Eq. (5.25) to the findings of conventional scattering theory, we display $J(E)$ in an entirely different, yet wholly equivalent fashion. Employing a complete orthonormal set of eigenfunctions $|\psi_{\mathrm{fi}}\rangle$ of the Hamiltonian H, $\delta(E - H)|\psi_{\mathrm{fi}}\rangle = \delta(E - E_{\mathrm{fi}})|\psi_{\mathrm{fi}}\rangle$ follows, and replacing $|\sigma\rangle = D(\mathbf{r})|\psi_{\mathrm{atom}}\rangle$ (5.23), we may formally decompose (5.26) into a sum over eigenfunctions:

$$J(E) = \frac{2\pi}{\hbar} \sum_{\mathrm{fi}} \delta(E - E_{\mathrm{fi}}) \left| \langle \psi_{\mathrm{fi}} | D(\mathbf{r}) | \psi_{\mathrm{atom}} \rangle \right|^2 . \qquad (5.28)$$

Thus, Fermi's golden rule is recovered. Another noteworthy consequence of (5.25) and (5.26) emerges in the limit of pointlike sources, $\sigma(\mathbf{r}) \sim C\delta(\mathbf{r} - \mathbf{R})$. We then find [17]

$$J(E) = -\frac{2}{\hbar} |C|^2 \Im [G(\mathbf{R}, \mathbf{R}; E)] = \frac{2\pi}{\hbar} |C|^2 n(\mathbf{R}; E) , \qquad (5.29)$$

where $n(\mathbf{R}; E) = \sum_{\mathrm{fi}} \delta(E - E_{\mathrm{fi}}) |\psi_{\mathrm{fi}}(\mathbf{R})|^2$ is the local density of states of H at the source position $\mathbf{R}$. Equation (5.29) forms the theoretical basis of the Tersoff–Hamann description of scanning tunneling microscopy [17, 19]. The advantage of the formulation in terms of quantum sources over the traditional Fermi's golden rule approach (which involves an integral over the final states) is that it emphasizes the dynamical aspects of the propagation in real space and opens the possibility to a semiclassical calculation of photocurrents with closed-orbit theories [20, 21].

5.3.4 Photodetachment and Wigner's threshold laws

To find out how we can use the formalism for real physics we continue our discussion of the photoeffect. Applying the photoelectric effect to negative ions means that the emitted electron only weakly interacts with the remaining neutral atom. Just as in Young's double-slit experiment, the fringe pattern in the current profile can be interpreted as interference between the two classical trajectories, here of a particle in a constant force field [24, 25]. From the interference pattern one can determine the kinetic energy of the electrons

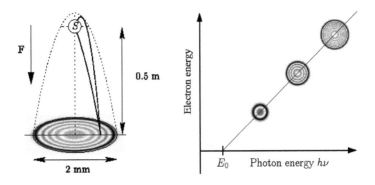

FIGURE 5.4: Near-threshold detachment of oxygen ions: $O^- \to O + e^-$ in the presence of a homogeneous electric force field $F = e\mathscr{E}$. The two possible classical trajectories for a photoelectron leading from the source (marked by S) to any destination will give rise to interference on a distant detector screen. The fringe pattern in the current distribution depends sensitively on the energy. By counting the number of fringes the binding energy E_0 of the outer electron can be determined from Einstein's law [22, 23, 24].

and plot it against the photon energy to check Einstein's law (right panel of Fig. 5.4). Near threshold, the photoelectron has very little kinetic energy and, hence, a large de Broglie wavelength. In the absence of a final-state Coulomb interaction the relevant Green function is that of a particle falling freely in a constant field [24, 26]. Experimental results are reported in Ref. [23], Fig. 5.4. They show a highly accurate verification of Einstein's law which can be used to obtain the binding energy of O^- with unprecedented accuracy.

As noted above, we can interpret $D(\mathbf{r})|\psi_{\text{atom}}\rangle$ as a source $|\sigma\rangle$ for the photoelectrons. Expanding the source in terms of multipoles $\sigma_{lm}(r)$ and, taking into account that for O^- the photoelectron leaves the atom near threshold in an s–wave continuum state, we retain only the $l = 0$ component of the source by writing $\sigma(\mathbf{r}) = \sigma_{00}(r)/\sqrt{4\pi}$. In the absence of external fields, the free Green function (5.11), an outgoing spherical wave, yields after multipole expansion

$$J(E) \propto k \left[\int_0^\infty \frac{r\,dr}{k} \sin(kr)\sigma_{00}(r) \right]^2. \tag{5.30}$$

For $E = \hbar^2 k^2/(2m) \to 0$ it follows that $J(E) \propto k$. This result is independent of the form of the atomic source and it reflects Wigner's threshold law [24, 27]. It provides an alternative way to determine the electron affinity of a negative ion.

A similar analysis applies to the more complicated Green function in an electric force field $F = e\mathscr{E}$, one of the few quantum problems in more than one dimension that have exact solutions:

$$G(\mathbf{r},\mathbf{o};E) = \frac{m}{2\hbar^2 r}\left[Ci(\alpha_+)Ai'(\alpha_-) - Ci'(\alpha_+)Ai(\alpha_-)\right]. \qquad (5.31)$$

Here, $\beta^3 = m/(2\hbar F)^2$, $\alpha_\pm = -\beta[2E + F(z \pm r)]$, and $Ai(u)$ and $Ci(u) = Bi(u) + iAi(u)$ denote Airy functions [13]. It leads to a modified Wigner law for the s–wave absorption cross section near threshold [18]:

$$J(E) \propto Ai'(-2\beta E)^2 + 2\beta E\, Ai(-2\beta E)^2. \qquad (5.32)$$

A static electric field opens up a sub-threshold ($E < 0$) tunneling regime that has been confirmed by experiment [28].

Emission of particles from pointlike sources has been considered in the literature [29] long before the advent of mesoscopic physics. It was Schwinger [30] who introduced sources as a means of describing quantum dynamics in the context of emission and absorption of light.

5.4   Spatially extended sources: the atom laser

Atomic electron sources are usually sufficiently small to be considered point-like. A different situation arises when particles are coherently emitted from an extended region in space. An example for such a "fuzzy" source is the continuous *atom laser*, a beam of ultracold atoms fed by a Bose–Einstein condensate (BEC) [31]. In the experiment, only atoms in a specific Zeeman substate ($m = -1$) are magnetically trapped and form a BEC. Application of a suitably tuned radiofrequency (RF) field will cause transitions into another magnetic substate of the atoms ($m = 0$) that is not subject to the trapping potential. Under the influence of gravity, these "outcoupled" atoms fall freely from the trap region and form a coherent, continuous atom laser "beam." In our language, the macroscopic BEC wave function $\psi_0(\mathbf{r})$ serves as the source $\sigma(\mathbf{r})$ and corresponds to the atomic bound state $\psi_{atom}(\mathbf{r})$ in Eq. (5.18), whereas the outcoupled beam $\psi(\mathbf{r})$ of accelerating atoms takes over the role of the scattered wave $\psi_{sc}(\mathbf{r})$, akin to photodetachment in an electric field (Sec. 5.3.4). From a theoretical viewpoint, the only essential difference is the macroscopic size of the source.

For ideal, non-interacting atoms, the ultracold cloud populates the ground state of the nearly parabolic trapping potential, leading to a Gaussian density

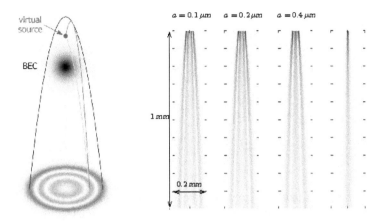

FIGURE 5.5: Left panel: A Gaussian source of freely falling particles can be replaced by a *virtual* point source of particles with the same energy, located upstream from the actual extended source. Right panels: Size dependence of the beam profile for Gaussian BEC sources with different widths a.

profile in the BEC. For simplicity, we assume an isotropic distribution:

$$\sigma(\mathbf{r}) = \hbar\Omega\psi_0(\mathbf{r}) = \hbar\Omega N_0 \exp(-r^2/(2a^2)). \tag{5.33}$$

Here, $\hbar\Omega$ denotes the strength of the transition-inducing oscillating RF field. The parameter a describes the width of the source (which is related to the field gradient in the trap), and $N_0 = a^{-3/2}\pi^{-3/4}$ denotes the proper normalization from the condition

$$\int d^3\mathbf{r}\,|\psi_0(\mathbf{r})|^2 = 1. \tag{5.34}$$

To obtain expressions for the currents generated by a Gaussian source, we work in the time-dependent propagator representation (see Eq. (5.9)). The beam wave function $\psi(\mathbf{r})$ then may be written

$$\psi(\mathbf{r}) = -i\Omega N_0 \int_0^\infty dt\, e^{iEt/\hbar} \int d^3\mathbf{r}'\, K_{\text{field}}(\mathbf{r},t|\mathbf{r}',0)\, e^{-r'^2/(2a^2)}. \tag{5.35}$$

It is possible to carry out the integration over the source volume. With negligible corrections, outside the source the integral (5.35) assumes the form

$$\psi(\mathbf{r}) = \hbar\Omega(2\sqrt{\pi}a)^{3/2} e^{-ma^2E/\hbar^2 + m^2F^2a^6/(3\hbar^4)}\, G\!\left(\mathbf{r}, -\frac{m\mathbf{F}}{2\hbar^2}a^4; E\right), \tag{5.36}$$

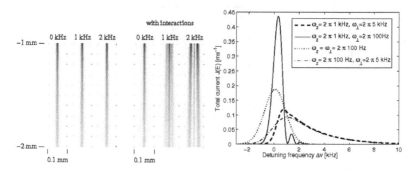

FIGURE 5.6: Left panels: Atom-laser beam profile from a Rb BEC of size $a = 0.8\,\mu$m at different detuning energies $\Delta v = E/h$ (5.36). The first series shows the beam profile for non-interacting particles, whereas the next series includes interactions due to 500 atoms, which lead to a transverse substructure [33]. Right panel: Anisotropic trapping frequencies cause a strong modulation of the total particle current $J(E)$.

where $G(\mathbf{r}, \mathbf{r}'; E)$ denotes the energy Green function for uniformly accelerated particles (5.31). This expression displays a remarkable feature of the beam wave function $\psi(\mathbf{r})$ originating from a Gaussian source: The extended source can be formally replaced by a virtual point source of the same energy, albeit at a location shifted by $\mathbf{r}' = -m\mathbf{F}a^4/(2\hbar^2)$ from the center of the Gaussian distribution (see Fig. 5.5). Expressions for the beam profile and currents are then conveniently found from the analogous expressions for a point source by performing the indicated shifts.

As an immediate, and somewhat surprising, consequence of the concept of a virtual source, the beam profile shows a sharp fringe pattern that results from the interference between the two virtual paths in Fig. 5.5. The number of fringes depends sensitively on the size a of the source, as displayed in Fig. 5.5. In the limit of *extended* Gaussian sources with $E < mF^2a^4/(2\hbar^2)$, the virtual source turns into a tunneling source (as discussed in greater detail in the following section), and the beam profile itself becomes Gaussian [26]. The spectrum of the total particle current $J(E)$ as a function of the detuning of the RF field $E = h\Delta v$ then may be written in the suggestive form

$$J(E) \approx \frac{2\pi}{\hbar} \int d^3\mathbf{r}\, |\sigma(\mathbf{r})|^2\, \delta(E + Fz), \qquad (5.37)$$

an expression that has a simple geometrical interpretation. For extended sources, the energy dependence of $J(E)$ reflects the source structure: By the resonance

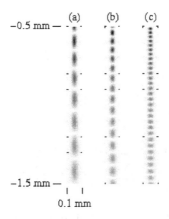

-0.5 mm —

-1.5 mm —

0.1 mm

FIGURE 5.7: Beam profile for simultaneous output coupling with two different radio frequencies. The outcoupling frequencies $\Delta v_{1,2} = E/h$ are (a) ± 0.5 kHz, (b) ± 1.0 kHz, and (c) ± 2.0 kHz. The number of longitudinal interference fringes is proportional to the difference in the detuning frequencies. Parameter: $a = 0.8$ μm, $F = m_{Rb} g$, with $g = 9.81$ m/s^2, and $m_{Rb} = 87$ u.

condition $E + Fz = 0$, the total current probes the density $|\psi_0(\mathbf{r})|^2$ of the BEC on different slices across the source. Finally, we note that the approximation (5.37) obeys the sum rule (5.27) for the total current $J(E)$.

In an actual atomic BEC, the repulsive interactions between atoms lead to a broadening of the condensate. For most cases, the inclusion of the interactions via a mean-field approach is sufficient. The repulsive forces of the much denser BEC act on the outcoupled atom beam and lead to a further splitting of the beam profile, as shown in Fig. 5.6. Also the total current is modified by the interactions [33]. Both effects have been observed experimentally. Non-isotropic trapping frequencies and currents from higher trapping modes allow to control the shape and rate of the atom laser [33].

Fig. 5.7 shows the "dripping quantum faucet", which is produced by superposition of two laser beams with slightly different energy that are outcoupled from the same BEC [32]. It is not surprising to see that rotating BECs which sustain vortices are described in terms of rotating Gaussian sources with nodal structures [24].

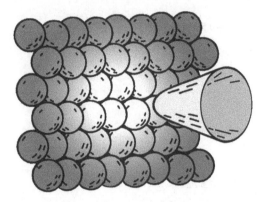

FIGURE 5.8: Electric field emission out of (or into) the sharp apex of a conducting wire. The bright spot symbolizes the Gaussian density profile along the central tunneling path between tip and surface of the sample.

5.5   Ballistic tunneling: STM

The quantum theory of scattering is not limited to asymptotic problems where particles are generated (and observed) far away from the scattering region. A prominent candidate for scattering at finite distances is the scanning tunneling microscope (STM). There, an electric current flows down a macroscopic wire that ends in a sharp tip. Its apex can be viewed as a source of electrons which leave the tip by tunneling due to the applied electric field between tip and sample surface. In some cases, the apex of the tip is ultra-sharp, consisting of a single atom. In an experiment the tip is slowly moved across the surface. In the constant current mode the tip is raised and lowered so as to keep the current constant. The raising and lowering process produces a computer-generated contour map of the surface [37, 38]. The method is capable of resolving individual atoms and works best with conducting materials. Electrons drawn from the apex of the tip (see Fig. 5.8) exhibit dynamically forbidden motion because the electron transfer between tip and surface occurs via field-driven tunneling, confining the current to a narrow filament with Gaussian profile that samples the surface.

It is straightforward to model the apex of an STM tip as a source (or sink) of electrons. To be specific, we consider here a conducting sample surface that harbors a two-dimensional electron gas; in practice, the band of surface states on the densely packed, smooth Cu(111) surface has been exploited for this

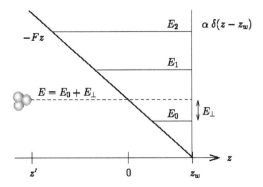

FIGURE 5.9: Sketch of the bouncing ball problem. An electron exits the three-dimensional tunnel at $z = 0$ with energy $E = 0$. It then undergoes multiple reflections between the exit of the tunnel and the surface barrier at $z = z_w$ before traveling into the solid ($z > z_w$). Adatoms (not drawn in the figure) are necessary to observe the resonance-induced ripples (see Fig. 5.11) with energy $E_\perp$.

purpose [35, 36]. The STM tip will emit a spreading surface electron wave that is scattered at adsorbed surface atoms (adatoms). Since the electrons are slow, s–wave scattering prevails that can be modelled by a short-range potential. In this case, one has an analytic solution for the scattering problem which forms the basis for the calculation of the corrugation (surface roughness). As a result of the analytic approach we will show that scattering resonances play an essential role for resolving atoms and detecting electron surface states.

The current flowing through the STM tip is proportional to the local density of surface states, and therefore the imaginary part of the Green function at the tip position, $\Im[G(\mathbf{R}, \mathbf{R}; E)]$ (5.29). A route that leads conveniently to the Green function in this problem consists of the following three steps:

5.5.0.0.2 Step 1. One-dimensional problem: We first calculate the one-dimensional Green function $G_1^\alpha(z, z'; E)$ that corresponds to the model potential of Fig. 5.9. In this one-dimensional problem, the electron is allowed to tunnel in direction of the electric field (z–direction). The energy E_0 corresponds to the bound (and unoccupied) surface state of Cu(111). The energy E of the tunneling electron is taken to lie in a band gap of the substrate (solid). As a result the electron faces a potential barrier at $z = z_w$ and bounces back and forth between tunnel exit ($z = 0$) and barrier. However, the electron can move

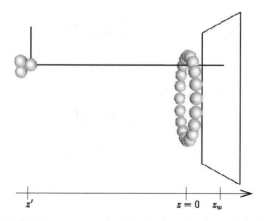

FIGURE 5.10: Schematic plot of tip (at $z = z'$), adsorbed atoms (at $z = 0$), and a strongly reflecting, clean surface (at $z = z_w$).

freely with energy $E_\perp = E - E_0$ in the surface plane orthogonal to $z = z_w$. Because of inelastic scattering with phonons the electron will finally disappear in the solid. By assuming a point source at $z = z'$ the Green function that belongs to the problem of Fig. 5.9 can be solved analytically in terms of Airy functions [34].

5.5.0.0.3 Step 2. Three-dimensional background Green function: Since the STM is a three-dimensional device we must calculate the Green function in three dimensions. The uncertainty principle for momentum and position applied in the lateral $(x - y)$ direction results in a tunneling spot (see Fig. 5.8) of finite width, and approximately Gaussian profile. The three-dimensional Green function $G_{\text{sym}}(\mathbf{r}, \mathbf{r}', z)$ for a particle moving in the potential $\tilde{U}(\mathbf{r}) = \tilde{U}(z)$ of Fig. 5.9 is obtained from its one-dimensional counterpart by integrating G_1^α over all momenta $\hbar k_\perp$ vertical to the field direction (i. e., parallel to the surface),

$$G_{\text{sym}}(\mathbf{r}, \mathbf{r}', E) = G_{\text{sym}}(z, z', \Delta\rho, E)$$
$$= \frac{1}{2\pi} \int_0^\infty dk\, k_\perp\, J_0(k_\perp \Delta\rho)\, G_1^\alpha\left(z, z', E - \frac{\hbar^2 k_\perp^2}{2M}\right), \quad (5.38)$$

with $J_0(\cdots)$ being the usual cylindrical Bessel function of degree zero. The lateral distance between $\mathbf{r}$ and $\mathbf{r}'$ is given by $\Delta\rho^2 = |\rho - \rho'|^2 = (x - x')^2 + (y - y')^2$. Obviously $G_{\text{sym}}(\mathbf{r}', \mathbf{r}', E)$ is independent of the lateral position (x', y') of

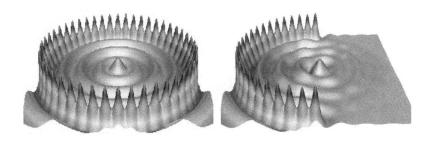

FIGURE 5.11: Interference of waves scattered by corral atoms: Model calculation [34] of a corrugation plot at constant electric current. The structure of the circular ripples both inside and outside the corral can be related to the quantum bounce problem illustrated in Fig. 5.9 and discussed in the text. For comparison with experimental results obtained by Eigler's group [35, 36] we show a similar setup with a quantum corral consisting of 48 iron atoms on a circle with radius 71.3 Å adsorbed on a Cu(111) surface (left panel). The corrugation of the adatoms is approximately 0.5 Å and corresponds to a conductivity of $\sigma_0 = 2.7 \cdot 10^{-8}$ A/V. Note that there is no adsorbed atom in the center of the corral.

the tip. Hence, for $z' = \mathrm{const}$, G_{sym} represents a constant background corrugation. The full solution for the Green function with the adatoms present, is then obtained from

5.5.0.0.4 Step 3. Dyson equation for $G(\mathbf{r}, \mathbf{r}', E)$: We must now take into account the adsorbed atoms (see Fig. 5.10). Using the appropriate Dyson equation, we obtain an algebraic equation for the full Green function,

$$G(\mathbf{r}, \mathbf{r}', E) = G_{\mathrm{sym}}(\mathbf{r}, \mathbf{r}', E) + \sum_{j,k=1}^{n} G_{\mathrm{sym}}(\mathbf{r}, \mathbf{r}_j, E) (\mathbf{T}(E))_{jk} G_{\mathrm{sym}}(\mathbf{r}_k, \mathbf{r}', E),$$

(5.39)

where the sum runs over all adatoms, and the T matrix describing the effects of the multiple scattering events between the adsorbed atoms can be expressed using the background Green function $G_{\mathrm{sym}}(\mathbf{r}_j, \mathbf{r}_k, E)$. Details of the calculation of the Green function and the experimentally observable tunneling current

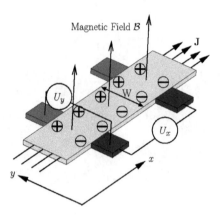

FIGURE 5.12: Schematic view of a Hall bar. A current $\mathbf{J}_x$ is flowing through a two-dimensional electron gas (2DEG) in the x-y–plane, which is orientated perpendicular to an external magnetic field $\mathcal{B}$. The deflected electrons at the sample edges produce a Hall voltage U_y over the sample width W, which is measured along with the longitudinal voltage drop U_x.

$J(\mathbf{R};E)$ (5.29) can be found in Ref. [34]. A zero-temperature plot obtained from such a calculation (which typically takes a few minutes on a personal computer) is shown in Fig. 5.11. We should point out that since G_{sym} gives rise only to a uniform background current, the observed roughness of the surface is entirely contained in the T matrix.

5.6 Electrons in electric and magnetic fields: the quantum Hall effect

In this section we explore the strange and fascinating ways of electrons in electric and magnetic fields. Of particular importance here is the *Hall configuration*, where the electrons are confined to an effectively two-dimensional conductor in the presence of orthogonal electric and magnetic fields. The Hall geometry is displayed in Fig. 5.12. Fig. 5.13 shows some of the classical paths followed by the electrons in the conducting plane. Notwithstanding the complicated pattern of motion, all trajectories share the same distinctive behavior, uniform *drift motion* perpendicular to both fields with a characteristic velocity

$v_D = \mathscr{E}/\mathscr{B}.$

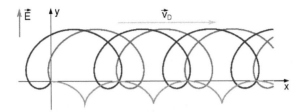

FIGURE 5.13: Drift motion in crossed fields: Trajectories drift with velocity $v_D = \mathscr{E}/\mathscr{B}$ perpendicular to the fields. The motion of the electrons in two dimensions is a superposition of cyclotron motion with drift motion, resulting in *trochoidal* (cycloidal) trajectories.

Because of the universal drift motion, the current density in the Hall bar is simply proportional to the local density of states (LDOS) $n(E)$ in the material, which in turn is related to the Green function in the corresponding external potentials (see Sec. 5.3.4). Hence, from a mathematical point of view we are interested in finding the energy-dependent density of states for the moving electrons. For a purely magnetic field, the two-dimensional LDOS has a spike-like structure [41, 42], formally written as a superposition of discrete δ–distributions positioned at the Landau levels at $E = (2k+1)\hbar\omega_L$, where $\omega_L = e\mathscr{B}/(2m)$ denotes the Larmor frequency.

$$n_{\mathscr{B}}^{(2D)}(E) = \frac{e\mathscr{B}}{2\pi\hbar}\sum_{k=0}^{\infty}\delta\left(E - \hbar\omega_L[2k+1]\right). \tag{5.40}$$

The addition of an electric field leads to important changes in the density of states for a purely magnetic field. As we know from Eq. (5.29), the local density of states $n(E)$ is always linked to the imaginary part of the energy-dependent retarded Green function $G(\mathbf{r}=\mathbf{o},\mathbf{r}'=\mathbf{o};E)$,

$$n(E) = -\frac{1}{\pi}\Im\left[G(\mathbf{o},\mathbf{o};E)\right], \tag{5.41}$$

which in turn can be expressed as the Laplace transform of the quantum propagator $K(\mathbf{o},t|\mathbf{o},0)$ (5.9):

$$G(\mathbf{o},\mathbf{o};E) = \frac{1}{i\hbar}\int_0^\infty dt\, e^{iEt/\hbar}K(\mathbf{o},t|\mathbf{o},0). \tag{5.42}$$

For $\mathbf{r} = 0$, the two-dimensional quantum propagator for the Hamiltonian of a (spinless) electron in crossed fields,

$$H^{(2D)}_{\mathscr{E}\times\mathscr{B}} = \frac{\mathbf{p}_x^2 + \mathbf{p}_y^2}{2m} + \frac{1}{2}m\omega_L^2\left(x^2 + y^2\right) - \mathbf{r}_\perp \cdot \mathbf{F}_\perp - \mathbf{p}_y x\omega_L + \mathbf{p}_x y\omega_L, \quad (5.43)$$

where $\mathbf{F}_\perp = -e\mathscr{E}_\perp$ denotes the electric force in the x-y–plane reads [11],

$$K^{(2D)}_{\mathscr{E}\times\mathscr{B}}(\mathbf{0},t|\mathbf{0},0) = -\frac{im\omega_L}{2\pi\hbar\sin(\omega_L t)}\exp\left\{\frac{iF_\perp^2 t}{8m\hbar\omega_L^2}\left[\omega_L t\cot(\omega_L t) - 1\right]\right\}. \quad (5.44)$$

To account for the effects of the electron spin, we note that its interaction with the magnetic field merely adds a spatially constant term to the Hamiltonian, and thus shifts the effective energy of the two spin populations by a fixed amount $\pm\frac{1}{2}g\mu_B\mathscr{B} = \pm\frac{1}{2}g\hbar\omega_L$ [41]. The spin-dependent densities of states become

$$n_{\uparrow,\downarrow}(E) = n\left(E \pm \frac{1}{2}g\hbar\omega_L\right) \quad (5.45)$$

and the total LDOS including spin can be mapped back to the LDOS without spin: $n_{\uparrow\downarrow}(E) = n_\uparrow(E) + n_\downarrow(E)$. Thus, it suffices to evaluate $n(E)$. For the discussion of the density of states in the Hall configuration it is useful to replace the trigonometric functions in the propagator (5.44) by a sum. This can be done using the identity

$$\frac{\exp[-\alpha\coth(z)]}{\sinh(z)} = 2e^{-\alpha}\sum_{k=0}^{\infty}L_k^{(0)}(2\alpha)e^{-2z(k+1/2)}, \quad (5.46)$$

which follows directly from the generating function of the Laguerre polynomials $L_k^{(0)}(z)$ [13]. The Laplace transform (5.42) then can be performed analytically, and we finally find for the density of states

$$n_{\mathscr{E}\times\mathscr{B}}(E) = \frac{1}{2\pi^{3/2}l^2\Gamma}\sum_{k=0}^{\infty}\frac{1}{2^k k!}e^{-E_k^2/\Gamma^2}\left[H_k\left(E_k/\Gamma\right)\right]^2, \quad (5.47)$$

where the level width parameter

$$\Gamma = F_\perp l \quad (5.48)$$

is related to the magnetic length $l = \sqrt{\hbar/(e\mathscr{B})}$. $H_k(z)$ denotes the kth Hermite polynomial [13] and E_k is the effective energy shift for the kth Landau level

$$E_k = E - \Gamma^2/(4\hbar\omega_L) - (2k+1)\hbar\omega_L. \quad (5.49)$$

Interestingly, the density of states can again be interpreted as a sum over Landau levels. However, they now appear spread in energy, with a distribution that is isomorphic to the probability density of the corresponding eigenstate of a one-dimensional harmonic oscillator

$$|u_k(\xi)|^2 = \frac{1}{2^k k! \sqrt{\pi}} e^{-\xi^2} [H_k(\xi)]^2. \tag{5.50}$$

The *total* contribution of the kth Landau level integrated over energy-space is readily available from the normalization of the oscillator eigenstates:

$$\int_{-\infty}^{\infty} dE \, n_{k,\mathscr{E}\times\mathscr{B}}(E) = \frac{e\mathscr{B}}{2\pi\hbar} \int_{-\infty}^{\infty} d\xi \, |u_k(\xi)|^2 = \frac{e\mathscr{B}}{2\pi\hbar}. \tag{5.51}$$

This result reflects the quantization of each Landau level in a purely magnetic field (5.40).

At low temperatures, only the *occupied states* with energies smaller than the Fermi energy E_F of the system will contribute to the Hall current. Thus, we expect that the current density will be proportional to the integrated density of states $N(E_F)$ with energies $E < E_F$,

$$N(E_F) = \int_{-\infty}^{E_F} dE \, n(E). \tag{5.52}$$

Within the Fermi gas model for a dilute gas of electrons in two dimensions it is straightforward to obtain resistivity plots which bear remarkable resemblance to actual quantum Hall data [40]. An example is shown in Fig. 5.14.

We finally mention the implications of our Green function model of the integer quantum Hall effect for the observed breakdown of the quantized conductivity at higher electric fields [43]. Kawaji [44, 45, 46] studied the width of the quantized resistivity "plateaus" as a function of the electric current and thus the Hall field in the sample. He finds a characteristic power law for the shrinking of the plateaus, which can be expressed in terms of a critical electric field:

$$\mathscr{E}_{\text{crit}} \propto B^{3/2}. \tag{5.53}$$

The non-perturbative inclusion of the electric field via the Green function formalism as presented here leads automatically to this power law (see [40, 47, 48, 49])

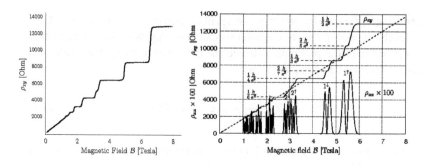

FIGURE 5.14: The quantum Hall effect. Left panel: Schematic sketch of the experimental results adapted from [39] for ρ_{xy} in GaAs/AlGaAs heterojunctions at $T = 50$ mK. The theoretical simulations [40] (right panel) are based on the density of states (5.47). The dotted straight line is the classical prediction.

5.7 The semiclassical method

It is usually rewarding to analyze quantum problems with the tools of classical physics. First, the semiclassical solution facilitates in many cases the understanding of the quantum solution properties. Second, semiclassics leads to approximate solutions ("WKB solutions") that are usually numerically much less expensive than *ab initio* quantum mechanical calculations. In context of the propagator methods discussed here, the semiclassical method can be described by the following steps:

1. Find *number* of paths N joining source $\mathbf{r}'$ and destination $\mathbf{r}$. This is an effective method for finding the caustics (turning surfaces).

2. Find all trajectories $\mathbf{r}_k(t)$ leading from $\mathbf{r}'$ to $\mathbf{r}$.

3. Establish *classical weight* $\rho_k(\mathbf{r})$, i.e., the local density of trajectories, for each path.

4. Determine its *semiclassical phase* $\Phi_k(\mathbf{r})$ from the reduced action along the path.

5. Create *semiclassical approximation* to the Green function $G_{sc}(\mathbf{r}, \mathbf{r}'; E)$.

6. Use *uniform approximations* to correct the divergence of the WKB solution near the caustics.

Therefore, the semiclassical method requires knowledge of the classical trajectories $\mathbf{r}_k(t)$, their local density $\rho_k(\mathbf{r})$, and the corresponding action fields $W_k(\mathbf{r},\mathbf{r}';E)$. A semiclassical treatment of the photodetachment problem (Sec. 5.3.4) can be found e.g., in Ref. [26]. Recently, the technique has been applied to the more complicated dynamics of electrons in parallel electric and magnetic fields [50, 51]. Here, we concentrate on the related problem of semiclassical motion in the Hall configuration (Sec. 5.6).

5.7.0.0.5 Classical orbits for the Hall effect: The propagation of electrons in crossed electric and magnetic fields has received much attention in classical physics. Indeed, the corresponding trajectory field of electrons emitted by a point source located at the origin (for convenience), plotted in Fig. 5.15, looks interesting by itself. Again assuming the Hall geometry displayed in Fig. 5.12, the motion of the charges is governed by the Hamiltonian (5.43), and we find the family of orbits:

$$
\mathbf{r}(t) = \begin{pmatrix} v_D t \\ 0 \end{pmatrix} + \frac{1}{2\omega_L} \begin{pmatrix} -v_{0y} & v_{0x}-v_D \\ v_{0x}-v_D & v_{0y} \end{pmatrix} \cdot \begin{pmatrix} \cos(2\omega_L t) \\ \sin(2\omega_L t) \end{pmatrix}
$$
$$
+ \frac{1}{2\omega_L} \begin{pmatrix} v_{0y} \\ v_D - v_{0x} \end{pmatrix} .
\tag{5.54}
$$

Here, the initial velocity vector depends on the emission angle θ and is given by

$$
\dot{\mathbf{r}}(t=0) = \mathbf{v}_0 = \begin{pmatrix} v_{0x} \\ v_{0y} \end{pmatrix} = \begin{pmatrix} v_0 \cos\theta \\ v_0 \sin\theta \end{pmatrix} .
\tag{5.55}
$$

It must fulfill $E = \frac{1}{2} m v_0^2$. Eq. (5.54) is conveniently interpreted as a sum of three terms: On average, only the first term contributes to the transport of the electron. The corresponding drift velocity (averaged over one cyclotron period $T = \pi/\omega_L$) reads

$$
\mathbf{v}_D = \frac{1}{T} \int_t^{t+T} dt'\, \dot{\mathbf{r}}(t') = (\mathscr{E} \times \mathscr{B})/\mathscr{B}^2.
\tag{5.56}
$$

In a "drift" reference frame that is moving with this velocity the otherwise trochoidal orbit becomes a circle with angle-dependent radius $R(\theta)$ whose center is shifted from the origin by a constant displacement $\mathbf{r}_c$

$$
R(\theta)^2 = \frac{v_0^2 - 2 v_0 v_D \cos\theta + v_D^2}{4\omega_L^2} , \qquad \mathbf{r}_c = \frac{1}{2\omega_L} \begin{pmatrix} v_0 \sin\theta \\ v_D - v_0 \cos\theta \end{pmatrix} .
\tag{5.57}
$$

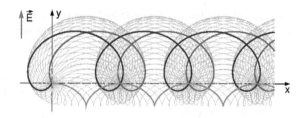

FIGURE 5.15: Caustics and the trajectory field (cf. Fig. 5.13): The trajectory field traces out the caustics. Electrons pass through focal points. The number of paths connecting the source with a given point increases with the magnetic field.

Some sample orbits are plotted in Fig. 5.13. Variation of the angle θ yields the trajectory field displayed in Fig. 5.15.

5.7.0.0.6 The Hall conductivity tensor: Noting that the velocity in turn is related to the classical current density $\mathbf{j}$

$$\mathbf{j} = Ne\mathbf{v}_D, \qquad (5.58)$$

where N denotes the electron density, one can extract the resistivity tensor ρ (or its inverse, the conductivity tensor σ) from Ohm's law in the two-dimensional (x,y)–plane:

$$\mathbf{j} = \rho^{-1} \cdot \mathcal{E} \quad \Rightarrow \quad \sigma = \rho^{-1} = \frac{Ne}{\mathcal{B}} \begin{pmatrix} 0 & -1 \\ 1 & 0 \end{pmatrix}. \qquad (5.59)$$

This remarkable equation predicts a finite conductivity even in the absence of scattering, which is usually invoked in theories of conduction in order to guarantee a finite carrier velocity. We note that the drift velocity is independent of the kinetic energy of the electrons.

5.7.0.0.7 Closed orbits in the classical picture: Eq. (5.58) shows that the current in the Hall conductor is determined by the density of states N. According to Sec. 5.3, the local density of states (LDOS) $n(E)$ within a narrow energy interval is related to the Green function $G(\mathbf{o}, \mathbf{o}; E)$ (5.41), which, from a semi-classical perspective, is governed by those trajectories that return to the source

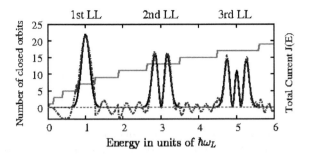

FIGURE 5.16: Density of states in crossed electric and magnetic fields. The curves show the quantum (solid) and semiclassical result (dashed line). The staircase structure denotes the count of closed orbits. Crossed electric field $\mathscr{E} = 4000$ V/m and magnetic field $\mathscr{B} = 5$ T.

$(\mathbf{r} = \mathbf{r}' = \mathbf{o})$. This is the basic motivation of the closed orbit theory [52, 53] for source processes.

These trajectories, which lead from the origin back to the origin, are best found by using the classical action. Here, it is convenient to start with the *time-dependent* action functional $S_{cl}(\mathbf{r},t|\mathbf{r}',0)$. In crossed electric and magnetic fields, this classical action is uniquely given by

$$S_{cl}(\mathbf{o},t|\mathbf{o},0) = -\frac{m}{2}v_D^2 t + \frac{m\omega_L}{2}\cot(\omega_L t)v_D^2 t^2. \qquad (5.60)$$

This expression describes the single closed orbit returning to the source in a predetermined time of flight t. However, we are rather interested in the energy E of the electron,

$$E(t) = -\frac{\partial S_{cl}(\mathbf{o},t|\mathbf{o},0)}{\partial t}. \qquad (5.61)$$

For fixed emission energy E, this is an implicit equation for the time of flight t, and generally several solutions t_k, pertaining to distinct classical trajectories $\mathbf{r}_k(t)$, exist. (To find their initial velocities $\mathbf{v}_0$, it is sufficient to set $\mathbf{r} = \mathbf{o}$ and $t = t_k$ in the equation of motion (5.54), and solve the ensuing linear equation system for v_{0x} and v_{0y}.) The reduced action for each contributing path then follows from the Legendre transform:

$$W_k(\mathbf{o},\mathbf{o};E) = S_{cl}(\mathbf{o},t_k|\mathbf{o},0) + Et_k. \qquad (5.62)$$

As shown in Fig. 5.16, the number of closed orbits increases with the magnetic field strength.

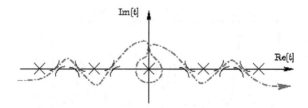

FIGURE 5.17: Principal structure of the classical action in the complex time plane for $N = 3$ saddle points (closed orbits). The dashed line denotes the integration path. Singularities are denoted by $\times$ and saddle points by $)($. Note that the singularity at the origin arises from the prefactor in the propagator and not from the classical action at $t = 0$.

5.7.0.0.8 Density of states and propagator: The classical action (5.60) is an important ingredient of the quantum-mechanical time-evolution operator. Using Eqs. (5.41) and (5.42), it is possible to relate the local density of states with the propagator via [40, 52, 54]

$$n^{(2D)}_{\mathscr{E} \times \mathscr{B}}(E) = \frac{1}{2\pi\,\hbar} \int_{-\infty}^{\infty} dT\, e^{iET/\hbar}\, K(\mathbf{o}, T | \mathbf{o}, 0), \qquad (5.63)$$

where the time-dependent propagator (5.44) is given by

$$K(\mathbf{o}, t | \mathbf{o}, 0) = \frac{m\omega_L}{2\pi i\,\hbar \sin(\omega_L t)} \exp\left\{ \frac{i}{\hbar} S_{cl}(\mathbf{o}, t | \mathbf{o}, 0) \right\}. \qquad (5.64)$$

As we have shown before, this expression can be evaluated analytically in terms of harmonic oscillator eigenstates in energy space (5.47).

5.7.0.0.9 Quantum result versus semiclassical approach: An asymptotic evaluation of the integral (5.63) provides the link between closed orbits and the density of states. The original path of integration follows the real time-axis. Analytic continuation of the propagator makes it possible to deform this path of integration to the one sketched in Fig. 5.17. This path passes through saddle points of the exponent (denoted by $)($ in the figure) using the paths of steepest descent. The singularities in the integrand at times $T = k\pi/\omega_L$ (denoted by $\times$) are avoided. The only contribution of a singularity comes from $t = 0$, which may be evaluated by the residue theorem:

$$I_{\text{origin}} = \frac{1}{2\pi\,\hbar} \oint dt\, e^{iEt/\hbar}\, K(\mathbf{o}, t | \mathbf{o}, 0) = \frac{m}{2\pi\,\hbar^2}. \qquad (5.65)$$

Comparison of Eqs. (5.63) and (5.64) with (5.61) shows that the saddle points of the integrand coincide with the classical times of flight for the various closed orbits. Adding their contributions yields the semiclassical result:

$$
n_{sc,\mathscr{E}\times\mathscr{B}}^{(2D)} = I_{\text{origin}} + 2\,\text{Re}\left[\frac{m\omega_L}{4\pi^2 \mathrm{i}\,\hbar^2}\sum_{k=1}^{N}\frac{\mathrm{e}^{\mathrm{i}W_k(\mathbf{o},\mathbf{o};E)/\hbar+\mathrm{i}\pi\,\text{sgn}[\ddot{S}_{cl}(\mathbf{o},t_k|\mathbf{o},0)]/4}}{\sin(\omega_L t_k)\,\sqrt{|\ddot{S}_{cl}(\mathbf{o},t_k|\mathbf{o},0)|/(2\pi\,\hbar)}}\right]
$$
(5.66)

Fig. 5.16 compares the semiclassical and quantum results. Despite their very different origins (sum over Landau levels vs. interfering classical trajectories), they are in striking agreement. In the "plateau regions" of the conductivity, destructive interference between the properly weighted classical trajectories strongly suppresses the LDOS, leading to a quantization into separated levels with a substructure. Note that the number of trajectories does not change in each Landau level. It is the relative phase that modulates the LDOS.

Acknowledgements

This work benefited greatly from discussions with W. Becker, M. Betz, B. Donner, E. Heller, M. Moshinsky and M. O. Scully. Partial financial support from the Deutsche Forschungsgemeinschaft is gratefully acknowledged.

References

[1] R. P. Feynman, *The Character of Physical Law*, The M. I. T. Press, Cambridge, MA, 1967.

[2] S. Datta, editor, *Electronic Transport in Mesoscopic Systems*, Cambridge University Press, Cambridge, MA, 1997.

[3] Y. Imry, *Introduction to Mesoscopic Physics* (2nd edition), Oxford University Press, Oxford, 2002.

[4] J. J. Sakurai, *Modern Quantum Mechanics*, Addison-Wesley, New York, 1994.

[5] R. P. Feynman and A. R. Hibbs, *Quantum Mechanics and Path Integrals*, McGraw-Hill, New York, 1965.

[6] L. S. Schulman, *Techniques and Applications of Path Integration*, Wiley, New York, 1981.

[7] H. Kleinert, *Path Integrals in Quantum Mechanics, Statistics and Polymer Physics*, World Scientific, Singapore, 1990.

[8] V. V. Dodonov, V. I. Man'ko, and D. E. Nikonov. Exact propagators for time-dependent Coulomb, delta and other potentials. *Physics Letters A*, 162:359–364, 1992.

[9] L. M. Nieto, Green's function for crossed time-dependent electric and magnetic fields. Phase-space quantum mechanics approach, *J. Math. Phys.*, 33:3402–3409, 1992.

[10] M. Kleber, Exact solutions for time-dependent phenomena in quantum mechanics, *Phys. Rep.*, 236:331–393, 1994.

[11] C. Grosche and F. Steiner, *Handbook of Feynman Path Integrals*, Vol. 145 of *Springer Tracts in Modern Physics*, Springer, Berlin, 1998.

[12] M. Moshinsky, Diffraction in time, *Phys. Rev.*, 88:625–631, 1952.

[13] M. Abramowitz and I. A. Stegun, *Handbook of Mathematical Functions*, Dover, New York, 1965.

[14] A. del Campo and J. G. Muga, Single-particle matter wave pulses, *J. Phys. A*, 38:9803–9819, 2005.

[15] E. N. Economou, *Green's Functions in Quantum Physics (Solid-State Sciences 7)*, Springer, Berlin, 1983.

[16] T. Kramer, C. Bracher, and M. Kleber, Four-path interference and uncertainty principle in photodetachment microscopy, *Europhys. Lett.*, 56:471–477, 2001.

[17] C. Bracher, M. Riza, and M. Kleber, Propagator theory in scanning tunneling microscopy, *Phys. Rev. B*, 56:7704–7715, 1997.

[18] T. Kramer, C. Bracher, and M. Kleber, Matter waves from quantum sources in a force field, *J. Phys. A*, 35:8361–8372, 2002.

[19] J. Tersoff and D. R. Hamann, Theory and application for the scanning tunneling microscope, *Phys. Rev. Lett.*, 50:1998–2001, 1983.

[20] D. Kleppner and J. B. Delos, Beyond quantum mechanics: Insights from the work of Martin Gutzwiller, *Found. Phys.*, 31:593–612, 2001.

[21] M. L. Du and J. B. Delos, Photodetachment of H$^-$ in an electric field, *Phys. Rev. A*, 38:5609–5616, 1988.

[22] C. Blondel, C. Delsart, and F. Dulieu, The photodetachment microscope, *Phys. Rev. Lett.*, 77:3755–3758, 1996.

[23] C. Blondel, C. Delsart, F. Dulieu, and C. Valli, Photodetachment microscopy of O$^-$, *Eur. Phys. J. D*, 5:207–216, 1999.

[24] C. Bracher, T. Kramer, and M. Kleber, Ballistic matter waves with angular momentum: Exact solutions and applications, *Phys. Rev. A*, 67:043601–1–20, 2003.

[25] Yu. N. Demkov, V. D. Kondratovich, and V. N. Ostrovskii, Interference of electrons resulting from the photoionization of an atom in an electric field, *JETP Lett.*, 34:403–405, 1982. [Pis'ma Zh. Eksp. Teor. Fiz. 34:425–427, 1981].

[26] C. Bracher, W. Becker, S. A. Gurvitz, M. Kleber, and M. S. Marinov, Three-dimensional tunneling in quantum ballistic motion, *Am. J. Phys.*, 66:38–48, 1998.

[27] E. P. Wigner, On the behavior of cross sections near thresholds, *Phys. Rev.*, 73:1002–1009, 1948.

[28] N. D. Gibson, M. D. Gasda, K. A. Moore, D. A. Zawistowski, and C. W. Walter, s–wave photodetachment from S$^-$ ions in a static electric field, *Phys. Rev. A*, 64:061403–1–10, 2001.

[29] L. S. Rodberg and R. M. Thaler, *Introduction to the Quantum Theory of Scattering*, Academic Press, New York, 1967.

[30] J. Schwinger, *Particles, Sources, and Fields*, Vol. 2, Addison-Wesley, New York, 1973.

[31] I. Bloch, T. W. Hänsch, and T. Esslinger, Atom laser with a cw output coupler, *Phys. Rev. Lett.*, 82:3008–3011, 1999.

[32] I. Bloch, T. W. Hänsch, and T. Esslinger, Measurement of the spatial coherence of a trapped Bose gas at the phase transition, *Nature*, 403:166–170, 2000.

[33] T. Kramer and M. Rodríguez. Quantum theory of an atom laser originating from a Bose-Einstein condensate or a Fermi gas in the presence of gravity. *Phys. Rev. A*, 74:013611, 2006.

[34] B. Donner, C. Bracher, M. Kleber, T. Kramer, and H. J. Kreuzer, Corrugation amplification in scanning tunneling microscopy, *Am. J. Phys.*, 73:690–700, 2005.

[35] M. F. Crommie, C. P. Lutz, and D. M. Eigler, Confinement of electrons to quantum corrals on a metal surface, *Science*, 262:218–220, 1993.

[36] M. F. Crommie, C. P. Lutz, D. M. Eigler, and E. J. Heller, Quantum corrals, *Physica D*, 83:98–108, 1995.

[37] G. Binnig and H. Rohrer, Scanning tunneling microscopy, *Helv. Phys. Acta*, 55:726–735, 1982.

[38] C. J. Chen, *Introduction to Scanning Tunneling Microscopy*, Oxford University Press, New York, 1993.

[39] M. A. Paalanen, D. C. Tsui, and A. C. Gossard, Quantized Hall effect at low temperatures, *Phys. Rev. B*, 25:5566–5569, 1982.

[40] T. Kramer, C. Bracher, and M. Kleber, Electron propagation in crossed magnetic and electric fields, *J. Opt. B*, 6:21–27, 2004.

[41] R. E. Prange and S. M. Girvin, editors, *The Quantum Hall Effect*, Springer, Berlin, 1987.

[42] G. Grosso and G. P. Parravicini, *Solid State Physics*, Academic Press, New York, 2000.

[43] K. v. Klitzing, G. Dorda, and M. Pepper, New method for high-accuracy determination of the fine-structure constant based on quantized Hall resistance, *Phys. Rev. Lett.*, 45:494–497, 1980.

[44] S. Kawaji, K. Hirakawa, and M. Nagata, Device-width dependence of plateau width in quantum Hall states, *Physica B*, 184:17–20, 1993.

[45] S. Kawaji, Breakdown of the integer quantum Hall effect at high currents in GaAs/AlGaAs heterostructures, *Semicond. Sci. Technol.*, 11:1546–1551, 1996.

[46] T. Shimada, T. Okamoto, and S. Kawaji, Hall electric field-dependent broadening of extended state bands in Landau levels and breakdown of the quantum Hall effect, *Physica B*, 249–251:107–110, 1998.

[47] T. Kramer, *Matter waves from localized sources in homogeneous force fields*, PhD thesis, Technische Universität München, 2003. (Online: http://tumb1.biblio.tu-muenchen.de/publ/diss/ph/2003/kramer.html.)

[48] T. Kramer and C. Bracher, Propagation in crossed magnetic and electric fields: The quantum source approach, in B. Gruber, G. Marmo, and N. Yoshinaga, editors, *Symmetries in Science XI*, Kluwer, Dordrecht, 2004, pp. 317–353. (Online: http://arxiv.org/abs/cond-mat/0309424.)

[49] T. Kramer, A heuristic quantum theory of the integer quantum Hall effect, *Int. J. Mod. Phys. B*, 20:1243–1260, 2006.

[50] C. Bracher and J. B. Delos, Motion of an electron from a point source in parallel electric and magnetic fields, *Phys. Rev. Lett.*, 96:100404–1–4, 2006.

[51] C. Bracher, T. Kramer, and J. B. Delos, Electron dynamics in parallel electric and magnetic fields, *Phys. Rev. A*, 73:062114–1–21, 2006.

[52] M. Berry and K. W. Mount, Semiclassical approximations in wave mechanics, *Rep. Prog. Phys.*, 35:315–397, 1972.

[53] A. D. Peters and J. B. Delos, Photodetachment cross-section of H^- in crossed electric and magnetic fields. 1. Closed-orbit theory, *Phys. Rev. A*, 47:3020–3035, 1993.

[54] M. Gutzwiller, *Chaos in Classical and Quantum Mechanics*, Springer, New York, 1990.

Chapter 6

The Probabilistic Nature of Quantum Mechanics

Leon Cohen[1]

Abstract   We show that within classical probability theory there are mathematical quantities which are similar to quantum mechanical wave functions and operators. This is shown by generalizing a theorem of Khinchin on the necessary and sufficient conditions for a function to be a characteristic function. We show that for the one dimensional case the methods of quantum mechanics for obtaining expectation values and distributions of observables follow simply. Particular difficulties arise for two non-commuting operators but nonetheless improper "quasi-distributions" can be defined and used with profit. Quasi-distributions can be thought of as two dimensional mappings of a one dimensional function. In the mathematical sense the distribution contains the same information as the wave function, since it is constructed from it and the wave function can be obtained from it uniquely. Nonetheless, an immense simplification occurs when one studies the wave function in a quasi-representation: the physical nature of the wave function becomes much clearer. A number of explicit examples are given.

[1]Work supported by the Air Force Office of Scientific Research and the NSA HBCU/MI program.

6.1 Introduction

Quantum mechanics is a probability theory but it is unlike any other probability theory. While the end results of quantum mechanics are standard probabilistic quantities, such as expectation values and probability distributions,
quantum mechanics calculates these quantities in very unusual ways compared
to standard probability theory. Clearly, whoever constructed quantum mechanics did not wish to use standard probability theory as exemplified by the classic
books of, for example, Feller or Doob, and also did not bother with the axioms
of probability as formulated by Kolmogorov and others. Quantum mechanical
ideas and methods are seemingly foreign to standard probability theory, as it
deals with operators, wave functions, transformation theory, and yet, as just
mentioned, at the end it gives standard probabilistic results. Why should this
be so? Why isn't standard probability theory good enough? Conversely, is it
possible that in standard probability theory there are things like wave functions
and operators but that we have not noticed them? That is not to ask whether
quantum mechanics can be derived from classical probability theory but to ask
why the mathematical methods of standard probability theory are not good
enough. It is our aim to consider the fundamental probabilistic structure of
quantum mechanics and classical probability theory and study the relationship
between the two. We will see that some of the strange ideas of quantum mechanics do exist in classical probability theory. In particular we will show that
operators and quantum mechanical-like wave functions can be defined in classical probability theory and used to calculate averages in the quantum mechanical manner. But as just mentioned one can derive other quantum mechanical
properties such as the Schrodinger's equation or the fact that wave functions
obey the superposition principle.

6.2 Are there wave functions in standard probability theory?

Seemingly, nothing like wave functions or operators exist in standard probability theory, however, we will argue that at least for one dimensional probability distributions, operators and wave functions do appear in standard probability. This may seem absurd at first thought since wave functions are complex and standard probability theory generally deals with real quantities. Are

there functions in standard probability theory that are complex? Without question the most important complex quantity is the characteristic function, $M(\theta)$, which is the Fourier transform of the probability density, $P(x)$,

$$M(\theta) = \int e^{i\theta x} P(x) \, dx. \tag{6.1}$$

The probability distribution can be calculated from it,

$$P(x) = \frac{1}{2\pi} \int M(\theta) e^{-i\theta x} d\theta. \tag{6.2}$$

The characteristic function has many uses and advantages and very often it is easier to use to get statistical quantities than the probability distribution itself. For example, it can be used to calculate the moments

$$\langle x^n \rangle = \frac{1}{i^n} \frac{d^n}{d\theta^n} M(\theta) \big|_{\theta=0}. \tag{6.3}$$

Also, if the moments are known they can be used to construct the distribution, since from the moments we can calculate the characteristic function

$$M(\theta) = \left\langle e^{i\theta x} \right\rangle = \sum_{n=0}^{\infty} \frac{(i\theta)^n}{n!} \langle x^n \rangle \tag{6.4}$$

and hence the distribution by way of Eq. (6.2).

The characteristic function is generally a complex function. What is important for our considerations is the fact that not every complex function is a characteristic function. Because a probability distribution is manifestly positive and because it has certain properties, e.g., the integral over all space has to be one, these properties get reflected on the characteristic function. It is easy to come up with many necessary conditions for a function to be a characteristic function but of importance is to determine necessary and sufficient conditions. Historically, the answer was given by Khinchin [14, 18], however, we believe a better answer can be given: $M(\theta)$ is a characteristic function if and only if it can be expressed in the following form [7]

$$M(\theta) = \int g^*(x) e^{i\theta \mathbf{A}} g(x) dx, \tag{6.5}$$

for some function $g(x)$ and for some Hermitian operator $\mathbf{A}$. Also, $g(x)$ is normalized to one,

$$\int |g(x)|^2 dx = 1. \tag{6.6}$$

We emphasize though that while the characteristic function is unique the pair $(g, \mathbf{A})$ is not and quite the contrary there is an infinite number of such pairs that will give the same characteristic function. As we will see shortly the Khinchin result is one such pair.

Expectation values: Let us now calculate the expectation value using the standard method, namely Eq. (6.3),

$$\langle x \rangle = \frac{1}{i} \frac{\partial}{\partial \theta} M(\theta) \Big|_{\theta=0} = \frac{1}{i} \frac{\partial}{\partial \theta} \int g^*(x) e^{i\theta \mathbf{A}} g(x) dx \Big|_{\theta=0}$$
$$= \int g^*(x) \mathbf{A} g(x) dx.$$

This is precisely the quantum mechanical procedure for getting expectation values. Therefore we argue that the $g(x)'s$ are analogous to the quantum mechanical wave functions.

Obtaining probability distributions. First, we review how one gets probabilities in quantum mechanics. One solves the eigenvalue problem for the Hermitian operator representing the physical quantity. The spectrum can be discrete or continuous depending on the operator. We first assume that it is discrete

$$\mathbf{A} u_n(x) = a_n u_n(x), \tag{6.7}$$

where $a_n, u_n(x)$ are the eigenvalues and eigenfunctions respectively and $\mathbf{A}$ is the operator. According to quantum mechanics the only random variables are the eigenvalues a_n and the probability for getting these values is calculated in the following way. Expand the wave function as

$$g(x) = \sum_n c_n u_n(x), \tag{6.8}$$

then, the probability of measuring a_n is

$$P(a_n) = |c_n|^2 \quad \text{quantum mechanical probability (discrete case)}. \tag{6.9}$$

For the continuous case we write

$$\mathbf{A} u(a, x) = a u(a, x), \tag{6.10}$$

where $u(a, x)$ are the eigenfunctions. One now expands as

$$g(x) = \int \varphi(a) u(a, x) dx, \tag{6.11}$$

where

$$\varphi(a) = \int g(x) u^*(a, x) da. \tag{6.12}$$

Then, the probability of measuring a is

$$P(a) = |\varphi(a)|^2 \text{ quantum mechanical probability (continuous case).} \quad (6.13)$$

Now we ask whether one can derive these standard quantum results using our form of the characteristic function, that is by combining Eq. (6.5) with Eq. (6.2). We consider first the continuous case. Substituting Eq. (6.5) into Eq. (6.2) we have

$$P(a) = \frac{1}{2\pi} \int M(\theta) e^{-i\theta x} d\theta \quad (6.14)$$

$$= \frac{1}{2\pi} \int \varphi^*(a') u^*(a',x) e^{i\theta A} \varphi(a'') u(a'',x) g(x)$$
$$e^{-i\theta a} dx d\theta da' da'' \quad (6.15)$$

$$= \frac{1}{2\pi} \int \varphi^*(a') u^*(a',x)$$
$$e^{i\theta a} \varphi(a'') u(a'',x) g(x) e^{-i\theta a} dx d\theta da' da'' \quad (6.16)$$

$$= |\varphi(a)|^2 \quad (6.17)$$

which is Eq. (6.13), the quantum mechanical result. Thus we see that, again, the g's defined by Eq. (6.5) act as "wave" functions.

One of the remarkable properties of quantum mechanics is discreteness of certain observables and that the only values that a random variable can take are the eigenvalues. To see how that comes in consider the probability when we have the discrete case. Substituting Eq. (6.8) into Eq. (6.2) we have

$$P(a) = \int \sum_{n,m} c_m^* u_m^*(x) e^{i\theta A} c_n u_n(x) e^{-i\theta a} dx d\theta \quad (6.18)$$

$$= \int \sum_{n,m} c_m^* u_m^*(x) e^{i\theta a_n} c_n u_n(x) e^{-i\theta a} dx d\theta \quad (6.19)$$

or

$$P(a) = \sum_n |c_n|^2 \delta(a - a_n) \quad (6.20)$$

which shows that the only values which the random variable can take with nonzero probability are the eigenvalues.

6.2.1   The Khinchin theorem

As mentioned above Khinchin derived necessary and sufficient conditions that a function be a characteristic function. His result is that a complex function, $M(\theta)$, is a characteristic function if and only if there exists the representation

$$M(\theta) = \int g^*(x)g(x+\theta)dx. \tag{6.21}$$

We now show that this is a special case of our result, Eq. (6.5). Take

$$\mathbf{A} = \frac{1}{i}\frac{d}{dx}, \tag{6.22}$$

and substitute into Eq. (6.5) to obtain

$$M(\theta) = \int g^*(x)e^{i\theta\mathbf{A}}g(x)dx \tag{6.23}$$

$$= \int g^*(x)e^{\theta\frac{d}{dx}}g(x)dx \tag{6.24}$$

$$= \int g^*(x)g(x+\theta)dx, \tag{6.25}$$

which is precisely the Khinchin theorem. In going from Eq. (6.24) to Eq. (6.25) we have used the fact that for any function [26]

$$e^{\theta\frac{d}{dx}}g(x) = g(x+\theta). \tag{6.26}$$

Thus the Khinchin theorem is a special case, it is one pair, $(g, \frac{1}{i}\frac{d}{dx})$, out of an infinite number. We note that $\mathbf{A}$ as defined by Eq. (6.22) in this example is essentially the momentum operator and is the generator of translations as per Eq. (6.26).

6.3   Two variables

We now consider the case of two variables and see if and how the above can be generalized. For two random variables, a and b the characteristic function, $M(\theta, \tau)$, and distribution, $P(a,b)$, are related by

$$M(\theta, \tau) = \int\int e^{i\theta a + i\tau b}P(a,b)dadb, \tag{6.27}$$

$$P(a,b) = \frac{1}{4\pi^2}\int\int M(\theta, \tau)e^{-i\theta a - i\tau b}d\theta d\tau. \tag{6.28}$$

Now just like in the one dimensional case the characteristic function is an expectation value, in this case it is the expectation value of $e^{i\theta a + i\tau b}$,

$$M(\theta, \tau) = \langle e^{i\theta a + i\tau b} \rangle, \tag{6.29}$$

Let us recall that in the last section we showed that the characteristic function is given by

$$M(\theta) = \int g^*(x) e^{i\theta \mathbf{A}} g(x) dx \qquad (6.30)$$

where $\mathbf{A}$ is the operator associated with the variable.

If we have two variables we would hope that we can calculate the characteristic function in a similar way. Suppose a and b are represented by the operators $\mathbf{A}$ and $\mathbf{B}$ then we expect to obtain the characteristic function by way of [4, 24]

$$M(\theta, \tau) = \int \psi^*(q) e^{i\theta \mathbf{A} + i\tau \mathbf{B}} \psi(q) dq. \qquad (6.31)$$

However, it is a remarkable fact that the above procedure does not work! It does not work because the characteristic function thus defined is improper, that is, it is not the Fourier transform of a positive density. Why it does not work is a long standing issue in quantum mechanics although most authors dismiss the issue by saying that it is because of the uncertainty principle or because the operators do not commute. Of central importance is that since the operators in general do not commute, one can take for the characteristic function many choices such as

$$M(\theta, \tau) = \int \psi^*(q) e^{i\theta \mathbf{A} + i\tau \mathbf{B}} \psi(q) dq \qquad (6.32)$$

$$\text{or} \int \psi^*(q) e^{i\theta \mathbf{A}} e^{i\tau \mathbf{B}} \psi(q) dq \qquad (6.33)$$

$$\text{or} \int \psi^*(q) e^{i\tau \mathbf{B}} e^{i\theta \mathbf{A}} \psi(q) dq \qquad (6.34)$$

$$\text{or} \int \psi^*(q) e^{i\theta \mathbf{A}/2} e^{i\tau \mathbf{B}} e^{i\theta \mathbf{A}/2} \psi(q) dq \qquad (6.35)$$

$$\text{or all possible arrangements} \qquad (6.36)$$

and clearly there is an infinite number of ways to rearrange the operators. A way to characterize all of these is the kernel method, which we will describe shortly. Even though one gets improper densities it is rather remarkable that these densities are nonetheless very useful and indeed quantum mechanics can be formulated in terms of these improper densities. These improper densities are called quasi-probability distributions and the formulation of quantum mechanics in terms of them is called the phase space formulation of quantum mechanics. Instead of considering operators in general we describe here the case when the operators are position and momentum. What we want is to define a "probability distribution," $C(q, p)$, so that we can calculate averages in the standard probabilistic way. For a quantum operator, $\mathbf{G}(\mathbf{q}, \mathbf{p})$, we want to associate a classical function, $g(q,p)$, so that its average calculated in the classical way agrees with the quantum way, that is

$$\iint g(q,p)\,P(q,p)\,dq\,dp = \int \psi^*(q)\,\mathbf{G}(\mathbf{q},\mathbf{p})\,\psi(q)\,dq. \tag{6.37}$$

Of course we also want the distribution to satisfy the marginal conditions,

$$\int C(q,p)\,dp = |\psi(q)|^2, \tag{6.38}$$

$$\int C(q,p)\,dq = |\varphi(p)|^2, \tag{6.39}$$

where $|\varphi(p)|^2$ is the probability distribution of momentum and where $\varphi(p)$ is the momentum wave function,

$$\varphi(p) = \frac{1}{\sqrt{2\pi}}\int e^{-ipq}\,\psi(q)\,dq. \tag{6.40}$$

Wigner was the first to give such a distribution [27]. It is possible to generate all distributions that satisfy Eqs. (6.38)–(6.39),

$$C(q,p) = \frac{1}{4\pi^2}\iiint \psi^*(u-\tfrac{1}{2}\tau)\,\psi(u+\tfrac{1}{2}\tau)\,\Phi(\theta,\tau)\,e^{-i\theta q - i\tau p + i\theta u}\,du\,d\tau\,d\theta, \tag{6.41}$$

where $\Phi(\theta,\tau)$ is a two-dimensional function called the kernel that characterizes the distribution [4, 15, 3, 13, 17, 28]. In Eq. (6.41) and subsequently we take $\hbar = 1$. In the Table 6.1 we list some distributions and their corresponding kernels.

Eq. (6.41) can be written as

$$C(q,p) = \frac{1}{4\pi^2}\iint M(\theta,\tau)\,e^{-i\theta q - i\tau p}\,d\theta\,d\tau, \tag{6.42}$$

where

$$M(\theta,\tau) = \Phi(\theta,\tau)\int \psi^*(u-\tfrac{1}{2}\tau)\,\psi(u+\tfrac{1}{2}\tau)\,e^{i\theta u}\,du \tag{6.43}$$

is called the quasi-characteristic function.

6.3.1   Generalized characteristic function

A central idea in understanding and deriving Eq. (6.41) is the notion of the characteristic function operator, an idea first introduced by Moyal [21] for the

Name	Kernel: $\phi(\theta,\tau)$	Distribution: $C(q,p)$
General class (Cohen)	$\phi(\theta,\tau)$	$\dfrac{1}{4\pi^2}\displaystyle\iiint e^{-i\theta q-i\tau p+i\theta u}\,\phi(\theta,\tau)$ $\psi^*(u-\tfrac{1}{2}\tau)\,\psi(u+\tfrac{1}{2}\tau)\,du\,d\tau\,d\theta$
Wigner	1	$\dfrac{1}{2\pi}\displaystyle\int e^{-i\tau p}\psi^*(q-\tfrac{1}{2}\tau)\,\psi(q+\tfrac{1}{2}\tau)\,d\tau$
Margenau–Hill	$\cos\tfrac{1}{2}\theta\tau$	$\text{Re }\dfrac{1}{\sqrt{2\pi}}\,\psi(q)\varphi^*(p)\,e^{-itp}$
Kirkwood	$e^{i\theta\tau/2}$	$\dfrac{1}{\sqrt{2\pi}}\,\psi(q)\varphi^*(p)\,e^{-iqp}$
Born–Jordan–Cohen	$\dfrac{\sin\tfrac{1}{2}\theta\tau}{\tfrac{1}{2}\theta\tau}$	$\dfrac{1}{2\pi}\displaystyle\int\dfrac{1}{\lvert\tau\rvert}e^{-i\tau p}\int_{q-\lvert\tau\rvert/2}^{q+\lvert\tau\rvert/2}$ $\psi^*(u-\tfrac{1}{2}\tau)\,\psi(u+\tfrac{1}{2}\tau)\,du\,d\tau$
Choi–Williams	$e^{-\theta^2\tau^2/\sigma}$	$\dfrac{1}{4\pi^{3/2}}\displaystyle\iint\dfrac{1}{\sqrt{\tau^2/\sigma}}e^{-\sigma(u-q)^2/\tau^2-i\tau p}$ $\psi^*(u-\tfrac{1}{2}\tau)\,\psi(u+\tfrac{1}{2}\tau)\,du\,d\tau$
Spectrogram	$\displaystyle\int h^*(u-\tfrac{1}{2}\tau)e^{-i\theta u}$ $h(u+\tfrac{1}{2}\tau)\,du$	$\left\lvert\dfrac{1}{\sqrt{2\pi}}\displaystyle\int e^{-ip\tau}\,\psi(\tau)h(\tau-q)\,d\tau\right\rvert^2$
Zhao–Atlas–Marks	$g(\tau)\lvert\tau\rvert\dfrac{\sin a\theta\tau}{a\theta\tau}$	$\dfrac{1}{4\pi a}\displaystyle\int g(\tau)e^{-i\tau p}\int_{q-\lvert\tau\rvert a}^{q+\lvert\tau\rvert a}$ $\psi^*(u-\tfrac{1}{2}\tau)\,\psi(u+\tfrac{1}{2}\tau)\,du\,d\tau$

Table 6.1 Some distributions and their kernels.

Wigner distribution. One defines a characteristic function operator, $M(\theta, \tau)$, so that the characteristic function is its expectation value,

$$M(\theta, \tau) = \int \psi^*(q) M \psi(q) dq. \tag{6.44}$$

As mentioned, the reason we have an infinite number of distributions is that we have an infinite number of characteristic function operators that correspond to a classical characteristic function as discussed above. What the kernel method does is characterize in a simple way all these possibilities [4, 5, 6]. The general characteristic function operator is given by

$$\mathbf{M}(\theta, \tau) = \Phi(\theta, \tau) e^{i\theta \mathbf{q} + i\tau \mathbf{p}}. \tag{6.45}$$

Putting this into Eq. (6.44) gives

$$\langle \mathbf{M}(\theta, \tau) \rangle = \int \psi^*(q) \Phi(\theta, \tau) e^{i\theta \mathbf{q} + i\tau \mathbf{p}} \psi(q) \, dq \tag{6.46}$$

$$= \Phi(\theta, \tau) \int \psi^*(u - \tfrac{1}{2}\tau) \, \psi(u + \tfrac{1}{2}\tau) \, e^{i\theta u} \, du. \tag{6.47}$$

Using Eq. (6.47) and Eq. (6.27) one obtains Eq. (6.41). We now give the correspondence between a quantum operator $\mathbf{G}(\mathbf{q}, \mathbf{p})$ and the classical function $g(q, p)$ so that

$$\langle g(q,p) \rangle = \iint g(q,p) C(q,p) \, dq \, dp \tag{6.48}$$

$$= \int \psi^*(q) \, \mathbf{G}(\mathbf{q}, \mathbf{p}) \, \psi(q) \, dq. \tag{6.49}$$

The relationship is given by [4]

$$\mathbf{G}(\mathbf{q}, \mathbf{p}) = \iint \gamma(\theta, \tau) \Phi(\theta, \tau) \, e^{i\theta \mathbf{q} + i\tau \mathbf{p}} \, d\theta \, d\tau, \tag{6.50}$$

where

$$\gamma(\theta, \tau) = \frac{1}{4\pi^2} \iint g(q,p) e^{-i\theta q - i\tau p} \, dq \, dp. \tag{6.51}$$

Equivalently,

$$\mathbf{G}(\mathbf{q}, \mathbf{p}) = \frac{1}{4\pi^2} \iiiint g(q,p) \Phi(\theta, \tau) e^{i\theta(\mathbf{q}-q)+ i\tau(\mathbf{p}-p)} \, d\theta \, d\tau \, dq \, dp. \tag{6.52}$$

Conversely, if we start with a quantum operator we can construct the classical function in the following way. Starting with the quantum operator $\mathbf{G}(\mathbf{q}, \mathbf{p})$

rearrange it, using the commutation relation between $\mathbf{q}$ and $\mathbf{p}$, so that all the $\mathbf{q}$ operators are to the left of the $\mathbf{p}$ operators. Afterwards replace the operators by classical variables p and q and call the resulting function $G_Q(q,p)$. Then, the classical function corresponding to $\mathbf{G}(\mathbf{q},\mathbf{p})$ is given by

$$\mathbf{G}(\mathbf{q},\mathbf{p}) \rightarrow g(q,p) = \frac{\exp(i\frac{\hbar}{2}\frac{\partial}{\partial q}\frac{\partial}{\partial p})}{\Phi(-i\frac{\partial}{\partial q},-i\frac{\partial}{\partial p})} G_Q(q,p). \tag{6.53}$$

Constraints on the kernel. The advantage of the kernel method is that one can readily obtain conditions on the kernel corresponding to properties of the distribution function [4, 5, 6, 15, 17, 8, 19] . For the satisfaction of the marginals

$$\int C(q,p)dp = |\psi(q)|^2, \tag{6.54}$$

$$\int C(q,p)dq = |\varphi(p)|^2, \tag{6.55}$$

one must have that

$$\Phi(\theta,0) = \Phi(0,\tau) = 1. \tag{6.56}$$

Particularly interesting is the first conditional moment. We define the conditional value of momentum, p, for a fixed value of position by way of

$$\langle p \rangle_q = \int pC(q,p)dp. \tag{6.57}$$

Substituting Eq. (6.41) into Eq. (6.57) one obtains that

$$\langle p \rangle_q = \int pC(q,p)dp = j(q), \tag{6.58}$$

where $j(q)$ is the quantum mechanical current (per unit mass)

$$j(q) = \frac{1}{2i}\left(\psi^* \frac{d\psi}{dq} - \psi \frac{d\psi^*}{dq} \right). \tag{6.59}$$

This is the case if the kernel satisfies

$$\frac{\partial}{\partial \tau}\Phi(\theta,\tau)|_{\tau=0} = 0, \tag{6.60}$$

in addition to the marginal conditions.

A more revealing way to write the current is to write the wave function in terms of amplitude and phase

$$\psi(q) = R(q)e^{iS(q)}, \tag{6.61}$$

in which case the quantum mechanical current is given by the derivative of the phase

$$j(q) = R^2(q)S'(q),$$ (6.62)

where the prime denotes differentiation. In obtaining the above we have used the fact that

$$\frac{d}{dq}R(q)e^{iS(q)/\hbar} = e^{iS(q)/\hbar}\left(R' + \frac{i}{\hbar}RS'\right)$$ (6.63)

$$= R(q)e^{iS(q)/\hbar}\left(\frac{R'}{R} + \frac{i}{\hbar}S'\right).$$ (6.64)

For future use we note that

$$\frac{d^2}{dq^2}R(q)e^{iS(q)/\hbar}$$

$$= e^{iS(q)/\hbar}\left(R'' - \frac{1}{\hbar^2}RS'^2 + \frac{i}{\hbar}\left[2R'S' + RS''\right]\right)$$ (6.65)

$$= R(q)e^{iS(q)/\hbar}\left(\frac{R''}{R} - \frac{1}{\hbar^2}S'^2 + \frac{i}{\hbar}\left[\frac{2R'S'}{R} + S''\right]\right)$$ (6.66)

$$= R(q)e^{iS(q)/\hbar}\left(\frac{1}{R^2}\frac{d}{dq}RR' - \left(\frac{R'}{R}\right)^2 - \frac{1}{\hbar^2}S'^2\right.$$

$$\left. + \frac{i}{\hbar}\left[\frac{2R'S'}{R} + S''\right]\right) - \frac{\hbar^2}{2m}\psi^*\frac{d^2}{dq^2}\psi$$ (6.67)

$$= \frac{\hbar^2}{2m}\left(\left(\frac{R'}{R}\right)^2 + \frac{1}{\hbar^2}S'^2 - \frac{1}{R^2}\frac{d}{dq}RR'\right.$$

$$\left. + \frac{i}{\hbar}\left[\frac{2R'S'}{R} + S''\right]\right)R^2.$$ (6.68)

6.4 Visualization of quantum wave functions

One can think of a quantum quasi-distribution as a two dimensional mapping of a one dimensional function. In the mathematical sense the distribution contains the same information as the wave function since it is constructed from it and the wave function can be obtained from it uniquely. However a dramatic thing happens when one plots or studies the distribution instead of the wave function: the physical nature of the wave function is much clearer. We show this by taking a few examples. Each figure we show has three parts. In the top part we plot the real part of the wave function, the left figure is the absolute square of the momentum wave function and in the center is plotted the two dimension distribution. In all cases we plot the Wigner distribution.

Example 6.1
Consider the wave function (unnormalized)

$$\psi(q) = e^{-\alpha q^2/2 + i\beta q^2/2 + ip_0 q}. \tag{6.69}$$

In Fig. 6.1 we plot the distribution. Notice that it is concentrated a-

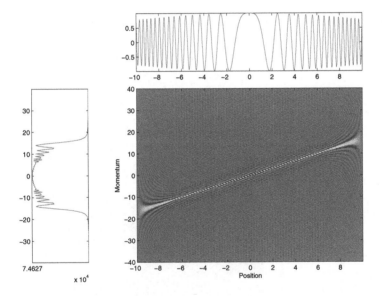

FIGURE 6.1: The Wigner distribution of $\psi(q) = e^{-\alpha q^2/2 + i\beta q^2/2 + ip_0 q}$ is shown in the central part. The top figure is the real part of the wave function and the left figure is the absolute square of the momentum wave function. The Wigner distribution is concentrated along the quantum mechanical current, $j(q) = p = p_0 + \beta q$.

long a trajectory and indeed the trajectory is the quantum current (the derivative of the phase)

$$j(q) = p = p_0 + \beta q. \tag{6.70}$$

☐

Example 6.2

Now consider

$$\psi(q) = e^{-\alpha q^2/2} \cos\left(\beta q^2/2 + i p_0 q\right) \tag{6.71}$$

and the distribution is shown in Fig. 6.2. We see that in some sense this real wave function was made into two complex functions

$$\psi(q) = e^{-\alpha q^2/2} \left(e^{i\beta q^2/2 + i p_0 q} + e^{-i\beta q^2/2 - i p_0 q} \right) \tag{6.72}$$

and gets concentrated along the currents of each complex part along the lines

$$p = p_0 + \beta q \quad ; \quad p = -p_0 - \beta q. \tag{6.73}$$

☐

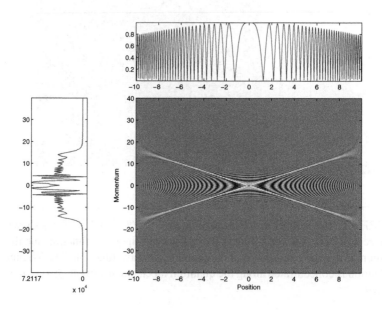

FIGURE 6.2: The Wigner distribution of Eq. (6.71).

Example 6.3

Now consider

$$\psi(q) = e^{-\alpha_1 q^2/2 + ip_1 q} + e^{-\alpha_2 q^2/2 + i\beta q^2/2 + ip_2 q} = Re^{iS}, \qquad (6.74)$$

where we have explicitly emphasized that the sum can be written as per the right hand side. We do so to emphasize that when this wave function is plotted it does not get concentrated along the derivative of the phase, S', but in phase space it seems to "know" that it consists of two parts. The distribution is plotted in Fig. 6.3. It is concentrated

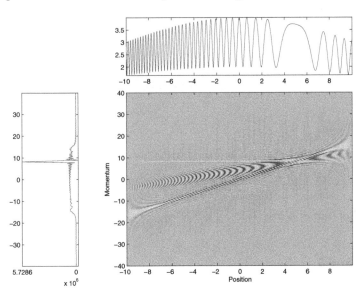

FIGURE 6.3: The Wigner distribution of Eq. (6.74).

along the phases of each part

$$p = p_0 + \beta q \quad ; \quad p = p_0. \qquad (6.75)$$

We call such wave functions multipart. □

Example 6.4

Consider

$$\psi(q) = e^{-\alpha_1 q^2/2 + i\beta_1 q^2/2 + ip_1 q} + e^{-\alpha q^2/2 + i\gamma q^3/3 + i\beta q^2/2 + ip_0 q}, \qquad (6.76)$$

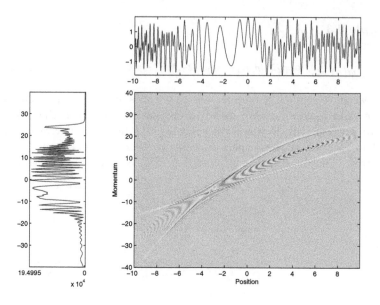

FIGURE 6.4: The Wigner distribution of Eq. (6.76).

this is illustrated in Fig. 6.4.　　☐

Example 6.5

As a last example consider

$$\psi(q) = (\alpha/\pi)^{1/4} e^{-\alpha q^2/2 + i\eta q^4/4 + i\gamma q^3/3 + i\beta q^2/2 + ip_0 q}, \qquad (6.77)$$

whose distribution is illustrated in Fig. 6.5. Again the concentration is along the current

$$p = \eta q^3 + \gamma q^2 + \beta q + p_0. \qquad (6.78)$$

☐

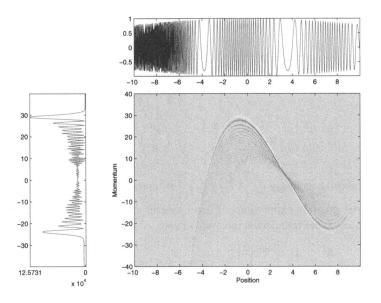

FIGURE 6.5: The Wigner distribution of Eq. (6.77).

6.5   Local kinetic energy

An interesting related issue is the question of local kinetic energy and the local virial theorem [1, 2, 9, 11, 12, 20, 22, 25, 23]. We use the notation $K(q)$ for local kinetic energy and we require that the integral of $K(q)$ give the global kinetic energy

$$\int K(q)\, dq = -\frac{\hbar^2}{2m} \int \psi^* \frac{d^2}{dq^2} \psi\, dq = \left\langle \frac{p^2}{2m} \right\rangle. \qquad (6.79)$$

The global kinetic energy can be written as

$$\begin{aligned} KE &= -\frac{\hbar^2}{2m} \int \psi^* \frac{d^2}{dq^2} \psi\, dq \\ &= \frac{\hbar^2}{2m} \int \left| \frac{d}{dq} \psi \right|^2 dq, \end{aligned}$$

and we note that

$$-\frac{\hbar^2}{2m}\psi^*\frac{d^2}{dq^2}\psi = \frac{\hbar^2}{2m}\left(S'^2 + \left(\frac{R'}{R}\right)^2 - \frac{1}{R^2}\frac{d}{dq}RR'\right.$$

$$\left. +i\left[\frac{2R'S'}{R} + S''\right]\right)R^2 dq, \tag{6.80}$$

$$\frac{\hbar^2}{2m}\left|\frac{d}{dq}\psi\right|^2 = \frac{\hbar^2}{2m}\left[S'^2 + \left(\frac{R'}{R}\right)^2\right]R^2. \tag{6.81}$$

We emphasize that only $\frac{\hbar^2}{2m}\left[S'^2 + \left(\frac{R'}{R}\right)^2\right]R^2$ contributes to the total kinetic energy. All possible expressions of local kinetic energy must contain these terms and any other terms must integrate to zero,

$$K(q) = \frac{\hbar^2}{2m}\left[S'^2 + \left(\frac{R'}{R}\right)^2\right]R^2 + \text{terms that integrate to zero}. \tag{6.82}$$

Some of the expressions that have been studied in the literature are:

$$K_A = -\frac{\hbar^2}{2m}\psi^*\frac{d^2}{dq^2}\psi$$

$$= \frac{\hbar^2}{2m}\left[\left(\frac{R'}{R}\right)^2 + S'^2 - \frac{1}{R^2}\frac{d}{dq}RR' + i\left[\frac{2R'S'}{R} + S''\right]\right], \tag{6.83}$$

$$K_B = \frac{\hbar^2}{2m}\left|\frac{d}{dq}\psi\right|^2 = \frac{\hbar^2}{2m}\left[S'^2 + \left(\frac{R'}{R}\right)^2\right]R^2, \tag{6.84}$$

$$K_C = -\frac{\hbar^2}{4m}\left(\psi\frac{d^2}{dq^2}\psi + \psi\frac{d^2}{dq^2}\psi^*\right)$$

$$= \frac{\hbar^2}{2m}\left[S'^2 + \left(\frac{R'}{R}\right)^2 - \frac{1}{R^2}\frac{d}{dq}RR'\right]R^2, \tag{6.85}$$

$$K_D = \frac{\hbar^2}{2m}\left|\frac{d}{dq}\nabla\psi\right|^2 - \frac{\hbar^2}{8m}\frac{d^2}{dq^2}|\psi^2|$$

$$= \frac{\hbar^2}{2m}\left[S'^2 + \left(\frac{R'}{R}\right)^2 - \frac{1}{4R^2}\frac{d^2}{dq^2}R^2\right]R^2. \tag{6.86}$$

All of these can be obtained by considering

$$K(q) = \frac{\hbar^2}{2m}\int p^2 C(q,p)\,dp, \tag{6.87}$$

and taking different kernels. Evaluating Eq. (6.87) for an arbitrary kernel one obtains

$$K(q) = K_D - \frac{1}{2\pi} \int e^{i\theta(u-q)} \left[\frac{1}{2m} |\psi(u)|^2 h_2(\theta) + ih_1(\theta) j(u) \right] du \, d\theta, \quad (6.88)$$

where $j(q)$ is the quantum mechanical current as defined by Eq. (6.59) and h_1 and h_2 are

$$h_1(\theta) = \frac{\partial}{\partial \tau} \Phi(\theta, \tau)|_{\tau=0} = 0, \qquad (6.89)$$

$$h_2(\theta) = \frac{\partial^2}{\partial \tau^2} \Phi(\theta, \tau)|_{\tau=0} = 0. \qquad (6.90)$$

For product kernels,
$$\Phi(\theta, \tau) = \Phi(\theta \tau). \qquad (6.91)$$

Eq. (6.88) reduces to a much simpler form

$$K(q) = K_D + \frac{\Phi''(0)}{2m} \frac{d^2}{dq^2} R^2 + \Phi'(0) \frac{d}{dq} j(q), \qquad (6.92)$$

where the primes denote differentiation with respect to the argument of Φ. Any number of forms can be obtained by choosing kernels with different values for h_1 and h_2. If we want to satisfy a local virial theorem

$$2K(q) = q \frac{\partial}{\partial q} V(q), \qquad (6.93)$$

it can be verified that one can take any $h_2(\theta)$ given by

$$h_2(\theta) = -2m \frac{\int \left[q \frac{\partial V}{\partial q} |\psi(q)|^2 - \frac{\hbar^2}{8m} \left(|\frac{d}{dq} \psi|^2 - \frac{d^2}{dq^2} |\psi^2| \right) e^{i\theta q} \right] dq}{\int |\psi(q)|^2 e^{i\theta q} dq}. \qquad (6.94)$$

6.6 Conclusion

We have shown that the unusual method of calculating expectation values in quantum mechanics exists within classical probability theory. This was done by generalizing a theorem of Khinchin on the necessary and sufficient conditions for a function to be a characteristic function. Also, we have discussed the issue of writing joint distributions for two non-commuting variables and have shown that the currently known distributions, while not having all the properties of true joint distributions, nonetheless are a powerful method for understanding and visualizing wave functions. In addition, we showed that these joint distributions may be used to define local quantities such as current and local kinetic energy and have shown that these local quantities do not behave as expected because they are not consistent with obtaining these quantities from a proper two dimensional distribution. We give an example. It is possible to construct wave functions that have a momentum distribution which is zero outside a band of momentum values, but nonetheless the current or local kinetic energy has nonzero values outside the band. The reason is that these types of conditional quantities are "obtained" with a distribution that is not manifestly positive. The consequences of these peculiar behaviors have not been fully explored or interpreted.

References

[1] R. F. W. Bader and J. T. Preston, "The kinetic energy of molecular charge distribution and molecular stability," *Int. J. Quantum Chem.* **3**, 327 (1969).

[2] R. F. W. Bader and P. M. Beddall, "Virial field relationship for molecular charge distributions and the spatial partitioning of molecular properties," *J. Chem Phys.* **56**, 3320 (1972); *J. Am. Chem. Soc.* **95**, 305 (1973).

[3] H. I. Choi and W. J. Williams, "Improved time–frequency representation of multicomponent signals using exponential kernels," *IEEE Trans. on Acoust., Speech, Signal Processing*, vol. 37, pp. 862–871, 1989.

[4] L. Cohen, "Generalized phase–space distribution functions," *Jour. Math. Phys.*, vol. 7, pp. 781–786, 1966.

[5] L. Cohen *Time-Frequency Analysis*, Prentice-Hall, 1995.

[6] L. Cohen, "Time-frequency distributions – A review," *Proc. of the IEEE*, vol. 77, pp. 941–981, 1989.

[7] L. Cohen, "Rules of Probability in Quantum Mechanics," *Found. Phys.*, **18**, 983 (1988).

[8] L. Cohen, "Local Values in Quantum Mechanics," *Physics Letters A*, vol. A 212, pp. 315–319, 1996."

[9] L. Cohen, "Local Kinetic Energy in Quantum Mechanics,"*J. Chem Phys.* **70**, 788 (1979).

[10] L. Cohen, "Can quantum mechanics be formulated as a classical probability theory?," *Philosophy of Science*, **33**, 317 (1966).

[11] L. Cohen, "Representable Local Kinetic Energy," *J. Chem. Phys.* **80**, 4277 (1984).

[12] S. K. Ghosh, M. Berkowitz, and R. G. Parr, *Proc. Natl. Acad. Sci. USA*, **81**, 8028 (1984).

[13] J. Jeong and W. Williams, "Kernel design for reduced interference distributions," *IEEE Trans. Sig. Process.*, vol. 40, no. 2, pp. 402–412, 1992.

[14] A. Khinchin, *Bull. Univ. Moscow*, 1, (1937).

[15] H. W. Lee, *Physics Reports*, 259, 147–211, 1995.

[16] P. Loughlin (ed.), *Proc. of the IEEE*, Special Issue on Applications of Time-Frequency Analysis, vol. 84, no. 9, 1996.

[17] P. Loughlin, J. Pitton and L. E. Atlas, "Bilinear time-frequency representations: new insights and properties," *IEEE Trans. Sig. Proc.*, vol. 41, pp. 750–767, 1993.

[18] E. Lukacs, *Characteristic Functions*, Charles Griffin and Company, London, 1970.

[19] H. Margenau and L. Cohen, "Probabilities in Quantum Mechanics," in: *Quantum Theory and Reality*, edited by Mario Bunge, Springer–Verlag (1967).

[20] A. Mazziotti, R. G. Parr, and G. Simons, *J. Chem Phys.* **59**, 939 (1973).

[21] J. E. Moyal, "Quantum mechanics as a statistical theory," *Proc. Cambridge. Philos. Soc.* **45**, 99, 1949.

[22] J. G. Muga, J. P. Palao and R. Sala "Average local values and local variances in quantum mechanics," *Phys. Lett.* A 238, 90–94 (1998).

[23] J. G. Muga, D. Seidel, and G. C. Hegerfeldt, "Quantum kinetic energy densities: an operational approach," *J. Chem. Phys.* **122** (2005), 154106.

[24] M. O. Scully and L. Cohen, "Quasi-Probability Distributions for Arbitrary Operators" in: *The Physics of Phase Space*, edited by Y. S. Kim and W. W. Zachary, Springer Verlag, New York, 1986.

[25] G. F. Thomas, F. Javor, and S. M. Rothstein, *J. Chem Phys.*, **64**, 1574 (1976).

[26] R. M. Wilcox, "Exponential Operators and Parameter Differentiation in Quantum Physics," *J. Math. Phys.* **8**, 962, 1967.

[27] E. P. Wigner, "On the quantum correction for thermodynamic equilibrium," *Physical Review*, vol. 40, pp. 749–759, 1932.

[28] Y. Zhao, L. E. Atlas, and R. J. Marks, "The use of cone–shaped kernels for generalized time–frequency representations of nonstationary signals," *IEEE Trans. Acoust., Speech, Signal Processing*, vol. 38, pp. 1084–1091, 1990.

Chapter 7

Superconducting Quantum Computing Devices

Zhigang Zhang and Goong Chen

Abstract   Superconductivity manifests macroscopic quantum phenomena. A superconducting circuit composed of Josephson junctions, Cooper pair boxes, and rf-/dc-SQUID, properly miniaturized, becomes quantized and demonstrates Rabi oscillations and entanglement. In this paper, we offer an overview of this approach. We begin with an introduction of the history and elementary theory of superconductivity. Then we describe the building blocks of superconducting classical circuits and derive their canonical quantizations. The set-up of qubits and superconducting quantum logic gates are then examined. Finally, ways to make measurements are discussed.

7.1   Introduction

Even though quantum effects are mostly observed in microscopic scales, they also manifest macroscopically. A particular case of such is *superconductivity*. Superconducting devices composed of Josephson junctions (JJ), Cooper-pair boxes and SQUID (superconducting quantum interference devices) have been developed since the 1980s as magnetometers, gradiometers, gyroscopes, sensors, transistors, voltmeters, etc., to perform measurements on small magnetic fields, and to demonstrate the quantum effects of tunneling, resonance and coherence [8, 13, 23, 36, 39, 44]. Many industrial and medical applications have also resulted: maglev trains, superconducting power generator, cables and transformers, MRI and NMR for medical scans, to mention a few. With the advances in solid-state lithography and thin-film technology, superconducting devices have the great advantage of being easily scalable and engineering-designable. For a bulk superconductor, if its size is reduced smaller and smaller, then the quasi-continuous electron conduction band therein turns into discrete energy levels. In principle, such energy levels can be used to constitute a qubit. The first demonstration of quantum-coherent oscillations of a Josephson "charge qubit" in a superposition of eigenstates was made by Nakamura et al. [27] in 1999. Ever since, theoretically and experimentally there has been steady progress. New proposals for qubits based on *charges, flux, phase* and *charge-flux* have been made, with observations of microwave-induced Rabi oscillations of two-level populations in those qubit systems [10, 11, 12, 46, 47].

In this paper, we will first introduce superconductivity in Subsection 7.2, the Josephson junction in Section 7.3, and the elementary superconducting circuits in Section 7.4. Superconducting quantum circuits and gates are studied in Sections 7.5 and 7.6, and conclude with measurements in Subsection 7.6.6.

An earlier version of this chapter has been adopted in the book by Chen et al. [7]. The present version of this article is more comprehensive than [7, Chap. 9]. We are grateful to Chapman & Hall/CRC for permission to reprint the overlapping material in order to make this chapter self-contained.

7.2 Superconductivity

Superconductivity was discovered in 1911 by the Dutch physicist Heike Kamerlingh Onnes (1853–1926), who dedicated his career to the exploration of extremely cold refrigeration. In 1908, he successfully liquefied helium by cooling it to $-452°$ F (4 K). In 1911, he began the investigation of the electrical properties of metals in extremely cold temperatures, using liquid helium. He noticed that for solid mercury at cryogenic temperature of 4.2 K, its electric resistivity abruptly disappeared (as if there were a jump discontinuity). This is the discovery of superconductivity, and Onnes was awarded the Nobel Prize for physics in 1913.

Subsequently, superconductivity was found in other materials. For example, lead was found to superconduct at 7 K, and (in 1941) niobium nitride was found to superconduct at 16 K.

Important understanding of superconductivity was made by Meissner and Ochsenfeld in 1933 who discovered that superconductors expelled applied magnetic fields, a phenomenon which has come to be known as the *Meissner effect*. In 1935, F. and H. London showed that the Meissner effect was a consequence of the minimization of the electromagnetic free energy carried by superconducting current. This causes the complete absence of electrical resistance and the exclusion of the interior magnetic field below some critical temperature T_c. As a consequence, the electric current density inside a superconductor must be zero. Shielding currents, which are confined on the surface of the superconducting body, are not damped and can circulate indefinitely.

In 1950, Russian scientists Ginzburg and Landau [14] developed a phenomenological theory of superconductivity which can successfully explain macroscopic properties of superconductors. From that theory, Abrikosov showed that the theory can predict the classification of superconductors into two types (see Type I and II superconductors below). Abrikosov and Ginzburg were awarded the Nobel Prize for their contributions to superconductivity in 2003.

The theory of superconductivity was further advanced in 1957 by three American physicists (then at the University of Illinois), J. Bardeen, L. Cooper, and J. Schrieffer, and called the *BCS Theory* [3]. The BCS theory explains superconductivity at temperatures close to absolute zero. Cooper theorized that atomic lattice vibrations were directly responsible for unifying and moderating the entire current. Such vibrations force the electrons to pair up into partners that enable them to pass all of the obstacles which cause resistance in the conductor. These partners of electrons are known as *Cooper pairs*. This electron coupling is viewed as an exchange of *phonons*, with phonons being the quanta of lattice vibration energy. The electron Cooper pairs are coupled over a range of hundreds of nanometers, three orders of magnitude larger than the lattice spacing. The effective net attraction between the normally repulsive electrons produces a pair binding energy on the order of milli-electron volts, enough to keep them paired at extremely low temperatures. Experimental corroboration of an interaction with the lattice was provided by the *isotope effect* on the superconducting transition temperature T_c. More on Cooper pairs in the next section.

Superconductivity phenomena have been found in metals, alloys, heavily doped semiconductor and ceramic materials at low temperatures. There are two types of superconductors, Type I and II. Twenty-nine metals together with their critical temperatures, are called Type I (or soft) superconductors. See Table 7.1. Superconductors made from alloys and ceramics of the high temperature kind are called Type II superconductors.

Mat.	$T_c(K)$
Be	0
Rh	0
W	0.015
Ir	0.1
Lu	0.1
Hf	0.1
U	0.2
Ti	0.39
Ru	0.5
Cd	0.56
Zr	0.61
Os	0.7
Mo	0.92
Zn	0.85
Ga	1.083

Mat.	$T_c(K)$
Al	1.2
Pa	1.4
Th	1.4
Re	1.4
Tl	2.39
In	3.408
Sn	3.722
Hg	4.153
Ta	4.47
V	5.38
La	6.00
Pb	7.193
Tc	7.77
Nb	9.46

Table 7.1   The twenty-nine Type I superconductors and their critical temperatures, excerpted and adapted from [17]. Some other reference sources [15, 16] also include Gd, Am or others as Type I superconductors, for which some additional physical treatment may be necessary.

Mat.	$T_c(K)$
NbTi	10
PbMoS	14.4
V_3Ga	14.8
NbN	15.7
V_3Si	16.9
Nb_3Sn	18.0
Nb_3Al	18.7
$Nb_3(AlGe)$	20.7
Nb_3Ge	23.2
MgB_2	39

Mat.	$T_c(K)$
YBaCuO	93
BiSrCaCuO	110
TlBaCaCuO	125
HgBaCaCuO	135

Table 7.2 Some Type II superconductors and their critical temperatures. (Data taken partly from [17].)

Among Type II, *cuprate* perovskite superconductors are certain ceramic compounds containing planes of copper and oxygen CuO_2 atoms. They can have much higher critical temperatures: $YBa_2Cu_3O_7$ (YBCO), one of the first cuprate superconductors to be discovered (by P.C.W. Chu and M.K. Wu in 1987), has a critical temperature of 93 K, and mercury-based cuprates have been found with critical temperatures in excess of 130 K. These are high temperature superconductors and so far, there is no explanation for their high critical temperatures. (The BCS theory explains superconductivity in conventional superconductors, but it does not explain superconductivity in the newer class of superconductors with high T_c.) Reports on materials with high T_c are constantly undergoing updating and verification. The highest figure of T_c known today is 134 K of some mercury compound which under high pressure is 164 K.

Interested readers may find more information in superconductivity textbooks [19, 34, 35, 42], for example.

7.3   More on Cooper pairs and Josephson junctions

In the preceding section, we briefly introduced Cooper pairs. For electrons in a metal at low temperature, despite the fact that the electron Coulomb forces repel each other, the lattice of positive ions in the metal can have phonon vibration energy that mediates the coupling or pairing of electrons to overcome the repelling force. It works as follows [32]: When one of the electrons that make up a Cooper pair passes close to an ion in the crystal lattice, the attraction between the negative electron and the positive ion causes a vibration (i.e., phonon) to pass from ion to ion until the other electron of the pair absorbs the vibration. The net effect is that the electron has emitted a phonon and the other electron has absorbed the phonon. It is this exchange that keeps the Cooper pairs together. It is important to understand, however, that the pairs are constantly breaking and re-forming. Because electrons are indistinguishable particles, it is easier to think of them as permanently paired. The composite entity, the Cooper pair, thus behaves as a single particle. These coupled electrons can take the character of a *boson* with charge twice that of an electron and zero spin. The first excited state of Cooper pairs has a minimum energy of 2 Δ, where Δ is what we had referred to earlier as the *superconducting gap*. See also Δ in (7.1) and (7.2) below. Cooper pairs carry the current in a superconductor.

Now, consider two superconductors. If they are kept apart and totally isolated from each other, then the phases of their wave functions will be independent. Bring them close together but separate by a thin non-conducting oxide barrier of tens of angströms thickness. Then Cooper pairs begin to tunnel stronger across the barrier as the separation decreases. This current is called the *Josephson current*. The "sandwich-like" arrangement is called the *Josephson junction*, see Fig. 7.1. Both were named after the British physicist B.D. Josephson (Nobel laureate in physics 1973).

The basic equations governing the dynamics of the Josephson tunneling are

$$V(t) = \frac{\hbar}{2e} \frac{\partial \phi(t)}{\partial t}, \qquad I(t) = I_c \sin(\phi(t)), \qquad (7.1)$$

where $V(t)$ and $I(t)$ are, respectively, the voltage and current across the JJ, $\phi(t)$ is the phase difference of the superconductors across the JJ, and I_c, a constant, is the *critical current*. In the microscopic theory of superconductivity [42], it is known that

$$I_c = \frac{\pi \Delta}{2eR_N} \tanh \frac{\Delta}{2T}, \qquad (7.2)$$

where Δ is the superconducting order parameter *energy gap*, T is the temperature, and R_N is a constant.

It follows from the equations in (7.1) that there are three major effects:

(1) The *DC Josephson effect*: This is the phenomenon of a direct current crossing the insulator in the absence of any external electromagnetic field due to tunneling. The second Eq. in (7.1) applies and the DC Josephson current is proportional to the sine of the phase difference across the insulator, and may take values between $-I_c$ and I_c.

(2) The *AC Josephson effect*: If the voltage U_{DC} is fixed across the junctions, the phase will vary linearly with time and the current will be an AC current with amplitude Ic and frequency $\frac{2e}{\hbar}U_{DC}$. Thus, a Josephson junction can act as a perfect *voltage-to-frequency converter*.

(3) The *inverse AC Josephson effect*: This works in a reverse way as (2) above, where for distinct DC voltages, the junction may carry a DC current and acts like a perfect *frequency-to-voltage converter*.

For example, one can apply (3) above to make the JJ a superfast voltage-switching device. JJ can perform voltage-switching functions approximately ten times faster than ordinary semiconducting circuits. This is a distinct and ideal advantage for building superfast electronic computers.

There are two general types of JJ: *overdamped* and *underdamped*, and they behave differently when $I(t) > I_c$ and $T \ll T_c$. The two types of junctions are distinguished by their Stewart–McCumber parameters which are defined as $\beta_c = \frac{2\pi I_c R^2 C}{\Phi_0}$, where C is the junction capacitance and Φ_0 is the magnetic flux quantum. When $\beta_C > 1$, the junction is underdamped, and it is called overdamped when $\beta_C < 1$. When β_C approaches zero, the time average voltage across the junction is defined uniquely by the through current with identity [26]

$$V = R(I^2 - I_c^2)^{1/2}.$$

The function changes smoothly from $V = 0$ when $I < I_c$ to $V = RI$ when $I \gg I_c$.

In contrast, for an underdamped junction, the barrier is an insulator. The junction's internal resistance (R in (7.21) below) will be maximum and the current-voltage curve is hysteretic near I_c.

A superconducting quantum interference device (SQUID) consists of one or more Josephson junctions included in a superconducting loop. SQUIDs are usually made of either a lead alloy (with 10% gold or indium) and/or niobium, often consisting of the tunnel barrier sandwiched between a base electrode of niobium and the top electrode of lead alloy.

There are two types of SQUID:

(1) dc-SQUID: It was invented by R. Jaklevic, J. Lambe, A. Silver, and J. Mercereau of Ford Research Labs in 1964. It consists of two JJ placed

in parallel such that electrons tunneling through the junctions manifest quantum interference, depending upon the strength of the magnetic field within a loop.

(2) rf-(or ac-) SQUID: It was invented by J. E. Zimmerman and A. Silver at Ford in 1965. It is made up of one Josephson junction, which is mounted on a superconducting ring. An oscillating current is applied to an external circuit, whose voltage changes as an effect of the interaction between it and the ring. The magnetic flux can then be measured.

The dc type is more difficult and expensive to fabricate, but they are much more sensitive. A SQUID can detect a change of energy as much as 100 billion times weaker than the electromagnetic energy that moves a compass needle. We will study dc and rf SQUID in more technical detail in the following sections.

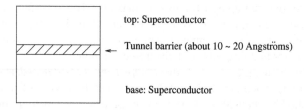

FIGURE 7.1: Schematic of a simple Josephson junction. It has a "sandwich" structure. The base is an electrode made of a very thin niobium layer, formed by deposition. The midlayer, the tunnel barrier, is oxidized onto the niobium surface. The top layer, also an electrode, made of lead alloy (with about 10% gold or indium) is then deposited on top of the other two.

7.4 Superconducting circuits: classical

There are about a half dozen major proposals for superconducting qubits. We will introduce some of them in this section. First, *classical* superconducting circuits characterized by their *Lagrangians* will be presented. Then we advance to their quantum versions through the *canonical quantization* procedure. For different ways of the setup of qubits, the number of electrons on the circuit needs not be small. (But if the qubit is a charge qubit, to be explained later in Subsection 7.6.2, the number of electrons involved is small.) Our discussions mainly follow the excellent tutorial paper by Wendin and Shumeiko [48].

In superconducting quantum computing applications, four basic types of circuits with JJ are commonly used as *building blocks*:

(1) single current-biased JJ;

(2) single Cooper-pair box (SCB);

(3) rf-SQUID;

(4) dc-SQUID.

We address each of them separately in the following subsection.

7.4.1 Current-biased JJ

This is the simplest superconducting circuit, consisting of a tunnel Josephson junction with superconducting electrodes connected to a current source. A schematic is given in Fig. 7.2.

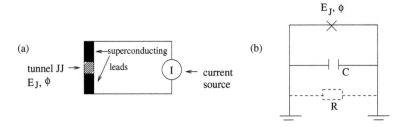

FIGURE 7.2: (a) A current biased Josephson junctions. (b) An equivalent lumped circuit, where × signifies the barrier of the JJ. (Adapted from [48]).

Let $\phi(t)$ be the phase difference between the wave functions in the two superconductors across the junction. Let $V(t)$ denote the voltage difference across the junction. Then by the first equation in (7.1),

$$\phi(t) = \frac{2e}{\hbar} \int_{t_0}^{t} V(\tau)d\tau + \phi_0. \tag{7.3}$$

(The superconducting phase $\phi(t)$ is also related to a magnetic flux $\Phi(t)$ as

$$\phi(t) = \frac{2e}{\hbar}\Phi(t) = 2\pi\frac{\Phi(t)}{\Phi_0}, \tag{7.4}$$

where $\Phi_0 = h/(2e)$ is the magnetic flux quantum.) As noted in the second equation of (7.1) in Section 7.3, the JJ current is proportional to the sine of $\phi(t)$ across the insulator:

$$I_J = I_c \sin\phi, \quad I_c \equiv \text{the critical Josephson current, (cf. (7.1)).} \tag{7.5}$$

Differentiating (7.3), we have

$$\dot{\phi}(t) = \frac{2e}{\hbar} V(t). \tag{7.6}$$

Refer to Fig. 7.2 (b). The current-voltage relations for the junction capacitance C and resistance R are given by the standard formulas

$$I_C = C\frac{dV}{dt}, \quad I_R = V/R. \tag{7.7}$$

From (7.6) and (7.7), by the Kirchhoff law of the circuit (see Fig. 7.2 (b)), we now have

$$\frac{\hbar}{2e}C\ddot{\phi} + \frac{\hbar}{2eR}\dot{\phi} + I_c\sin\phi = I_e, \tag{7.8}$$

where I_e is the *bias current*. Eq. (7.8) takes the form of a *damped forced pendulum*.

The damping term $\frac{\hbar}{2eR}\dot{\phi}$ in (7.8) determines the lifetime of the (future superconducting quantum circuit) qubit. Thus, the dissipation must be extremely small. Ideally, we assume that it is zero. So we consider an undamped Eq. (7.8):

$$\frac{\hbar}{2e}C\ddot{\phi} + I_c\sin\phi = I_e. \tag{7.9}$$

REMARK 7.1   It is necessary to emphasize that dropping the damping term $\hbar\dot{\phi}/(2eR)$ in (7.8) constitutes a reasonable approximation only under the following conditions of superconductivity:

(i) low temperature, i.e., T is small;

(ii) $|\dot{\phi}|$ is very small;

(iii) $T, \hbar\omega \ll \Delta$, where Δ is the energy gap in (7.2).

∎

For the undamped Eq. (7.9), Lagrangian and Hamiltonian variational forms can now be obtained by kinetic and potential energies:

$$\text{kinetic energy } K = K(\dot{\phi}) = \left(\frac{\hbar}{2e}\right)^2 \frac{C}{2}\dot{\phi}^2, \tag{7.10}$$

$$\text{potential energy } U = U(\phi) = \frac{\hbar}{2e}\int [I_c \sin\phi - I_e]d\phi$$

$$= \frac{\hbar}{2e}I_c(1 - \cos\phi) - \frac{\hbar}{2e}I_e\phi, \tag{7.11}$$

where the kinetic energy is proportional to the electrostatic energy of the junction capacitor (corresponding to the first term in (7.9)), while the potential energy consists of the energy of the Josephson current and the magnetic energy of the bias current (corresponding to the last two terms in (7.9)).

For future quantum superconducting circuit applications, we introduce several useful constants. The first is the *charging energy* of the junction capacitor charged with a single Cooper pair (of electrons)

$$E_C \equiv \frac{(2e)^2}{2C}. \tag{7.12}$$

The second,

$$E_J \equiv \frac{\hbar}{2e}I_c \tag{7.13}$$

is called the *Josephson energy*. The third constant,

$$\omega_J \equiv \sqrt{\frac{2eI_c}{\hbar C}}, \tag{7.14}$$

is called the *plasma frequency* of the JJ. This is the frequency of the small-amplitude oscillation of the unforced pendulum (i.e., Eq. (7.9) with $I_e = 0$). With (7.12) and (7.13), we can write (7.10) and (7.11) as

$$K = \frac{\hbar^2\dot{\phi}^2}{4E_C}, \quad U = E_J(1 - \cos\phi) - \frac{\hbar}{2e}I_e\phi.$$

Thus, we obtain the Lagrangian

$$L(\phi, \dot{\phi}) = K - U = \frac{\hbar^2\dot{\phi}^2}{4E_C} - E_J(1 - \cos\phi) + \frac{\hbar}{2e}I_e\phi,$$

whose Lagrangian variational equation

$$\frac{d}{dt}\frac{\partial L}{\partial \dot{\phi}} - \frac{\partial L}{\partial \phi} = 0$$

is exactly (7.9).

The Hamiltonian H is related to the Lagrangian L through

$$H(p,\phi) = p\dot{\phi} - L, \quad \text{where} \quad p = \frac{\partial L}{\partial \dot{\phi}} = \frac{\hbar^2}{2E_C}\dot{\phi}, \tag{7.15}$$

with p being the canonical momentum operator conjugate to ϕ. Then

$$H(p,\phi) = \frac{E_C}{\hbar^2}p^2 + E_J(1 - \cos\phi) - \frac{\hbar}{2e}I_e\phi, \tag{7.16}$$

and the Hamiltonian equations of motion

$$\dot{\phi} = \frac{\partial H}{\partial p}, \quad \dot{p} = -\frac{\partial H}{\partial \phi} \tag{7.17}$$

are again equivalent to (7.9).

7.4.2 Single Cooper-pair box (SCB)

An SCB is driven by an applied voltage Vg through capacitance Cg to induce an offset charge. The circuit consists of a small superconducting "island" connected via a Josephson tunnel junction to a large superconducting reservoir. See a schematic in Fig. 7.3.

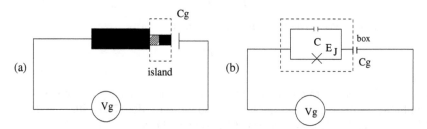

FIGURE 7.3: (a) A single Cooper-pair box. (b) An equivalent lumped circuit, where $\times$ signifies the barrier of JJ. (Adapted from [48] and [52].)

The electrostatic energy of the SCB is the sum

$$K = \frac{CV^2}{2} + \frac{Cg(Vg - V)^2}{2},$$

which, after using (7.6) and completing the square, gives

$$K = \frac{(C + Cg)}{2}\left(\frac{\hbar}{2e}\dot{\phi} - \frac{Cg}{C + Cg}Vg\right)^2 + \frac{1}{2}\left(Cg - \frac{Cg^2}{C + Cg}\right)Vg^2. \tag{7.18}$$

Dropping the (last) constant term in (7.18) and denoting $C_\Sigma \equiv C + Cg$, we have

$$K = K(\dot\phi) = \frac{C_\Sigma}{2}\left(\frac{\hbar}{2e}\dot\phi - \frac{Cg}{C_\Sigma}Vg\right)^2.$$

The potential energy U from (7.11) (by dropping the bias current I_e as it is no longer present) is

$$U = U(\phi) = E_J(1 - \cos\phi).$$

Therefore, we obtain the Lagrangian

$$L(\phi,\dot\phi) = \frac{C_\Sigma}{2}\left(\frac{\hbar}{2e}\dot\phi - \frac{Cg}{C_\Sigma}Vg\right)^2 - E_J(1 - \cos\phi). \qquad (7.19)$$

The Hamiltonian, according to (7.15), is

$$H(\phi,p) = \frac{1}{2C_\Sigma}\left(\frac{2e}{\hbar}\right)^2 p^2 + E_J(1 - \cos\phi). \qquad (7.20)$$

7.4.3 rf- or ac-SQUID

The rf-SQUID, also called an ac-SQUID or a magnetic-flux box, is depicted in Fig. 7.4. It is the magnetic analogue of the (electrostatic) SCB discussed in Subsection 7.4.2. It consists of a tunnel JJ inserted in a superconducting loop.

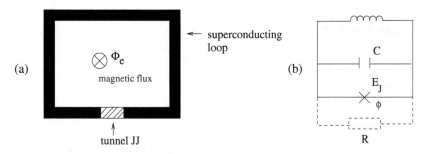

FIGURE 7.4: (a) An rf-SQUID. (b) An equivalent lumped circuit. (Adapted from [48, Fig. 8].)

Let I_L denote the current associated with the inductance L of the superconducting leads. Then by (7.4),

$$I_L = \frac{\hbar}{2eL}(\phi - \phi_e), \quad \phi_e = \frac{2e}{\hbar}\Phi_e,$$

where Φ_e is the external magnetic flux piercing the rf-SQUID loop. Using the same arguments as in Subsection 7.4.1, by the Kirchhoff circuit law we arrive at

$$\frac{\hbar}{2e}C\ddot{\phi} + \frac{\hbar}{2eR}\dot{\phi} + I_c\sin\phi + \frac{\hbar}{2eL}(\phi - \phi_e) = 0, \qquad (7.21)$$

where in (7.8) the bias current I_e is replaced by $-I_L$.

If the damping is very small, then the term containing $\dot{\phi}$ can again be dropped and the Lagrangian of the rf-SQUID is

$$L(\phi, \dot{\phi}) = \frac{\hbar^2\dot{\phi}^2}{4E_C} - E_J(1 - \cos\phi) - E_L\frac{(\phi - \phi_e)^2}{2}, \qquad \left(E_L \equiv \frac{\hbar^2}{(2e)^2L}\right). \quad (7.22)$$

The Hamiltonian is then obtained as

$$H(\phi, p) = \frac{E_C}{\hbar^2}p^2 + E_J(1 - \cos\phi) + E_L\frac{(\phi - \phi_e)^2}{2}. \qquad (7.23)$$

7.4.4   dc-SQUID

A dc-SQUID consists of two JJ in parallel coupling to a current source. It has some similarity to the current-biased single junction (Fig. 7.2), except that there is an additional magnetic flux piercing the SQUID loop, which serves as a control on the effective Josephson energy of the double JJ. See Fig. 7.5 for a schematic of a dc-SQUID.

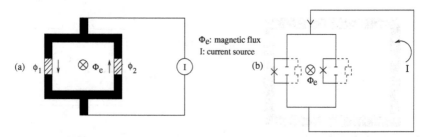

FIGURE 7.5: (a) Schematic of a dc-SQUID. (b) An equivalent (nominal) lumped circuit.

Let ϕ_1 and ϕ_2 be superconducting phase differences across the JJ 1 and 2, respectively. Assume that the inductance of the SQUID loop is small so that the magnetic energy of the circulating currents can be neglected. Then the total voltage drop over the two JJ is zero:

$$V_1 + V_2 = 0.$$

From (7.6),

$$\dot{\phi}_1 + \dot{\phi}_2 = 0,$$

and, thus $\phi_1 + \phi_2$ is a constant, and

$$\phi_1 + \phi_2 = \phi_e, \tag{7.24}$$

where ϕ_e is the biasing superconducting phase related to the biasing magnetic flux. Define

$$\phi_\pm = \frac{\phi_1 \pm \phi_2}{2}.$$

Then

$$\phi_+ = \frac{\phi_1 + \phi_2}{2} = \frac{1}{2}\phi_e, \quad \phi_- = \frac{\phi_1 - \phi_2}{2},$$

which leads to

$$\phi_1 = \phi_- + \frac{\phi_e}{2}, \quad \phi_2 = \frac{\phi_e}{2} - \phi_-. \tag{7.25}$$

For the symmetric case, the two JJ have the same E_J, C and I_c. Assume that there is no dissipation, thus we can neglect the $\dot{\phi}$ term. The equation (7.24) can be rewritten using ϕ_+ as

$$2\phi_+ - \phi_e = 0.$$

The Kirchhoff circuit law requires

$$\frac{\hbar}{2e}C\ddot{\phi}_1 + I_c \sin\phi_1 - \frac{\hbar}{2e}C\ddot{\phi}_2 - I_c \sin\phi_2 - I_e = 0,$$

or

$$\frac{\hbar}{e}C\ddot{\phi}_- + 2I_c \cos\phi_+ \sin\phi_- - I_e = 0,$$

by using trigonometric identities. Thus the dynamic equation for the system of ϕ_+ and ϕ_- can be obtained as

$$\begin{cases} \frac{\hbar}{e}C\ddot{\phi}_- + 2I_c \cos\phi_+ \sin\phi_- - I_e = 0 \\ \frac{\hbar}{eL}(2\phi_+ - \phi_e) = 0. \end{cases} \tag{7.26}$$

The system has in fact only one degree of freedom since $2\phi_+ = \phi_e$. By substituting ϕ_+ by $\phi_e/2$ and comparing (7.26) with

$$\frac{d}{dt}\frac{\partial L}{\partial \dot{\phi}_-} - \frac{\partial L}{\partial \phi_-} = 0, \tag{7.27}$$

we can obtain the Lagrangian of the dc-SQUID as

$$L = \left(\frac{\hbar}{2e}\right)^2 C\dot{\phi}_-^2 + \frac{\hbar}{2e}2I_c \cos\phi_+ \cos\phi_- + \frac{\hbar}{2e}I_e\phi_-.$$

Its Hamiltonian, in turn, is

$$H = \left(\frac{\hbar}{2e}\right)^2 C\dot{\phi}_-^2 - \frac{\hbar}{2e}2I_c \cos\phi_+ \cos\phi_- - \frac{\hbar}{2e}I_e\phi_-.$$

The kinetic energy of the dc-SQUID can be obtained from the Lagrangian as

$$K(\phi_-) = \left(\frac{\hbar}{2e}\right)^2 2C\frac{\dot{\phi}_-^2}{2}. \tag{7.28}$$

It has a simple interpretation as the charging energy of the two junction capacitances (cf. Fig. 7.5 (b)) by looking at identity:

$$2C\tfrac{\hbar}{2e}\dot{\phi}_- = C\tfrac{\hbar}{2e}(\dot{\phi}_1 - \dot{\phi}_2)$$
$$= C(V_1 - V_2)$$
$$= q.$$

By setting $E_C \equiv \frac{(2e)^2}{2 \cdot 2C}$ and define E_J and E_L as before, we can rewrite the Hamiltonian as

$$H = \frac{\hbar^2}{4E_C}\dot{\phi}_-^2 - 2E_J \cos\frac{\phi_e}{2}\cos\phi_- - \frac{\hbar}{2e}I_e\phi_-,$$

and in terms of $p = \frac{\hbar^2}{2E_C}\dot{\phi}_-$,

$$H = \frac{E_C}{\hbar^2}p^2 - 2E_J \cos\frac{\phi_e}{2}\cos\phi_- - \frac{\hbar}{2e}I_e\phi_-. \tag{7.29}$$

7.5   Superconducting circuits: quantum

We know that the quantization of the electromagnetic field gives a simple harmonic oscillator. A classical superconducting circuit may be viewed as an antenna. It can thus radiate electromagnetic waves. From this analogue, we see that superconducting circuits can be quantized as well when the JJ becomes microscopically small, and the continuous electric current becomes discretely charged.

We now formalize the above argument by following the standard approach of *canonical quantization*. From the classical Lagrangian L, and then $p = \partial L/\partial \dot{\phi}$ we have the Hamiltonian H just as in (7.15). Now consider the simplest case of a single junction (Subsection 7.4.1, in particular Fig. 7.2). From (7.15),

$$p = \frac{\partial L}{\partial \dot{\phi}} = \frac{\hbar^2}{2E_C}\dot{\phi}, \quad \text{(cf. (7.12) for } E_C\text{)}. \tag{7.30}$$

From the first equation in (7.1),

$$V = \frac{1}{2e}\frac{h}{2\pi}\dot{\phi} = \frac{\hbar}{2e}\dot{\phi}. \tag{7.31}$$

Thus,

$$\begin{aligned}
p &= \frac{\hbar^2}{2E_C}\dot{\phi} = \left(\frac{\hbar}{2e}\right)^2 C\dot{\phi} \\
&= \left(\frac{\hbar}{2e}\right)^2 C\left(\frac{2e}{\hbar}\right)V = \frac{\hbar}{2e}CV \\
&= \frac{\hbar}{2e}q \qquad (q = CV \text{ on the junction capacitor}) \\
&= \hbar\frac{q}{2e} = \hbar n, \tag{7.32}
\end{aligned}$$

where $q/(2e)$ is n, the number of Cooper pairs. Therefore, the momentum p has a simple interpretation that it is proportional to the number of Cooper pairs n on the junction capacitor. Substituting (7.32) into (7.16), we obtain the (quantum) Hamiltonian for the current-biased JJ:

$$H = E_c n^2 - E_J\cos\phi - \frac{\hbar}{2e}I_e\phi, \tag{7.33}$$

where the constant E_J in (7.16) has been dropped.

For the SCB, from (7.19) we have the conjugated momentum

$$p = \frac{\partial L}{\partial \dot{\phi}} = \frac{\hbar C_\Sigma}{2e}\left(\frac{\hbar}{2e}\dot{\phi} - \frac{C_g}{C_\Sigma}Vg\right), \tag{7.34}$$

and by using (7.32) and (7.34) in (7.20), we have

$$H = \overline{E}_c(n - n_g)^2 - E_J\cos\phi, \tag{7.35}$$

where

$$\bar{E}_c \equiv (2e)^2/(2C_\Sigma), \quad n_g = C_g V_g/(2e), \tag{7.36}$$

and n_g is the number of Cooper pairs on the gate capacitor. This n_g is tunable through different designs of Cg and Vg.

For the dc-SQUID, according to the derivations of (7.29), we obtain the Hamiltonian

$$H = E_C n_-^2 - 2E_J \cos\frac{\phi_e}{2}\cos\phi_- - \frac{\hbar}{2e}I_e\phi_-, \tag{7.37}$$

where $E_C = \frac{(2e)^2}{4C}$ and $n_- = 2C\frac{\hbar}{(2e)^2}\dot{\phi}_-$.

In quantization, the classical momentum p in (7.30) becomes the *differential operator*

$$\hat{p} = -i\hbar\frac{\partial}{\partial\phi}, \tag{7.38}$$

where using ϕ we mean ϕ_- for the dc-SQUID. From (7.32), we thus also have the *operator of the pair number*

$$\hat{n} = -i\frac{\partial}{\partial\phi}, \tag{7.39}$$

and the commutator relation

$$[\phi,\hat{n}] = i. \tag{7.40}$$

The time evolution of the wave function $\psi = \psi(\phi,t)$ satisfies the Schrödinger equation

$$i\hbar\frac{\partial}{\partial t}\psi(\phi,t) = H\psi(\phi,t) = H\left(\phi,\frac{\hbar}{i}\frac{\partial}{\partial\phi}\right)\psi(\phi,t), \tag{7.41}$$

where $H = H(\phi,p) = H(\phi,\hbar\hat{n})$ is the Hamiltonian derived in (7.33) through (7.37).

7.6   Quantum gates

We begin the discussion by using CPB as a major reference model for this section. Recall from (7.36), that the Hamiltonian for a CPB is given by

$$H = E_C(\hat{n} - n_g)^2 - E_J\cos\phi. \tag{7.42}$$

Here we assume that

$$E_C \gg E_J. \tag{7.43}$$

The pair-number operator $\hat{n}$ is defined by

$$\hat{n}|n\rangle = n|n\rangle, \qquad n = \text{an integer}, \tag{7.44}$$

where $|n\rangle$ is called the number state. From (7.39), we see that the wave function $\psi = \psi(\phi)$ of $|n\rangle$ satisfies the differential equation

$$-i\frac{\partial}{\partial\phi}\psi = n\psi. \tag{7.45}$$

To allow only integer n in (7.45) for consideration in solving ψ, a periodic constraint must be imposed:

$$\psi(\phi + 2\pi) = \psi(\phi). \tag{7.46}$$

(Without such a constraint, the number of electrons on the island may be odd, or n could be a real value number. But here the electrode is miniaturized small enough that such cases would not happen as only a finite number of Cooper pairs can exist on the island.) Therefore, from (7.45) and (7.46), we obtain

$$\psi(\phi) = \frac{1}{\sqrt{2\pi}}e^{in\phi}, \quad \text{for} \quad n = 0, \pm 1, \pm 2, \cdots, \tag{7.47}$$

where $1/\sqrt{2\pi}$ is the normalization factor with respect to the $L^2(0, 2\pi)$-norm. From (7.42), we see that for the lowest energy eigenstate $|0\rangle$ and $|1\rangle$ of $\hat{n}$, when (7.43) holds, the states $|0\rangle$ and $|1\rangle$ are nearly degenerate when $n_g = 0.5$:

$$
\begin{aligned}
H|0\rangle &= [E_C(0-0.5)^2 - E_J\cos\phi]|0\rangle \approx \tfrac{1}{4}E_C|0\rangle, \\
H|1\rangle &= [E_C(1-0.5)^2 - E_J\cos\phi]|1\rangle \approx \tfrac{1}{4}E_C|1\rangle.
\end{aligned}
\tag{7.48}
$$

This is a favorable situation. (Normally, if two states $|0\rangle$ and $|1\rangle$ differ much in energy levels, then even though they discriminate better, the higher lying state $|1\rangle$ is *less* stable, and the system tends to decohere and lie more often in $|0\rangle$ than in $|1\rangle$, an unbalanced situation in quantum computing to be avoided.)

Similarly, if $n_g = n + 1/2$, then the two states $|n\rangle$ and $|n+1\rangle$ are nearly degenerate for any integer n. For simplicity, let us just consider $n_g \approx 0.5$.

THEOREM 7.1
Assume that (7.43) holds, and that $n_g \approx 0.5$. Let

$$V = span\{|0\rangle, |1\rangle\}. \tag{7.49}$$

Then the projection of the Hamiltonian H in (7.42) with respect to the ordered basis in (7.49) satisfies

$$P_H = \begin{bmatrix} E_C[\frac{1}{4} + (n_g - 0.5)] & -\frac{1}{2}E_J \\ -\frac{1}{2}E_J & E_C[\frac{1}{4} - (n_g - 0.5)] \end{bmatrix} + \mathcal{O}(|n_g - 0.5|^2). \quad (7.50)$$

PROOF   The projection matrix P_H of H on V is easily evaluated as

$$P_H = \begin{bmatrix} a_0 & b \\ c & a_1 \end{bmatrix}, \quad (7.51)$$

where

$$a_j = \langle j|H|j \rangle \quad \text{for } j = 0, 1, \quad (7.52)$$

and

$$b = \langle 0|H|1 \rangle, c = \langle 1|H|0 \rangle. \quad (7.53)$$

Using (7.47) for $|0\rangle$ and $|1\rangle$, we compute, e.g.,

$$a_1 = \langle 1|H|1 \rangle$$
$$= \int_0^{2\pi} \left(\frac{1}{\sqrt{2\pi}} e^{-i\phi} \right) \left(E_C(-i\frac{\partial}{\partial\phi} - n_g)^2 - E_J \cos\phi \right) \left(\frac{1}{\sqrt{2\pi}} e^{i\phi} \right) d\phi$$
$$= \frac{1}{2\pi} \int_0^{2\pi} \{E_C(1 - n_g)^2 - E_J \cos\phi\} d\phi$$
$$= \frac{E_C}{2\pi} \cdot 2\pi [(1 - 0.5) + (0.5 - n_g)]^2$$
$$= E_C [0.5^2 + 2(0.5)(0.5 - n_g) + (0.5 - n_g)^2]$$
$$= E_C \left[\frac{1}{4} - (n_g - 0.5) \right] + \mathcal{O}(|n_g - 0.5|^2). \quad (7.54)$$

Similarly, the entries a_0, b and c can be computed. We obtain (7.50).
∎

As

$$P_H = \frac{1}{4} E_C \begin{bmatrix} 1 & 0 \\ 0 & 1 \end{bmatrix} + \begin{bmatrix} E_C(n_g - 0.5) & -\frac{1}{2}E_J \\ -\frac{1}{2}E_J & -E_C(n_g - 0.5) \end{bmatrix}$$
$$+ \mathcal{O}(|n_g - 0.5|^2), \quad (7.55)$$

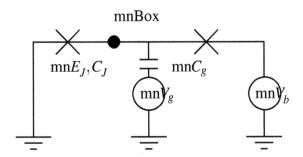

FIGURE 7.6: Schematic of a charge qubit constructed with a Cooper pair box. The box is denoted by a black dot and the two Josephson junctions are denoted by two crosses. The pulse gate voltage V_g can change the offset charge of the junction. The other junction is connected to a voltage V_b used for measurement, and the gate is called the probe gate.

we can just use the effective Hamiltonian

$$
\bar{P}_H = \begin{bmatrix} E_C(n_g - 0.5) & -\frac{1}{2}E_J \\ -\frac{1}{2}E_J & -E_C(n_g - 0.5) \end{bmatrix}
$$

$$
= E_C(n_g - 0.5)\sigma_z - \frac{1}{2}E_J\sigma_x, \tag{7.56}
$$

as an approximate Hamiltonian in the subsequent discussion. The state $|0\rangle$ and $|1\rangle$ constitute a charge-qubit system. In addition, a probe gate may be coupled to the box through a junction to perform measurement, shown in Fig. 7.6.

7.6.1   Some basic facts about SU(2) and SO(3)

We need to recall some basic facts about rotation matrices $R \in \mathbf{SO}(3)$ and $U \in \mathbf{SU}(2)$ (see [30], e.g.), where

$$
\begin{aligned}
\mathbf{SO}(3) &= \text{the special orthogonal group on } \mathbf{R}^3, \\
\mathbf{SU}(2) &= \text{the special unitray group on } \mathbf{C}^2.
\end{aligned} \tag{7.57}
$$

Every $R \in \mathbf{SO}(3)$ is a rotation with axis n and angle θ, where n is a unit vector in $\mathbf{R}^3$ and $\theta \in [0, 2\pi)$. So we denote

$$
R = R(\theta, n). \tag{7.58}
$$

We have the following [30]:

- Let $\rho \in \mathbf{SO}(3)$ be any rotation. Then

$$\rho R(\theta,n)\rho^{-1} = R(\theta,\rho n). \tag{7.59}$$

- For every $U \in \mathbf{SU}(2)$, there exists an angle $\theta \in [0,2\pi)$ and a unit vector $n = (n_1,n_2,n_3) \in \mathbf{R}^3$ such that

$$\begin{aligned}
U(\theta,n) &= e^{-i\frac{\theta}{2}n\cdot\sigma} \\
&= \begin{bmatrix} \cos\frac{\theta}{2} - in_3\sin\frac{\theta}{2} & -\sin\frac{\theta}{2}(n_2 + in_1) \\ \sin\frac{\theta}{2}(n_2 - in_1) & \cos\frac{\theta}{2} + in_3\sin\frac{\theta}{2} \end{bmatrix} \\
&= \cos\frac{\theta}{2}I_2 - i\sin\frac{\theta}{2}n\cdot\sigma,
\end{aligned} \tag{7.60}$$

where $\sigma = [\sigma_x,\sigma_y,\sigma_z]$, and I_2 is the 2×2 identity matrix.

- The map

$$\begin{aligned}
\mathscr{R} &: \mathbf{SU}(2) \to \mathbf{SO}(3) \\
\mathscr{R}(U(\theta,n)) &\equiv R(\theta,n)
\end{aligned} \tag{7.61}$$

is a 2-to-1 homomorphism, where

$$\begin{aligned}
\mathscr{R}(U(\theta_1,n_1)U(\theta_2,n_2)) &= \mathscr{R}(U(\theta_1,n_1))\mathscr{R}(U(\theta_2,n_2)), \\
\mathscr{R}(I_2) &= \mathscr{R}(-I_2) = I_3.
\end{aligned} \tag{7.62}$$

Specifically, for $U \in \mathbf{SU}(2)$ and $R \in \mathbf{SO}(3)$ and

$$U = (u_{ij})_{2\times2}, \quad R = (r_{ij})_{3\times3}, \quad R = \mathscr{R}(U), \tag{7.63}$$

then

$$r_{ij} = Tr(\frac{1}{2}\sigma_i U\sigma_j U^\dagger); \quad i,j = 1,2,3, \tag{7.64}$$

where σ_1, σ_2 and σ_3 are, respectively, the Pauli matrices σ_x, σ_y and σ_z.

7.6.2   One qubit operations (I): charge-qubit

There are various methods to manipulate the information encoded in the CP-B system, and the essence is to know how to control the time-varying Hamiltonian. In the constrained linear subspace V (cf. (7.49)) spanned by number states $|0\rangle$ and $|1\rangle$, the system Hamiltonian has been obtained in Eq. (7.56). We assume that n_g is nearly equal to 0.5 and E_C is far smaller than the supercon-ducting gap Δ. The evolution matrix of this system in time duration τ can be easily computed using results from NMR:

$$e^{-i\bar{P}_H\tau/\hbar} = e^{-i(E_C(n_g-0.5)\sigma_z - \frac{1}{2}E_J\sigma_x)\tau/\hbar}, \qquad (7.65)$$

which is a rotation around the following axis:

$$\frac{1}{\sqrt{E_J^2/4 + E_C^2(n_g-0.5)^2}} \left(-\frac{1}{2}E_Je_x + E_C(n_g-0.5)e_z \right)$$

with angle $\tau\sqrt{E_J^2/4 + E_C^2(n_g-0.5)^2}/\hbar$.

In this section, our main objective is to show that we can derive the Rabi (1-qubit) rotation gate

$$U_{\theta,\alpha} = \begin{bmatrix} \cos(\theta) & -i\sin(\theta)e^{-i\alpha} \\ -i\sin(\theta)e^{i\alpha} & \cos(\theta) \end{bmatrix}, \qquad (7.66)$$

by using the evolution matrix (7.65) with different choices of the parameter n_g and time duration τ. Note that the only tunable parameter is n_g. So we signify the dependence of $\bar{P}_H$ on n_g from (7.56) as

$$\bar{P}_H = \bar{P}_H(n_g). \qquad (7.67)$$

LEMMA 7.1

 We have the x-rotation matrix

$$R_{x,\psi} = e^{-i\bar{P}_H(\bar{n}_g)\tau/\hbar} = \begin{bmatrix} \cos(\psi/2) & -i\sin(\psi/2) \\ -i\sin(\psi/2) & \cos(\psi/2) \end{bmatrix} \qquad (7.68)$$

where $\bar{n}_g = 0.5$ and $\psi = -E_J\tau/\hbar$. In particular,

$$R_{x,\pi} = \begin{bmatrix} 0 & -i \\ -i & 0 \end{bmatrix} = -i\sigma_x. \qquad (7.69)$$

PROOF   When $\bar{n}_g = 0.5$, we have from (7.56)

$$\bar{P}_H(\bar{n}_g) = \begin{bmatrix} 0 & -\frac{1}{2}E_J \\ -\frac{1}{2}E_J & 0 \end{bmatrix}. \qquad (7.70)$$

The rest follow immediately from taking the exponential matrix $e^{-i\bar{P}_H(\bar{n}_g)\tau/\hbar}$. ∎

REMARK 7.2   The operation in Lemma 7.1 is achieved through several steps. First, the offset charge $n_g = C_g V_g/(2e)$ as controlled by V_g is abruptly switched to the degeneration point $n_g = 0.5$, kept for duration τ, and then abruptly switched back. Time duration τ is in the order of $10^{-10}s$, and the switching must be fast enough to avoid any adiabatic transition. ∎

LEMMA 7.2
   Define

$$R_{+,\theta} = e^{-i\bar{P}_H(\bar{n}_g^1)\tau/\hbar}, \quad R_{-,\phi} = e^{-i\bar{P}_H(\bar{n}_g^2)\tau/\hbar}, \tag{7.71}$$

where $\bar{n}_g^1$ and $\bar{n}_g^2$ satisfy, respectively,

$$E_C(\bar{n}_g^1 - 0.5) = -\frac{1}{2}E_J \equiv -\delta, \quad E_C(\bar{n}_g^2 - 0.5) = \frac{1}{2}E_J = \delta, \tag{7.72}$$

and

$$\theta = \phi = 2\sqrt{2}\tau\delta/\hbar. \tag{7.73}$$

Then we obtain the y-rotation and z-rotation matrices as

$$R_{y,\theta} = \begin{bmatrix} \cos\frac{\theta}{2} & -\sin\frac{\theta}{2} \\ \sin\frac{\theta}{2} & \cos\frac{\theta}{2} \end{bmatrix} = -R_{-,3\pi/2}R_{+,\theta}R_{-,\pi/2}, \tag{7.74}$$

$$R_{z,\phi} = \begin{bmatrix} e^{-i\phi/2} & 0 \\ 0 & e^{i\phi/2} \end{bmatrix} = -R_{x,\pi/2}R_{y,\phi}R_{x,3\pi/2}. \tag{7.75}$$

PROOF   With the choice of $\bar{n}_g^1$, $\bar{n}_g^2$, δ and τ given in (7.72) and (7.73), we have

$$R_{+,\theta} = e^{i\frac{\theta}{2\sqrt{2}}(\sigma_x+\sigma_z)}, \quad R_{-,\phi} = e^{i\frac{\phi}{2\sqrt{2}}(\sigma_x-\sigma_z)}. \tag{7.76}$$

Note that $R_{+,\theta}$ and $R_{-,\theta}$ are rotations with respect to axes $-\frac{1}{\sqrt{2}}(e_x+e_z)$, $-\frac{1}{\sqrt{2}}(-e_x+e_z)$, respectively. According to the properties (7.6.1)–(7.62), we have

$$R_{-,\pi/2}R_{+,-\theta}R_{-,3\pi/2} = -R_{y,\theta}$$

$$= -\begin{bmatrix} \cos\frac{\theta}{2} & -\sin\frac{\theta}{2} \\ \sin\frac{\theta}{2} & \cos\frac{\theta}{2} \end{bmatrix},$$

$$R_{x,\pi/2}R_{y,\phi}R_{x,3\pi/2} = -R_{z,\phi}$$

$$= -\begin{bmatrix} e^{-i\phi/2} & 0 \\ 0 & e^{i\phi/2} \end{bmatrix}. \tag{7.77}$$

The negative sign comes from the fact that $R_{n,2\pi} = -I_2$ for any unit vector n. ∎

COROLLARY 7.1
We have the Rabi rotation gate

$$U_{\theta/2,\alpha} = e^{-i\frac{\theta}{2}(\cos\alpha\,\sigma_x + \sin\alpha\,\sigma_y)}$$
$$= -R_{x,\pi/2}R_{y,-\alpha}R_{x,\theta}R_{y,\alpha}R_{x,3\pi/2}, \tag{7.78}$$

through the cascading of quantum operations $e^{-i\bar{P}_H(n_g)\tau/\hbar}$ by tuning the parameter n_g and time duration τ.

Next, we construct the Rabi rotation gate $U_{\theta,\phi}$ in an alternative approach which is perhaps easier to implement. From (7.56), if we let the voltage V_g be oscillating (called a phase gate [27]) such that

$$E_C(n_g - 0.5) = \varepsilon\cos(\omega t + \alpha), \tag{7.79}$$

where ε is the amplitude, then (7.56) gives (an approximate Hamiltonian)

$$H = \varepsilon\cos(\omega t + \alpha)\sigma_z - \frac{1}{2}E_J\sigma_x. \tag{7.80}$$

The above Hamiltonian is with reference to the ordered basis $\{|0\rangle, |1\rangle\}$. Now define a new basis

$$|\uparrow\rangle \equiv \frac{1}{\sqrt{2}}(|0\rangle + |1\rangle), \quad |\downarrow\rangle \equiv \frac{1}{\sqrt{2}}(|0\rangle - |1\rangle). \tag{7.81}$$

Then, with respect to the above ordered basis, the Hamiltonian (7.6.2) becomes

$$\tilde{H} = \varepsilon\cos(\omega t + \alpha)\sigma_x + \frac{1}{2}E_J\sigma_z, \tag{7.82}$$

where we rename E_J to $-E_J$ just for simplicity.

We now utilize a standard procedure in NMR by transforming the system into a rotating frame, namely, for the original wave function $|\chi(t)\rangle$ with Hamiltonian (7.82), let

$$|\psi(t)\rangle = e^{i\omega t \sigma_z/2}|\chi(t)\rangle. \tag{7.83}$$

The $|\psi(t)\rangle$ satisfies the Schrödinger equation:

$$i\hbar \frac{d}{dt}|\psi(t)\rangle = (-\frac{\hbar\omega}{2}\sigma_z + e^{i\omega t \sigma_z/2}\tilde{H}e^{-i\omega t \sigma_z/2})|\psi(t)\rangle, \tag{7.84}$$

which, by using

$$\begin{aligned} e^{i\omega t \sigma_z/2}\sigma_z e^{-i\omega t \sigma_z/2} &= \sigma_z, \\ e^{i\omega t \sigma_z/2}\sigma_x e^{-i\omega t \sigma_z/2} &= \sigma_x \cos(\omega t) - \sigma_y \sin(\omega t), \end{aligned} \tag{7.85}$$

gives

$$\begin{aligned} i\hbar \frac{d}{dt}|\psi(t)\rangle = &(\varepsilon(\sigma_x \cos(\omega t) - \sigma_y \sin(\omega t))\cos(\omega t + \alpha) \\ &+ (\tfrac{1}{2}E_J - \tfrac{1}{2}\hbar\omega)\sigma_z)|\psi(t)\rangle. \end{aligned} \tag{7.86}$$

We choose $\hbar\omega = E_J$, the *resonance* case, and obtain

$$\begin{aligned} i\hbar \frac{d}{dt}|\psi(t)\rangle &= (\sigma_x \varepsilon \cos(\omega t)\cos(\omega t + \alpha) - \sigma_y \varepsilon \sin(\omega t)\cos(\omega t + \alpha))|\psi(t)\rangle \\ &= (\sigma_x(\frac{\varepsilon}{2}(\cos(2\omega t + \alpha) + \cos\alpha)) \\ &\quad - \sigma_y(\frac{\varepsilon}{2}(\sin(2\omega t + \alpha) - \sin\alpha)))|\psi(t)\rangle. \end{aligned} \tag{7.87}$$

We now invoke the *rotating-wave approximation* by dropping the high frequency terms $\cos(2\omega t + \alpha)$ and $\sin(2\omega t + \alpha)$. (These represent high frequency oscillations which either can not be observed in laboratory conditions or contribute little to the measurement data.) Then the Schrödinger equation (7.87) is further simplified to

$$i\hbar \frac{d}{dt}|\psi(t)\rangle = \frac{\varepsilon}{2}(\cos\alpha\,\sigma_x + \sin\alpha\,\sigma_y)|\psi(t)\rangle, \tag{7.88}$$

whose revolution matrix is

$$\begin{aligned} U_{\theta/2,\alpha} &= e^{-i\frac{\varepsilon t}{2\hbar}(\cos\alpha\,\sigma_x + \sin\alpha\,\sigma_y)} \\ &= \begin{bmatrix} \cos(\frac{\theta}{2}) & -i\sin(\frac{\theta}{2})e^{-i\alpha} \\ -i\sin(\frac{\theta}{2})e^{i\alpha} & \cos(\frac{\theta}{2}) \end{bmatrix}, \end{aligned} \tag{7.89}$$

where $\theta = \varepsilon t / \hbar$. This is a Rabi rotation with respect to the ordered basis $\{|\uparrow\rangle, |\downarrow\rangle\}$. A Rabi rotation with respect to the ordered basis $\{|0\rangle, |1\rangle\}$ can be obtained by using a similarity transformation using the Walsh–Hadamard gate.

The density matrix of the system, according to the Boltzmann distribution, is given by

$$e^{\frac{-H}{k_B T}},$$

where $k_B = 1.381 \times 10^{-23}$J/K is the Boltzmann constant and T is the absolute temperature. When $k_B T \ll E_C$, and $n_g \neq 0.5$, the Coulomb energy dominates the Hamiltonian and the system is initialized to its ground state, and this initializes the system.

7.6.3 One qubit operations (II): flux-qubit

In an rf-SQUID, the magnetic flux Φ through the loop is quantized and must satisfy

$$(\Phi_0/2\pi)\phi + \Phi_{ext} + \Phi_{ind} = m\Phi_0, \qquad (7.90)$$

where $\Phi_0 = 2.07 \times 10^{-15}$Wb, and as before, m is an integer, Φ_{ext} is the external magnetic field and Φ_{ind} is induced by a current through the loop as in Fig. 7.7. That surface current through the loop is induced to compensate Φ_{ext} and its direction can be either clockwise or counterclockwise. If we denote the two surface current states as $|\uparrow\rangle$ and $|\downarrow\rangle$, then they form a basis and the qubit is called a *flux qubit*. The main references for this qubit setup are [37, 45, 53]. When Φ_{ext} is near one half of Φ_0, the current can be either clockwise or counterclockwise and the system behaves like a Cooper pair box when n_g is near 0.5. Recall the Hamiltonian of an rf-SQUID in (7.23). When the self-inductance L is large enough such that $\beta_0 = E_J 4\pi^2 L / \Phi_0^2 = E_J / E_L > 1$ and Φ_{ext} is near $\Phi_0/2$ (this means $\phi_{ext} = 2\pi\Phi_{ext}/\Phi_0$ is near π), the Hamiltonian has a shape of a double-well near $\Phi = \Phi_0/2$ ($\phi = 2\pi\Phi/\Phi_0 = \pi$), see Fig. 7.8. The two lowest states at the bottom of each well are well separated from other excited levels in low temperature and suitable for quantum computation. When $\Phi_{ext} = \Phi_0/2$, the two states are degenerate and the two eigenstates of the system are maximally superposed states of $|\uparrow\rangle$ and $|\downarrow\rangle$ [53]. When Φ_{ext} is away from $\Phi_0/2$, they approach $|\uparrow\rangle$ and $|\downarrow\rangle$, respectively. The Hamiltonian of this two level system has a simplified form as

$$H = -\frac{1}{2}B_z \sigma_z - \frac{1}{2}B_x \sigma_x,$$

where B_z can be tuned by Φ_{ext} and B_x is a function of E_J which is also tunable if the junction is replaced by a dc-SQUID. Thus, any 1-qubit operation can

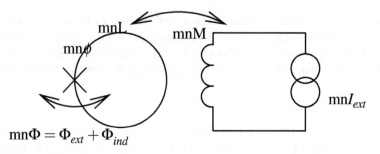

FIGURE 7.7: Schematic of a flux qubit constructed with a Josephson junction in a loop.

be realized through combinations of different choices of Φ_{ext} and E_J. When $\Phi_{ext} = \Phi_0/2$, $B_z = 0$.

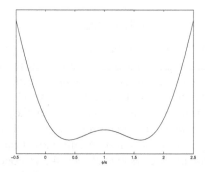

FIGURE 7.8: The double-well shape potential of a flux qubit with Hamiltonian (7.23). We take $\Phi_{ext} = \Phi_0/2$ and plot the potential curve near $\phi = \pi$.

A shortcoming of the simple rf-SQUID design is that its size is large in order to obtain high self-inductance and that makes it very susceptible to external noise. A better design uses more junctions in the loop and makes the size smaller [25, 31]. A three junction flux-qubit is shown in Fig. 7.9. Two of the junctions are designed to be the same, $E_J^1 = E_J^2 = E_J$ and $C_1 = C_2 = C$, while the third junction has different parameters, $E_J^3 = \alpha E_J$ and $C_3 = \alpha C_J$. The set up has been experimentally examined [18, 21].

To obtain the Hamiltonian of the system, we first find its Josephson energy and electronic charge energy terms. The quantum constraint (7.90) still applies, implying that

$$\phi_1 - \phi_2 + \phi_3 = -\phi_{ext} = -2\pi\Phi_{ext}/\Phi_0 = -2\pi f, \qquad (7.91)$$

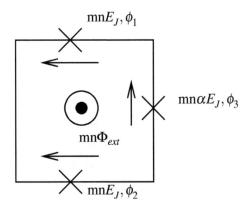

FIGURE 7.9: A three junction superconducting loop serving as a flux qubit. Compared with the simple design of rf-SQUID in Fig. 7.7, it has a smaller size and better coherence performance.

where we let $m = 0$ in (7.90) and define $f = \Phi_{ext}/\Phi_0$. When f is fixed, ϕ_3 is a function of ϕ_1 and ϕ_2. The total Josephson energy of the device, which is the sum of the Josephson energy of the three junctions, is [31, 53]

$$\begin{aligned}
\frac{U}{E_J} &= 1 - \cos\phi_1 + 1 - \cos\phi_2 + \alpha(1 - \cos\phi_3) \\
&= 2 + \alpha - \cos\phi_1 - \cos\phi_2 - \alpha\cos(2\pi f + \phi_1 - \phi_2).
\end{aligned} \tag{7.92}$$

It is a function of both ϕ_1 and ϕ_2. When $\alpha > 0.5$, it has two stable minima at $(\phi^*, -\phi^*)$ and $(-\phi^*, \phi^*)$, respectively. When $f = 0.5$, ϕ^* satisfies equation $\cos\phi^* = \frac{1}{2\alpha}$. Thus it shows a double-well pattern in a $2\pi \times 2\pi$ cell, and this pattern repeats itself with period 2π, as shown in Fig. 7.10.

Similarly, as in previous discussion, the electronic charge term is a function of $\dot{\phi}_1$ and $\dot{\phi}_2$:

$$\begin{aligned}
T &= \tfrac{1}{2}CV_1^2 + \tfrac{1}{2}CV_2^2 + \tfrac{1}{\alpha}CV_3^2 \\
&= \tfrac{C}{2}(\tfrac{\hbar}{2e})^2(\dot{\phi}_1^{\,2} + \dot{\phi}_2^{\,2} + \alpha\dot{\phi}_3^{\,2}) \\
&= \tfrac{C}{2}(\tfrac{\hbar}{2e})^2(\dot{\phi}_1^{\,2} + \dot{\phi}_2^{\,2} + \alpha(\dot{\phi}_1 - \dot{\phi}_2)^2) \\
&= \tfrac{C}{2}(\tfrac{\hbar}{2e})^2((1+\alpha)\dot{\phi}_1^{\,2} + (1+\alpha)\dot{\phi}_2^{\,2} - 2\alpha\dot{\phi}_2\dot{\phi}_1).
\end{aligned} \tag{7.93}$$

Combining (7.92) and (7.93), we obtain the classical Lagrangian $\mathscr{L}$ of the device:

$$\mathscr{L} = T - U. \tag{7.94}$$

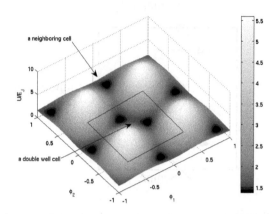

FIGURE 7.10: A repeating double-well pattern of the Josephson energy of a 3 junction flux qubit on the ϕ_1 and ϕ_2 plane. The potential function is given by (7.92) where $\alpha = 0.8$ and $f = 0.5$. A double well pattern is shown in a framed cell. The lowest levels in the two wells form two basic states of the flux qubit.

The classical Hamiltonian, H, is then derived as

$$
\begin{aligned}
H &= \Sigma_{i=1}^{2} \frac{\partial \mathscr{L}}{\partial \dot{\phi}_i} \cdot \dot{\phi}_i - \mathscr{L} \\
&= \frac{C}{2}(\frac{\hbar}{2e})^2[(1+\alpha)\dot{\phi}_1{}^2 + (1+\alpha)\dot{\phi}_2{}^2 - 2\alpha\dot{\phi}_1\dot{\phi}_2] + U.
\end{aligned}
\tag{7.95}
$$

To eliminate the off-diagonal term $-\alpha\dot{\phi}_1\dot{\phi}_2$, we rotate the system by defining $\phi_p = \frac{\phi_1+\phi_2}{2}$ and $\phi_m = \frac{\phi_1-\phi_2}{2}$. Then the Hamiltonian H becomes

$$
\begin{aligned}
H &= C(\frac{\hbar}{2e})^2(\dot{\phi}_p{}^2 + (1+2\alpha)\dot{\phi}_m{}^2) \\
&\quad +E_J(2 + \alpha - 2\cos\phi_p\cos\phi_m - \alpha\cos(2\pi f + 2\phi_m)),
\end{aligned}
\tag{7.96}
$$

while the corresponding Lagrangian is now

$$
\mathscr{L} = C(\frac{\hbar}{2e})^2(\dot{\phi}_p{}^2 + (1+\alpha)\dot{\phi}_m{}^2) - U.
$$

The canonical momenta are defined by

$$
\begin{aligned}
P_p &= \frac{\partial \mathscr{L}}{\partial \dot{\phi}_p} = 2C(\frac{\hbar}{2e})^2\dot{\phi}_p, \\
P_m &= \frac{\partial \mathscr{L}}{\partial \dot{\phi}_m} = 2C(\frac{\hbar}{2e})^2(1+2\alpha)\dot{\phi}_m,
\end{aligned}
\tag{7.97}
$$

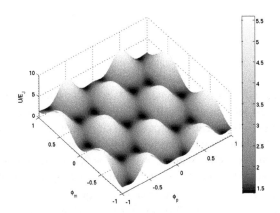

FIGURE 7.11: A repeating double-well pattern of the Josephson energy of a three junction flux qubit on the ϕ_p and ϕ_m plane.

and H can be rewritten in terms of P_p and P_m:

$$H = \frac{(2e)^2}{4C\hbar^2}P_p^2 + \frac{(2e)^2}{4C\hbar^2(1+2\alpha)}P_m^2 + U, \tag{7.98}$$

while the potential energy is

$$U = E_J(2 + \alpha - 2\cos\phi_p\cos\phi_m - \alpha\cos(2\pi f + 2\phi_m)). \tag{7.99}$$

After this rotation, the same double-well pattern still appears for the potential energy on the two dimensional ϕ_p and ϕ_m plane, as shown in Fig. 7.11. The quantization of the device can be realized by replacing P_p and P_m with operators $-i\hbar\frac{\partial}{\partial P_p}$ and $-i\hbar\frac{\partial}{\partial P_m}$, respectively.

Computation shows that the probability for the system leaking from one stable state to a neighboring cell can be sufficiently suppressed by appropriately choosing f and α, which changes the barrier between the two stable states in one cell and the barrier between two stable states in neighboring cells. Thus only the two lowest levels at the bottom of the two wells in one cell are considered, and denoted as $|\uparrow\rangle$ and $|\downarrow\rangle$, respectively. After ignoring the constant term, the Hamiltonian of the device with respect to the ordered basis $\{|\uparrow\rangle, |\downarrow\rangle\}$, H, can be put in a matrix form:

$$H = \begin{pmatrix} F & -t \\ -t & -F \end{pmatrix}, \tag{7.100}$$

where matrix entries F and $-t$ are defined as follows. Let

$$\varepsilon_\uparrow = \langle \uparrow | H | \uparrow \rangle,$$
$$\varepsilon_\downarrow = \langle \downarrow | H | \downarrow \rangle. \tag{7.101}$$

The matrix element F can be computed as $F = \frac{\varepsilon_\uparrow - \varepsilon_\downarrow}{2}$, which is the energy change of the lowest level in each well with respect to the energy level at the degeneracy point ($f = 0.5$). The other parameter, t, which can be computed as $t = -\langle \uparrow | H | \downarrow \rangle$, is the tunneling matrix element for the system to tunnel through the barrier between two wells. When $f = 0.5$, $F = 0$, and the eigenvectors of the system are $\frac{1}{\sqrt{2}}(1,1)^T$ and $\frac{1}{\sqrt{2}}(-1,1)^T$ with respect to the ordered basis $\{| \uparrow \rangle, | \downarrow \rangle\}$. When f is slightly away from 0.5, F quickly dominates t since E_J is much large than E_C for the flux qubit design. Thus H becomes almost diagonal and the eigenvectors are approximated by $| \uparrow \rangle$ and $| \downarrow \rangle$. In either case, H can be diagonalized to

$$H = -\sqrt{F^2 + t^2}\,\sigma_z \tag{7.102}$$

by rotating the coordinate system.

By changing Φ_{ext}, the magnetic field piecing through the junction loop, we can manipulate the Hamiltonian of the device and control the evolution of the system. To achieve this purpose, f is moved slightly below the degeneracy point and an additional resonant field is applied. Let f_0 be the new work point where the system Hamiltonian before diagonalization, cf. (7.100), is

$$H = F_0 \sigma_z - t_0 \sigma_x. \tag{7.103}$$

The additional resonant field results in a small change on f_0, $f = f_0 + \delta(s)$, where s is time. Since δ is so small, F and t can be approximated well by their first order approximations:

$$F \approx F_0 + r_F \delta(s),$$
$$t \approx t_0 + r_t \delta(s). \tag{7.104}$$

Thus, the time varying system Hamiltonian at time s can be approximated by

$$H = (F_0 + r_F \delta(s))\sigma_z - (t_0 + r_t \delta(s))\sigma_x. \tag{7.105}$$

LEMMA 7.3

Within the rotating wave approximation, a small oscillating signal $\delta(s)$ at certain frequency yields a Rabi rotation of the system described by Hamiltonian (7.103).

PROOF   We first rotate the coordinate system using the same rotation to diagonalize (7.103). The rotation matrix D is defined as

$$D = \begin{pmatrix} \cos \frac{\theta}{2} & -\sin \frac{\theta}{2} \\ \sin \frac{\theta}{2} & \cos \frac{\theta}{2} \end{pmatrix}, \tag{7.106}$$

where $\theta = \tan^{-1}(t_0/F_0)$. We also assume that $F_0 < 0$, which happens when f_0 is below 0.5. This results in a Hamiltonian without constant terms in its off-diagonal elements:

$$\begin{aligned} H_D &= DHD^T \\ &= -\sqrt{F_0^2 + t_0^2}\,\sigma_z \\ &\quad + (r_F \delta \cos \theta + r_t \delta \sin \theta)\sigma_z + (r_F \delta \sin \theta - r_t \delta \cos \theta)\sigma_x. \end{aligned} \tag{7.107}$$

As desired, δ is an oscillating signal with frequency ω. After collecting all the constants, we can rewrite H_D as

$$H_D = (B_1 + \varepsilon_1 \cos(\omega t + \alpha))\sigma_z + \varepsilon_2 \cos(\omega t + \alpha)\sigma_x, \tag{7.108}$$

where B_1, ε_1, ε_2, and α are some constants. Different from the NMR case and charge qubit case, the coefficient of σ_z is also time varying. Fortunately, since the Rabi frequency, which is proportional to ε_2 is much smaller than the resonant frequency (Larmor frequency), this time varying term causes no problem.

As in Subsection 7.6.2, we transform the system into a rotating frame by defining

$$|\psi(t)\rangle = e^{i\omega t \sigma_z/2}|\xi(t)\rangle, \tag{7.109}$$

where $|\xi(t)\rangle$ is the wave function of the system. Thus (7.108) leads to

$$\begin{aligned} i\hbar \tfrac{d}{dt}|\psi(t)\rangle &= (-\tfrac{\hbar\omega}{2}\sigma_z + (B_1 + \varepsilon_1 \cos(\omega t + \alpha))\sigma_z \\ &\quad + \varepsilon_2(\sigma_x \cos(\omega t) - \sigma_y \sin(\omega t))\cos(\omega t + \alpha))|\psi(t)\rangle. \end{aligned} \tag{7.110}$$

We assume that $B_1 = \tfrac{\hbar}{2}\omega$, the resonant case, and the Schrödinger equation (7.110) changes to

$$\begin{aligned} i\hbar \tfrac{d}{dt}|\psi(t)\rangle &= (\varepsilon_1 \cos(\omega t + \alpha)\sigma_z + \tfrac{1}{2}\varepsilon_2\sigma_x(\cos(2\omega t + \alpha) + \cos\alpha) \\ &\quad - \tfrac{1}{2}\sigma_y(\sin(2\omega t + \alpha) - \sin(\alpha)))|\psi(t)\rangle. \end{aligned} \tag{7.111}$$

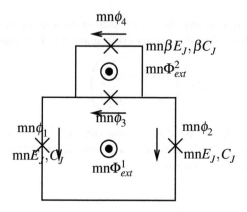

FIGURE 7.12: Schematic of a four junction superconducting flux qubit. Junctions 1 and 2 are symmetric with junction energy E_J and capacitance C_J. The top loop with junctions 3 and 4 forms a DC-SQUID, and these two junctions are also symmetric but different from the first two, $E_J^3 = E_J^4 = \beta E_J$ and $C_J^3 = C_J^4 = \beta C_J$.

The high frequency terms $\cos(2\omega t + \alpha)$, $\sin(2\omega t + \alpha)$, and $\cos(\omega t + \alpha)$, can be dropped. Thus, (7.111) is further simplified to

$$i\hbar \frac{d}{dt}|\psi(t)\rangle = \frac{1}{2}\varepsilon_2(\sigma_x \cos\alpha + \sigma_y \sin(\alpha))|\psi(t)\rangle. \qquad (7.112)$$

This is the same equation as (7.88), and the evolution of the system in time duration τ constitutes a Rabi rotation. ∎

If the third junction is replaced by a DC-SQUID as in Fig. 7.12, it is called a four-junction flux qubit. This design has two loops, and both loops have a magnetic field piercing through them. This gives us an extra degree of freedom for control purposes. The total Josephson energy of this four-junction design is now [31]

$$U/E_J = 2 + 2\beta - 2\cos\phi_p \cos\phi_m - 2\beta\cos(\pi f_a)\cos(2\pi f_b + 2\phi_m), \qquad (7.113)$$

where ϕ_m and ϕ_p are defined as before, while $f_a = f_2 = \Phi_{ext}^2/\Phi_0$ and $f_b = f_1 + f_2/2$. It is still a function of ϕ_1 and ϕ_2 when f_1 and f_2 are both fixed. We note that $2\beta\cos(\pi f_a)$ takes the place of α in (7.92). By regulating f_a and thus the flux piercing through the top loop, we change the shape of the double well pattern and the evolution of the quantum state.

7.6.4   Charge-flux qubit and phase qubit

In this subsection, we briefly describe two other ways of setting up qubits in a superconducting circuit.

For a flux qubit, E_J is much larger than E_C. When E_J is almost equal to E_C, both the Coulomb and JJ terms are important, and the qubit is called the *charge-flux qubit* [9, 43]. Neither ϕ nor n is a good quantum number and the lowest energy states are superpositions of several charge states. A typical design is shown in Fig. 7.13, which is developed from that of a Cooper pair box with a dc-SQUID. A larger junction is inserted in the loop for measurement, which is shunted by capacitors to reduce phase fluctuations. An external flux Φ_{ext} is also imposed as in the dc-SQUID case. Normally, the qubit works near $n_g = 1/2$, and the two lowest eigenstates are superpositions of number states $|0\rangle$ and $|1\rangle$. Denoted by $|+\rangle$ and $|-\rangle$, the two states have an energy difference E_J and the system Hamiltonian can be written as $H = \frac{1}{2}E_J\sigma_z$ when $n_g = 0.5$ exactly. Control signal with resonant frequency can be applied on the gate to manipulate the system. After putting the system in a "rotating frame" as before, the system Hamiltonian changes to

$$H = v(\sigma_x \cos\alpha + \sigma_y \sin\alpha),$$

when the control signal is $\Delta n_g \cos(\omega t + \alpha)$, while $v = 2E_C\Delta n_g\langle +|\hat{n}|-\rangle$. The system behaves like an NMR spin and all technologies, such as composite pulses can be used to increase the accuracy and robustness of the operation [9]. Charge-flux qubit shows better decoherence than charge- or flux-qubit in experiments.

Readout of the charge-flux qubit is realized through the current in the loop instead of the charge on the island. When a biased current I_b slightly below the critical current I_c of the large junction is applied, the large junction is switched into a finite voltage state depending on the qubit state. In theory, the measurement efficiency $p_+ - p_- = 0.95$ holds, where p_i is the probability to obtain a voltage in the readout when the qubit is in state $|i\rangle$.

Lastly, we address the *phase-qubit* setup, which is a current-biased Josephson junction. Its special feature is that the junction energy E_J is much larger than the Coulomb energy E_C. See Fig. 7.2. Here, our references are [24, 40, 55]. For such, the Coulomb term is neglected, so its Hamiltonian can be obtained from (7.33) as

$$H = -E_J \cos\phi - \frac{\hbar}{2e}I_e\phi,$$

and the potential is a periodic function of ϕ offset by $I_e\phi$, with shape appearing like a "washboard", see Fig. 7.14. Normally, the JJ is undamped and we choose I_e not too large so that there are a series of wells on the potential curve. In every

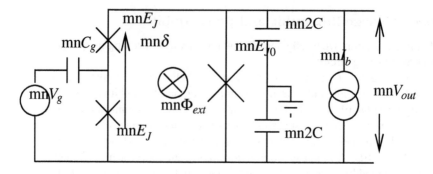

FIGURE 7.13: Circuit of a charge-flux qubit, with two small junctions and one large junction in a loop. An external flux Φ_{ext} penetrates the loop and a voltage V_g is applied through capacitor C_g to control the bias charge n_g. A bias current I_b is used for measurement.

well formed by $\cos\phi$, it is well-known that the energy is quantized and has different levels. Besides the lowest two states serving as qubit states $|0\rangle$ and $|1\rangle$, sometimes there are one or more other states in the well. The extra level or levels may be used for measurement. Transitions between $|0\rangle$ and $|1\rangle$, in the form of Rabi rotation, are realized by applying a resonant electromagnetic field with $\omega = E_{10}/\hbar$, where E_{10} is the energy difference between $|0\rangle$ and $|1\rangle$. Measurement is accomplished by inspecting the tunneling probability of states through the well. There are two methods. One is to use a microwave field resonant with E_{21} (i.e. the energy difference between $|1\rangle$ and $|2\rangle$) to pump $|1\rangle$ to the second excited state $|2\rangle$, which has a higher tunneling probability. The other is to tilt the washboard by increasing I_e so that $|1\rangle$ can tunnel through the barrier with high probability.

7.6.5   Two qubit operations: charge and flux qubits

Various proposals have been suggested to couple two qubits for different kinds of superconducting qubits.

7.6.5.1   Charge qubit case

Capacitors, for example, can be used to couple two charge qubits. Experiments have shown two-qubit oscillations using this scheme [33], and a conditional gate operation has also been demonstrated using the same device [49]. One disadvantage of the capacitor coupling is that it is not switchable, which

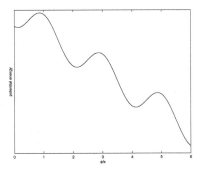

FIGURE 7.14: A "washboard" shape potential energy curve of a phase qubit. It is obtained by tilting the cosine function of ϕ by $-\frac{\hbar}{2e}I_e\phi$. When $I_e > \frac{2e}{\hbar}E_J$, there will be no well on the curve.

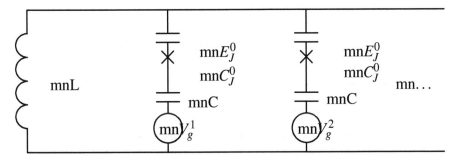

FIGURE 7.15: A simple design to couple charge qubits with inductance. The inductance and the effective capacitance of the charge qubits configured in parallel form a weak coupling among the qubits.

makes the pulse design inflexible. It is also difficult to couple two qubits far away from each other because only the neighboring qubit coupling is convenient.

Inductance, instead, seems more promising. The simplest design is to construct a weak coupling between the qubits through the CL (capacitance-inductance) oscillation, see Fig 7.15 [41]. But the coupling is still not switchable and thus lacks engineering flexibility. An improved design embeds a dc-SQUID into the qubit circuit with the advantage that the Josephson energy can be controlled [22]. The junction in Fig. 7.15 is replaced by a dc-SQUID. See Fig. 7.16. An external magnetic field Φ_e^i penetrates the SQUID and changes the term of the Josephson Hamiltonian to $-2E_J^0\cos(\pi\Phi_e/\Phi_0)\cos\phi$, where the

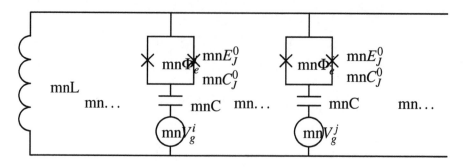

FIGURE 7.16: A design for coupling charge qubits with inductance where the junctions in the charge qubits are replaced by a dc-SQUID. All qubits are coupled through an inductor L, and an external field Φ_e^i penetrates every dc-SQUID. This changes the effective Josephson term in the Hamiltonian to $-2E_J^0 \cos(\pi\Phi_e/\Phi_0)\cos\phi$ and makes E_J tunable by Φ_e^i.

effective phase difference ϕ equals half of the difference of the two phase drops at the two junctions and $\frac{2\pi\Phi_e}{\Phi_0} = \phi_e$, cf. Section 7.4.4. This means that we replace E_J in equation (7.42) by a tunable $E_J(\Phi_e)$:

$$E_J(\Phi_e) = 2E_J^0 \cos(\pi\Phi_e/\Phi_0).$$

In this configuration, the additional effective interaction Hamiltonian induced by the oscillation in the LC-circuit can be given in the form of Pauli matrices as

$$H_{int} = -\sum_{i<j} \frac{E_J(\Phi_e^i)E_J(\Phi_e^j)}{E_L}\sigma_y^i\sigma_y^j,$$

where $E_L = [\Phi_0^2/(\pi^2 L)](C_J/C_{qb})^2$, while C_{qb} is the capacitor of the qubit defined by $C_{qb}^{-1} = C_J^{-1} + C^{-1}$.

Assume that we can still constrain every qubit in the projected subspace spanned by $|0\rangle$ and $|1\rangle$, see V in (7.49), and note that the whole Hamiltonian of the n-qubit system can be written as

$$H = \sum_{i=1}^{n} (\varepsilon(V_g^i)\sigma_z^i - \frac{1}{2}E_J(\Phi_e^i)\sigma_x^i) - \sum_{i<j} \frac{E_J(\Phi_e^i)E_J(\Phi_e^j)}{E_L}\sigma_y^i\sigma_y^j, \qquad (7.114)$$

where we collect all parameters before σ_z in $\varepsilon(V_g^i)$ for simplicity. If we let all $\Phi_e^j = \Phi_0/2$ and $n_g^j = 0.5$ when $j \neq i$, the whole system Hamiltonian changes to

$$H = \varepsilon(V_g^i)\sigma_z^i - \frac{1}{2}E_J(\Phi_e^i)\sigma_x^i,$$

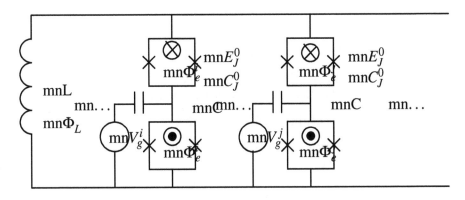

FIGURE 7.17: An improved design to couple charge qubits with inductance. The top and bottom magnetic fluxes piercing through each of the two SQUIDs are designed to have the same amplitude but different directions. Similar to the design in Fig.7.16, the JJ term is tunable through the magnetic fluxes, and the interaction term now has the form of $\sigma_x^i \sigma_x^j$, which is more preferable.

and all the other terms are turned off. We can perform any single qubit operation through the approximation offered by (7.56).

Similarly, a two qubit operation between qubits i and j can be performed by turning off all $E_J^k(\Phi_e^k)$ and n_g^k except qubits i and j. By doing this, now the Hamiltonian becomes

$$H = \varepsilon(V_g^i)\sigma_z^i + \varepsilon(V_g^j)\sigma_z^j - \frac{1}{2}E_J(\Phi_e^i)\sigma_x^i - \frac{1}{2}E_J(\Phi_e^j)\sigma_x^j + \Pi_{ij}\sigma_y^i\sigma_y^j.$$

If we also move the two qubits to their degenerate state, i.e., $n_g^i = n_g^j = 0.5$, the Hamiltonian is simplified to

$$H = -\frac{1}{2}E_J(\Phi_e^i)\sigma_x^i - \frac{1}{2}E_J(\Phi_e^j)\sigma_x^j + \Pi_{ij}\sigma_y^i\sigma_y^j.$$

Because σ_x does not commute with σ_y, the computation of the evolution matrix is tedious and the design of the CNOT gate and conditional phase change gate is complicated. Although it provides a mechanism to realize any qubit gates in combination with one qubit gates, more simplification is helpful.

You et al. [50] improved this design further and obtained a simpler pulse sequence for two-qubit operations. In fact, the conditional phase gate can be achieved with just one two-qubit pulse combined with several one-qubit operations, leading to a much more efficient scheme. The improved design has two dc-SQUIDs instead of one, see Fig. 7.17. Similar to the previous design, the

JJ term is tunable through the magnetic field:

$$H_J^i = -E_J^i(\Phi_e^i)(\cos(\phi_A^i) + \cos(\phi_B^i)),$$

where ϕ_A^i and ϕ_B^i are the effective phase drops of the top and bottom SQUIDs, respectively, in Fig. 7.17. The new effective junction energy is given by

$$E_J^i(\Phi_e^i) = 2E_J^0 \cos(\pi\Phi_e^i/\Phi_0)$$

as previously. The inductance couples all qubits and the whole system Hamiltonian of n qubits now is

$$H = \sum_{k=1}^{n} H_k + \frac{1}{2}LI^2,$$

where $H_k = E_C^k(\hat{n}_k - n_{gk})^2 - E_J^k(\Phi_e^k)(\cos(\phi_A^i) + \cos(\phi_B^i))$, E_C^k is the Coulomb energy of qubit k, and I is the persistent current through the superconducting inductance. Written in Pauli matrices form, the new overall Hamiltonian is

$$H = \sum_{k=i,j} [\varepsilon_k(V_g^k)\sigma_z^k - \bar{E}_J^k(\Phi_e^k,\Phi_L,L)\sigma_x^k] + \Pi_{ij}\sigma_x^i\sigma_x^j. \qquad (7.115)$$

The $\sigma_x^i\sigma_x^j$ forms the interaction term which brings the advantage that it commutes with the Josephson term, and we will show later that it make the two-qubit gate design much more straightforward and simple. Also, note that the effective junction energy $\bar{E}_J^k$ in (7.115) is not the same as the E_J in (7.114) and also depends on the inductance L and its magnetic flux Φ_L, although it is still tunable through Φ_e^k. Similarly, the interaction coefficients Π_{ij} are also functions of Φ_L, Φ_e^i and Φ_e^j. Thus all terms are *switchable*.

By setting $\Phi_e^k = \frac{1}{2}\Phi_0$ and $n_g^k = 0.5$ for all qubits, we can let all terms vanish and obtain $H = 0$. The system state will not change. If we need to perform an operation on qubit i, we change the corresponding Φ_e^i from $\frac{1}{2}\Phi_0$ and n_g^i from 0.5, and then the Hamiltonian becomes

$$H = \varepsilon_i(V_g^i)\sigma_z^i - \bar{E}_J^i(\Phi_e^i,\Phi_L,L)\sigma_x^i.$$

Because both n_g^i and Φ_e^i can be tuned separately, the 1-qubit operators $e^{i\alpha\sigma_z^i}$ and $e^{i\beta\sigma_x^i}$ can be obtained easily by choosing $\bar{E}_J^i = 0$ or $\varepsilon_i(V_g^i) = 0$, with an appropriate time duration. Any other one-qubit operations can be constructed by combining these two operators.

Two-qubit operations can now be performed by tuning Φ_e^i and Φ_e^j away from $\Phi_0/2$. Then the Hamiltonian becomes

$$H = -\bar{E}_J^i\sigma_x^i - \bar{E}_J^j\sigma_x^j + \Pi_{ij}\sigma_x^i\sigma_x^j. \qquad (7.116)$$

THEOREM 7.2

For the Hamiltonian (7.116) with tunable coefficients $\bar{E}_J^i$, $\bar{E}_J^j$ and Π_{ij}, we can construct the two-bit quantum phase gate Q_π and the CNOT gate U_{CNOT} in conjunction with one-bit Rabi gate $U_{\theta,\phi}$ (cf. (7.66), as warranted by Corollary 7.1), where

$$Q_\pi = \begin{bmatrix} 1 & 0 & 0 & 0 \\ 0 & 1 & 0 & 0 \\ 0 & 0 & 1 & 0 \\ 0 & 0 & 0 & -1 \end{bmatrix}, \quad U_{CNOT} = \begin{bmatrix} 1 & 0 & 0 & 0 \\ 0 & 1 & 0 & 0 \\ 0 & 0 & 0 & 1 \\ 0 & 0 & 1 & 0 \end{bmatrix}. \tag{7.117}$$

PROOF   We choose the control parameters such that $\bar{E}_J^i = \bar{E}_J^j = \Pi_{ij} = \delta$. Then the evolution matrix for the Hamiltonian (7.116) becomes

$$U = e^{-iH\tau/\hbar} = e^{-(i\delta\tau/\hbar)(-\sigma_x^i - \sigma_x^j + \sigma_x^i\sigma_x^j)}. \tag{7.118}$$

It is easy to check that the eigenvalue equations for H now are:

$$\begin{aligned} H|++\rangle &= -\delta|++\rangle, \quad H|+-\rangle = -\delta|+-\rangle, \\ H|-+\rangle &= -\delta|-+\rangle, \quad H|--\rangle = 3\delta|--\rangle, \\ (|\pm\rangle &= \tfrac{1}{\sqrt{2}}(|0\rangle \pm |1\rangle)). \end{aligned} \tag{7.119}$$

By choosing $\delta\tau/\hbar = \pi/4$ in (7.118), we see that (7.118) gives the evolution matrix

$$\tilde{U} = e^{i\pi/4}\begin{bmatrix} 1 & & & \\ & 1 & & \\ & & 1 & \\ & & & -1 \end{bmatrix} \tag{7.120}$$

with respect to the ordered basis $\{|++\rangle, |+-\rangle, |-+\rangle, |--\rangle\}$. We can convert the matrix representation (7.120) to a representation with respect to the standard ordered basis $\{|00\rangle, |01\rangle, |10\rangle, |11\rangle\}$ by

$$Q_\pi = H_i^\dagger H_j^\dagger \tilde{U} H_i H_j,$$

where H_i and H_j are, respectively, the Walsh–Hadamard gate for the i-th and j-th qubit. Since the Walsh–Hadamard gate satisfies

$$H_i = H_j = \frac{1}{\sqrt{2}} \begin{bmatrix} 1 & 1 \\ 1 & -1 \end{bmatrix} = e^{-i\pi/2} R_{y,\pi/2} R_{z,\pi}, \qquad (7.121)$$

we have obtained Q_π as promised.

From Q_π, we have

$$U_{CNOT} = U_{\pi/4,\pi/2}^2 \tilde{U} U_{\pi/4,-\pi/2}^2, \qquad (7.122)$$

we also have the CNOT-gate. ∎

Corollary 6.5 Superconducting one-bit gates $U_{\theta,\phi}$ obtained in Corollary 7.1 together with two-bit gates Q_π or U_{CNOT} obtained in Theorem 7.2 are universal.

PROOF   This is a consequence of a result of J. Brylinski and R. Brylinski [6]. ∎

7.6.5.2   Flux qubit case

Two flux qubits can be coupled with a direct inductance coupling or LC circuit [22, 25, 54]. In Fig. 7.18, a closed superconducting loop couples two flux qubits through the mutual inductance M. The coupled Hamiltonian of this two qubit system is then

$$H = H^A(\tilde{f}_1^A) + H^B(\tilde{f}_1^B) + M I_1^A I_1^B, \qquad (7.123)$$

where I_1^A and I_1^B are the persistent currents circulating in the bottom loops of qubit A and qubit B, respectively, and H_1^A and H_1^B are the single qubit Hamiltonian of qubit A and qubit B, respectively, in the form of (7.113), while f_1^A is replaced by $\tilde{f}_1^A = f_1^A + M I^B / \Phi_0$ and f_1^B is replaced by $\tilde{f}_1^B = f_1^B + M I^A / \Phi_0$. The coupling is very weak, and the induced magnetic field from the current of the other qubit is much smaller than the external magnetic field applied through the qubit itself. Put in Pauli matrices, the interaction Hamiltonian has the form of

$$H^{int} = k_1 \sigma_z^A \sigma_z^B + k_2 \sigma_x^A \sigma_z^B + k_3 \sigma_z^A \sigma_x^B, \qquad (7.124)$$

where k_1, k_2, and k_3 are some constants. The coupling can be turned off by inserting another DC-SQUID in the coupling loop.

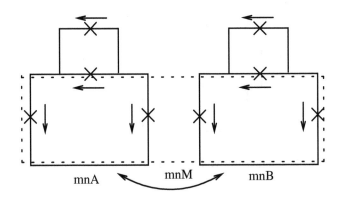

FIGURE 7.18: Schematics of two flux qubits coupled with mutual inductance.

The coupling between the two qubits results in different energy split for qubit B when qubit A is different. Let E_{00}, E_{01}, E_{10}, and E_{11} be the energy levels of state $|00\rangle$, $|01\rangle$, $|10\rangle$, and $|11\rangle$, respectively, and let $\delta E_0^B = E_{01} - E_{00}$ and $\delta E_1^B = E_{11} - E_{10}$. Then $\delta E_0^B \neq \delta E_1^B$ when the coupling is turned on. A laser field resonant to δE_1^B applied on qubit B causes a Rabi rotation only when qubit A is in state $|1\rangle$. A controlled-NOT gate is then achieved when the time duration of the pulse is properly controlled such that the state of qubit B is flipped.

Two flux qubits can also be coupled by an LC circuit as in Fig. 7.19. The interaction Hamiltonian is given by

$$H^{int} = \varepsilon_{AB}\sigma_y^A\sigma_y^B, \tag{7.125}$$

where ε_{AB} represents the coupling strength. This is called a $\sigma_y\sigma_y$ coupling. The coupling strength is a function of the mutual inductance and the tunneling amplitude, among other parameters, and the coupling can be turned off by adjusting these parameters.

To check the evolution of the system with this coupling, we investigate a general Hamiltonian with $\sigma_y\sigma_y$ coupling:

$$\begin{aligned} H &= H^A + H^B - \varepsilon\sigma_y^A\sigma_y^B \\ &= \tfrac{E}{2}\sigma_z^A + \tfrac{E}{2}\sigma_z^B - \varepsilon_{AB}\sigma_y^A\sigma_y^B. \end{aligned} \tag{7.126}$$

We assume that the two qubits are symmetric, and

$$H^A = \tfrac{E}{2}\sigma_A$$
$$H^B = \tfrac{E}{2}\sigma_B.$$

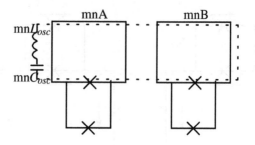

FIGURE 7.19: Schematics of two flux qubits coupled with a LC oscillator.

LEMMA 7.4

The evolution matrix of the system with the above Hamiltonian within time duration τ, which is a unitary gate on the two qubit system, is imprimitive [6] when τ is properly chosen.

PROOF   As before, we transform the Hamiltonian into the interaction picture by defining a new wave function $|\psi(t)\rangle = e^{i\omega(\sigma_z^A + \sigma_z^B)t/2}|\xi(t)\rangle$, where $|\xi(t)\rangle$ is the original wave function of the two qubit system. Thus, the Schrödinger equation can be derived as

$$i\hbar \tfrac{\partial}{\partial t}|\psi(t)\rangle = ((\tfrac{E}{2} - \tfrac{\hbar\omega}{2})(\sigma_z^A + \sigma_z^B)$$
$$-\varepsilon_{AB}(\sigma_x^A \sin(\omega t) + \sigma_y^A \cos(\omega t)) \qquad (7.127)$$
$$(\sigma_x^B \sin(\omega t) + \sigma_y^B \cos(\omega t)))|\psi(t)\rangle.$$

We choose $\hbar\omega = E$, and the above Schrödinger equation can be simplified to

$$i\hbar \tfrac{\partial}{\partial t}|\psi(t)\rangle = -\varepsilon_{AB}(\tfrac{1-\cos 2\omega t}{2}\sigma_x^A \sigma_x^B + \tfrac{1+\cos 2\omega t}{2}\sigma_y^A \sigma_y^B$$
$$+\tfrac{\sin 2\omega t}{2}(\sigma_x^A \sigma_y^B + \sigma_y^A \sigma_x^B))|\psi(t)\rangle. \qquad (7.128)$$

Normally, the interaction is very weak, and ε_{AB} is much smaller than E. Thus we can drop the high frequency terms and further simplify (7.128) into

$$\hbar i \frac{\partial}{\partial t}|\psi(t)\rangle = -\varepsilon_{AB}\frac{1}{2}(\sigma_x^A \sigma_x^B + \sigma_y^A \sigma_y^B)|\psi(t)\rangle. \qquad (7.129)$$

Put in matrix form, the associated Hamiltonian is

$$\mathcal{H} = -\varepsilon_{AB} \begin{pmatrix} 0 & 0 & 0 & 0 \\ 0 & 0 & 1 & 0 \\ 0 & 1 & 0 & 0 \\ 0 & 0 & 0 & 0 \end{pmatrix}. \tag{7.130}$$

The evolution matrix after time duration τ can be obtained by straight-forward computation:

$$e^{-i\mathcal{H}\tau/\hbar} = \begin{pmatrix} 1 & 0 & 0 & 0 \\ 0 & \cos(\varepsilon_{AB}\tau/\hbar) & i\sin(\varepsilon_{AB}\tau/\hbar) & 0 \\ 0 & i\sin(\varepsilon_{AB}\tau/\hbar) & \cos(\varepsilon_{AB}\tau/\hbar & 0 \\ 0 & 0 & 0 & 1 \end{pmatrix}. \tag{7.131}$$

When $\varepsilon_{AB}\tau/\hbar = \pi/2$, we obtain a gate similar to the SWAP gate:

$$U_1 = \begin{pmatrix} 1 & 0 & 0 & 0 \\ 0 & 0 & i & 0 \\ 0 & i & 0 & 0 \\ 0 & 0 & 0 & 1 \end{pmatrix}. \tag{7.132}$$

When $\varepsilon_{AB}\tau/\hbar = \pi/4$, its square root is obtained as

$$U_1 = \begin{pmatrix} 1 & 0 & 0 & 0 \\ 0 & \frac{\sqrt{2}}{2} & i\frac{\sqrt{2}}{2} & 0 \\ 0 & i\frac{\sqrt{2}}{2} & \frac{\sqrt{2}}{2} & 0 \\ 0 & 0 & 0 & 1 \end{pmatrix}. \tag{7.133}$$

This unitary gate is imprimitive. It maps a factorizable state $|01\rangle$ to $\frac{\sqrt{2}}{2}(|01\rangle + i|10\rangle)$, which is nonfactorizable. ∎

7.6.6   Measurement of charge qubit

The energy level of the first excited state $|1\rangle$ changes with the offset charge; when it is higher than the superconducting gap, the Cooper pair is broken apart into two quasi-particles. In Fig. 7.6, a read pulse applied on the probe gate will break the pair and let them tunnel through the junction. Repeating the experiment and measurement at frequency v and assuming that the probability of observing the qubit at state $|1\rangle$ is P_1, we can obtain a classical current through the probe gate which is proportional to P_1:

$$I = 2eP_1 v.$$

This measurement is *destructive*. Although state $|0\rangle$ is kept unchanged, state $|1\rangle$ is destroyed after measurement. Nakamura has used this method to observe the coherence in a SCB and quantum oscillation in two coupled charge qubits [28, 29, 33].

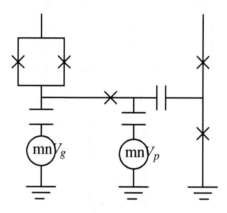

FIGURE 7.20: Schematic of a circuit for measuring a charge qubit using low frequency SET. The charge qubit is coupled capacitively to an SET through a charge trap which is connected to the Cooper pair box with a tunnel junction. To reduce dissipation, the junction has high resistance. The SET is in Coulomb blockade state and there is no current through the junctions when there is no charge in the trap. When a read pulse moves extra charges from the charge qubit to the trap, the SET is biased and a current is observed through the SET.

The above method is easy to apply, but it requires many repeated experiments and measurements. A single shot measurement requires only one measurement and would save much time. One example was realized by a group in Japan [2] using single electron transistor (SET), a sensitive electrometer, and similar setups were also investigated by other groups. See Fig. 7.20. When an appropriate pulse V_p is applied to the probe gate, such as mentioned in the preceding paragraph, the extra Cooper pair in the box is broken into two quasi-particles and tunnels into the trap. If the box is originally in state $|0\rangle$,

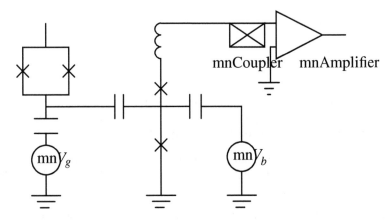

FIGURE 7.21: Schematic of circuit for measuring a charge qubit using rf-SET. Different from the low frequency SET where it is the current from the source to the drain to be measured, the rf-SET measures the conductance and this makes it faster and more sensitive. A radio frequency (rf) signal resonant to the SET, referred to as a "carrier" but not shown in this figure, is launched toward the SET though the coupler. Then a conductance change of the SET due to the extra charge in the charge qubit results in the change of the damping of the SET circuit, and it is reflected in the output of the amplifier.

no electron will tunnel through the junction. Then the extra charge in the trap may be detected by the SET. This completes the measurement. During normal operations, the trap junction is kept unbiased and the charge qubit is isolated from the trap and SET.

The qubit may be coupled to the SET directly through a capacitor without the trap and junction, but this may induce more decoherence to the qubit. The above low frequency SET can be replaced by an rf-SET [1, 38], a more sensitive and fast electrometer, see Fig. 7.21. Different from the low frequency SET where it is the current from the source to the drain to be measured, the rf-SET measures the conductance.

There is a worrisome aspect of measurement due to the effects of noise in hight T_c superconductors as Kish and Svedlindh [20] and others have reported excessively strong magnetic and conductance noise on such superconductors.

References

[1] A. Aassime, G. Johansson, G. Wendin, R.J. Schoelkopf, and P. Delsing, Radio-frequency single-electron transistor as readout device for qubits: charge sensitivity and backaction, *Phys. Rev. Lett.* **86** (2001), 3376.

[2] O. Astafiev, Y.A. Pashkin, T. Yamamoto, Y. Nakamura, and J.S. Tsai, Single-shot measurement of Josephson charge qubit, *Phys. Rev. B* **69** (2004), 180507.

[3] J. Bardeen, L.N. Cooper and J.R. Schrieffer, Theory of Superconductivity, *Phys. Rev.* **108** (1957), 1175.

[4] F.J. Blatt, *Modern Physics*, McGraw–Hill, New York, 1992.

[5] V. Bouchiat, D. Vion, P. Joyez, D. Esteve and M.H. Devoret, Quantum coherence with a single Cooper pair, *Physica Scripta* **T76** (1998), 165–170.

[6] J.L. Brylinski and R. Brylinski, Universal quantum gate, in *Mathematics of Quantum Computation*, R. Brylinski and G. Chen, (eds.), Chapman & Hall/CRC, Boca Raton, Florida, 2002, 101–116.

[7] G. Chen, D.A. Church, B.G. Englert, C. Henkel, B. Rohwedder, M.O. Scully, and M.S. Zubairy, *Quantum Computing Devices: Principles, Designs, and Analysis*, Chapman & Hall/CRC, Boca Raton, Florida, 2006.

[8] C. Cosmelli, P. Carelli, M.G. Castellano, F. Chiarello, Palazzi G. Diambrini, R. Leoni, and G. Torrioli, *Phys. Rev. Lett.* **82** (1999), 5357.

[9] E. Collin, G. Ithier, A. Aassime, P. Joyez, D. Vion, and D. Esteve, NMR-like control of a quantum bit superconducting circuit, *Phys. Rev. Lett.* **93** (2004), 157005.

[10] M.H. Devoret and J.M. Martinis, Implementing qubits with superconducting circuits, *Quantum Information Processing* **3** (2004), 163–203.

[11] M.H. Devoret, A. Wallraff, and J.M. Martinis, Superconducting qubits: a short review, (2004); cond-mat/0411174.

[12] D. Esteve and D. Vion, Solid state quantum bit circuits, *Les Houches Summer School-Session LXXXI on Nanoscopic Quantum Physics*(2004), Elsevier Science, Amsterdam, 2005.

[13] J.R. Friedman, V. Patel, W. Chen, S.K. Tolpygo, and J.E. Lukens, *Nature* **406** (2000), 43.

[14] V.L. Ginzburg and L.D. Landau, On the theory of superconductivity, *Zh. Eksperim. i. Teor. Fiz.* **20** (1950), 1064.

[15] http://ffden-2.phys.uaf.edu/113.web.stuff/Travis/what_is.html

[16] http://www.superconductors.org/Type1.htm

[17] http://hyperphysics.phy-astr.gsu.edu/hbase/solids/scond.html#c4

[18] A. Izmalkov, M. Grajcar, E. Il'ichiv, Th. Wagner, H. Meyer, A. Smirnov, M. Amin, A. Brink, and A. Zagoskin, *Phys. Rev. Lett* **93** (2004), 037003.

[19] J.B. Ketterson and S.N. Song, *Superconductivity*, Cambridge University Press, Cambridge, U.K., 1999.

[20] L.B. Kish and P. Svedlindh, Noise in hight T_c superconductors, *IEEE Trans. Electron Devices* **41** (1994), 2112–2122.

[21] J. Majer, F. Paauw, A. Haar, C. Harmans, and J. Mooij, *Phys. Rev. Lett* **94** (2005), 090501.

[22] Y. Makhin, G. Schön, and A. Shnirman, Josephson-junction qubits with controlled coupling, *Nature* **398** (Mar. 1999), 305–307.

[23] J.M. Martinis, M.H. Devoret, and J. Clark, *Phys. Rev. B* **35** (1987), 4682.

[24] J.M. Martinis, S. Nam, and J. Aumentado, Rabi oscillations in a large Josephson-junction qubit, *Phys. Rev. Lett.* **89** (2002), 117901.

[25] J. Mooij, T. Orlando, L. Levitov, L. Tian, C. Wal, and S. Lloyd, *Science* **285** (1999), 1036–1039.

[26] M. Tinkham, *Introduction to Superconductivity*, McGraw-Hill, New York, 1996.

[27] Y. Nakamura, Y.A. Pashkin, and J.S. Tsai, Coherent control of macroscopic quantum states in a single-Cooper-pair box, *Nature* **398** (1999), 786–788.

[28] Y. Nakamura, Y.A. Pashkin, and J.S. Tsai, Quantum coherence in a single-Cooper-pair box: experiments in the frequency and time domains, *Physica B* **280** (2000), 405–409.

[29] Y. Nakamura and J.S. Tsai, Quantum-state control with a single-Cooper-pair box, *J. of Low Temperature Physics* **118** (2000), 765–779.

[30] J. Normand, *A Lie Group: Rotations in Quantum Mechanics*, North-Holland, New York, 1980.

[31] T. Orlando, J. Mooij, L. Tian, C. Wal, L. Levitov, S. Lloyd, and J. Mazo, *Phys. Rev. B* **60** (1999), 15398.

[32] http://www.ornl.gov/info/reports/m/ornlm3063r1/pt2.html

[33] Y. Pashkin, T. Yamamoto, O. Astaflev, Y. Nakamura, D.V. Averin, and J.S. Tsai, Quantum oscillations in two coupled charge qubits, *Nature* **421** (Feb. 2003), 823–826.

[34] C.P. Poole, H.A. Farach, and R.J. Creswick, *Superconductivity*, Academic Press, New York, 1996.

[35] A.C. Rose-Innes and E.H. Rhoderick, *Introduction to Superconductivity*, Second Edition. Pergamon Press, Oxford, 1978.

[36] R. Rouse, S. Han, and J.S. Lukens, *Phys. Rev. Lett.* **75** (1995), 514.

[37] S. Saito, M. Thorwart, H. Tanaka, M. Ueda, H. Nakano, K. Semba, and H. Takayan, Multiphoton transitions in a macroscopic quantum two-state system, *Phys. Rev. Lett.* **93** (2004), 037001.

[38] R.J. Schoelkof, P. Wahlgren, A.A. Kozhevnikov, P. Delsing, and D.E. Prober, The radio-frequency single-electron transistor(RF-SET): a fast and ultrasensitive electrometer, *Science* **280** (May 1998), 1238–1242.

[39] P. Silvestrini, V.G. Palmieri, B. Ruggiero, and M. Russo, *Phys. Rev. Lett.* **79** (1997), 3046.

[40] R.W. Simmonds, K.M. Lang, D.A. Hite, S. Nam, D.P. Pappas, and J.M. Martinis, Decoherence in Josephson junction phase qubits from junction resonators, *Phys. Rev. Lett.* **93** (2004), 077003.

[41] A. Shnirman, G. Schön, and Z. Hermon, Quantum manipulation of small Josephson junctions, *Phys. Rev. Lett* **79** (1997), 2317.

[42] M. Tinkham, *Introduction To Superconductivity*, McGraw-Hill, Singapore, 1996.

[43] D. Vion, A. Aassime, A. Cottet, P. Joyez, H. Pothier, C. Urbina, D. Esteve, and M.H. Devoret, Manipulating the quantum state of an electrical circuit, *Science* **296** (2002), 886–889.

[44] R.F. Voss and R.A. Webb, *Phys. Rev. Lett.* **47** (1981), 265.

[45] Caspar H. van der Wal, A.C.J. ter Haar, F.K. Wilhelm, R.N. Schouten, C.J.P.M. Harmans, T.P. Orlando, Seth Lloyd, J.E. Mooij, Quantum superposition of macroscopic persistent-current states, *Science* **290** (2000), 773–776.

[46] G. Wendin, Scalable solid state qubits: challenging decoherence and readout, *Phil. Trans. R. Soc. Lond. A* **361** (2003), 1323.

[47] G. Wendin, Superconducting quantum computing, *Physics World*, May 2003.

[48] G. Wendin and V.S. Shumeiko, Superconducting quantum circuits, qubits and computing, arXiv:cond-mat/0508729v1, Aug. 30, 2005.

[49] T. Yamamoto, Y.A. Pashkin, O. Astaflev, Y. Nakamura, and J.S. Tsai, Demonstration of conditional gate operation using superconducting charge qubits, *Nature* **425** (Oct. 2003), 941–944.

[50] J.Q. You, J.S. Tsai, and F. Nori, Scalable quantum computing with Josephson charge qubits, *Phys. Rev. Lett.* **89** (2002), 197902.

[51] J.Q. You, J.S. Tsai, and F. Nori, Controllable manipulation and entanglement of macroscopic quantum states in coupled charge qubits, *Phys. Rev. B* **68** (2003), 024510.

[52] J.Q. You and F. Nori, Quantum information processing with superconducting qubits in a microwave field, *Phys. Rev. B* **68** (2003), 064509.

[53] J.Q. You and F. Nori, Superconducting circuits and quantum information, *Physics Today* (Nov. 2005), 42–47.

[54] J.Q. You, Y. Nakamura, and F. Nori, Fast two-bit operations in inductively coupled flux qubits, *Phys. Rev. B* **71** (2005), 024532.

[55] Y. Yu, S. Han, X. Chu, S. Chu, and Z. Wang, Coherent temporal oscillations of macroscopic quantum states in a Josephson junction, *Science* **296** (2002), 889–892.

Chapter 8

Nondeterministic Logic Gates in Optical Quantum Computing

Federico M. Spedalieri, Jonathan P. Dowling, and Hwang Lee

Abstract   We present a detailed description of nondeterministic quantum logic gates: the crucial component of linear optical quantum computing. We use the qubits that are encoded in the polarization degrees of freedom, as opposed to the original dual-rail encoding. Employing the polarization-encoding scheme, we can avoid the undetected errors due to the inefficiency of the photodetectors and, therefore, achieve high gate fidelity, independent of the quantum efficiency of the detectors. Furthermore, we show that such a high-fidelity gate operation can be performed with no need of number-resolving detectors.

8.1   Introduction

Suppose we have a calculator that gives correct answers from time to time, but not always. Generally, it can hardly be considered as a useful device. However, it will be a fine machine if it says "yes" and stops whenever the output is correct—when not correct, it should say "no" and repeat the calculation. Of course, it would be perfect if it computes fast and its saying yes happens more frequently than not. Nondeterministic logic gates are supposed to do a similar job at the individual gate-operation level. In optical quantum computing, due to the lack of efficiency in the photon-photon interaction, the required two-qubit logic gates rely on the *effective nonlinearity* based on photodetection [1, 2]. In

223

doing so, the gate operation is not always successful—there must always be a nonvanishing probability of an unwanted outcome of photodetection. If there were no such unwanted outcome, we would not have the detection! The good news is that we would know if it is successful whenever the outcome is the desired one—if not, abort and repeat.

When the gate operation is probabilistic as described above, however, the number of gates must increase to counter the errors, resulting in an exponential slowdown as the size of computation increases. Therefore, the probability of success of the individual gate operation must be boosted, up to the point where the failure gets below the error threshold required by the fault-tolerant computation. In essence for the linear optical quantum computing proposed by Knill, Laflamme, and Miburn (KLM) [1], the boost of the success probability is made by adding more ancilla qubits. For example, the probability of success of the two-qubit CNOT gate is given by $n^2/(n+1)^2$, where n is the number of ancilla qubits. Thereby these are called "near-deterministic" gates. What is the price to pay? It is that the ancilla qubits are to be entangled in a certain form [3]. The KLM scheme utilizes the gate-teleportation technique, first suggested by Gottesman and Chuang [4], in order that the ancilla states are prepared outside the main stream of the gate operations—by the so-called off-line production.

On the other hand, the high probability of success of the nondeterministic gates basically assumes perfect photodetection with unit efficiency and the number-resolving capability. The KLM scheme, in its basic form, is extremely fragile against errors in photodetection. This is not manifested in a way that the machine says "yes" less frequently. Rather, the machine would say "yes, go ahead," but it was a wrong answer and we would never know it was wrong. Thereby there are errors that are not detected. So the output of the computation quickly becomes unreliable as the quality of the photodetectors is far off being "ideal". The tolerable efficiency of the detector required for reliable LOQC is above 99%, which is far beyond what is currently available.

In a recent article, we have proposed a polarization encoding for the KLM scheme [5], as opposed to its original dual-rail encoding. The aim of the polarization encoding is to prevent the *undetected errors* associated with the imperfect photodetection. As a result, it requires twice more ancilla photons and naturally twice more photodetectors than the dual-rail encoding. Using the polarization encoding, whereas the fidelity of the implemented gate is increased (by avoiding the undetected errors), the probability of having successful gate operations decreases when non-ideal detectors are employed (by the increase of the number of detectors).

Sacrificing the success probability of gate operation is bad news. However, as opposed to the usual quantum circuit model, there is an alternative way of doing quantum computation, proposed by Raussendorf and Briegel [6], the so-called cluster-state approach. In the cluster-state model, the computation is carried out by a series of single-qubit measurements on a set of highly entangled qubits, the cluster states. The construction of optical cluster states using the nondeterministic two-qubit gates was proposed by Nielsen [7]. In essence, lower probability of success can be tolerated while still being able to construct the desired cluster states. Hence, with its high-fidelity gate operation the polarization encoding scheme can be used to construct high-fidelity optical cluster states.

Here we describe the basic ideas of LOQC using the polarization encoding scheme and its advantages in optical cluster-state quantum computing. Sections to follow consist of Section 8.2 the representation of the qubit in the optical quantum computing, Section 8.3 non-deterministic quantum teleportation, Sections 8.4 and 8.5 non-deterministic two-qubit gate operation and construction of the ancilla states, and Section 8.6 the application to optical cluster states, the issue of gate fidelity, and the optimal ancilla state. Section 8.7 contains the conclusions and Appendices A and B are devoted to the descriptions of basic single-quit gates and the nonlinear sign gate.

8.2 Photon as a qubit

Typically, in the optical approach to quantum computation, the states of polarization of a single photon are used to define the qubit. In this polarization encoding the logical qubit $|0\rangle$ is represented, for example, by a single photon with the horizontal polarization $|H\rangle$ and $|1\rangle$ by one with the vertical polarization $|V\rangle$.

$$|0\rangle \equiv |H\rangle, \qquad |1\rangle \equiv |V\rangle. \tag{8.1}$$

Another convenient choice of the logical qubit can be the utilization of two physical paths containing a single photon:

$$|0\rangle \equiv |1\rangle_1 |0\rangle_2$$
$$|1\rangle \equiv |0\rangle_1 |1\rangle_2, \tag{8.2}$$

where the subscripts $1, 2$ represent the relevant two modes. The interchange between the polarization qubit to this so-called dual-rail qubit can be simply made by polarizing beam splitters followed by a polarization rotation. The polarizing beam splitters (oriented in a certain direction, i.e., H-V basis) always transmit horizontally polarized light (H-photons) and reflect vertically polarized light (V-photons), and therefore allow the following transformation:

$$|H\rangle \rightarrow |0\rangle_1 |1\rangle_2,$$
$$|V\rangle \rightarrow |1\rangle_1 |0\rangle_2, \tag{8.3}$$

with the additional polarization rotation (H↔V) in one of the two paths.

In order to do universal quantum computation, it is required to have the ability to perform all one-qubit gates and the two-qubit CNOT gate, based on the universality theorem that all unitary operations on arbitrarily many qubits can be made out of a combination of these gates [8]. Any one-qubit logic operation can be achieved by the manipulation of the polarization of a single photon using simple linear optical devices such as beam splitters and phase shifters.

Normally a beam splitter is described as follows: Taking $\hat{a}$ and $\hat{b}$ as the input modes, the output modes is given by

$$\begin{pmatrix} \hat{a}' \\ \hat{b}' \end{pmatrix} = \begin{pmatrix} r & t \\ t & r \end{pmatrix} \begin{pmatrix} \hat{a} \\ \hat{b} \end{pmatrix} \tag{8.4}$$

where for lossless beam splitters the reflection coefficient r and the transmission coefficient t need to satisfy $|r|^2 + |t|^2 = 1$, and $rt^* + tr^* = 0$. We may define the beam splitter as

$$\begin{pmatrix} \hat{a}' \\ \hat{b}' \end{pmatrix} = \begin{pmatrix} \cos\frac{\theta}{2} & i\sin\frac{\theta}{2} \\ i\sin\frac{\theta}{2} & \cos\frac{\theta}{2} \end{pmatrix} \begin{pmatrix} \hat{a} \\ \hat{b} \end{pmatrix}. \tag{8.5}$$

Polarizing beam splitters, however, allow complete transmission for one polarization and total reflection for the other orthogonal polarization. The effect of the phase shifters on the mode operators is described as $\hat{a}' = e^{-i\hat{a}^\dagger \hat{a}\phi}\hat{a}$. Physical implementations of the various one-qubit gates necessary for optical quantum computing are described in Appendix A.

The two-qubit gate, on the other hand, requires certain interactions between the photons, which can be envisioned through nonlinear optical processes as the Kerr effect [9]. However, the fact that the efficiency of these nonlinear interactions is very small at the single-photon level poses the main difficulty in optical quantum computing. The idea of linear-optical quantum computing is to replace the requirement of nonlinear interaction with corrections to the output of gate operations (made of linear-optical devices) based on the results of single-photon detectors.

A typical two-qubit gate, controlled-NOT (CNOT) is represented as follows:

$$
\begin{aligned}
|H\rangle|H\rangle &\rightarrow |H\rangle|H\rangle \\
|H\rangle|V\rangle &\rightarrow |H\rangle|V\rangle \\
|V\rangle|H\rangle &\rightarrow |V\rangle|V\rangle \\
|V\rangle|V\rangle &\rightarrow |V\rangle|H\rangle,
\end{aligned}
\tag{8.6}
$$

where the second qubit (target) flips when the first qubit (control) is in the state $|V\rangle$ (the logical value 1), yielding the controlled σ_x operation. The CNOT gate plays an important role in that any n-qubit gates can be decomposed into the CNOT gates and one-qubit gates and thus form a universal set of gates.

Another frequently used two-qubit gate is the conditional sign-flip gate (C-SIGN). The CSIGN gate is equivalent to CNOT gate in that they can be transformed to each other by using one-qubit gates only. The transformation by the CSIGN gate is written as

$$
\begin{aligned}
|H\rangle|H\rangle &\rightarrow |H\rangle|H\rangle \\
|H\rangle|V\rangle &\rightarrow |H\rangle|V\rangle \\
|V\rangle|H\rangle &\rightarrow |V\rangle|H\rangle \\
|V\rangle|V\rangle &\rightarrow -|V\rangle|V\rangle.
\end{aligned}
\tag{8.7}
$$

Note that the CSIGN gate is the controlled σ_z operation so that the CNOT gate is then simply constructed by using CSIGN and two one-qubit gates (e.g., Hadamard on the target, followed by the CSIGN and another Hadamard gate on the target).

The CSIGN gate is perhaps more familiar to the optics community as it can be viewed as a third-order nonlinear optical process. The interaction caused by the Kerr nonlinearity can be described by a Hamiltonian [10]

$$
\mathscr{H}_{\text{Kerr}} = \hbar\kappa\hat{a}^\dagger\hat{a}\hat{b}^\dagger\hat{b},
\tag{8.8}
$$

where κ is a coupling constant depending on the third-order nonlinear susceptibility, and $\hat{a}^\dagger$, $\hat{b}^\dagger$ and $\hat{a}$, $\hat{b}$ are the creation and annihilation operators for two optical modes.

We assign modes 1, 2 for the control qubit, and 3, 4 for the target qubit and suppose now only the modes 2, 4 are coupled under the interaction given by Eq. (8.8). For a given interaction time τ, the transformation (after we put the dual rail back to the polarization encoding) can be written as

$$
|H\rangle|H\rangle \rightarrow |H\rangle|H\rangle
$$

$$|H\rangle|V\rangle \;\rightarrow\; |H\rangle|V\rangle$$
$$|V\rangle|H\rangle \;\rightarrow\; |V\rangle|H\rangle$$
$$|V\rangle|V\rangle \;\rightarrow\; e^{i\varphi}|V\rangle|V\rangle. \qquad (8.9)$$

where $\varphi \equiv \kappa n_a n_b \tau$ and $n_a = \langle \hat{a}^\dagger \hat{a} \rangle, n_b = \langle \hat{b}^\dagger \hat{b} \rangle$. This operation yields a conditional phase shift. When $\varphi = \pi$, we have the CSIGN gate. In order to have φ of the order of π at the single-photon level, however, a huge third-order nonlinear coupling is required [11]. In the next section we discuss how to avoid such difficulties by using the quantum teleportation technique on the one hand, and effective nonlinearities produced by projective measurements on the other.

8.3 Linear optical quantum computing

The application of quantum gates with linear optical elements is based on an approach developed by Gottesman and Chuang [4] that relies on quantum teleportation. Quantum teleportation transmits information encoded in a qubit to another location without sending the qubit itself. As depicted in Fig. 8.1, this is done by

i) transmitting (via a quantum channel) one of the EPR pair,

ii) Bell-state measurement (BM) of the qubit and the other one of the EPR pair,

iii) transmission (via a classical channel) of the result of the BM, and

iv) making a correction ($\sigma_j, j = 0,1,2,3$, corresponding to the identity operation I, and the three Pauli operations $\sigma_x, i\sigma_y = \sigma_z \sigma_x, \sigma_z$) to the one transmitted in the process (i).

The EPR pair can be any one of the four Bell states ($\phi^\pm, \psi^\pm$), where

$$\phi^{(+)} = (|HH\rangle + |VV\rangle)/\sqrt{2},$$
$$\phi^{(-)} = (|HH\rangle - |VV\rangle)/\sqrt{2},$$
$$\psi^{(+)} = (|HV\rangle + |VH\rangle)/\sqrt{2},$$
$$\psi^{(-)} = (|HV\rangle - |VH\rangle)/\sqrt{2}. \qquad (8.10)$$

Gate teleportation, as shown in Fig. 8.2, is carried out simply by i) applying the desired gate (say U_1) before the transmission of one of the EPR pair and

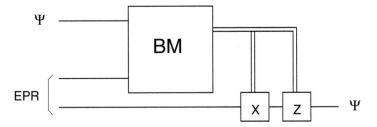

FIGURE 8.1: Quantum teleportation: Ψ is the quantum state of the qubit. EPR stands for the EPR pair, which can be any one of the Bell states ($\phi^{\pm}, \psi^{\pm}$). One of the EPR pair is transmitted to another location and Bell-state measurement (BM) is carried out for the qubit and the other one of the EPR pair. The one of the four possible results of the BM is then transmitted through the classical channel. Finally, the receiver needs to make a correcting operation ($X = \sigma_x$, $Z = \sigma_z$).

ii) making the correction as $U\sigma_j U^{-1}$. The usual quantum teleportation can be thought as a special case in which the unitary applied is the identity operation. Obviously, the gate teleportation is trivial and of no use for a one-qubit gate.

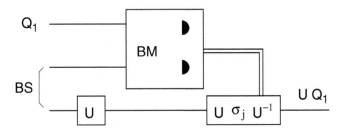

FIGURE 8.2: One-qubit gate teleportation: The desired gate operation (say U_1) is done on one part of the Bell state (BS: the EPR pair) before the transmission to the receiver. The correction operation is then given by $U\sigma_j U^{-1}$.

Generalization to n-qubit gates is straightforward. For two-qubit operations (say U_2), there are two EPR pairs, and the two-qubit gate is applied between the ones from each pair, and two separate BMs are needed. Then, the correcting operation after the two separate BMs becomes $U_2(\sigma_j \otimes \sigma_k)U_2^{-1}$ (see Fig. 8.3). In particular when $U_2 = $ CNOT, the operation CNOT $(\sigma_j \otimes \sigma_k)$ CNOT can be written as direct product of Pauli matrices and the identity matrix (such as $\pm\sigma_l \otimes \sigma_m$). Consequently, we do not have to perform the two-qubit operation at the correcting stage—only one-qubit operations are needed. This property alone does not help much since we would have to apply the two-qubit gate to the ones from each of the EPR pairs.

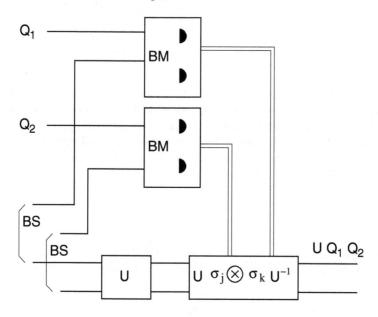

FIGURE 8.3: Two-qubit gate teleportation: There are now two EPR pairs (Bell states, BS), and the two-qubit gate is applied between the ones from each pair, and two separate BMs are needed. Now the correcting operation after the two Bell measurements is given by $U_2(\sigma_j \otimes \sigma_k)U_2^{-1}$.

The big advantage of the gate teleportation appears when the two-qubit gate is not deterministic. When the attempted gate operation is not always successful, it is hardly useful unless its probability of success is high enough within the fault tolerant range. Using the gate teleportation, one can move the desired two-qubit gate to the stage of preparing the auxiliary qubit (namely the EPR pairs here) and avoid applying the probabilistic gates directly on to the main qubits.

The idea of gate teleportation is to replace the usual Bell pair used to teleport the state of a qubit by a different entangled state. However, the teleportation procedure requires that we are able to perform the Bell measurement (a projective measurement on the basis composed by the four Bell states). Roughly speaking, the two-qubit Bell measurement consists of a CNOT gate and Hadamard gate (on one of the two qubits) followed by the measurement in the computational basis. This is again a problem for a linear-optical implementation, since in this case a complete Bell measurement is not possible without the capability of CNOT operation. To circumvent this problem we need to relax the requirement of a complete Bell measurement.

It is well known that linear-optical elements can be used to implement an incomplete Bell measurement. If we use the teleportation scheme of Gottesman and Chuang to apply our quantum gates, these gates will only work half of the time. There is, however, a way to improve the probability of success of the teleportation step. This is accomplished by using a more complex version of the Bell state than employing in the usual qubit teleportation. As we shall see, this allows us to increase the probability of success arbitrarily close to 1.

We will use the notation $|V\rangle^j$ to represent a state of j modes, each one occupied by a vertically polarized photon. For the most part we will omit the explicit numbering of the modes unless it is needed to avoid confusion. For example, $|V\rangle^2|H\rangle$ represents the state $|VV\rangle_{12}|H\rangle_3$. Now consider the state

$$|t_n\rangle = \frac{1}{\sqrt{n+1}} \sum_{j=0}^{n} |V\rangle^j|H\rangle^{n-j}|H\rangle^j|V\rangle^{n-j}. \qquad (8.11)$$

This is a state of $2n$ modes that contains $2n$ photons, exactly one photon per mode (see Fig. 8.4).

To simplify the calculations it is useful to write the state of the modes in terms of creation operators acting on the vacuum state. We will call $a_k^\dagger$ the creation operator of a *vertically* polarized photon in mode k, and $b_k^\dagger$ the creation operator of a *horizontally* polarized photon in mode k. Then we have

$$a_k^\dagger|\mathrm{vac}\rangle = |V\rangle_k$$
$$b_k^\dagger|\mathrm{vac}\rangle = |H\rangle_k, \qquad (8.12)$$

where $|\mathrm{vac}\rangle$ represents the vacuum state. We will write $|\mathrm{vac}\rangle_{1...n}$ to represent the vacuum state of modes 1 to n.

We would like to teleport the state of a qubit that is encoded in a mode of the field that we will call mode 0. This state can be written as

$$|\psi\rangle = \alpha|H\rangle_0 + \beta|V\rangle_0 = (\alpha b_0^\dagger + \beta a_0^\dagger)|\mathrm{vac}\rangle_0. \qquad (8.13)$$

The first step consists in applying a Fourier transform to the set of $n+1$ modes formed by mode 0 and the first n modes of the state (8.11). This Fourier transform has a very simple mathematical expression when given in terms of its action on the creation operators,

$$\hat{F}_n(a_k^\dagger) = \frac{1}{\sqrt{n+1}} \sum_{l_k=0}^{n} \omega^{kl_k} a_{l_k}^\dagger$$

$$\hat{F}_n(b_k^\dagger) = \frac{1}{\sqrt{n+1}} \sum_{l_k=0}^{n} \omega^{kl_k} b_{l_k}^\dagger, \tag{8.14}$$

where $\omega = \exp[2\pi i/(n+1)]$. One very important fact is that this Fourier transform can be implemented with linear optical elements such as mirrors, beam splitters, and phase shifters [13]. Another important point is that this operation does not mix the polarizations of the photon, which can be seen from the E-qs. in (8.14) through the fact that the creation operators for each polarization transform among themselves.

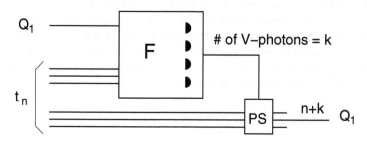

FIGURE 8.4: Nondeterministic quantum teleportation: Improvement of the success probability of the incomplete Bell measurement is achieved by using a more complex version of the Bell state employed in the usual qubit teleportation. This can increase the probability of success arbitrarily close to 1.

We can now rewrite the teleporting state $|t_n\rangle$ using the creation operators for horizontally and vertically polarized photons, and get

$$|t_n\rangle = \frac{1}{\sqrt{n+1}} \sum_{j=0}^{n} a_1^\dagger \dots a_j^\dagger b_{j+1}^\dagger \dots b_n^\dagger b_{n+1}^\dagger \dots b_{n+j}^\dagger a_{n+j+1}^\dagger \dots a_{2n}^\dagger |vac\rangle_{1\dots 2n}.$$

$$\tag{8.15}$$

Before applying the Fourier transform, consider the joint state formed by our qubit in state (8.13) together with the state $|t_n\rangle$. Expanding this, we have

$$|\psi\rangle|t_n\rangle = \frac{1}{\sqrt{n+1}} \sum_{j=0}^{n} \left\{ \alpha b_0^\dagger \left(\prod_{k=1}^{j} a_k^\dagger \right) \left(\prod_{k=1}^{n} b_{j+k}^\dagger \right) \left(\prod_{k=1}^{n-j} a_{n+j+k}^\dagger \right) + \right.$$

$$\left. + \beta a_0^\dagger \left(\prod_{k=1}^{j} a_k^\dagger \right) \left(\prod_{k=1}^{n} b_{j+k}^\dagger \right) \left(\prod_{k=1}^{n-j} a_{n+j+k}^\dagger \right) \right\} |vac\rangle_{0\dots 2n}. \tag{8.16}$$

This is a state of $2n+1$ modes. Note that the difference between the two terms inside the curly brackets is, besides the values of α and β, that the first term has a creation operator for a horizontally polarized photon in mode 0, while the second has a creation operator for a vertically polarized photon in that mode. The next step is to apply the Fourier transform to the first $n+1$ modes (i.e., modes 0 to n). So what we have to do is replace the first $n+1$ creation operators appearing in the two terms in (8.16) by the corresponding sum of operators that can be read from (8.14). Note that, since the two terms have different numbers of creation operators of each type (H or V), and since the Fourier transform does not mix polarizations, the same will hold after the transformation is applied.

The state after the Fourier transform is

$$\Sigma_{j=0}^{n} \left\{ \sum_{0 \leq l_0,\ldots,l_n \leq n} \omega^{\Sigma_{k=0}^{n} k l_k} \left(\alpha b_{l_0}^\dagger a_{l_1}^\dagger \ldots a_{l_j}^\dagger b_{l_{j+1}}^\dagger \ldots b_{l_n}^\dagger + \right. \right.$$

$$\left. \left. + \beta a_{l_0}^\dagger a_{l_1}^\dagger \ldots a_{l_j}^\dagger b_{l_{j+1}}^\dagger \ldots b_{l_n}^\dagger \right) |vac\rangle_{0\ldots n} \right\} \underbrace{|H\rangle^j |V\rangle^{n-j}}_{\text{modes}(n+1,\ldots,2n)} , \quad (8.17)$$

where we have written the state of the last n modes as $|H\rangle^j |V\rangle^{n-j}$ to simplify the expression, and have omitted the normalization factor $(n+1)^{-\frac{n+2}{2}}$. Note that the α terms have j V-photons and $(n-j+1)$ H-photons, while the β terms have $(j+1)$ V-photons and $(n-j)$ H-photons. This difference will be responsible for transferring the superposition that was present in mode 0 to one of the last n modes.

The next step consists in measuring the first $n+1$ modes. The idea of this measurement is to collapse the state (8.17) in such a way that the state of our qubit (originally encoded in mode 0) is transferred to one the last n modes. What we need to do then is to measure how many photons of each polarization are present in each of the first $n+1$ modes. To do this we first need to send each of these modes through a polarization beamsplitter, that sends vertically and horizontally polarized photons through different paths, and then measure the number of photons present using a number-resolving photodetector.

Assume that we have performed this measurement, and we have obtained that in mode j; there are r_j of the V-photons and h_j of the H-photons. Note that since there was one photon per mode in the first $n+1$ modes of (8.16) and the Fourier transform cannot create or destroy photons, the total number of photons measured at the output *must remain* $n+1$. First, consider the two simplest cases. If $\Sigma_{j=0}^{n} r_j = n+1$, then all the photons detected are V-photons (i.e., $h_j = 0$, $\forall j$). Looking at (8.17), we see that the only term that has $n+1$ of the V-photons in the first $n+1$ modes corresponds to the β term with $j = n$. Any other term in (8.17) has at least one H-photon. Then the state corresponding to that measurement result is:

$$|V\rangle^{n+1}|H\rangle^{n}. \tag{8.18}$$

It is clear that in this case the original superposition present in mode 0 has been lost, and after the measurement all modes are left in known states with either a V-photon or an H-photon. In this case the teleportation has failed. This particular outcome occurs with probability $|\beta|^2/(n+1)$. Similarly, if we measure that $\sum_{j=0}^{n} r_j = 0$, that means $\sum_{j=0}^{n} h_j = n+1$, and we can repeat the reasoning above by replacing V-photons with H-photons. So again, the result is a projective measurement that destroys the superposition. The probability of this event occurring is $|\alpha|^2/(n+1)$, and so the total probability of failure of the teleportation is $1/(n+1)$ independent of the input state. Franson and coworkers suggested that the probability of failure can be reduced to the order of $1/n^2$ by tailoring the probability amplitudes of the ancilla state [12].

Let us study the more interesting case in which the measurement result is such that $\sum_{j=0}^{n} r_j \neq n+1, 0$, and write $\sum_{j=0}^{n} r_j = k$. Then we also have $\sum_{j=0}^{n} h_j = n-k+1$, since the total number of photons detected is always $n+1$. The state corresponding to that measurement result is

$$\left\{ \sum_{\mathscr{S}} \omega^{\sum_{p=0}^{n} p l_p} \alpha b_{l_0}^{\dagger} a_{l_1}^{\dagger} \dots a_{l_k}^{\dagger} b_{l_{k+1}}^{\dagger} \dots b_{l_n}^{\dagger} |\text{vac}\rangle_{0\dots n} |H\rangle^{k} |V\rangle^{n-k} + \right.$$

$$\left. + \sum_{\mathscr{S}'} \omega^{\sum_{p=0}^{n} p l_p} \beta a_{l_0}^{\dagger} a_{l_1}^{\dagger} \dots a_{l_{k-1}}^{\dagger} b_{l_k}^{\dagger} \dots b_{l_n}^{\dagger} |\text{vac}\rangle_{0\dots n} |H\rangle^{k-1} |V\rangle^{n-k+1} \right\},$$

$$\tag{8.19}$$

with

$$\mathscr{S} = \left\{ (l_0, \dots, l_n) / \{l_1, \dots, l_k\} \text{ contains the value } j, r_j \text{ times, and} \right.$$

$$\{l_0, l_{k+1}, \dots, l_n\} \text{ contains the value } j, h_j \text{ times, } j \in \{0, \dots, n\} \right\}$$

$$\mathscr{S}' = \left\{ (l_0, \dots, l_n) / \{l_0, \dots, l_{k-1}\} \text{ contains the value } j, r_j \text{ times, and} \right.$$

$$\{l_k, \dots, l_n\} \text{ contains the value } j, h_j \text{ times, } j \in \{0, \dots, n\} \right\}. \tag{8.20}$$

By looking at the two sums in (8.19) we can see that these two terms have the same state for the first $n+1$ modes since they have the same number of V-photons and H-photons in each of the first $n+1$ modes (fixed by the result of the measurement). The only difference is given by the state of the last n modes and by the factors introduced by the two sums

$$\sum_{\mathscr{S}} \omega^{\sum_{p=0}^{n} p l_p} \quad \text{and} \quad \sum_{\mathscr{S}'} \omega^{\sum_{p=0}^{n} p l_p}. \tag{8.21}$$

Since the sums are over two different sets of $(n+1)$-tuples, it is not clear if this will change the relative weights in the superposition given by α and β. We will now show that actually the two factors in (8.21) differ only by a phase.

To see this, first let us note that the sets $\mathscr{S}$ and $\mathscr{S}'$ are isomorphic. Let $\vec{l}'$ be an element of $\mathscr{S}'$. Then $\vec{l}'$ is related to a unique element $\vec{l}$ of $\mathscr{S}$ by a function $f : \mathscr{S}' \to \mathscr{S}$ defined by

$$\vec{l} = f[\vec{l}'] = f[(l'_0, \ldots, l'_n)] = (l'_n, l'_0, \ldots, l'_{n-1}), \tag{8.22}$$

so we have

$$l_0 = l'_n$$
$$l_1 = l'_0$$
$$\vdots$$
$$l_n = l'_{n-1}. \tag{8.23}$$

It follows that

$$\sum_{\mathscr{S}'} \omega^{\sum_{p=0}^{n} p\, l'_p} = \sum_{\mathscr{S}} \omega^{\sum_{p=0}^{n-1} p\, l_{p+1}}\, \omega^{n l_0}$$
$$= \sum_{\mathscr{S}} \omega^{\sum_{p=1}^{n} (p-1)\, l_p}\, \omega^{n l_0}$$
$$= \sum_{\mathscr{S}} \omega^{\sum_{p=1}^{n} p\, l_p}\, \omega^{-\sum_{p=1}^{n} l_p}\, \omega^{n l_0}$$
$$= \sum_{\mathscr{S}} \omega^{\sum_{p=0}^{n} p\, l_p}\, \omega^{-\sum_{p=0}^{n} l_p}\, \omega^{(n+1) l_0}. \tag{8.24}$$

In the last equality we have replaced $\sum_{p=1}^{n} p\, l_p$ by $\sum_{p=0}^{n} p\, l_p$ in the first exponent, since extending the lower limit doesn't change the sum (the term with $p=0$ is zero). Now recall that $\omega = \exp[2\pi i/(n+1)]$, so we have $\omega^{(n+1) l_0} = 1$. As a consequence,

$$\sum_{\mathscr{S}'} \omega^{\sum_{p=0}^{n} p\, l'_p} = \sum_{\mathscr{S}} \omega^{\sum_{p=0}^{n} p\, l_p}\, \omega^{-\sum_{p=0}^{n} l_p}. \tag{8.25}$$

Let us look at the factor $\omega^{-\sum_{p=0}^{n} l_p}$ in more detail. In particular, look at the exponent. It is the sum of the values of all the components of an $(n+1)$-tuple belonging to $\mathscr{S}$. But all $(n+1)$-tuples of $\mathscr{S}$ have exactly the same components, up to a permutation of components 1 to k among themselves, and a permutation of components $0, k+1, \ldots, n$ among themselves. Thus the sum

$$\sum_{p=0}^{n} l_p, \tag{8.26}$$

is a constant over $\mathscr{S}$, and so this phase factor can be extracted from the sum over $\mathscr{S}$, and we get

$$\sum_{\mathscr{S}'}\omega^{\Sigma_{p=0}^{n}p\,l_p'} = \omega^{-\Sigma_{p=0}^{n}l_p}\sum_{\mathscr{S}}\omega^{\Sigma_{p=0}^{n}p\,l_p}, \qquad (8.27)$$

where the values of l_p used in the prefactor are the values of any element of $\mathscr{S}$. Furthermore, we can compute that value in terms of the measurement results, by writing

$$\sum_{p=0}^{n}l_p = \sum_{p=1}^{k}l_p + \sum_{p=k+1}^{n}l_p + l_0. \qquad (8.28)$$

Remember that the set of indices $l_1,\ldots,l_k$ tells us where the V-photons were measured. For example, if 4 of these indices take the value "3", this means that there were 4 V-photons measured in mode 3. Now it is not difficult to see that

$$\sum_{p=1}^{k}l_p = \sum_{j=0}^{n}j r_j, \qquad (8.29)$$

where we are summing over all the modes, counting how many V-photons were measured on that mode, and multiplying that by the number of the mode. Then, it is clear also that

$$\sum_{p=k+1}^{n}l_p + l_0 = \sum_{j=0}^{n}j h_j, \qquad (8.30)$$

and finally we have

$$\sum_{p=0}^{n}l_p = \sum_{j=0}^{n}j(r_j + h_j), \qquad (8.31)$$

where $(r_j + h_j)$ is actually the total number of photons measured in mode j. In summary, we have

$$\sum_{\mathscr{S}'}\omega^{\Sigma_{p=0}^{n}p\,l_p'} = \omega^{-\Sigma_{j=0}^{n}j(r_j+h_j)}\sum_{\mathscr{S}}\omega^{\Sigma_{p=0}^{n}p\,l_p}. \qquad (8.32)$$

Taking all of these into account we can rewrite the state after the measurement (8.19) as

$$|\Phi\rangle_{0\ldots n}|H\rangle^{k-1}\left(\alpha|H\rangle + \beta\,\omega^{-\Sigma_{j=0}^{n}j(r_j+h_j)}|V\rangle\right)|V\rangle^{n-k+1}, \qquad (8.33)$$

where $|\Phi\rangle_{0\ldots n}$ is a state of the first $n+1$ modes fixed by the result of the measurement. We can see that the superposition that was originally encoded in mode 0 was teleported to the mode $n + k$, up to a known relative phase. But since we know exactly the value of this phase, we can correct it by using linear optical elements such as polarization beamsplitters and phase-shifters. The final state is then

$$|\Phi\rangle_{0...n}|H\rangle^{k-1}(\alpha|H\rangle + \beta|V\rangle)|V\rangle^{n-k+1}. \tag{8.34}$$

The last n modes, with the exception of mode $n + k$ of course, are left in a known state and can be reused later. The teleportation succeeds with probability $n/(n+1)$. By increasing the value of n (i.e., increasing the size of the teleporting state $|t_n\rangle$) we can make this probability arbitrarily close to 1.

8.4 Nondeterministic two-qubit gate

The near-deterministic teleportation step we have described is the basis of the application of a linear optical C-SIGN gate. The idea is to use the Gottesman and Chuang approach for applying unitary gates through teleportation. To this end we will need a particular entangled state given by

$$|t'_n\rangle = \sum_{i,j=0}^{n}(-1)^{(n-j)(n-i)}|V\rangle^j|H\rangle^{n-j}|H\rangle^j|V\rangle^{n-j} \times |V\rangle^i|H\rangle^{n-i}|H\rangle^i|V\rangle^{n-i}.$$
$$\tag{8.35}$$

This is a state of $4n$ modes with $4n$ photons. It is nothing but two copies of the state $|t_n\rangle$ with CSIGN gates applied between each one of the last n modes of one copy and each of the last n modes of the other copy (see Fig. 8.5).

The application of the CSIGN between two optical modes that encode the control and target qubits will proceed by teleporting the control qubit using the first $2n$ modes of (8.35) as we discussed before, and then by teleporting the target qubit using the last $2n$ modes. The two output modes into which the qubits are teleported will be in a state that corresponds to applying the CSIGN gate.

Before we start teleporting the first qubit, the state of the system is given by

$$|\Psi\rangle = \alpha|H\rangle|t'_n\rangle + \beta|V\rangle|t'_n\rangle$$
$$= \sum_{i=0}^{n}\sum_{j=0}^{n}(-1)^{(n-j)(n-i)}\left(\alpha|H\rangle|V\rangle^j|H\rangle^{n-j}|H\rangle^j|V\rangle^{n-j}\right.$$
$$\left. + \beta|V\rangle|V\rangle^j|H\rangle^{n-j}|H\rangle^j|V\rangle^{n-j}\right) \times |V\rangle^i|H\rangle^{n-i}|H\rangle^i|V\rangle^{n-i}, \tag{8.36}$$

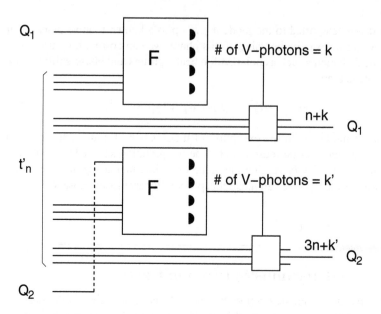

FIGURE 8.5: Nondeterministic two-qubit CSIGN gate: The ancilla state consists of $4n$ modes with $4n$ photons. Simply, it is two copies of the state $|t_n\rangle$ (for the teleportation) with CSIGN gates applied between the two halves of the two copies. The CSIGN operation between two optical modes that encode the control and target qubits is achieved by teleporting the control qubit using the first $2n$ modes, and then teleporting the target qubit using the last $2n$ modes.

where we have left out the state of the second qubit to make the notation less cumbersome. Now we need to apply the Fourier transform to the first $n+1$ modes (those modes labeled from 0 to n) to perform the teleportation of the first qubit

$$(\hat{F}_n \otimes \mathbf{1})|\Psi\rangle = \sum_{i=0}^{n} \sum_{j=0}^{n} \left(\alpha \hat{F}_n \left[|H\rangle |V\rangle^j |H\rangle^{n-j} \right] + \beta \hat{F}_n \left[|V\rangle |V\rangle^j |H\rangle^{n-j} \right] \right)$$
$$\times |H\rangle^j |V\rangle^{n-j} \times |V\rangle^i |H\rangle^{n-i} |H\rangle^i |V\rangle^{n-i} (-1)^{(n-j)(n-i)}.$$

$$(8.37)$$

Note that for each value of i we have basically the same state we had when we described the near-deterministic teleportation, except for the phase factor $(-1)^{(n-j)(n-i)}$. Keeping that in mind, we can proceed as before. We measure the $n+1$ output modes of the Fourier transform, distinguishing the number of V-photons and H-photons in each mode. If the total number of V-photons is k, then the α term in (8.37) collapses to the value $j = k$ while the β term collapses to $j = k - 1$. Everything is the same as before except for the extra phase factor, that depends on k. The state after the measurement is

$$\sum_{i=0}^{n} |H\rangle^{k-1} \left(\alpha(-1)^{(n-k)(n-i)}|H\rangle + \beta(-1)^{(n-k+1)(n-i)}\omega^{-\Sigma_{j=0}^{n} j(r_j+h_j)}|V\rangle \right)$$

$$\times |V\rangle^{n-k} \times |V\rangle^{i}|H\rangle^{n-i}|H\rangle^{i}|V\rangle^{n-i}, \tag{8.38}$$

where we have omitted the final state of the first $n+1$ modes after the measurement. Just as before, the mode into which the state was teleported is mode $n+k$. From now on we will designate that mode as mode A to avoid confusion with the remaining modes. Again, since the phase factor $\omega^{-\Sigma_{j=0}^{n} j(r_j+h_j)}$ is known, we can correct it using a phase shifter. We can rewrite the state of the system after the measurement and the phase correction as

$$|H\rangle_{n+1,\dots,n+k-1}|V\rangle_{n+k+1,\dots,2n} \Sigma_{i=0}^{n} \left(\alpha(-1)^{(n-k)(n-i)}|H\rangle_A + \right.$$

$$\left. \beta(-1)^{(n-k+1)(n-i)}|V\rangle_A \right) |V\rangle^{i}|H\rangle^{n-i}|H\rangle^{i}|V\rangle^{n-i}, \tag{8.39}$$

where we have written in a slightly different way the state of the first $2n$ modes. Except for mode A, which we have singled out in the notation, these modes are of no use anymore for the application of the CSIGN gate (although they can be physically reused), and so we will drop them from the notation.

Now we need to teleport the target qubit of our CSIGN gate using the remaining $2n$ modes of the state (8.35). Note that these modes are now entangled with the survived mode A from the first teleportation. We will rename the modes such that this target qubit is labeled mode 0 and the remaining ones are labeled from 1 to $2n$. In this way, this second teleportation will have the same mathematical form as the first one, with the addition of mode A. If we write the state of the target qubit as $\gamma|H\rangle_0 + \delta|V\rangle_0$, then we have

$$|\Psi'\rangle = \sum_{i=0}^{n} \left(\alpha(-1)^{(n-k)(n-i)}|H\rangle_A + \beta(-1)^{(n-k+1)(n-i)}|V\rangle_A \right)$$

$$\times (\gamma|H\rangle_0 + \delta|V\rangle_0) |V\rangle^{i}|H\rangle^{n-i}|H\rangle^{i}|V\rangle^{n-i}. \tag{8.40}$$

Now we apply the Fourier transform to the first $n+1$ modes as usual and get

$$(\hat{F}_n \otimes 1)|\Psi'\rangle = \sum_{i=0}^{n} \left(\alpha(-1)^{(n-k)(n-i)}|H\rangle_A + \beta(-1)^{(n-k+1)(n-i)}|V\rangle_A \right)$$

$$\times \left(\gamma \hat{F}_n \left[|H\rangle_0 |V\rangle^i |H\rangle^{n-i}\right] + \delta \hat{F}_n \left[|V\rangle_0 |V\rangle^i |H\rangle^{n-i}\right]\right) |H\rangle^i |V\rangle^{n-i}.$$

$$(8.41)$$

This state has exactly the same form we encountered before when we first discussed the near-deterministic teleportation, except that the state of mode A is acting as an extra factor in the coefficients γ and δ, but does not affect the procedure. Now we measure the number of photons in each of the first $n+1$ modes as before, and if the total number of V-photons is k', then the γ terms in (8.41) collapse to the value $i = k'$ while the δ terms collapse to the value $i = k' - 1$. Note that this will have an effect on the phases multiplying the coefficients α and β. The state after the measurement is

$$|\Phi'\rangle_{0...n} |H\rangle_{n+1,...,n+k'-1} |V\rangle_{n+k'+1,...,2n}$$
$$\times \left\{ \left(\alpha(-1)^{(n-k)(n-k')}|H\rangle_A + \beta(-1)^{(n-k)(n-k+1)}|V\rangle_A\right) \gamma|H\rangle_{k'}\right.$$
$$\left. + \left(\alpha(-1)^{(n-k'+1)(n-k)}|H\rangle_A + \beta(-1)^{(n-k'+1)(n-k+1)}|V\rangle_A\right) \delta|V\rangle_{k'}\right\}.$$

$$(8.42)$$

Finally, the state of the teleported qubits which we will refer to as modes A and B is given by

$$|\psi\rangle_{AB} = \alpha\gamma(-1)^{(n-k)(n-k')}|HH\rangle_{AB} + \beta\gamma(-1)^{(n-k')(n-k+1)}|VH\rangle_{AB} +$$
$$+ \alpha\delta(-1)^{(n-k'+1)(n-k)}|HV\rangle_{AB} + \beta\delta(-1)^{(n-k'+1)(n-k+1)}|VV\rangle_{AB}.$$

$$(8.43)$$

By noting that

$$\begin{aligned}
(n-k)(n-k') &= n^2 - nk' - nk + kk', \\
(n-k+1)(n-k') &= n^2 - nk - nk' + n - k' + kk', \\
(n-k)(n-k'+1) &= n^2 - nk' - nk + n - k + kk', \\
(n-k+1)(n-k'+1) &= n^2 - nk - nk' + kk' + 2n - k - k' + 1, \quad (8.44)
\end{aligned}$$

we can extract an overall phase factor and get

$$|\psi\rangle_{AB} = (-1)^{n^2 - n(k+k') + kk'} \left(\alpha\gamma|HH\rangle_{AB} + \beta\gamma(-1)^{(n-k')}|VH\rangle_{AB}\right.$$
$$\left. + \alpha\delta(-1)^{(n-k)}|HV\rangle_{AB} + \beta\delta(-1)^{2n-(k+k')+1}|VV\rangle_{AB}\right).$$

$$(8.45)$$

Now we can use beam splitters, mirrors and phase shifters to apply a phase shift $(-1)^{n-k'}$ to V-photons in mode A, and a phase shift $(-1)^{n-k}$ to V-photons in mode B. The resulting state is

$$|\psi\rangle_{AB} = \left(\alpha\gamma|HH\rangle_{AB} + \beta\gamma|VH\rangle_{AB} + \alpha\delta|HV\rangle_{AB} + \right.$$

$$\left. + \beta\delta\underbrace{(-1)^{(n-k)+(n-k')}(-1)^{2n-(k+k')+1}}_{(-1)}|VV\rangle_{AB}\right)$$

$$= \alpha\gamma|HH\rangle_{AB} + \beta\gamma|VH\rangle_{AB} + \alpha\delta|HV\rangle_{AB} - \beta\delta|VV\rangle_{AB}, \quad (8.46)$$

which is just the result of a C-SIGN gate applied to the two-qubit state

$$\left(\alpha|H\rangle_A + \beta|V\rangle_A\right)\left(\gamma|H\rangle_B + \delta|V\rangle_B\right)$$

$$= \alpha\gamma|HH\rangle_{AB} + \beta\gamma|VH\rangle_{AB} + \alpha\delta|HV\rangle_{AB} + \beta\delta|VV\rangle_{AB}. \quad (8.47)$$

Since each teleportation step succeeds independently with probability $n/(n+1)$, the total success probability of this non-deterministic CSIGN gate is $[n/(n+1)]^2$. Again, by increasing the value of n this probability can be made arbitrarily close to 1. Once again, the CSIGN gate can be simply converted to the CNOT gate with additional one-qubit Hadamard gate on the target, followed by the CSIGN and another Hadamard on the target.

8.5 Ancilla-state preparation

We have seen that linear optical elements and photodetection are sufficient to perform an entangling operation (CSIGN or CNOT) between two qubits encoded in two optical modes. But this is only possible provided that we have access to a certain entangled ancilla state. We have not yet discussed how this entangled state is created and it is natural to ask whether it can also be built using only linear optical elements and photodetection. One of the main results of the seminal work by Knill, Laflamme and Milburn is that this is in fact possible [1]. And the essential building block for this construction is the nonlinear phase shift gate.

Let us first look at the so-called nonlinear sign (NS) gate while deferring the reason why we need this gate. The NS gate performs the following transformation on the state of an optical mode:

$$\alpha_0|0\rangle + \alpha_1|1\rangle + \alpha_2|2\rangle \rightarrow \alpha_0|0\rangle + \alpha_1|1\rangle - \alpha_2|2\rangle, \quad (8.48)$$

where $|0\rangle$, $|1\rangle$ and $|2\rangle$ represent states of the optical mode with 0, 1 and 2 photons respectively. This transformation cannot be implemented with only linear optical elements since it is clear that its effect does not scale linearly with the number of photons. However, complementing linear optics with photodetection and postselection allow us to implement this transformation. The price we have to pay (as before) is that the transformation will become nondeterministic.

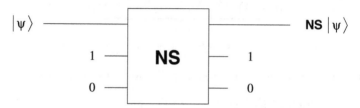

FIGURE 8.6: Nonlinear sign gate: For a given input ($\psi\rangle = \alpha_0|0\rangle + \alpha_1|1\rangle + \alpha_2|2\rangle$) the output state is given by $NS|\psi\rangle = \alpha_0|0\rangle + \alpha_1|1\rangle - \alpha_2|2\rangle$. The numbers one the left side of the NS box represent the number of photons input in each mode (one in mode 2 and zero in mode 3). The numbers to the right represent the measurement outcome corresponding to the success of the gate.

The schematic for the NS gate is shown in Fig. 8.6. The box NS implements a transformation among the three optical modes that acts on the creation operators of each mode according to

$$a_k^\dagger \to \sum_{i=1}^{3} u_{ij} a_j^\dagger. \tag{8.49}$$

The u_{ij} are the elements of a unitary matrix U given by

$$U = \begin{pmatrix} 1 - 2^{1/2} & 2^{-1/4} & (3/2^{1/2} - 2)^{1/2} \\ 2^{-1/4} & 1/2 & 1/2 - 1/2^{1/2} \\ (3/2^{1/2} - 2)^{1/2} & 1/2 - 1/2^{1/2} & 2^{1/2} - 1/2 \end{pmatrix}. \tag{8.50}$$

This unitary transformation among modes can be implemented with linear optical elements [13]. The nonlinear phase shift gate succeeds with probability 1/4. A simple setup for the NS gate with three beam splitters is given in Appendix B.

Now we discuss the reason why this NS gate is an essential element for performing the CSIGN gate and preparing the entangled state of ancilla photons. To make the CSIGN gate, from Eq. (8.7) all we need is to change the sign of

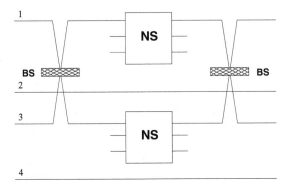

FIGURE 8.7: Setup for an entangling four mode gate using two nonlinear sign gates. The modes 1 and 2 represent the control qubit, and 3 and 4 are for the target qubit. The NS gate makes the sign changes (α_2 to $-\alpha_2$) only for the case when the input contains two photons. With the 50:50 beam splitter there is a sign change only when there is one photon in mode 1 and one photon in mode 3 ($|1\rangle_1|1\rangle_3 \rightarrow |2,0\rangle_{1,3} + |0,2\rangle_{1,3} \rightarrow -(|2,0\rangle_{1,3} + |0,2\rangle_{1,3}) \rightarrow |1\rangle_1|1\rangle_3$). All other cases of inputs in the modes 1 and 3 remain intact.

the input when the input is corresponding to $|V\rangle|V\rangle$, i.e., when the qubits have logical value of one ($|1\rangle_L|1\rangle_L$ state). Given the two input qubits (suppose these are the smallest ancilla states for the teleportation)

$$
\begin{aligned}
|Q_1\rangle &= \frac{1}{\sqrt{2}}(|0\rangle_L + |1\rangle_L) = \frac{1}{\sqrt{2}}(|0\rangle_1|1\rangle_2 + |1\rangle_1|0\rangle_2), \\
|Q_2\rangle &= \frac{1}{\sqrt{2}}(|0\rangle_L + |1\rangle_L) = \frac{1}{\sqrt{2}}(|0\rangle_3|1\rangle_4 + |1\rangle_3|0\rangle_4),
\end{aligned} \qquad (8.51)
$$

the transformation by applying the CSIGN gate can be written as (by ignoring the normalization factor)

$$
\begin{aligned}
|Q_1\rangle|Q_2\rangle &\Rightarrow |0\rangle_L|0\rangle_L + |0\rangle_L|1\rangle_L + |1\rangle_L|0\rangle_L - |1\rangle_L|1\rangle_L \\
&= |0,1,0,1\rangle + |0,1,1,0\rangle + |1,0,0,1\rangle - |1,0,1,0\rangle, \quad (8.52)
\end{aligned}
$$

where the modes 1, 2 are designated for the control qubit, and 3, 4 are for the target qubit, and the number state representation was given in this order (see Fig. 8.7). We can immediately see that there is a sign change only when there is one photon in mode 1 and one photon in mode 3.

At this point let us take a look at the first beam splitter in Fig. 8.7. When there is one photon in mode 1 and one photon in mode 3 impinging upon

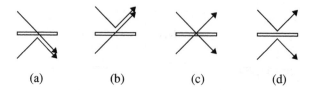

(a)   (b)   (c)   (d)

FIGURE 8.8: Four possibilities obtained by sending a $|1,1\rangle$ state through a 50:50 beam splitter. The diagrams (c) and (d) lead to the same final state, and interfere destructively. (c) Transmission-transmission $(i)(i) = -1$. (d) Refection-reflection: $(-1)(-1) = 1$, where we have used the reflected mode acquiring a phase -1 while the transmitted mode acquires a phase of i, respectively.

a 50:50 beam splitter, the output should be either two photons in one mode or two photons in the other. Formally the 50:50 beam splitter transforms the input $|1,1\rangle$ into the output $|2,0\rangle + |0,2\rangle$ in that the probability amplitude for having $|1,1\rangle$ at the output of the beam splitter vanishes [14]. Combining this well-known Hong–Ou–Mandel effect and the NS gate allows us to entangle states of two or more photons nondeterministically. From the state $\frac{1}{2}(|10\rangle_{12} + |01\rangle_{12})(|10\rangle_{34} + |01\rangle_{34})$ we can generate the state $\frac{1}{2}(|1010\rangle + |1001\rangle + |0101\rangle - |0101\rangle)$ with the setup of Fig. 8.8.

Using polarization beam splitters and regular beam splitters we can transform this state into a polarization Bell pair $\frac{1}{\sqrt{2}}(|HH\rangle + |VV\rangle)$. With two of these Bell pairs we can use the setup of Fig. 8.9 to nondeterministically generate the state $|t_1'\rangle$.

We can construct the more complex state $|t_n'\rangle$ using the state $|t_1'\rangle$ to apply C-SIGN gates nondeterministically and entangle more and more photons together. The resources required to construct these more complex states scale in principle exponentially with n. However, there are clever ways of carrying out this construction that manage to reuse the states leftover from failures due to the nondeterministic nature of the process. In this case, the scaling can be lowered to subexponential [1].

Two important points are: this construction can be made offline, and that a nonzero probability of failure can be tolerated in the application of a CSIGN gate during a quantum computation. This means that we do not need to construct states with arbitrarily high values of n. However, the resources required are still too high for this approach to be practical in the usual quantum circuit model of quantum computing.

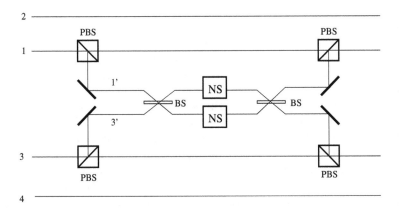

FIGURE 8.9: Setup to nondeterministically generate $|t_1'\rangle$ from two polarization Bell pairs.

8.6   Cluster-state approach and gate fidelity

Even though the KLM scheme allows us to implement a CSIGN gate with an arbitrarily high success probability, and hence to perform an arbitrarily long quantum computation, in practice this approach turns out to be not very useful. The main reason is that the resources required to implement a CSIGN gate with high enough probability of success are completely unrealistic for present day technology. On the one hand, number-resolving photodetectors with an efficiency above 99% are required to prevent photon-loss errors, and on the other hand an order of 10^4 optical modes is required to implement a single CSIGN gate [1]. The best detector currently available with the required photon counting characteristics has an efficiency of around 80% [15], and dealing with such an enormous number of modes is well beyond the realm of present quantum optics technology.

Fortunately, this is not a dead end for LOQC. The way around this road block is to abandon the implementation of the KLM techniques within the quantum circuit model of quantum computation, and instead work with a different approach known as the cluster state model. The cluster state model was introduced in the seminal work of Raussendorf and Briegel [6], and belongs to the class of measurement-based quantum computing models. The basic idea is to carry the computation by performing single-qubit measurements on a certain entangled state. The entangled states that allow a quantum computation to be performed through measurements are known as cluster states.

The application of the KLM techniques to the cluster state model was first suggested by Nielsen [7], who realized that this union could lead to a drastic reduction of the number of optical modes required for each quantum gate. This is due to the fact that a cluster state can be constructed by applying an entangling gate between pairs of qubits. The gate does not have to be successful all the time, as long as the success probability of the gate and the sequence of entangling attempts is such that the size of the cluster grows on average. It turns out that the probability of success required for this entangling gate can be in principle any number greater than zero, and can still allow an efficient construction of the cluster state [16, 17]. Thus an optical CSIGN gate implemented using the techniques of the KLM scheme that requires only a small number of optical modes can be used in this approach. However, there are still some important issues associated with the fidelity of the gate that need to be taken into account.

The original KLM scheme relied on *dual-rail encoding*, in which the state of a qubit is represented by the state of a pair of optical modes. This essentially encodes the information in the number of photons present in the modes. Since very high efficiency detectors are not available, and the implementation of the CSIGN gate depends crucially on measuring the exact number of photons present in a subset of the modes, this approach can introduce *undetected errors* in the gate. If a detector fails to register a photon, the total number of photons measured will not be the same as the total number of photons present, and thus, for instance, the wrong mode will be post-selected and an error will be introduced by the gate. The worst part is that this error is not detected by this procedure.

An imperfect detector is characterized by its quantum efficiency η, representing the probability of the detector registering a photon when one is actually present, and its dark-count rate λ. Ignoring the dark counts and assuming that the quantum efficiency η does not depend on the intensity of light, the probability of the detector registering k photons when l were actually present is given by

$$P(k|l) = \binom{l}{k} \eta^k (1-\eta)^{l-k}. \tag{8.53}$$

So far we have assumed that even though our gate was nondeterministic, whenever it was signaled as "successful", the gate was also perfect. However, when we consider the errors introduced by imperfect detectors we realize that this assumption is not justified. Furthermore, if we consider imperfect detectors with the highest efficiency available today, the probability of undetected errors introduced by our gate can be as high as 30% [5]. This is unacceptable even for the cluster state approach, since due to this low fidelity of the gate, the cluster state we build will also have a low fidelity and so the result of any computation performed with it cannot be trusted.

By using *polarization encoding* we are able to have a high-fidelity gate without requiring high-efficiency detectors. This is due to the fact that with polarization encoding the crucial quantity that needs to be measured is the number of vertically polarized photons present in a set of optical modes, while the total number of photons remains fixed. The information is not encoded in total photon number, which is more susceptible to detection errors. This provides us with an independent way of checking for detector failure—if the total number of photons detected is different from the number expected, some of the detectors have malfunctioned and the result of the gate cannot be trusted. Note that if the information was encoded in total photon number these errors would not be detected, since different values of total photon number are also valid outcomes.

Even with polarization encoding it is possible for certain errors to occur and go undetected. The dark count rate λ quantifies the rate of false positives, i.e., the rate at which the detector registers the presence of a photon when none is actually present. These dark counts are usually assumed to have a Poisson distribution and so the probability of having d of them during the measurement interval τ is

$$D(d) = e^{-\lambda\tau} \frac{(\lambda\tau)^d}{d!}. \tag{8.54}$$

It follows that the probability of the detector registering k photons when l were actually present is given by [18]

$$P_D(k|l) = \sum_{d=0}^{k} D(k-d) \binom{l}{d} \eta^d (1-\eta)^{l-d}. \tag{8.55}$$

This requires us to carefully analyze any protocol that depends crucially on a perfect measurement of the photon number to prevent unwanted and undetected errors. For example, if a detector fails to register a photon while another detector registers a dark count at the same time, the total number of photons registered does not change but its distribution among vertically and horizontally polarized photons may be different. However, since typically the dark count rates of currently available detectors are extremely low [19], these events have very low probability (of the order of 10^{-7} per gate). And since it is this probability of undetected errors that affects the fidelity of the gate, very high fidelities can be achieved.

Given that the gate fidelity decreases as the undetected error and the probability of having this undetected error increase and as the number of detectors increases, the highest fidelity of the gate is to be the scheme of employing the smallest number of detectors. This, in turn, implies the optimal ancilla state should have the smallest number of photons. From Eq. (8.35) the smallest such ancilla state is given by

$$|t_1'\rangle = \frac{1}{2} \sum_{i,j=0}^{1} (-1)^{(1-j)(1-i)} |V\rangle^j |H\rangle^{1-j} |H\rangle^j |V\rangle^{1-j} \times |V\rangle^i |H\rangle^{1-i} |H\rangle^i |V\rangle^{1-i}.$$

$$= \frac{1}{2} \left(-|HVHV\rangle + |HVVH\rangle + |VHHV\rangle + |VHVH\rangle \right). \tag{8.56}$$

Besides the fact that this smallest ancilla state yields the highest gate fidelity, it also has several other merits:

(1) The whole setup for the discrete Fourier transform (shown in Fig. 8.5) is now just one beam splitter.

(2) The number of possible output modes is just one for each teleportation (the control and target qubit) so that we do not need to post-select the output mode. Only the necessary phase shift needs to be applied.

(3) The successful teleportation is signaled by the detection of one H-photon and one V-photon, so that two out of four detectors should fire and we do not know how many photons are detected at each detector.

Of course, items (1) and (2) mean great simplification of the implementation. And most importantly, item (3) says that we do not necessarily need detectors that have the number-resolving capability. Therefore, employing the smallest ancilla state of (8.56), we achieve the highest fidelity of the two-qubit CSIGN gate (limited only by the dark-count rate) and remove the requirement of number resolution for the detectors.

The main purpose of the chapter is to provide a comprehensive description of the quantum gate teleportation and nondeterministic two-qubit gates, the crucial elements in linear optical quantum computing. Here the desired non-linearities come from projective measurements and post-selection. Simply, the measurements over some part of the total quantum system and the selection of only the correct outcomes project the rest of the system into a desired quantum state.

The qubits in our description are encoded in the *polarization* degrees of freedom as opposed to the original *dual-rail* encoding of the KLM scheme. In doing so, we needed twice more photons for the ancilla state as well as twice more detectors for a given probability of success of the gate. However, when it comes to the issue of gate fidelity, the polarization encoding avoids the *undetected errors* due to the finite quantum efficiency of the detectors and consequently yields extremely high fidelity. The reduction of undetected errors by employing the polarization encoding is independent of the efficiency of the detectors and requires only the small dark-count rate of the detectors. This property of the polarization-encoding scheme is particularly useful in the *optical cluster-state quantum computing* where the building the cluster states does not require near-unity success probability of the entangling operations.

In such a circumstance where the highest gate fidelity is desired, the optimal ancilla state is found to be the smallest in the number of photons. This leads to a huge reduction of the number of optical elements in the interferometer design for the gate operation and there is no need for the mode selection for the output—only the feed forward for the phase correction. At the same time the entangling gate using the smallest ancilla states does not require the number-resolving capability of the detectors.

We wish to close by pointing out that, although the use of the number-resolving detectors can be avoided in the entangling-gate operation, they still need to be used in the preparation of the ancilla state (even the smallest one). The requirement of the number-resolving detectors in the state preparation comes from the utilization of the NS gate as in Sec. 8.5. It turns out that we may be able to eliminate the use of NS gate in the ancilla-state preparation, too. The smallest ancilla of the form Eq. (8.56) can be generated without relying on the number-resolving detectors if we have high-fidelity Bell states [5]. The requirement for the photodetectors to be number resolving, which is considered to be the biggest obstacle so far in the realization of optical quantum computing, can then be completely eliminated.

Acknowledgments

We would like to acknowledges supports from the Hearne Institute of Theoretical Physics, NASA Code Y, the National Security Agency, the Disruptive Technologies Office, and the Army Research Office.

8.7   Appendices

Appendix A: One-qubit gates

An arbitrary one-qubit operation can be made if we have an arbitrary phase gate and any two of x-rotation, y-rotation, and z-rotation. The phase gate is given by

$$P(\theta) = \begin{pmatrix} e^{i\theta} & 0 \\ 0 & e^{-i\theta} \end{pmatrix}, \tag{8.57}$$

which, acting upon a qubit in a state $\alpha|0\rangle + \beta|1\rangle$, leads to a change of relative phase as $\alpha|0\rangle + \beta e^{-i2\theta}|1\rangle$.

Now using σ_j as the usual notation of the Pauli operators, the x-rotation, $R_x(\theta)$ is formally written as

$$R_x(\theta) = e^{i\sigma_x \frac{\theta}{2}} = \cos\frac{\theta}{2} + i\sigma_x \sin\frac{\theta}{2} = \begin{pmatrix} \cos\frac{\theta}{2} & i\sin\frac{\theta}{2} \\ i\sin\frac{\theta}{2} & \cos\frac{\theta}{2} \end{pmatrix}. \tag{8.58}$$

Similarly, y-rotation and z-rotation are given by

$$R_y(\theta) = e^{i\sigma_y \frac{\theta}{2}} = \cos\frac{\theta}{2} + i\sigma_x \sin\frac{\theta}{2} = \begin{pmatrix} \cos\frac{\theta}{2} & \sin\frac{\theta}{2} \\ -\sin\frac{\theta}{2} & \cos\frac{\theta}{2} \end{pmatrix},$$

$$R_z(\theta) = e^{i\sigma_z \frac{\theta}{2}} = \cos\frac{\theta}{2} + i\sigma_z \sin\frac{\theta}{2} = \begin{pmatrix} e^{\frac{i\theta}{2}} & 0 \\ 0 & e^{-\frac{i\theta}{2}} \end{pmatrix}. \tag{8.59}$$

Here we examine the linear optical devices to construct these one-qubit gates.

A1. Phase Gate

Using the polarization encoding of Eq. (8.1), the phase gate, then, can be realized by a waveplate and a phase shifter. The wave plate is a doubly refracting (birefringent) transparent crystal where the index of refraction is larger for the slow axis than the one for the fast axis. We assume that the slow axis is the vertical direction and the fast is the horizontal direction. After propagating by a distance d, the vertically polarized light acquires a phase shift of $\exp[in_1 \frac{2\pi}{\lambda}d]$. Similarly, the horizontally polarized light acquires $\exp[in_2 \frac{2\pi}{\lambda}d]$ ($n_1 > n_2$). The relative phase shift acquired by the vertically polarized light can be described as

$$|V\rangle \Rightarrow e^{i(n_1 - n_2)\frac{\lambda}{2}d}|V\rangle \equiv e^{i\phi}|V\rangle. \tag{8.60}$$

Now choosing the thickness of the crystal $d = \lambda/4(n_1 - n_2)$, the relative phase shift $\phi = \pi/2$ is acquired. This corresponds to the *quarter wave plate* as we have

$$|H\rangle + |V\rangle \Rightarrow |H\rangle + e^{i\pi/2}|V\rangle. \tag{8.61}$$

Using the convention for right and left circularly polarized light, $\hat{e}_\pm = \frac{1}{\sqrt{2}}(\hat{e}_1 \pm i\hat{e}_2)$, one can see that the linearly polarized light ($|H\rangle + |V\rangle$) becomes a left circularly polarized light.

In general, an arbitrary phase angle ϕ can be obtained by choosing the thickness d as

$$d = \frac{\lambda}{n_1 - n_2}\frac{\phi}{2\pi}. \tag{8.62}$$

The overall phase factor then depends on the value of ϕ as it is given by

$$e^{in_2\frac{2\pi}{\lambda}d} = e^{in_2\frac{2\pi}{\lambda}\frac{\lambda}{2\pi}\frac{\phi}{n_1-n_2}} = e^{i\frac{n_2}{n_1-n_2}\phi}. \tag{8.63}$$

A2. z-Rotation

For the z-rotation, for Eq. (8.59) we have

$$R_z(\theta)\left[|0\rangle + |1\rangle\right] = e^{i\frac{\theta}{2}}|0\rangle + e^{-i\frac{\theta}{2}}|1\rangle = e^{i\frac{\theta}{2}}\left[|0\rangle + e^{-i\theta}|1\rangle\right]. \tag{8.64}$$

One can have this operation with a wave plate by taking $\phi = -\theta$. Then, the overall phase of Eq. (8.63) becomes

$$\frac{n_2}{n_1 - n_2}\phi = -\frac{n_2}{n_1 - n_2}\theta \equiv -r\theta \tag{8.65}$$

where we defined $r = \frac{n_2}{n_1-n_2}$. In order to match the overall phase factor $e^{i\frac{\theta}{2}}$ of the z-rotation we need to have a phase shifter such that $\phi' - r\theta = \theta/2$, which yields

$$\phi' = \frac{\theta}{2}(1 + 2r) = \frac{\theta}{2}\frac{n_1 + n_2}{n_1 - n_2} \equiv s\theta. \tag{8.66}$$

Therefore, the z-rotation is made by a wave plate of $-\theta$ and a phase shifter of $s\theta$, where $s = [(n_1 + n_2)/2(n_1 - n_2)]$. Symbolically, we may write this relation as

$$R_z(\theta) :=: \text{PS}(s\theta)\text{WP}(-\theta). \tag{8.67}$$

A3. x-Rotation

Suppose the wave plate we described above is now rotated around the propagation axis by an amount of α. Then the operation by the wave plate can be given as

$$\text{WP}(\phi, \alpha = 0) = e^{ir\phi} \begin{pmatrix} 1 & 0 \\ 0 & e^{i\phi} \end{pmatrix},$$

$$\text{WP}(\phi, \alpha) = e^{ir\phi} \begin{pmatrix} \cos\alpha & -\sin\alpha \\ \sin\alpha & \cos\alpha \end{pmatrix} \begin{pmatrix} 1 & 0 \\ 0 & e^{i\phi} \end{pmatrix} \begin{pmatrix} \cos\alpha & \sin\alpha \\ -\sin\alpha & \cos\alpha \end{pmatrix},$$

$$= e^{ir\phi} \begin{pmatrix} \cos^2\alpha + e^{i\phi}\sin^2\alpha & \cos\alpha\sin\alpha(1 - e^{i\phi}) \\ \cos\alpha\sin\alpha(1 - e^{i\phi}) & \sin^2\alpha + e^{i\phi}\cos^2\alpha \end{pmatrix}.$$

$$\tag{8.68}$$

If we set the rotation angle α equal to $\pi/4$, we obtain

$$\text{WP}(\phi, \alpha = \pi/4) = \frac{e^{ir\phi}}{2} \begin{pmatrix} 1 + e^{i\phi} & 1 - e^{i\phi} \\ 1 - e^{i\phi} & 1 + e^{i\phi} \end{pmatrix}. \tag{8.69}$$

We can see that the x-rotation $R_x(\theta)$ given in Eq. (8.58), can be achieved by setting $\phi = -\theta$ as

$$\text{WP}(\phi = -\theta, \alpha = \pi/4) = e^{-ir\theta} e^{-i\frac{\theta}{2}} \begin{pmatrix} \cos\frac{\theta}{2} & i\sin\frac{\theta}{2} \\ i\sin\frac{\theta}{2} & \cos\frac{\theta}{2} \end{pmatrix}. \tag{8.70}$$

As we did for the case of $R_z(\theta)$, we compensate the overall phase factor by a phase shifter of $e^{is\theta}$ and then the x-rotation is given by

$$R_x(\theta) :=: \text{PS}(s\theta)\text{WP}(-\theta, \alpha = \pi/4). \tag{8.71}$$

When combined together, the basic gate operations given above, the phase gate, z-rotation, and x-rotation are sufficient to build any arbitrary one qubit gate. For example, one can then construct the y-rotation by using x-rotation and z-rotation

$$R_y(\theta) = R_z(-\frac{\pi}{2})R_x(\theta)R_z(\frac{\pi}{2}). \tag{8.72}$$

As another example, the Hadamard gate, one of the most frequently used one-qubit gates in the literature, can be built as follows:

$$H = \frac{1}{\sqrt{2}} \begin{pmatrix} 1 & 1 \\ 1 & -1 \end{pmatrix} = P(-\frac{\pi}{2})R_z(\pi)R_y(\frac{\pi}{2})$$

$$= P(-\frac{\pi}{2})R_z(\frac{\pi}{2})R_x(\frac{\pi}{2})R_z(\frac{\pi}{2}). \tag{8.73}$$

Appendix B: Nonlinear sign gate

Together with quantum teleportation the nonlinear sign (NS) gate serves the basic element in the architecture of linear optics quantum computation. The NS gate applies to photon number state consists of zero, one and two photons as is defined by

$$\alpha_0|0\rangle + \alpha_1|1\rangle + \alpha_2|2\rangle \rightarrow \alpha_0|0\rangle + \alpha_1|1\rangle - \alpha_2|2\rangle. \tag{8.74}$$

The NS gate can be implemented non-deterministically by three beam splitters, two photo-detectors, and one ancilla photon as depicted in Fig.8.10.

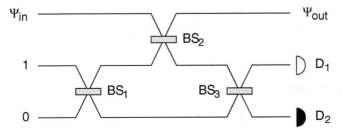

FIGURE 8.10: Nonlinear sign gate: The input state of $|\Psi_{in}\rangle = \alpha_0|0\rangle + \alpha_1|1\rangle + \alpha_2|2\rangle$ transforms to $|\Psi_{out}\rangle = \alpha_0|0\rangle + \alpha_1|1\rangle - \alpha_2|2\rangle$ upon detection of one photon at D_1 and non-detection at D_2.

We then fix the reflection coefficients of the three beam splitters for the desired operation with a certain detection event. Following Ref. [20], a set of beam splitters that can perform the nonlinear sign gate is given by

$$BS_1 = BS_3 :=: \begin{pmatrix} \sqrt{\eta} & \sqrt{1-\eta} \\ \sqrt{1-\eta} & -\sqrt{\eta} \end{pmatrix}, \qquad \eta = \frac{1}{(4-2\sqrt{2})} \approx 0.854,$$

$$BS_2 := \begin{pmatrix} -\sqrt{\eta_2} & \sqrt{1-\eta_2} \\ \sqrt{1-\eta_2} & \sqrt{eta_2} \end{pmatrix}, \qquad \eta_2 = (\sqrt{2}-1)^2 \approx 0.172. \quad (8.75)$$

These beam splitters are phase asymmetric, which can be achieved by ordinary beam splitters with additional phase shifters so we have

$$BS_1 = BS_3 := \begin{pmatrix} 1 & 0 \\ 0 & i \end{pmatrix} \begin{pmatrix} \sqrt{\eta} & -i\sqrt{1-\eta} \\ -i\sqrt{1-\eta} & \sqrt{\eta} \end{pmatrix} \begin{pmatrix} 1 & 0 \\ 0 & i \end{pmatrix}. \quad (8.76)$$

This corresponds to two phase shifters of $e^{i\pi/2}$ in the lower path before and after the ordinary beam splitters. Similarly, for BS_2 we have

$$BS_2 := \begin{pmatrix} i & 0 \\ 0 & 1 \end{pmatrix} \begin{pmatrix} \sqrt{\eta_2} & -i\sqrt{1-\eta_2} \\ -i\sqrt{1-\eta_2} & \sqrt{\eta_2} \end{pmatrix} \begin{pmatrix} i & 0 \\ 0 & 1 \end{pmatrix}, \quad (8.77)$$

which corresponds to two phase shifters of $e^{i\pi/2}$ in the upper path before and after the ordinary beam splitters with the reflection coefficient η_2. Conditioned upon a specific detector outcome (detection of one photon at D_1 and non-detection at D_2), we can have the desired output state with probability of 1/4. The gate operation succeeds only once in four times on average. But, the merit is that we know it was successful whenever it was successful.

References

[1] E. Knill, R. Laflamme, and G. J. Milburn, Nature **409**, 46 (2001).

[2] T. B. Pittman, B. C. Jacobs, and J. D. Franson, Phys. Rev. A **64**, 062311 (2001).

[3] J. D. Franson, M. M. Donnegan, and B. C. Jacobs, Phys. Rev. A **69**, 052328 (2004).

[4] D. Gottesman and I. Chuang, Nature **402**, 390 (1999).

[5] F. M. Spedalieri, H. Lee, and J. P. Dowling, Phys. Rev. A **73**, 012334 (2006).

[6] R. Raussendorf and H. J. Briegel, Phys. Rev. Lett. **86**, 5188 (2001).

[7] M. A. Nielsen, Phys. Rev. Lett. **93**, 040503 (2004).

[8] A. Barenco *et al.*, Phys. Rev. A **52**, 3457 (1995).

[9] G. J. Milburn, Phys. Rev. Lett. **62**, 2124 (1989).

[10] M. O. Scully and M. S. Zubairy, *Quantum Optics*, (Cambridge University Press, Cambridge, UK 1997).

[11] R. W. Boyd, *Nonlinear Optics*, (Academic Press, San Diego, CA, 1991).

[12] J. D. Franson *et al.*, Phys. Rev. Lett. **89**, 137901 (2002).

[13] M. Reck, A. Zeilinger, H. J. Bernstein, and P. Bertani, Phys. Rev. Lett. **73**, 58 (1994).

[14] C. K. Hong, Z. Y. Ou, and L. Mandel, Phys. Rev. Lett. **59**, 2044 (1987).

[15] D. Rosenberg, A. E. Lita, A. J. Miller, and S. W. Nam, Phys. Rev. A **71**, 061803(R) (2005).

[16] L.-M. Duan and R. Raussendorf, Phys. Rev. Lett. **95**, 080503 (2005).

[17] S. D. Barrett and P. Kok, Phys. Rev. A **71**, 060310(R) (2005).

[18] H. Lee *et al.*, J. of Mod. Opt. **51**, 1517 (2004).

[19] A. J. Miller, A. W. Nam, J. M. Martinis, and A. V. Sergienko, Applied Physics Letters **83**, 791 (2003).

[20] T. C. Ralph, A. G. White, W. J. Munro, and G. J. Milburn, Phys. Rev. A **65** 012314 (2001).

Quantum Information

Chapter 9

Exploiting Entanglement in Quantum Cryptographic Probes

Howard E. Brandt

Abstract   The mathematical physical bases are given for quantum cryptographic probes which exploit entanglement to eavesdrop on quantum key distribution. The quantum circuits and designs are presented for two different optimized entangling probes for attacking the BB84 Protocol of quantum key distribution (QKD) and yielding maximum information to the probes. Probe photon polarization states become optimally entangled with the signal states on their way between the legitimate transmitter and receiver. The designs are based on different optimum unitary transformations, each yielding the same maximum information on the pre-privacy amplified key. In each of the designs, the main quantum circuit consists of a single CNOT gate which produces the optimum entanglement between the BB84 signal states and the correlated states of the probe. For each design a different well-defined single-photon polarization state serves as the initial state of the probe, and in each case the probe is projectively measured. Symmetric projective measurements of the probe yield maximum information on the pre-privacy amplified key once basis information becomes available during basis reconciliation.

9.1 Introduction

In cryptography, the key is a random binary sequence used for encryption. An encrypted message can be produced by adding the key to the message (also written in binary), and decrypted by subtracting the key from the encryption. The key must be secure. In the BB84 protocol [1] of quantum key distribution [2] , the ones and zeros of a potential key can be encoded in four different single-photon linear polarization states. Those four states $|u\rangle$, $|\bar{u}\rangle$, $|v\rangle$, and $|\bar{v}\rangle$ all lie in a real two-dimensional Hilbert space such that $\langle u|\bar{u}\rangle = 0$, $\langle v|\bar{v}\rangle = 0$, and $\langle u|v\rangle = 2^{-1/2}$. The states $|u\rangle$ and $|v\rangle$ can be chosen to encode binary number 1, and the states $|\bar{u}\rangle$ and $|\bar{v}\rangle$ encode the number 0. The states $\{|u\rangle, |\bar{u}\rangle\}$ form one basis, the states $\{|v\rangle, |\bar{v}\rangle\}$ form the other basis, and the states $|u\rangle$ and $|v\rangle$ are nonorthogonal with angle $\pi/4$ between them. For analytical purposes it is convenient to choose two orthonormal basis states $|e_0\rangle$ and $|e_1\rangle$ oriented symmetrically with respect to the four signal states, the state $|e_0\rangle$ making an angle of magnitude $\pi/8$ with the states $|u\rangle$ and $|\bar{v}\rangle$, and the state $|e_1\rangle$ making an angle of magnitude $\pi/8$ with the states $|\bar{u}\rangle$ and $|v\rangle$ [3]. In the BB84 protocol, the transmitter (also known as Alice) randomly chooses a basis $\{|u\rangle, |\bar{u}\rangle\}$ or $\{|v\rangle, |\bar{v}\rangle\}$ and then randomly picks one of the two states in the chosen basis and then sends it to the receiver (also known as Bob). The receiver randomly chooses one of the two measurement bases $\{|u\rangle, |\bar{u}\rangle\}$ or $\{|v\rangle, |\bar{v}\rangle\}$, and during reconciliation publicly announces which measurement basis is chosen. This procedure is repeated for each photon in a train of single photons transmitted from the transmitter to the receiver. In those cases in which the basis choices by the transmitter and the receiver are the same, that bit is kept and contributes to the potential key. Bits resulting from differing basis selections by the transmitter and receiver, as well as any erroneous bits identified by block checksums and bisective search, are discarded and do not contribute to the key. Also during reconciliation the relative order of selected and discarded bits along with the respective basis choices are in principle publicly available to an eavesdropper.

Numerous analyses of various eavesdropping procedures have appeared in the literature [2], allowing quantitative comparisons of various protocols. These analyses are far too numerous to cite individually. A recent review with extensive references is given in Ref. [2]. Analyses of the protocol addressed here are given in [3], [4], [5].

The present work is limited to an individual attack in which each transmitted photon is measured by an independent probe. In particular, the eavesdropping attack addressed here is accomplished using a so-called 'quantum cryptographic entangling probe' (see Fig. 9.1). The probe becomes optimally entangled with the signal photon on its way from the transmitter to the receiver, extracting maximum possible Renyi information on the key. [See Appendices A and B for reviews of Renyi information gain and the optimization procedure.] The incident signal photon enters the control port of a quantum controlled-NOT gate (CNOT gate) and becomes optimally entangled with a probe photon in a specific polarization state determined by the desired error rate to be induced by the probe. The signal photon leaves the exit port of the CNOT gate and goes on to the legitimate key receiver. The gated probe photon leaves the exit port of the CNOT gate, and its state (optimally correlated with the state of the signal photon) is projectively measured using a polarization beamsplitter (Wollaston prism), aligned to separate two appropriate polarization basis states, together with two photodetectors. The details and justification are given in Sections 9.2–9.4.

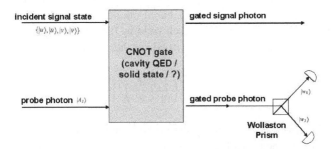

FIGURE 9.1: Quantum cryptographic entangling probe: Signal photon enters control port of CNOT gate. Probe photon in state $|A_2\rangle$ enters target port. CNOT gate might be implemented using cavity QED, solid state, etc. Gated signal photon goes on to legitimate receiver. Gated probe photon, optimally entangled with the signal, is measured using a Wollaston prism (polarization beamsplitter) and two photodetectors aligned to separate polarization states $|w_0\rangle$ and $|w_3\rangle$.

In addition to the individual attack, other approaches include: coherent collective attacks in which the eavesdropper entangles a separate probe with each transmitted photon and measures all probes together as one system; and also coherent joint attacks in which a single probe is entangled with the entire set of carrier photons. However, these approaches require maintenance of coherent superpositions of large numbers of states, and this is not currently feasible.

State storage and decoherence present major issues.

For the case of an individual attack, eavesdropping probe optimizations have been performed [3], [4], [6], [7], which on average yield the most information to the eavesdropper for a given error rate caused by the probe. The most general possible probes consistent with unitarity were considered [3], [4], [5], [6], [7], [8] in which each individual transmitted bit is made to interact with the probe so that the carrier and the probe are left in an optimum entangled state, and measurement of the probe then yields information about the carrier state. The probe optimizations are based on maximizing the order-two Renyi information gain by the probe on corrected data for a set error rate induced by the probe in the legitimate receiver. It is well to recall that a security proof was given in Ref. [3] for such an individual attack. However, the attack addressed here has been shown to be very taxing in terms of the amount of key that must be sacrificed during privacy amplification to make the key secure [5].

In the following sections, recent designs are summarized for some optimum quantum cryptographic entangling probes to be used in attacking the BB84 protocol [9], [10], [11], [12]. They are based on alternative optimal unitary transformations, $U^{(1)}$, $U^{(2)}$, and $U^{(3)}$, each yielding the same maximum Renyi information I_{opt}^R on the pre-privacy-amplified key, namely (see Appendix B),

$$I_{opt}^R = \log_2 \left[2 - \left(\frac{1-3E}{1-E} \right)^2 \right],\qquad(9.1)$$

where E is the error rate induced by the probe in the legitimate receiver. Equation (1) displays the tradeoff between information gained and the error rate induced by the probe in the legitimate receiver. The unitary transformation $U^{(1)}$ represents an entangling probe having a two-dimensional Hilbert space of states, while the transformations $U^{(2)}$ and $U^{(3)}$ represent probes having, in general, four-dimensional Hilbert spaces of states [9]. In Sections 9.2, 9.3, and 9.4, probe designs are described, faithfully representing the transformations $U^{(1)}$, $U^{(2)}$, and $U^{(3)}$ respectively. In each of the designs, the main quantum circuit consists of a single CNOT gate which produces the optimum entanglement between the BB84 signal states and the correlated probe states. The BB84 signal states enter the control port of the CNOT gate, and for each design a different particular initial single-photon polarization state enters the target port of the probe. In each case the probe is projectively measured.

9.2 Probe designs based on $U^{(1)}$

The effect of the optimum unitary transformation $U^{(1)}$ is to produce the following entanglements between the BB84 signal states $|u\rangle$, $|\bar{u}\rangle$, $|v\rangle$, and $|\bar{v}\rangle$ and probe states $|\alpha_+\rangle$, $|\alpha_-\rangle$, and $|\alpha\rangle$ [9], [10]:

$$|u\rangle \otimes |w\rangle \longrightarrow \frac{1}{4}\left(|u\rangle \otimes |\alpha_+\rangle + |\bar{u}\rangle \otimes |\alpha\rangle\right), \tag{9.2}$$

$$|\bar{u}\rangle \otimes |w\rangle \longrightarrow \frac{1}{4}\left(|u\rangle \otimes |\alpha\rangle + |\bar{u}\rangle \otimes |\alpha_-\rangle\right), \tag{9.3}$$

$$|v\rangle \otimes |w\rangle \longrightarrow \frac{1}{4}\left(|v\rangle \otimes |\alpha_-\rangle - |\bar{v}\rangle \otimes |\alpha\rangle\right), \tag{9.4}$$

$$|\bar{v}\rangle \otimes |w\rangle \longrightarrow \frac{1}{4}\left(-|v\rangle \otimes |\alpha\rangle + |\bar{v}\rangle \otimes |\alpha_+\rangle\right), \tag{9.5}$$

where $|w\rangle$ is the initial state of the probe, and the probe states $|\alpha_+\rangle$, $|\alpha_-\rangle$, and $|\alpha\rangle$ are given by

$$|\alpha_+\rangle = \left[\left(2^{1/2}+1\right)(1\pm\eta)^{1/2} + \left(2^{1/2}-1\right)(1\mp\eta)^{1/2}\right]|w_0\rangle$$
$$+ \left[\left(2^{1/2}+1\right)(1\mp\eta)^{1/2} + \left(2^{1/2}-1\right)(1\pm\eta)^{1/2}\right]|w_3\rangle, \tag{9.6}$$

$$|\alpha_-\rangle = \left[\left(2^{1/2}-1\right)(1\pm\eta)^{1/2} + \left(2^{1/2}+1\right)(1\mp\eta)^{1/2}\right]|w_0\rangle$$
$$+ \left[\left(2^{1/2}-1\right)(1\mp\eta)^{1/2} + \left(2^{1/2}+1\right)(1\pm\eta)^{1/2}\right]|w_3\rangle, \tag{9.7}$$

$$|\alpha\rangle = \left[-(1\pm\eta)^{1/2} + (1\mp\eta)^{1/2}\right]|w_0\rangle + \left[-(1\mp\eta)^{1/2} + (1\pm\eta)^{1/2}\right]|w_3\rangle, \tag{9.8}$$

expressed in terms of orthonormal basis states $|w_0\rangle$ and $|w_3\rangle$, and also the set error rate E induced by the probe in the legitimate receiver, where

$$\eta \equiv [8E(1-2E)]^{1/2}. \tag{9.9}$$

For this optimization the error rate is restricted to $0 \leq E \leq 1/4$. According to Eq. (9.9), η is monotonically increasing only for $0 \leq E \leq 1/4$. Note in Eqs. (9.6)–(9.8) that the Hilbert space of the probe is two-dimensional, depending on the two probe basis states, $|w_0\rangle$ and $|w_3\rangle$.

According to Eq. (9.2), the projected probe state $|\psi_{uu}\rangle$ correlated with the correct received signal state (in the notation of [3], [7]), in which the state $|u\rangle$ is sent by the transmitter, and is also received by the legitimate receiver, is

$|\alpha_+\rangle$. Analogously, from Eq. (9.3), it follows that the correlated probe state $|\psi_{\bar{u}\bar{u}}\rangle$ is $|\alpha_-\rangle$. The two states $|\alpha_+\rangle$ and $|\alpha_-\rangle$ are to be distinguished by the measurement of the probe. Also, according to Eqs. (9.4) and (9.5), the same two probe states $|\alpha_+\rangle$ and $|\alpha_-\rangle$ are the appropriate correlated states $|\psi_{\bar{v}\bar{v}}\rangle$ and $|\psi_{vv}\rangle$, respectively. This is consistent with the assumption in Section II of [3] that only two probe states must be distinguished by the probe.

In the following, this two-dimensional optimized unitary transformation is used to show that a simple quantum circuit representing the optimal entangling probe consists of a single quantum controlled-NOT gate (CNOT gate) in which the control qubit consists of two polarization basis states of the signal, the target qubit consists of two probe basis states, and the initial state of the probe is set in a specific way by the error rate. A method is given for measuring the appropriate correlated states of the probe. Finally, a design for the entangling probe is described.

One proceeds by exploiting the quantum circuit model of quantum computation [13] to determine the quantum circuit corresponding to the optimum unitary transformation, Eqs. (9.2)–(9.5). It was shown in [9] that the tensor products of the initial state $|w\rangle$ of the probe with the orthonormal basis states $|e_0\rangle$ and $|e_1\rangle$ of the signal transform as follows under the optimal unitary transformation:

$$|e_0 \otimes w\rangle \longrightarrow |e_0\rangle \otimes |A_1\rangle \qquad (9.10)$$

and

$$|e_1 \otimes w\rangle \longrightarrow |e_1\rangle \otimes |A_2\rangle, \qquad (9.11)$$

expressed in terms of probe states $|A_1\rangle$ and $|A_2\rangle$, where

$$|A_1\rangle = a_1 |w_0\rangle + a_2 |w_3\rangle, \qquad (9.12)$$
$$|A_2\rangle = a_2 |w_0\rangle + a_1 |w_3\rangle, \qquad (9.13)$$

in which

$$a_1 = 2^{-1/2}(1 \pm \eta)^{1/2}, \qquad (9.14)$$

$$a_2 = 2^{-1/2}(1 \mp \eta)^{1/2}, \qquad (9.15)$$

and η is given by Eq. (9.9).

Next, consider a quantum controlled-NOT gate (CNOT gate), in which the control qubit consists of the two signal basis states $\{|e_0\rangle, |e_1\rangle\}$, and the target qubit consists of the probe basis states $\{|w_0\rangle, |w_3\rangle\}$, and such that when $|e_0\rangle$ enters the control port then $\{|w_0\rangle, |w_3\rangle\}$ becomes $\{|w_3\rangle, |w_0\rangle\}$ at the target output port, or when $|e_1\rangle$ enters the control port then $\{|w_0\rangle, |w_3\rangle\}$ remains unchanged. It then follows that a simple quantum circuit affecting the transformations (9.10) and (9.11), and thereby faithfully representing the entangling probe, consists of this CNOT gate with the state $|A_2\rangle$ always entering the target port, and $\{|e_0\rangle, |e_1\rangle\}$ entering the control port. When $|e_0\rangle$ enters the control port, then $|A_2\rangle$ becomes $|A_1\rangle$, or when $|e_1\rangle$ enters the control port then $|A_2\rangle$ remains unchanged, in agreement with Eqs. (9.10) and (9.11) with $|w\rangle = |A_2\rangle$. According to the quantum circuit model of quantum computation, it is known that at most three CNOT gates, together with single-qubit gates, are in general necessary and sufficient in order to implement an arbitrary number of unitary transformations of two qubits [14]. Evidently in the present case, a single CNOT gate suffices to faithfully represent the optimized unitary transformation.

Next expanding the signal state $|u\rangle$ in terms of the signal basis states, using [3, Eq. (1)], one has

$$|u\rangle = \cos\frac{\pi}{8}|e_0\rangle + \sin\frac{\pi}{8}|e_1\rangle, \tag{9.16}$$

$$|\bar{u}\rangle = -\sin\frac{\pi}{8}|e_0\rangle + \cos\frac{\pi}{8}|e_1\rangle, \tag{9.17}$$

$$|v\rangle = \sin\frac{\pi}{8}|e_0\rangle + \cos\frac{\pi}{8}|e_1\rangle, \tag{9.18}$$

$$|\bar{v}\rangle = \cos\frac{\pi}{8}|e_0\rangle - \sin\frac{\pi}{8}|e_1\rangle. \tag{9.19}$$

It then follows from Eqs. (9.10), (9.11), and (9.16) that the CNOT gate affects the following transformation when the signal state $|u\rangle$ enters the control port:

$$|u\rangle \otimes |A_2\rangle \longrightarrow \cos\frac{\pi}{8}|e_0\rangle \otimes |A_1\rangle + \sin\frac{\pi}{8}|e_1\rangle \otimes |A_2\rangle. \tag{9.20}$$

Using Eqs. (9.16) and (9.17), one also has

$$|e_0\rangle = \cos\frac{\pi}{8}|u\rangle - \sin\frac{\pi}{8}|\bar{u}\rangle, \tag{9.21}$$

$$|e_1\rangle = \sin\frac{\pi}{8}|u\rangle + \cos\frac{\pi}{8}|\bar{u}\rangle. \tag{9.22}$$

Next substituting Eqs. (9.21) and (9.22) in Eq. (9.20), one has

$$|u\rangle \otimes |A_2\rangle \longrightarrow \cos\frac{\pi}{8}\left(\cos\frac{\pi}{8}|u\rangle - \sin\frac{\pi}{8}|\bar{u}\rangle\right) \otimes |A_1\rangle$$
$$+ \sin\frac{\pi}{8}\left(\sin\frac{\pi}{8}|u\rangle + \cos\frac{\pi}{8}|\bar{u}\rangle\right) \otimes |A_2\rangle, \tag{9.23}$$

and using

$$\sin\frac{\pi}{8} = \frac{1}{2}(2 - 2^{1/2})^{1/2}, \tag{9.24}$$

$$\cos\frac{\pi}{8} = \frac{1}{2}(2+2^{1/2})^{1/2}, \tag{9.25}$$

then Eq. (9.23) becomes

$$|u\rangle \otimes |A_2\rangle \longrightarrow \frac{1}{4}[(2+2^{1/2})|u\rangle \otimes |A_1\rangle + (2-2^{1/2})|u\rangle \otimes |A_2\rangle$$
$$-2^{1/2}|\bar{u}\rangle \otimes |A_1\rangle + 2^{1/2}|\bar{u}\rangle \otimes |A_2\rangle]. \tag{9.26}$$

Then substituting Eqs. (9.12)–(9.15) in Eq. (9.26), and using Eqs. (9.6) and (9.8), one obtains Eq. (9.2). Analogously, one also obtains Eqs. (9.3)–(9.5).

Thus the quantum circuit consisting of the CNOT gate does in fact faithfully represent the action of the optimum unitary transformation in entangling the signal states $|u\rangle$, $|\bar{u}\rangle$, $|v\rangle$, and $|\bar{v}\rangle$ with the probe states $|\alpha_+\rangle$, $|\alpha_-\rangle$, and $|\alpha\rangle$. It is to be emphasized that the initial state of the probe must be $|A_2\rangle$, given by Eqs. (9.13)–(9.15). (A sign choice in Eqs. (9.14) and (9.15) is made below following Eqs. (9.35) and (9.36), consistent with the measurement procedure defined there.)

Thus, according to Eqs. (9.2)–(9.5) and the above analysis, the probe produces the following entanglements for initial probe state $|w\rangle = |A_2\rangle$ and incoming BB84 signal states $|u\rangle$, $|\bar{u}\rangle$, $|v\rangle$, or $|\bar{v}\rangle$, respectively:

$$|u\rangle \otimes |A_2\rangle \longrightarrow \frac{1}{4}\left(|u\rangle \otimes |\alpha_+\rangle + |\bar{u}\rangle \otimes |\alpha\rangle\right), \tag{9.27}$$

$$|\bar{u}\rangle \otimes |A_2\rangle \longrightarrow \frac{1}{4}\left(|u\rangle \otimes |\alpha\rangle + |\bar{u}\rangle \otimes |\alpha_-\rangle\right), \tag{9.28}$$

$$|v\rangle \otimes |A_2\rangle \longrightarrow \frac{1}{4}\left(|v\rangle \otimes |\alpha_-\rangle - |\bar{v}\rangle \otimes |\alpha\rangle\right), \tag{9.29}$$

$$|\bar{v}\rangle \otimes |A_2\rangle \longrightarrow \frac{1}{4}\left(-|v\rangle \otimes |\alpha\rangle + |\bar{v}\rangle \otimes |\alpha_+\rangle\right). \tag{9.30}$$

Then, according to Eqs. (9.27) and (9.28), if, following the public reconciliation phase of the BB84 protocol, the signal basis mutually selected by the legitimate transmitter and receiver is publicly revealed to be $\{|u\rangle, |\bar{u}\rangle\}$, then the probe measurement must distinguish the projected probe state $|\alpha_+\rangle$, when the signal state $|u\rangle$ is both sent and received, from the projected probe state $|\alpha_-\rangle$, when the signal state $|\bar{u}\rangle$ is both sent and received. In this case one has the correlations:

$$|u\rangle \Longleftrightarrow |\alpha_+\rangle, \tag{9.31}$$

$$|\bar{u}\rangle \Longleftrightarrow |\alpha_-\rangle. \tag{9.32}$$

The same two states $|\alpha_+\rangle$ and $|\alpha_-\rangle$ must be distinguished, no matter which basis is chosen during reconciliation. Thus, according to Eqs. (9.29) and (9.30), if, following the public reconciliation phase of the BB84 protocol, the signal basis mutually selected by the legitimate transmitter and receiver is publicly revealed to be $\{|v\rangle,|\bar{v}\rangle\}$, then the probe measurement must distinguish the projected probe state $|\alpha_-\rangle$, when the signal state $|v\rangle$ is both sent and received, from the projected probe state $|\alpha_+\rangle$, when the signal state $|\bar{v}\rangle$ is both sent and received. In this case one has the correlations:

$$|v\rangle \Longleftrightarrow |\alpha_-\rangle, \tag{9.33}$$

$$|\bar{v}\rangle \Longleftrightarrow |\alpha_+\rangle. \tag{9.34}$$

Next, one notes that the correlations of the projected probe states $|\alpha_+\rangle$ and $|\alpha_-\rangle$ with the probe's two orthogonal basis states $|w_0\rangle$ and $|w_3\rangle$ are indicated, according to Eqs. (9.6) and (9.7), by the following probabilities:

$$\frac{|\langle w_0|\alpha_+\rangle|^2}{|\alpha_+|^2} = \frac{|\langle w_3|\alpha_-\rangle|^2}{|\alpha_-|^2} = \frac{1}{2} \pm \frac{[E(1-2E)]^{1/2}}{(1-E)}, \tag{9.35}$$

$$\frac{|\langle w_0|\alpha_-\rangle|^2}{|\alpha_-|^2} = \frac{|\langle w_3|\alpha_+\rangle|^2}{|\alpha_+|^2} = \frac{1}{2} \mp \frac{[E(1-2E)]^{1/2}}{(1-E)}. \tag{9.36}$$

At this point one can make a choice of the positive sign in Eq. (9.35), and correspondingly the negative sign in Eq. (9.36). This choice serves to define the Hilbert space orientation of the probe basis states, in order that the probe basis state $|w_0\rangle$ be dominantly correlated with the signal states $|u\rangle$ and $|\bar{v}\rangle$, and that the probe basis state $|w_3\rangle$ be dominantly correlated with the signal states $|\bar{u}\rangle$ and $|v\rangle$. With this sign choice, then Eqs. (9.35) and (9.36) become

$$\frac{|\langle w_0|\alpha_+\rangle|^2}{|\alpha_+|^2} = \frac{|\langle w_3|\alpha_-\rangle|^2}{|\alpha_-|^2} = \frac{1}{2} + \frac{[E(1-2E)]^{1/2}}{(1-E)}, \tag{9.37}$$

$$\frac{|\langle w_0|\alpha_-\rangle|^2}{|\alpha_-|^2} = \frac{|\langle w_3|\alpha_+\rangle|^2}{|\alpha_+|^2} = \frac{1}{2} - \frac{[E(1-2E)]^{1/2}}{(1-E)}, \tag{9.38}$$

and one then has the following state correlations:

$$|\alpha_+\rangle \Longleftrightarrow |w_0\rangle, \tag{9.39}$$

$$|\alpha_-\rangle \Longleftrightarrow |w_3\rangle. \tag{9.40}$$

Next combining the correlations (31)-(34), (39), and (40), one then establishes the following correlations:

$$\{|u\rangle,|\bar{v}\rangle\} \Longleftrightarrow |\alpha_+\rangle \Longleftrightarrow |w_0\rangle, \tag{9.41}$$

$$\{|\bar{u}\rangle, |v\rangle\} \iff |\alpha_-\rangle \iff |w_3\rangle, \tag{9.42}$$

to be implemented by the probe measurement method. This can be simply accomplished by a von Neumann-type projective measurement of the orthogonal probe basis states $|w_0\rangle$ and $|w_3\rangle$, implementing the probe projective measurement operators $\{|w_0 \rangle\langle w_0|, |w_3 \rangle\langle w_3|\}$ (see Appendix A). The chosen geometry in the two-dimensional Hilbert space of the probe is such that the orthogonal basis states $|w_0\rangle$ and $|w_3\rangle$ make equal angles with the states $|\alpha_+\rangle$ and $|\alpha_-\rangle$, respectively, and the sign choice is enforced in Eqs. (9.6) and (9.7), namely,

$$|\alpha_+\rangle = \left[\left(2^{1/2}+1\right)(1+\eta)^{1/2} + \left(2^{1/2}-1\right)(1-\eta)^{1/2}\right]|w_0\rangle \tag{9.43}$$
$$+ \left[\left(2^{1/2}+1\right)(1-\eta)^{1/2} + \left(2^{1/2}-1\right)(1+\eta)^{1/2}\right]|w_3\rangle,$$

$$|\alpha_-\rangle = \left[\left(2^{1/2}-1\right)(1+\eta)^{1/2} + \left(2^{1/2}+1\right)(1-\eta)^{1/2}\right]|w_0\rangle \tag{9.44}$$
$$+ \left[\left(2^{1/2}-1\right)(1-\eta)^{1/2} + \left(2^{1/2}+1\right)(1+\eta)^{1/2}\right]|w_3\rangle,$$

where

$$\eta = [8E(1-2E)]^{1/2}, \tag{9.45}$$

as in Eq. (9.9). Such a symmetric von Neumann test is known to be optimal [15], [16], and is an essential part of the optimization [3].

Next, based on the above results, one arrives at a simple entangling probe design [9], [10], [17] (see Fig. 9.1). An incident photon coming from the legitimate transmitter is received by the probe in one of the four signal-photon linear-polarization states $|u\rangle$, $|\bar{u}\rangle$, $|v\rangle$, or $|\bar{v}\rangle$ in the BB84 protocol. The signal photon enters the control port of the CNOT gate. The initial state of the probe is a photon in linear-polarization state $|A_2\rangle$ and entering the target port of the CNOT gate. The probe photon is produced by a single-photon source and is appropriately timed with reception of the signal photon by first sampling a few successive signal pulses to determine the repetition rate of the transmitter. The linear-polarization state $|A_2\rangle$, according to Eqs. (9.13)–(9.15) and (9.9), with the sign choice made above, is given by

$$|A_2\rangle = \left[\frac{1}{2}\{1 - [8E(1-2E)]^{1/2}\}\right]^{1/2}|w_0\rangle$$
$$+ \left[\frac{1}{2}\{1 + [8E(1-2E)]^{1/2}\}\right]^{1/2}|w_3\rangle, \tag{9.46}$$

and can be simply set for an error rate $E \leq 1/4$ by means of a polarizer. In this way the device can be tuned to the chosen error rate to be induced by the probe. The outgoing gated signal photon is relayed on to the legitimate receiver, and the gated probe photon enters a Wollaston prism, oriented to separate photon orthogonal-linear-polarization states $|w_0\rangle$ and $|w_3\rangle$, and the photon is then detected by one of two photodetectors. If the basis, revealed during the public basis-reconciliation phase of the BB84 protocol, is $\{|u\rangle, |\bar{u}\rangle\}$, then the photodetector located to receive the polarization state $|w_0\rangle$ or $|w_3\rangle$, respectively, will indicate, in accord with the correlations (41) and (42), that a state $|u\rangle$ or $|\bar{u}\rangle$, respectively, was most likely measured by the legitimate receiver. Alternatively, if the announced basis is $\{|v\rangle, |\bar{v}\rangle\}$, then the photodetector located to receive the polarization state $|w_3\rangle$ or $|w_0\rangle$, respectively, will indicate, in accord with the correlations (41) and (42), that a state $|v\rangle$ or $|\bar{v}\rangle$, respectively, was most likely measured by the legitimate receiver. By comparing the record of probe photodetector triggering with the sequence of bases revealed during reconciliation, then the likely sequence of ones and zeroes constituting the key, prior to privacy amplification, can be assigned. In any case the net effect is to yield, for a set error rate E, the maximum information gain to the probe, which is given by Eq. (9.1).

The geometry of the initial and shifted probe polarization states $|A_2\rangle$ and $|A_1\rangle$, respectively, and the orthogonal probe basis states, $|w_0\rangle$ and $|w_3\rangle$, in the two-dimensional Hilbert space of the probe, is such that the angle δ_0 between the probe state $|A_1\rangle$ and the probe basis state $|w_0\rangle$ is given by

$$\delta_0 = \cos^{-1}\left(\frac{\langle w_0|A_1\rangle}{|A_1|}\right), \tag{9.47}$$

or, substituting $|A_1\rangle$, given by Eqs. (9.12), (9.14), and (9.15) with the above sign choice, namely,

$$|A_1\rangle = \left[\frac{1}{2}\{1 + [8E(1-2E)]^{1/2}\}\right]^{1/2}|w_0\rangle$$
$$+ \left[\frac{1}{2}\{1 - [8E(1-2E)]^{1/2}\}\right]^{1/2}|w_3\rangle, \tag{9.48}$$

in Eq. (9.47), one obtains

$$\delta_0 = \cos^{-1}\left(\frac{1}{2}\{1 + [8E(1-2E)]^{1/2}\}\right)^{1/2}. \tag{9.49}$$

This is also the angle between the initial linear-polarization state $|A_2\rangle$ of the probe and the probe basis state $|w_3\rangle$. Also, the shift δ in polarization between the initial probe states $|A_2\rangle$ and the state $|A_1\rangle$ (the angle between $|A_1\rangle$ and $|A_2\rangle$) is given by

$$\delta = \cos^{-1}\left(\frac{\langle A_1|A_2\rangle}{|A_1||A_2|}\right), \tag{9.50}$$

or, substituting Eqs. (9.48) and (9.46), one obtains

$$\delta = \cos^{-1}(1 - 4E). \tag{9.51}$$

Possible implementations of the CNOT gate are under consideration, including ones based on cavity-QED, solid state, and/or linear optics [18]. However, a sufficiently robust high-fidelity CNOT gate, for control and target qubits based on single-photon orthogonal polarization states, is not yet available.

Recently a generalization [11] was given to include the full range of error rates, $0 \leq E \leq 1/3$. In this case, the CNOT target state $|A_2\rangle$ and the probe states $|\alpha_+\rangle$, $|\alpha_-\rangle$, and $|\alpha\rangle$ in Eqs. (9.2)–(9.5) become:

$$|A_2\rangle = \text{sgn}(1 - 4E)\left[\frac{1}{2}(1 - \eta)\right]^{1/2}|w_0\rangle + \left[\frac{1}{2}(1 + \eta)\right]^{1/2}|w_3\rangle, \tag{9.52}$$

$$\begin{aligned}|\alpha_+\rangle = &\left[\left(2^{1/2} + 1\right)(1 + \eta)^{1/2} + \text{sgn}(1 - 4E)\left(2^{1/2} - 1\right)(1 - \eta)^{1/2}\right]|w_0\rangle \\ &+ \left[\text{sgn}(1 - 4E)\left(2^{1/2} + 1\right)(1 - \eta)^{1/2}\right. \\ &\left. + \left(2^{1/2} - 1\right)(1 + \eta)^{1/2}\right]|w_3\rangle,\end{aligned} \tag{9.53}$$

$$\begin{aligned}|\alpha_-\rangle = &\left[\left(2^{1/2} - 1\right)(1 + \eta)^{1/2} + \text{sgn}(1 - 4E)\left(2^{1/2} + 1\right)(1 - \eta)^{1/2}\right]|w_0\rangle \\ &+ \left[\text{sgn}(1 - 4E)\left(2^{1/2} - 1\right)(1 - \eta)^{1/2}\right. \\ &\left. + \left(2^{1/2} + 1\right)(1 + \eta)^{1/2}\right]|w_3\rangle,\end{aligned} \tag{9.54}$$

$$\begin{aligned}|\alpha\rangle = &\left[\text{sgn}(1 - 4E)(1 - \eta)^{1/2} - (1 + \eta)^{1/2}\right]|w_0\rangle \\ &+ \left[(1 + \eta)^{1/2} - \text{sgn}(1 - 4E)(1 - \eta)^{1/2}\right]|w_3\rangle,\end{aligned} \tag{9.55}$$

respectively, where

$$\text{sgn}(x) \equiv \begin{cases} 1, & x > 0 \\ 0, & x = 0 \\ -1, & x < 0 \end{cases}. \tag{9.56}$$

Note that Eqs. (9.52)–(9.56) are consistent with Eqs. (9.46) and (9.6)–(9.9) (with the upper sign choice) for $0 \leq E \leq 1/4$, as must be the case. It then follows that Eqs. (9.2)–(9.5), along with Eqs. (9.53)–(9.56) above, now apply for $0 \leq E \leq 1/3$. (Note that according to Eq. (9.1), $E = 1/3$ corresponds to complete information gain by the quantum cryptographic entangling probe.) Also the probe and measurement method remain the same as in the above with the initial state of the probe now given by Eq. (9.52).

In obtaining the maximum Renyi information gain I_{opt}^R by the probe, one first uses Eq. (9.130) of Appendix A, namely,

$$I_{opt}^R = \log_2(2 - Q^2). \tag{9.57}$$

Next, to obtain the minimum overlap Q of correlated probe states (which yields the maximum I_{opt}^R), one substitutes Eqs. (9.53) and (9.54), with $|\psi_{uu}\rangle = |\alpha_+\rangle$ and $|\psi_{\bar{u}\bar{u}}\rangle = |\alpha_-\rangle$, in Eq. (9.147) of Appendix A to obtain:

$$Q = \frac{\langle \alpha_+|\alpha_-\rangle}{|\alpha_+||\alpha_-|} = \frac{1 + 3\mathrm{sgn}(1 - 4E)(1 - \eta^2)^{1/2}}{3 + \mathrm{sgn}(1 - 4E)(1 - \eta^2)^{1/2}}. \tag{9.58}$$

Then substituting Eq. (9.45) in Eq. (9.58), one obtains

$$Q = \frac{1 + 3\mathrm{sgn}(1 - 4E)((1 - 4E)^2)^{1/2}}{3 + \mathrm{sgn}(1 - 4E)((1 - 4E)^2)^{1/2}}, \tag{9.59}$$

in which the positive square root is meant, namely,

$$((1 - 4E)^2)^{1/2} = |1 - 4E|. \tag{9.60}$$

On noting that

$$\mathrm{sgn}(1 - 4E)|1 - 4E| = 1 - 4E, \tag{9.61}$$

and substituting Eqs. (9.60) and (9.61) in Eq. (9.59), one obtains

$$Q = \frac{1 - 3E}{1 - E}. \tag{9.62}$$

Finally, substituting Eq. (9.62) in Eq. (9.57), one obtains Eq. (9.1) for the full range of error rates, $0 \leq E \leq 1/3$, as required.

9.3 Probe designs based on $U^{(2)}$

The effect of the unitary transformation $U^{(2)}$ is to produce the following entanglements between the signal states $|u\rangle$, $|\bar{u}\rangle$, $|v\rangle$, and $|\bar{v}\rangle$ and the probe states $|\beta_+\rangle$, $|\beta_-\rangle$, and $|\beta\rangle$ [9]:

$$|u\rangle \otimes |w\rangle \longrightarrow \frac{1}{4}\left(|u\rangle \otimes |\beta_+\rangle + |\bar{u}\rangle \otimes |\beta\rangle\right), \qquad (9.63)$$

$$|\bar{u}\rangle \otimes |w\rangle \longrightarrow \frac{1}{4}\left(|u\rangle \otimes |\beta\rangle + |\bar{u}\rangle \otimes |\beta_-\rangle\right), \qquad (9.64)$$

$$|v\rangle \otimes |w\rangle \longrightarrow \frac{1}{4}\left(|v\rangle \otimes |\beta_-\rangle - |\bar{v}\rangle \otimes |\beta\rangle\right), \qquad (9.65)$$

$$|\bar{v}\rangle \otimes |w\rangle \longrightarrow \frac{1}{4}\left(-|v\rangle \otimes |\beta\rangle + |\bar{v}\rangle \otimes |\beta_+\rangle\right), \qquad (9.66)$$

in which

$$
\begin{aligned}
|\beta_\pm\rangle = &\left(1-r^2\right)^{1/2}\left[\left(2 \mp 2^{1/2}\right)\sin\mu + \left(2 \pm 2^{1/2}\right)\cos\mu\right]|w_0\rangle \\
&+\left(1-r^2\right)^{1/2}\left[\left(2 \pm 2^{1/2}\right)\sin\mu + \left(2 \mp 2^{1/2}\right)\cos\mu\right]|w_3\rangle \\
&+e_\lambda e_\theta r\left[\left(2 \mp 2^{1/2}\right)\sin\phi + \left(2 \pm 2^{1/2}\right)\cos\phi\right]|w_1\rangle \\
&+e_\lambda e_\theta r\left[\left(2 \pm 2^{1/2}\right)\sin\phi + \left(2 \mp 2^{1/2}\right)\cos\phi\right]|w_2\rangle, \qquad (9.67)
\end{aligned}
$$

$$
\begin{aligned}
|\beta\rangle = &2^{1/2}\left(1-r^2\right)^{1/2}\left(\sin\mu - \cos\mu\right)\left(|w_0\rangle - |w_3\rangle\right) \\
&+2^{1/2}e_\lambda e_\theta r(\sin\phi - \cos\phi)\left(|w_1\rangle - |w_2\rangle\right), \qquad (9.68)
\end{aligned}
$$

where $|w_0\rangle$, $|w_1\rangle$, $|w_2\rangle$, and $|w_3\rangle$ are orthonormal basis states of the probe, and r is defined by

$$r \equiv \left(\frac{\sin 2\mu - 1 + 4E}{\sin 2\mu - \sin 2\phi}\right)^{1/2}. \qquad (9.69)$$

Also, ϕ and μ are probe parameters constrained by

$$\sin 2\phi < \sin 2\mu \geq 1 - 4E \ \cup \ \sin 2\phi > \sin 2\mu \leq 1 - 4E, \qquad (9.70)$$

but otherwise free, and e_λ and e_θ are sign parameters defined by

$$e_\lambda \equiv \pm 1, \quad e_\theta \equiv \pm 1. \qquad (9.71)$$

Generally, because Eqs. (9.67) and (9.68) depend on four probe basis states $|w_0\rangle$, $|w_1\rangle$, $|w_2\rangle$, and $|w_3\rangle$, the Hilbert space of this probe is four-dimensional. However if one makes the following free choices for the parameters μ, ϕ, e_λ, and e_θ [19]:

$$\sin 2\mu = 0, \qquad (9.72)$$

$$\sin 2\phi = 1 - 4E, \qquad (9.73)$$

$$e_\lambda \, e_\theta \equiv 1, \qquad (9.74)$$

then Eqs. (9.67) and (9.68), for $0 \leq E \leq 1/4$ and with appropriate sign choices, become

$$|\beta_+\rangle = \left[\left(2^{1/2}+1\right)(1+\eta)^{1/2} + \left(2^{1/2}-1\right)(1-\eta)^{1/2}\right]|w_1\rangle \qquad (9.75)$$
$$+ \left[\left(2^{1/2}+1\right)(1-\eta)^{1/2} + \left(2^{1/2}-1\right)(1+\eta)^{1/2}\right]|w_2\rangle,$$

$$|\beta_-\rangle = \left[\left(2^{1/2}-1\right)(1+\eta)^{1/2} + \left(2^{1/2}+1\right)(1-\eta)^{1/2}\right]|w_1\rangle \qquad (9.76)$$
$$+ \left[\left(2^{1/2}-1\right)(1-\eta)^{1/2} + \left(2^{1/2}+1\right)(1+\eta)^{1/2}\right]|w_2\rangle,$$

$$|\beta\rangle = \left[-(1+\eta)^{1/2} + (1-\eta)^{1/2}\right]|w_1\rangle + \left[-(1-\eta)^{1/2} + (1+\eta)^{1/2}\right]|w_2\rangle. \qquad (9.77)$$

In this case, the probe Hilbert space reduces from four to two dimensions. Evidently, Eqs. (9.75)–(9.77) have the same form as Eqs. (9.6)–(9.8), can also be generalized to allow $0 \leq E \leq 1/3$, and will again lead to the probe design of Section 9.2.

9.4 Probe designs based on $U^{(3)}$

In this section, an implementation is presented for the optimum unitary transformation $U^{(3)}$ represented by [9, Eqs. (158)–(164)], however with restricted parameters such that the corresponding Hilbert space of the probe reduces from four to two dimensions. In particular, the free parameters μ and θ are here restricted to

$$\sin\mu = \cos\mu = 2^{-1/2}, \quad \cos\theta = 1. \qquad (9.78)$$

In this case, the entangling probe states $|\sigma_+\rangle, |\sigma_-\rangle, |\sigma\rangle, |\delta_+\rangle, |\delta_-\rangle, |\delta\rangle$, given by [9, Eqs. (159)–(164)], become [12]

$$|\sigma_+\rangle = |\delta_-\rangle = 4[(1-2E)^{1/2}|w_a\rangle - E^{1/2}|w_b\rangle], \qquad (9.79)$$

$$|\sigma_-\rangle = |\delta_+\rangle = 4[(1-2E)^{1/2}|w_a\rangle + E^{1/2}|w_b\rangle], \qquad (9.80)$$

$$|\sigma\rangle = -|\delta\rangle = 4E^{1/2}|w_b\rangle, \qquad (9.81)$$

in which the upper sign choices are chosen in [9, Eqs. (159)–(164)], E is the error rate induced by the probe, and the orthonormal probe basis vectors $|w_a\rangle$ and $|w_b\rangle$ are defined by

$$|w_a\rangle = 2^{-1/2}\left(|w_0\rangle + |w_3\rangle\right), \qquad (9.82)$$

$$|w_b\rangle = 2^{-1/2}\left(|w_1\rangle - |w_2\rangle\right), \qquad (9.83)$$

expressed in terms of the orthonormal basis vectors $|w_0\rangle$, $|w_3\rangle$, $|w_1\rangle$, and $|w_2\rangle$ of [9]. Thus, the optimum unitary transformation, [9, Eq. (158)], produces in this case the following entanglements for initial probe state $|w\rangle$ and incoming BB84 signal photon-polarization states $|u\rangle$, $|\bar{u}\rangle$, $|v\rangle$, or $|\bar{v}\rangle$, respectively:

$$|u\rangle \otimes |w\rangle \longrightarrow \frac{1}{4}\left(|u\rangle \otimes |\sigma_+\rangle + |\bar{u}\rangle \otimes |\sigma\rangle\right), \qquad (9.84)$$

$$|\bar{u}\rangle \otimes |w\rangle \longrightarrow \frac{1}{4}\left(|u\rangle \otimes |\sigma\rangle + |\bar{u}\rangle \otimes |\sigma_-\rangle\right), \qquad (9.85)$$

$$|v\rangle \otimes |w\rangle \longrightarrow \frac{1}{4}\left(|v\rangle \otimes |\sigma_-\rangle - |\bar{v}\rangle \otimes |\sigma\rangle\right), \qquad (9.86)$$

$$|\bar{v}\rangle \otimes |w\rangle \longrightarrow \frac{1}{4}\left(-|v\rangle \otimes |\sigma\rangle + |\bar{v}\rangle \otimes |\sigma_+\rangle\right). \qquad (9.87)$$

Here, the probe states $|\sigma_+\rangle$, $|\sigma_-\rangle$, $|\sigma\rangle$ are given by Eqs. (9.79)–(9.81). In the present case, the maximum information gain by the probe is again given by Eq. (9.1).

Using the same methods presented in Section 9.2, it can be shown that a quantum circuit consisting again of a single CNOT gate suffices to produce the optimum entanglement, Eqs. (9.84)–(9.87). Here, the control qubit entering the control port of the CNOT gate consists of the two signal basis states $\{|e_0\rangle, |e_1\rangle\}$. The target qubit entering the target port of the CNOT gate consists of the two orthonormal linearly-polarized photon polarization computational-basis states $2^{-1/2}(|w_a\rangle \pm |w_b\rangle)$. This choice of computational basis states is determined by the requirement that the probe measurement be a symmetric von-Neumann test [see argument preceding Eqs. (9.92) and (9.93)]. It follows that when $|e_0\rangle$ enters the control port, $\{|w_a\rangle, |w_b\rangle\}$ becomes $\{|w_a\rangle, -|w_b\rangle\}$, and when $|e_1\rangle$ enters the control port, $\{|w_a\rangle, |w_b\rangle\}$ remains the same. The initial target state of the probe can, in this case, be shown to be given by:

$$|A_2\rangle = (1-2E)^{1/2}|w_a\rangle + (2E)^{1/2}|w_b\rangle, \qquad (9.88)$$

and the transition state is given by

$$|A_1\rangle = (1-2E)^{1/2}|w_a\rangle - (2E)^{1/2}|w_b\rangle. \qquad (9.89)$$

Next, by arguments directly paralleling those of Section 9.2, using Eqs. (9.84)–(9.87), one has the following correlations between the signal states and the projected probe states, $|\sigma_+\rangle$ and $|\sigma_-\rangle$:

$$|u\rangle \Longleftrightarrow |\sigma_+\rangle, \quad |\bar{u}\rangle \Longleftrightarrow |\sigma_-\rangle, \tag{9.90}$$

and

$$|v\rangle \Longleftrightarrow |\sigma_-\rangle, \quad |\bar{v}\rangle \Longleftrightarrow |\sigma_+\rangle. \tag{9.91}$$

The measurement basis for the symmetric von Neumann projective measurement of the probe must be orthogonal and symmetric about the correlated probe states, $|\sigma_+\rangle$ and $|\sigma_-\rangle$, in the two-dimensional Hilbert space of the probe [3]. Thus, consistent with Eqs. (9.79) and (9.80), one defines the following orthonormal measurement basis states:

$$|w_+\rangle = 2^{-1/2}(|w_a\rangle + |w_b\rangle), \tag{9.92}$$

$$|w_-\rangle = 2^{-1/2}(|w_a\rangle - |w_b\rangle). \tag{9.93}$$

Next, one notes that the correlations of the projected probe states $|\sigma_+\rangle$ and $|\sigma_-\rangle$ with the measurement basis states $|w_+\rangle$ and $|w_-\rangle$ are indicated, according to Eqs. (9.79), (9.80), (9.92), and (9.93), by the following probabilities:

$$\frac{|\langle w_+|\sigma_+\rangle|^2}{|\sigma_+|^2} = \frac{|\langle w_-|\sigma_-\rangle|^2}{|\sigma_-|^2} = \frac{1}{2} - \frac{E^{1/2}(1-2E)^{1/2}}{(1-E)}, \tag{9.94}$$

$$\frac{|\langle w_+|\sigma_-\rangle|^2}{|\sigma_-|^2} = \frac{|\langle w_-|\sigma_+\rangle|^2}{|\sigma_+|^2} = \frac{1}{2} + \frac{E^{1/2}(1-2E)^{1/2}}{(1-E)}, \tag{9.95}$$

consistent with Eqs. (9.37) and (9.38), and implying the following dominant state correlations:

$$|\sigma_+\rangle \Longleftrightarrow |w_-\rangle, \quad |\sigma_-\rangle \Longleftrightarrow |w_+\rangle. \tag{9.96}$$

Next combining the correlations (9.90), (9.91), and (9.96), one thus establishes the following correlations:

$$\{|u\rangle, |\bar{v}\rangle\} \Longleftrightarrow |\sigma_+\rangle \Longleftrightarrow |w_-\rangle, \tag{9.97}$$

$$\{|\bar{u}\rangle, |v\rangle\} \Longleftrightarrow |\sigma_-\rangle \Longleftrightarrow |w_+\rangle, \tag{9.98}$$

to be implemented by the projective measurement of the probe, much as in Section 9.2.

One therefore arrives at the following alternative entangling probe design. An incident photon coming from the legitimate transmitter is received by the probe in one of the four signal-photon linear-polarization states $|u\rangle$, $|\bar{u}\rangle$, $|v\rangle$, or $|\bar{v}\rangle$ in the BB84 protocol. The signal photon enters the control port of a CNOT gate. The initial state of the probe is a photon in linear-polarization state $|A_2\rangle$ entering the target port of the CNOT gate. The photon linear-polarization state $|A_2\rangle$, according to Eq. (9.88), is given by

$$|A_2\rangle = (1 - 2E)^{1/2}|w_a\rangle + (2E)^{1/2}|w_b\rangle, \qquad (9.99)$$

and can be simply set for an error rate E by means of a polarizer. Note that this initial probe state has a simpler algebraic dependence on error rate than that in Eq. (9.46) of Section 9.2. In accord with Eq. (9.99), the entangling probe can be tuned to the chosen error rate to be induced by the probe. The outgoing gated signal photon is relayed on to the legitimate receiver, and the gated probe photon enters a Wollaston prism, oriented to separate photon orthogonal-linear-polarization states $|w_+\rangle$ and $|w_-\rangle$, and the photon is then detected by one of two photodetectors. This is an ordinary symmetric von Neumann projective measurement. If the basis, revealed during the public basis-reconciliation phase of the BB84 protocol, is $\{|u\rangle, |\bar{u}\rangle\}$, then the photodetector located to receive the polarization state $|w_-\rangle$ or $|w_+\rangle$, respectively, will indicate, in accord with the correlations (9.97) and (9.98), that a state $|u\rangle$ or $|\bar{u}\rangle$, respectively, was most likely measured by the legitimate receiver. Alternatively, if the announced basis is $\{|v\rangle, |\bar{v}\rangle\}$, then the photodetector located to receive the polarization state $|w_+\rangle$ or $|w_-\rangle$, respectively, will indicate, in accord with the correlations (9.97) and (9.98), that a state $|v\rangle$ or $|\bar{v}\rangle$, respectively, was most likely measured by the legitimate receiver. By comparing the record of probe photodetector triggering with the sequence of bases revealed during reconciliation, then the likely sequence of ones and zeroes constituting the key, prior to privacy amplification, can be assigned. In any case the net effect is to yield, for a set induced error rate E, the maximum Renyi information gain to the probe, which is given by Eq. (9.1).

9.5   Conclusion

Exploiting the quantum circuit model of quantum computation, the quantum circuits needed to implement optimum unitary transformations, Eqs. (9.2)–(9.5), or Eqs. (9.84)–(9.87), representing entangling probes attacking the B-

B84 protocol, were obtained and shown to yield the correct entangled states. The quantum circuits, faithfully representing the optimum entangling probes, consist of a single CNOT gate in which the control qubit consists of two photon-polarization basis states of the signal, the target qubit consists of the two probe-photon polarization basis states, and the probe photon is prepared in the initial linear-polarization state, Eqs. (9.46), (9.52), or (9.99), set by the induced error rate. The initial polarization state of the probe photon can be produced by a single-photon source together with a linear polarizer. The gated probe photon, optimally entangled with the signal, enters a Wollaston prism which separates the appropriate correlated states of the probe photon to trigger one or the other of two photodetectors. Basis selection, revealed on the public channel during basis reconciliation in the BB84 protocol, is exploited to correlate photodetector clicks with the signal transmitting the key, and to assign the most likely binary numbers, 1 or 0, such that the information gain by the probe of the key, prior to privacy amplification, is maximal. Explicit design parameters for the entangling probe are analytically specified, including: (1) the explicit initial polarization state of the probe photon; (2) the transition state of the probe photon; (3) the probabilities that one or the other photodetector triggers corresponding to a 0 or 1 of the key; (4) the relative angles between the various linear-polarization states in the Hilbert space of the probe; and (5) the information gain by the probe. The probes are simple special-purpose quantum information processors that will improve the odds for an eavesdropper in gaining access to the pre-privacy-amplified key, as well as impose a potentially severe sacrifice of key bits during privacy amplification [5]. The successful implementation of the probe awaits the development of a single robust high-fidelity CNOT gate, a practical single-photon source, and high-efficiency single-photon photodetectors.

Appendix A Renyi information gain

The expectation value of the Renyi information gain by the probe for each single-photon transmission is given by the expected value of the decrease in Renyi entropy of the probe, namely [3],

$$I^R = \sum_{\mu} P_{\mu} \left(R_0 - R_{\mu} \right), \tag{9.100}$$

where P_{μ} is the a priori probability of measurement outcome μ, R_0 is the initial Renyi entropy, and R_{μ} is the a posteriori Renyi entropy concerning the probe state for outcome μ. The initial Renyi entropy R_0 is given by [20], [21]

$$R_0 = -\log_2 \sum_i p_i^2, \tag{9.101}$$

where p_i is the a priori probability that the probe is in state i, and the a posteriori entropy R_μ is given by

$$R_\mu = -\log_2 \sum_i q_{i\mu}^2, \tag{9.102}$$

where $q_{i\mu}$ is the a posteriori probability that the probe is in state i for a measurement outcome μ (Here $i = 1$ or 2, and $\mu = 1$ or 2, corresponding to two probe states). The a priori probability P_μ of measurement outcome μ is given by

$$P_\mu = \mathrm{Tr}\left(E_\mu \sum_i p_i \rho^{(i)}\right), \tag{9.103}$$

in which $\rho^{(i)}$ is the density operator for the projected state $|\psi_i\rangle$ of the probe, correlated with the correct measurement outcome of the legitimate receiver, namely,

$$\rho^{(i)} = |\psi_i\rangle\langle\psi_i|, \tag{9.104}$$

and E_μ is the probe measurement operator,

$$E_\mu = |\bar{w}_\mu\rangle\langle\bar{w}_\mu|, \tag{9.105}$$

where $|\bar{w}_\mu\rangle$ are the orthonormal measurement basis vectors. The a posteriori probability $q_{i\mu}$ that the probe is in state i for measurement outcome μ is given by

$$q_{i\mu} = \frac{1}{P_\mu}\mathrm{Tr}\left(E_\mu \rho^{(i)}\right)p_i. \tag{9.106}$$

From Eqs. (9.106) and (9.103), it follows that

$$\sum_i q_{i\mu} = 1. \tag{9.107}$$

For the case in which only two probe states must be distinguished (which is the case in the present work), it follows from the symmetry of the protocol that one has equal a priori probabilities for the probe to be in state 1 or 2, namely,

$$p_1 = p_2 = \frac{1}{2}. \tag{9.108}$$

For the case of two probe states, $|\psi_1\rangle$ and $|\psi_2\rangle$, correlated with the correct measurements made by the legitimate receiver, it is known that the information gain is maximized when a simple two-dimensional von Neumann projective measurement is made in which the orthonormal measurement vectors $|\bar{w}_1\rangle$ and $|\bar{w}_2\rangle$ are located symmetrically with respect to the two correlated state vectors $|\psi_1\rangle$ and $|\psi_2\rangle$ of the probe in the real two-dimensional Hilbert space of the probe [3], [15], [16]. From this geometry, it then follows, for example, that the angle between $|\psi_1\rangle$ and $|\bar{w}_1\rangle$ is half the complement of the angle between $|\psi_1\rangle$ and $|\psi_2\rangle$. Thus one has

$$\langle\psi_1|\bar{w}_1\rangle = \langle\psi_2|\bar{w}_2\rangle \equiv \cos\xi = \cos\left(\frac{1}{2}\left[\frac{\pi}{2} - \cos^{-1}Q\right]\right), \qquad (9.109)$$

and

$$\langle\psi_1|\bar{w}_2\rangle = \langle\psi_2|\bar{w}_1\rangle \equiv \sin\xi = \sin\left(\frac{1}{2}\left[\frac{\pi}{2} - \cos^{-1}Q\right]\right), \qquad (9.110)$$

in which one defines the overlap of the states $|\psi_1\rangle$ and $|\psi_2\rangle$, correlated with the measurements of the signal, by [3]

$$Q \equiv \frac{\langle\psi_1|\psi_2\rangle}{|\psi_1||\psi_2|}. \qquad (9.111)$$

Next substituting Eqs. (9.105), (9.104), and (9.108) in Eq. (9.103), one has

$$P_\mu = \sum_i \mathrm{Tr}\frac{1}{2}\left(|\bar{w}_\mu\rangle\langle\bar{w}_\mu|\psi_i\rangle\langle\psi_i|\right), \qquad (9.112)$$

or equivalently,

$$P_\mu = \frac{1}{2}\left[|\langle\psi_1|\bar{w}_\mu\rangle|^2 + |\langle\psi_2|\bar{w}_\mu\rangle|^2\right]. \qquad (9.113)$$

Substituting Eqs. (9.109) and (9.110) in Eq. (9.113), one obtains

$$P_1 = \frac{1}{2}\left[\cos^2\xi + \sin^2\xi\right] = \frac{1}{2}, \qquad (9.114)$$

$$P_2 = \frac{1}{2}\left[\sin^2\xi + \cos^2\xi\right] = \frac{1}{2}. \qquad (9.115)$$

Next substituting Eqs. (9.114), (9.105), (9.104), and (9.108) in Eq. (9.106), one has

$$q_{11} = 2\mathrm{Tr}\left(\frac{1}{2}|\bar{w}_1\rangle\langle\bar{w}_1|\psi_1\rangle\langle\psi_1|\right) = |\langle\psi_1|\bar{w}_1\rangle|^2, \qquad (9.116)$$

and substituting Eq. (9.109) in Eq. (9.116), one has

$$q_{11} = \cos^2\left(\frac{1}{2}\left[\frac{\pi}{2} - \cos^{-1}Q\right]\right),$$
(9.117)

or equivalently, using a trigonometric identity, one has

$$q_{11} = \frac{1}{2} + \frac{1}{2}\cos\left(\frac{\pi}{2} - \cos^{-1}Q\right) = \frac{1}{2}\left(1 + \sin\cos^{-1}Q\right),$$
(9.118)

or

$$q_{11} = \frac{1}{2}\left[1 + \left(1 - Q^2\right)^{1/2}\right].$$
(9.119)

Analogously, one obtains

$$q_{12} = \frac{1}{2}\left[1 - \left(1 - Q^2\right)^{1/2}\right],$$
(9.120)

$$q_{21} = \frac{1}{2}\left[1 - \left(1 - Q^2\right)^{1/2}\right],$$
(9.121)

$$q_{22} = \frac{1}{2}\left[1 + \left(1 - Q^2\right)^{1/2}\right].$$
(9.122)

Next substituting Eq. (9.108) in Eq. (9.101), one has

$$R_0 = -\log_2\left[\left(\frac{1}{2}\right)^2 + \left(\frac{1}{2}\right)^2\right] = 1.$$
(9.123)

Also, according to Eq. (9.102), one has

$$R_1 = -\log_2\left[q_{11}^2 + q_{21}^2\right],$$
(9.124)

and substituting Eqs. (9.119) and (9.121) in Eq. (9.124), one obtains

$$R_1 = -\log_2\left(1 - \frac{1}{2}Q^2\right).$$
(9.125)

Analogously,

$$R_2 = -\log_2\left[q_{12}^2 + q_{22}^2\right],$$
(9.126)

or

$$R_2 = -\log_2\left(1 - \frac{1}{2}Q^2\right).$$
(9.127)

Next substituting Eqs. (9.114), (9.115), (9.123), (9.125), and (9.127) in Eq. (9.100), one obtains

$$I^R = \frac{1}{2}\left[1 + \log_2\left(1 - \frac{1}{2}Q^2\right)\right] + \frac{1}{2}\left[1 + \log_2\left(1 - \frac{1}{2}Q^2\right)\right],$$
(9.128)

or

$$I^R = 1 + \log_2\left(1 - \frac{1}{2}Q^2\right) = \log_2\left[2\left(1 - \frac{1}{2}Q^2\right)\right], \qquad (9.129)$$

or equivalently [3],

$$I^R = \log_2\left(2 - Q^2\right). \qquad (9.130)$$

It is evident from Eq. (9.130) that to maximize the Renyi information gain I^R, one can equivalently minimize Q, the overlap of correlated probe states. It is well to mention here that the maximum Renyi information gain by an eavesdropper is not only useful for probe optimization, but it is also important for determining how many bits of key must be sacrificed during privacy amplification, in order to insure that the key is secure [22], [5].

Appendix B Maximum Renyi information gain

Because the signal states $|u\rangle$, $|\bar{u}\rangle$, $|v\rangle$, and $|\bar{v}\rangle$ in the two-dimensional real Hilbert space of the signal can be expanded in terms of the signal basis states [as in Eqs. (9.16)–(9.19)], then by the linearity of quantum mechanics, the action of the probe on the signal states is fully described by the general unitary transformation U, representing the probe, acting on the signal basis states $|e_0\rangle$ and $|e_1\rangle$, which is given by [3], [8]

$$|e_m \otimes w\rangle \to U|e_m \otimes w\rangle = |e_0\rangle|\Phi_{0m}\rangle + |e_1\rangle|\Phi_{1m}\rangle, \qquad (9.131)$$

where $|w\rangle$ is the initial state of the probe, and the $|\Phi_{nm}\rangle$ are the transformed states of the probe (unnormalized and not orthogonal) and are functions of probe parameters $\{\lambda, \mu, \theta, \phi\}$,

$$|\Phi_{nm}\rangle = |\Phi_{nm}(\lambda, \mu, \theta, \phi)\rangle. \qquad (9.132)$$

(In general, $0 \le (\lambda, \mu, \theta, \phi) \le 2\pi$.) Specifically, unitarity and symmetry arguments determine the probe states $|\Phi_{nm}\rangle$ to be given by [3]

$$|\Phi_{00}\rangle = X_0|w_0\rangle + X_1|w_1\rangle + X_2|w_2\rangle + X_3|w_3\rangle, \qquad (9.133)$$
$$|\Phi_{01}\rangle = X_5|w_1\rangle + X_6|w_2\rangle, \qquad (9.134)$$
$$|\Phi_{10}\rangle = X_6|w_1\rangle + X_5|w_2\rangle, \qquad (9.135)$$
$$|\Phi_{11}\rangle = X_3|w_0\rangle + X_2|w_1\rangle + X_1|w_2\rangle + X_0|w_3\rangle, \qquad (9.136)$$

where

$$X_0 = \sin\lambda\cos\mu, \tag{9.137}$$
$$X_1 = \cos\lambda\cos\theta\cos\phi, \tag{9.138}$$
$$X_2 = \cos\lambda\cos\theta\sin\phi, \tag{9.139}$$
$$X_3 = \sin\lambda\sin\mu, \tag{9.140}$$
$$X_5 = \cos\lambda\sin\theta\cos\phi, \tag{9.141}$$
$$X_6 = -\cos\lambda\sin\theta\sin\phi. \tag{9.142}$$

The two states $|\psi_1\rangle$ and $|\psi_2\rangle$ which must be distinguished by the probe, for the basis $\{|u\rangle, |\bar{u}\rangle\}$, are

$$|\psi_1\rangle = |\psi_{uu}\rangle, \tag{9.143}$$
$$|\psi_2\rangle = |\psi_{\bar{u}\bar{u}}\rangle, \tag{9.144}$$

where $|\psi_{uu}\rangle$ is the projected state of the probe when state $|u\rangle$ is transmitted by the transmitter and also received by the legitimate receiver, namely,

$$|\psi_{uu}\rangle = \langle u|U|u\otimes w\rangle, \tag{9.145}$$

and analogously,

$$|\psi_{\bar{u}\bar{u}}\rangle = \langle \bar{u}|U|\bar{u}\otimes w\rangle. \tag{9.146}$$

Substituting Eqs. (9.143) and (9.146) in Eq. (9.111), then

$$Q = \frac{\langle \psi_{uu}|\psi_{\bar{u}\bar{u}}\rangle}{|\psi_{uu}||\psi_{\bar{u}\bar{u}}|}. \tag{9.147}$$

By the symmetry of the BB84 protocol, the value of Q given by Eq. (9.147) also holds in the $\{|v\rangle, |\bar{v}\rangle\}$ basis. Expanding the signal states in terms of the signal basis states, and using Eqs. (9.145)–(9.147), (9.131)–(9.142), it can be shown that [3], [4]

$$Q = \frac{\frac{1}{2}(d+a)+b}{\left\{\left[1+\frac{1}{2}(d+a)\right]^2 - \frac{1}{2}c^2\right\}^{1/2}}, \tag{9.148}$$

where

$$a = \sin^2\lambda\sin 2\mu + \cos^2\lambda\cos 2\theta\sin 2\phi, \tag{9.149}$$
$$b = \sin^2\lambda\sin 2\mu + \cos^2\lambda\sin 2\phi, \tag{9.150}$$
$$c = \cos^2\lambda\sin 2\theta\cos 2\phi, \tag{9.151}$$
$$d = \sin^2\lambda + \cos^2\lambda\cos 2\theta. \tag{9.152}$$

What is needed is the maximum Renyi information (or minimum Q in Eq. (9.130)) for any fixed error rate E chosen by the eavesdropper and induced in the legitimate receiver. Evidently the induced error rate is [3]

$$E = \frac{P_{u\bar{u}} + P_{\bar{u}u}}{P_{u\bar{u}} + P_{\bar{u}u} + P_{uu} + P_{\bar{u}\bar{u}}}, \tag{9.153}$$

where P_{ij} is the probability that if a photon in polarization state $|i\rangle$ is transmitted by the transmitter in the presence of the disturbing probe, the polarization state $|j\rangle$ is detected by the legitimate receiver. One has

$$P_{ij} = \left| \psi_{ij} \right|^2, \tag{9.154}$$

where $\left| \psi_{ij} \right\rangle$ is the projected state of the probe when polarization state $|i\rangle$ is transmitted, and polarization state $|j\rangle$ is detected by the legitimate receiver in the presence of the probe. The states $|\psi_{uu}\rangle$ and $|\psi_{\bar{u}\bar{u}}\rangle$ are given by Eqs. (9.145) and (9.146). Analogously one has

$$|\psi_{u\bar{u}}\rangle = \langle \bar{u}|U|u \otimes w\rangle, \tag{9.155}$$

and

$$|\psi_{\bar{u}u}\rangle = \langle u|U|\bar{u} \otimes w\rangle. \tag{9.156}$$

Using Eqs. (9.154), (9.145), (9.146), (9.155), (9.156), (9.149), (9.152) in Eq. (9.153), it can be shown that the induced error rate is given by [3]

$$E = \frac{1}{2}\left[1 - \frac{1}{2}(d+a)\right]. \tag{9.157}$$

Next substituting Eq. (9.157) in Eq. (9.148), one obtains

$$Q = \frac{1 - 2E + b}{\left[(2 - 2E)^2 - \frac{1}{2}c^2\right]^{1/2}}. \tag{9.158}$$

The optimization then becomes that of finding the probe parameters $\{\lambda, \mu, \theta, \phi\}$ such that Q in Eq. (9.158) is minimum for fixed E. The complete optimization was performed in [4], [6], [7]. To accomplish this, the critical points of Q were found by analytically determining the values of λ, μ, θ, and ϕ such that for fixed E,

$$\frac{\partial Q}{\partial \lambda}\bigg|_E = \frac{\partial Q}{\partial \mu}\bigg|_E = \frac{\partial Q}{\partial \theta}\bigg|_E = \frac{\partial Q}{\partial \phi}\bigg|_E = 0, \tag{9.159}$$

and then distinguishing minima from saddle points and maxima by numerically calculating Q in the neighborhood of the critical points. Three sets of optimum probe parameters were found, each yielding the identical minimum Q and maximum Renyi information gain for set error rate E. The three sets of optimum probe parameters are [4], [6], [7]

$$S^{(1)} \equiv \{\lambda, \mu, \theta, \phi; \cos\lambda = 0, \ \sin 2\mu = 1 - 4E\}, \qquad (9.160)$$

$$S^{(2)} \equiv \{\lambda, \mu, \theta, \phi; \sin 2\mu \sin^2 \lambda$$
$$= 1 - 4E - \cos^2 \lambda \sin 2\phi, \ \cos 2\theta = 1\}, \qquad (9.161)$$

$$S^{(3)} \equiv \{\lambda, \mu, \theta, \phi; \sin 2\phi = -1, \ \sin 2\mu \sin^2 \lambda$$
$$= 1 - 4E + \cos^2 \lambda\}. \qquad (9.162)$$

The minimum Q for all three sets, Eqs. (9.160)–(9.162), is [3], [4]

$$Q_{\min} = \frac{1 - 3E}{1 - E}, \qquad (9.163)$$

and finally substituting Eq. (9.163) in Eq. (9.130), the maximum Renyi information gain is

$$I_{opt}^R = \log_2\left[2 - \left(\frac{1-3E}{1-E}\right)^2\right], \qquad (9.164)$$

as in Eq. (9.1). Corresponding to each set of optimum probe parameters, $S^{(1)}$, $S^{(2)}$, and $S^{(3)}$ is an optimum unitary transformation $U^{(1)}$, $U^{(2)}$, and $U^{(3)}$, respectively, obtained by evaluating the transformations, Eqs. (9.131)–(9.142), for the optimum values of the probe parameters, Eqs. (9.160)–(9.162), respectively, yielding Eqs. (9.2)–(9.5), (9.63)–(9.66), and (9.84)–(9.87), respectively [9].

Acknowledgements

This work was supported by the U.S. Army Research Laboratory and the Defense Advanced Research Projects Agency. The author wishes to thank Goong Chen, Samuel Lomonaco, and Louis Kauffman for the invitation to present this paper at the conference, *Mathematics of Quantum Computation and Quantum Technology*, held 13-16 November 2005 at Texas A&M University, Department of Mathematics, College Station, Texas.

References

[1] C. H. Bennett and G. Brassard, "Quantum cryptography: public key distribution and coin tossing," in Proceedings of the IEEE International Conference on Computers, Systems, and Signal Processing, Bangalore, India (IEEE, New York, 1984), pp. 175–179.

[2] N. Gisin, G. Ribordy, W. Tittel, and H. Zbinden, "Quantum Cryptography," Rev. Mod. Phys. **74**, 145–195 (2002).

[3] B. A. Slutsky, R. Rao, P. C. Sun, and Y. Fainman, "Security of quantum cryptography against individual attacks," Phys. Rev. A **57**, 2383–2398 (1998).

[4] H. E. Brandt, "Probe Optimization in four-state protocol of quantum cryptography," Phys. Rev. A **66**, 032303 (16), (2002).

[5] H. E. Brandt, "Secrecy capacity in the four-state protocol of quantum key distribution," J. Math. Phys. **43**, 4526–4530 (2002).

[6] H. E. Brandt, "Optimization problem in quantum cryptography," J. Optics B **5**, S557–S560 (2003).

[7] H. E. Brandt, "Optimum probe parameters for entangling probe in quantum key distribution," Quantum Information Processing **2**, 37–79 (2003).

[8] C. A. Fuchs and A. Peres, "Quantum-state disturbance versus information gain: uncertainty relations for quantum information," Phys. Rev. A **53**, 2038–2045 (1996).

[9] H. E. Brandt, "Quantum cryptographic entangling probe," Phys. Rev. A **71**, 042312 (14), (2005).

[10] H. E. Brandt, "Design for a quantum cryptographic entangling probe," J. Modern Optics **52**, 2177–2185 (2005).

[11] H. E. Brandt and J. M. Myers, "Expanded quantum cryptographic entangling probe," J. Mod. Optics **53**, 1927–1930 (2006).

[12] H. E. Brandt, "Alternative design for quantum cryptographic entangling probe," J. Mod. Optics **53**, 1041–1045 (2006).

[13] M. A. Nielsen and I. L. Chuang, *Quantum Computation and Information* (Cambridge University Press, 2000).

[14] G. Vidal and C. M. Dawson, "Universal quantum circuit for two qubit transformation with three controlled-NOT gates," Phys. Rev. A **69**, 010301-1-4 (2004).

[15] C. W. Helstrom, *Quantum Detection and Estimation Theory* (Academic Press, New York, 1976).

[16] L. B. Levitan, "Optimal quantum measurement for two pure and mixed states," in *Quantum Communications and Measurement*, edited by V. P. Belavkin, O. Hirota, and R. L. Hudson (Plenum, New York, 1995), pp. 439–448.

[17] H. E. Brandt, Invention Disclosure: *Quantum Cryptographic Entangling Probe* (U.S. Army Research Laboratory, Adelphi, MD, 2004).

[18] See, e.g., L.-M. Duan and H. J. Kimble, Phys. Rev. Lett. **92**, 127902 (2004); T. B. Pittman, M. J. Fitch, B. C. Jacobs and J. D. Franson, Phys. Rev. A **68**, 032316 (2003); J. L. O'Brien, G. J. Pryde, A. G. White, T. C. Ralph and D. Branning, Nature **426**, 264 (2003); M. Fiorentino and F. N. C. Wong, Phys. Rev. Lett. **93**, 070502 (2004); S. Gasparoni, J. W. Pan, P. Walther, T. Rudolph, and A. Zeilinger, Phys. Rev. Lett. **93**, 020504 (2004).

[19] H. E. Brandt, "Entangled eavesdropping in quantum key distribution," J. Mod. Optics **53**, 2251–2257 (2006).

[20] C. H. Bennett, G.Brassard, C. Crepeau, and U. M. Maurer, "Generalized privacy amplification," IEEE Trans. Inform. Theory **41**, 1915–1923 (1995).

[21] A. Renyi, "On measures of entropy and information," in *Proc. 4th Berkeley Symp. on Mathematical Statistics and Probability*, Vol. 1, 1961, pp. 547–561.

[22] B. Slutsky, R. Rao, P. C. Sun, L. Tancevski, and S. Fainman, "Defense frontier analysis of quantum cryptographic systems," Appl. Optics **37**, 2869–2878 (1998).

Chapter 10

Nonbinary Stabilizer Codes

Pradeep Kiran Sarvepalli, Salah A. Aly, and
Andreas Klappenecker

Abstract   Recently, the field of quantum error-correcting codes has rapidly emerged as an important discipline. As quantum information is extremely sensitive to noise, it seems unlikely that any large scale quantum computation is feasible without quantum error-correction. In this paper we give a brief exposition of the theory of quantum stabilizer codes. We review the stabilizer formalism of quantum codes, establish the connection between classical codes and stabilizer codes and the main methods for constructing quantum codes from classical codes. In addition to the expository part, we include new results that cannot be found elsewhere. Specifically, after reviewing some important bounds for quantum codes, we prove the nonexistence of pure perfect quantum stabilizer codes with minimum distance greater than 3. Finally, we illustrate the general methods of constructing quantum codes from classical codes by explicitly constructing two new families of quantum codes and conclude by showing how to construct new quantum codes by shortening.

10.1   Introduction

Quantum error-correcting codes were introduced by Shor [54] in the wake of serious doubts cast over the practical implementation of quantum algorithms. Since then the field has made rapid progress and the pioneering works of

Gottesman [22] and Calderbank et al., [10] revealed a rich structure underlying the theory of quantum stabilizer codes. Their work spurred many researchers to study binary quantum codes, see [5, 6, 7, 11, 9, 15, 14, 16, 17, 21, 26, 24, 27, 23, 32, 29, 30, 28, 37, 38, 40, 45, 47, 50, 54, 58, 57, 59, 56]. The theory was later extended to the nonbinary case [1, 2, 3, 8, 12, 13, 18, 19, 25, 33, 31, 36, 39, 48, 49, 51, 53, 52]. This paper surveys the theory of nonbinary stabilizer codes – arguably, the most important class of quantum codes. There exists sufficient machinery to describe them compactly and make useful connections with classical coding theory. Moreover, they are very amenable to fault-tolerant implementation which makes them very attractive from a practical point of view.

We aim to provide an accessible introduction to the theory of nonbinary quantum codes. Section 10.2 gives a brief overview of the main ideas of stabilizer codes while Section 10.3 reviews the relation between quantum stabilizer codes and classical codes. This connection makes it possible to reduce the study of quantum stabilizer codes to the study of self-orthogonal classical codes, though the definition of self-orthogonality is a little broader than the classical one. Further, it allows us to use all the tools of classical codes to derive bounds on the parameters of good quantum codes. Section 10.4 gives an overview of the important bounds for quantum codes. Finally, Section 10.5 illustrates the general ideas behind quantum code construction by constructing the quantum Hamming codes, some cyclic quantum codes and codes from projective geometry.

While this paper is primarily an exposition of the theory of nonbinary stabilizer codes, we also included new results. For instance, we prove the nonexistence of pure perfect quantum codes with distance greater than 3. Furthermore, we derive two new families of quantum codes, the quantum projective Reed–Muller codes and the quantum m-adic residue codes. Finally, we illustrate the key ideas of shortening quantum codes by taking the newly introduced quantum projective Reed–Muller codes as an example.

We tried to keep the prerequisites to a minimum, though we assume that the reader has a minimal background in quantum computing. Some familiarity with classical coding theory will help; we recommend [34] and [46] as references. In general, we omitted long proofs of basic material – readers interested in more details should consult [36]. However, we made an effort to keep the overlap with [36] to a minimum, although some material is repeated here to make this chapter reasonably self-contained.

Notations. The finite field with q elements is denoted by $\mathbf{F}_q$, where $q = p^m$ and p is assumed to be a prime. The trace function from $\mathbf{F}_{q^l}$ to $\mathbf{F}_q$ is defined as $\mathrm{tr}_{q^l/q}(x) = \sum_{k=0}^{l-1} x^{q^k}$, and we may omit the subscripts if $\mathbf{F}_q$ is the prime field.

The center of a group G is denoted by $Z(G)$ and the centralizer of a subgroup S in G by $C_G(S)$. We denote by $H \leq G$ the fact that H is a subgroup of G. The trace $\text{Tr}(M)$ of a square matrix M is the sum of the diagonal elements of M.

10.2 Stabilizer codes

In this chapter, we use q-ary quantum digits, shortly called qudits, as the basic unit of quantum information. The state of a qudit is a nonzero vector in the complex vector space $\mathbf{C}^q$. This vector space is equipped with an orthonormal basis whose elements are denoted by $|x\rangle$, where x is an element of the finite field $\mathbf{F}_q$. The state of a system of n qudits is then a nonzero vector in $\mathbf{C}^{q^n}$. In general, quantum codes are just nonzero subspaces[1] of $\mathbf{C}^{q^n}$. A quantum code that encodes k qudits of information into n qudits is denoted by $[[n,k]]_q$, where the subscript q indicates that the code is q-ary. More generally, an $((n,K))_q$ quantum code is a K-dimensional subspace encoding $\log_q K$ qudits into n qudits.

As the codes are subspaces, it seems natural to describe them by giving a basis for the subspace. However, in case of quantum codes this turns out to be an inconvenient description.[2] An alternative description of the quantum error-correcting codes that are discussed in this chapter relies on error operators that act on $\mathbf{C}^{q^n}$. If we make the assumption that the errors are independent on each qudit, then each error operator E can be decomposed as $E = E_1 \otimes \cdots \otimes E_n$. Furthermore, linearity of quantum mechanics allows us to consider only a discrete set of errors. The quantum error-correcting codes that we consider here can be described as the joint eigenspace of an abelian subgroup of error operators. The subgroup of error operators is called the stabilizer of the code (because it leaves each state in the code unaffected) and the code is called a

---

[1]The more recent concept of an operator quantum error-correcting code generalizes this notion, where additional structure is imposed on the subspaces.
[2]For instance, a basis for the $[[7,1]]_2$ code is

$$|0_L\rangle = |0000000\rangle + |1010101\rangle + |0110011\rangle + |1100110\rangle$$
$$+ |0001111\rangle + |0111100\rangle + |1011010\rangle + |1101001\rangle,$$
$$|1_L\rangle = |0000000\rangle + |1010101\rangle + |0110011\rangle + |1100110\rangle$$
$$+ |0001111\rangle + |0111100\rangle + |1011010\rangle + |1101001\rangle.$$

stabilizer code.

10.2.1   Error bases

In general, we can regard any error as being composed of an amplitude error and a phase error. Let a and b be elements in $\mathbf{F}_q$. We can define unitary operators $X(a)$ and $Z(b)$ on $\mathbf{C}^q$ that generalize the Pauli X and Z operators to the q-ary case; they are defined as

$$X(a)|x\rangle = |x+a\rangle, \qquad Z(b)|x\rangle = \omega^{\mathrm{tr}(bx)}|x\rangle,$$

where tr denotes the trace operation from $\mathbf{F}_q$ to $\mathbf{F}_p$, and $\omega = \exp(2\pi i/p)$ is a primitive pth root of unity.

Let $\mathscr{E} = \{X(a)Z(b)\,|\,a,b \in \mathbf{F}_q\}$ be the set of error operators. The error operators in $\mathscr{E}$ form a basis of the set of complex $q \times q$ matrices as the trace $\mathrm{Tr}(A^\dagger B) = 0$ for distinct elements A, B of $\mathscr{E}$. Further, we observe that

$$X(a)Z(b)X(a')Z(b') = \omega^{\mathrm{tr}(ba')}X(a+a')Z(b+b'). \qquad (10.1)$$

The error basis for n q-ary quantum systems can be obtained by tensoring the error basis for each system. Let $\mathbf{a} = (a_1,\ldots,a_n) \in \mathbf{F}_q^n$. Let us denote by $X(\mathbf{a}) = X(a_1) \otimes \cdots \otimes X(a_n)$ and $Z(\mathbf{a}) = Z(a_1) \otimes \cdots \otimes Z(a_n)$ for the tensor products of n error operators. Then we have the following result whose proof follows from the definitions of $X(\mathbf{a})$ and $Z(\mathbf{b})$.

LEMMA 10.1
  The set $\mathscr{E}_n = \{X(\mathbf{a})Z(\mathbf{b})\,|\,\mathbf{a},\mathbf{b} \in \mathbf{F}_q^n\}$ is an error basis on the complex vector space $\mathbf{C}^{q^n}$.

10.2.2   Stabilizer codes

Consider the error group G_n defined as

$$G_n = \{\omega^c X(\mathbf{a})Z(\mathbf{b})\,|\,\mathbf{a},\mathbf{b} \in \mathbf{F}_q^n, c \in \mathbf{F}_p\}.$$

G_n is simply a finite group of order pq^{2n} generated by the matrices in the error basis $\mathscr{E}_n$.

Let S be an abelian subgroup of G_n, then a *stabilizer code* Q is a non-zero subspace of $\mathbf{C}^{q^n}$ defined as

$$Q = \bigcap_{E \in S} \{v \in \mathbf{C}^{q^n} \mid Ev = v\}. \qquad (10.2)$$

Alternatively, Q is the joint +1 eigenspace of S. A stabilizer code contains *all* joint eigenvectors of S with eigenvalue 1, as equation (10.2) indicates. If the code is smaller and does not contain all the joint eigenvectors of S with eigenvalue 1, then it is not a stabilizer code for S.

10.2.3   Stabilizer and error correction

Now that we have a handle on the quantum code through its stabilizer, we next need to be able to describe the performance of the code, that is, we should be able to tell how many errors it can detect (or correct) and how the error correction is done.

The central idea of error detection is that a detectable error acting on Q should either act as a scalar multiplication on the code space (in which case the error did not affect the encoded information) or it should map the encoded state to the orthogonal complement of Q (so that one can set up a measurement to detect the error). Specifically, we say that Q is able to detect an error E in the unitary group $U(q^n)$ if and only if the condition $\langle c_1|E|c_2\rangle = \lambda_E\langle c_1|c_2\rangle$ holds for all $c_1, c_2 \in Q$, see [43].

We can show that a stabilizer code Q with stabilizer S can detect all errors in G_n that are scalar multiples of elements in S or that do not commute with some element of S, see Lemma 10.2. In particular, an undetectable error in G_n has to commute with all elements of the stabilizer.

Let $S \leq G_n$ and $C_{G_n}(S)$ denote the centralizer of S in G_n,

$$C_{G_n}(S) = \{E \in G_n \mid EF = FE \text{ for all } F \in S\}.$$

Let $SZ(G_n)$ denote the group generated by S and the center $Z(G_n)$. We need the following characterization of detectable errors.

LEMMA 10.2
Suppose that $S \leq G_n$ is the stabilizer group of a stabilizer code Q of dimension $\dim Q > 1$. An error E in G_n is detectable by the quantum code Q if and only if either E is an element of $SZ(G_n)$ or E does not belong to the centralizer $C_{G_n}(S)$.

PROOF   See [36]. See also [3]; the interested reader can find a more general approach in [42, 41].   ∎

Since detectability of errors is closely associated to commutativity of error operators, we will derive the following condition on commuting elements in G_n:

LEMMA 10.3

Two elements $E = \omega^c X(\mathbf{a})Z(\mathbf{b})$ and $E' = \omega^{c'} X(\mathbf{a}')Z(\mathbf{b}')$ of the error group G_n satisfy the relation

$$EE' = \omega^{\mathrm{tr}(\mathbf{b}\cdot\mathbf{a}' - \mathbf{b}'\cdot\mathbf{a})} E'E.$$

In particular, the elements E and E' commute if and only if the trace symplectic form $\mathrm{tr}(\mathbf{b}\cdot\mathbf{a}' - \mathbf{b}'\cdot\mathbf{a})$ vanishes.

PROOF   We can easily verify that $EE' = \omega^{\mathrm{tr}(\mathbf{b}\cdot\mathbf{a}')}X(\mathbf{a}+\mathbf{a}')Z(\mathbf{b}+\mathbf{b}')$ and $E'E = \omega^{\mathrm{tr}(\mathbf{b}'\cdot\mathbf{a})}X(\mathbf{a}+\mathbf{a}')Z(\mathbf{b}+\mathbf{b}')$ using equation (10.1). Therefore, $\omega^{\mathrm{tr}(\mathbf{b}\cdot\mathbf{a}' - \mathbf{b}'\cdot\mathbf{a})}E'E$ yields EE', as claimed.   ∎

10.2.4   Minimum distance

The *symplectic weight* swt of a vector $(\mathbf{a}|\mathbf{b})$ in $\mathbf{F}_q^{2n}$ is defined as

$$\mathrm{swt}((\mathbf{a}|\mathbf{b})) = |\{k \,|\, (a_k, b_k) \neq (0,0)\}|.$$

The weight $\mathrm{w}(E)$ of an element $E = \omega^c E_1 \otimes \cdots \otimes E_n = \omega^c X(\mathbf{a})Z(\mathbf{b})$ in the error group G_n is defined to be the number of nonidentity tensor components i.e., $\mathrm{w}(E) = |\{E_i \neq I\}| = \mathrm{swt}((\mathbf{a}|\mathbf{b}))$.

A quantum code Q is said to have *minimum distance* d if and only if it can detect all errors in G_n of weight less than d, but cannot detect some error of weight d. We say that Q is an $((n,K,d))_q$ code if and only if Q is a K-dimensional subspace of $\mathbf{C}^{q^n}$ that has minimum distance d. An $((n,q^k,d))_q$ code is also called an $[[n,k,d]]_q$ code.

Due to the linearity of quantum mechanics, a quantum error-correcting code that can detect a set $\mathscr{D}$ of errors, can also detect all errors in the linear span of $\mathscr{D}$. A code of minimum distance d can correct all errors of weight $t = \lfloor (d-1)/2 \rfloor$ or less.

10.2.5   Pure and impure codes

We say that a quantum code Q is *pure to* t if and only if its stabilizer group S does not contain non-scalar error operators of weight less than t. A quantum code is called pure if and only if it is pure to its minimum distance. We will follow the same convention as in [10], that an $[[n,0,d]]_q$ code is pure. Impure codes are also referred to as degenerate codes. Degenerate codes are of interest because they have the potential for passive error correction.

10.2.6 Encoding quantum codes

Stabilizer also provides a means for encoding quantum codes. The essential idea is to encode the information into the code space through a projector. For an $((n,K,d))_q$ quantum code with stabilizer S, the projector P is defined as

$$P = \frac{1}{|S|} \sum_{E \in S} E.$$

It can be checked that P is an orthogonal projector onto a vector space Q. Further, we have

$$K = \dim Q = \mathrm{Tr}P = q^n/|S|.$$

The stabilizer allows us to derive encoded operators, so that we can operate directly on the encoded data instead of decoding and then operating on them. These operators are in $C_{G_n}(S)$. See [24] and [33] for more details.

10.3 Quantum codes and classical codes

In this section we show how stabilizer codes are related to classical codes (additive codes over $\mathbf{F}_q$ or over $\mathbf{F}_{q^2}$). The central idea behind this relation is the fact insofar as the detectability of an error is concerned the phase information is irrelevant. This means we can factor out the phase defining a map from G_n onto $\mathbf{F}_q^{2n}$ and study the images of S and $C_{G_n}(S)$. We will denote a classical code $C \le \mathbf{F}_q^n$ with K codewords and distance d by $(n,K,d)_q$. If it is linear then we will also denote it by $[n,k,d]_q$ where $k = \log_q K$. We define the Euclidean inner product of $x,y \in \mathbf{F}_q^n$ as $x \cdot y = \sum_{i=1}^n x_i y_i$. The dual code $C^{\perp}$ is the set of vectors in $\mathbf{F}_q^n$ orthogonal to C i.e., $C^{\perp} = \{x \in \mathbf{F}_q^n \mid x \cdot c = 0 \text{ for all } c \in C\}$. For more details on classical codes see [34] or [46].

10.3.1 Codes over $\mathbf{F}_q$

If we associate with an element $\omega^c X(\mathbf{a})Z(\mathbf{b})$ of G_n an element $(\mathbf{a}|\mathbf{b})$ of $\mathbf{F}_q^{2n}$, then the group $SZ(G_n)$ is mapped to the additive code

$$C = \{(\mathbf{a}|\mathbf{b}) \mid \omega^c X(\mathbf{a})Z(\mathbf{b}) \in SZ(G_n)\} = SZ(G_n)/Z(G_n).$$

To relate the images of the stabilizer and its centralizer, we need the notion of a trace-symplectic form of two vectors $(\mathbf{a}|\mathbf{b})$ and $(\mathbf{a}'|\mathbf{b}')$ in $\mathbf{F}_q^{2n}$,

$$\langle (\mathbf{a}|\mathbf{b}) \,|\, (\mathbf{a}'|\mathbf{b}') \rangle_s = \mathrm{tr}_{q/p}(\mathbf{b} \cdot \mathbf{a}' - \mathbf{b}' \cdot \mathbf{a}).$$

Let $C^{\perp_s}$ be the trace-symplectic dual of C defined as

$$C^{\perp_s} = \{x \in \mathbf{F}_q^{2n} \mid \langle x \mid c \rangle_s = 0 \text{ for all } c \in C\}.$$

The centralizer $C_{G_n}(S)$ contains all elements of G_n that commute with each element of S; thus, by Lemma 10.3, $C_{G_n}(S)$ is mapped onto the trace-symplectic dual code $C^{\perp_s}$ of the code C,

$$C^{\perp_s} = \{(\mathbf{a}|\mathbf{b}) \mid \omega^c X(\mathbf{a}) Z(\mathbf{b}) \in C_{G_n}(S)\}.$$

The next theorem crystallizes this connection between classical codes and stabilizer codes and generalizes the well-known connection to symplectic codes [10, 22] of the binary case.

THEOREM 10.1
An $((n,K,d))_q$ stabilizer code exists if and only if there exists an additive code $C \leq \mathbf{F}_q^{2n}$ of size $|C| = q^n/K$ such that $C \leq C^{\perp_s}$ and $\mathrm{swt}(C^{\perp_s} \setminus C) = d$ if $K > 1$ (and $\mathrm{swt}(C^{\perp_s}) = d$ if $K = 1$).

PROOF   See [3] or [36] for the proof.   ∎

In 1996, Calderbank and Shor [11] and Steane [58] introduced the following method to construct quantum codes. It is perhaps the simplest method to build quantum codes via classical codes over $\mathbf{F}_q$.

LEMMA 10.4
[CSS Code Construction] Let C_1 and C_2 denote two classical linear codes with parameters $[n,k_1,d_1]_q$ and $[n,k_2,d_2]_q$ such that $C_2^{\perp} \leq C_1$. Then there exists a $[[n, k_1 + k_2 - n, d]]_q$ stabilizer code with minimum distance $d = \min\{\mathrm{wt}(c) \mid c \in (C_1 \setminus C_2^{\perp}) \cup (C_2 \setminus C_1^{\perp})\}$ that is pure to $\min\{d_1, d_2\}$.

PROOF   Let $C = C_1^{\perp} \times C_2^{\perp} \leq \mathbf{F}_q^{2n}$. Clearly $C \leq C_2 \times C_1$. If $(c_1 \mid c_2) \in C$ and $(c_1' \mid c_2') \in C_2 \times C_1$, then we observe that

$$\mathrm{tr}(c_2 \cdot c_1' - c_2' \cdot c_1) = \mathrm{tr}(0 - 0) = 0.$$

Therefore, $C \leq C_2 \times C_1 \leq C^{\perp_s}$. Since $|C| = q^{2n-k_1-k_2}$, $|C^{\perp_s}| = q^{2n}/|C| = q^{k_1+k_2} = |C_2 \times C_1|$. Therefore, $C^{\perp_s} = C_2 \times C_1$. By Theorem 10.1 there exists an $((n,K,d))_q$ quantum code with $K = q^n/|C| = q^{k_1+k_2-n}$. The claim about the minimum distance and purity of the code is obvious from the construction. ∎

COROLLARY 10.1
If C is a classical linear $[n,k,d]_q$ code containing its dual, $C^\perp \leq C$, then there exists an $[[n, 2k-n, \geq d]]_q$ stabilizer code that is pure to d.

10.3.2 Codes over F_{q^2}

Sometimes it is more convenient to extend the connection of the quantum codes to codes over $\mathbf{F}_{q^2}$, especially as it allows us the use of codes over quadratic extension fields. The binary case was done in [10] and partial generalizations were done in [48, 39] and [49]. We provide a slightly alternative generalization using a trace-alternating form. Let (β, β^q) denote a normal basis of $\mathbf{F}_{q^2}$ over $\mathbf{F}_q$. We define a trace-alternating form of two vectors v and w in $\mathbf{F}_{q^2}^n$ by

$$\langle v|w\rangle_a = \mathrm{tr}_{q/p}\left(\frac{v \cdot w^q - v^q \cdot w}{\beta^{2q} - \beta^2}\right). \tag{10.3}$$

The argument of the trace is an element of $\mathbf{F}_q$ as it is invariant under the Galois automorphism $x \mapsto x^q$.

Let $\phi : \mathbf{F}_q^{2n} \to \mathbf{F}_{q^2}^n$ take $(\mathbf{a}|\mathbf{b}) \mapsto \beta\mathbf{a} + \beta^q\mathbf{b}$. The map ϕ is isometric in the sense that the symplectic weight of $(\mathbf{a}|\mathbf{b})$ is equal to the Hamming weight of $\phi((\mathbf{a}|\mathbf{b}))$. This map allows us to transform the trace-symplectic duality into trace-alternating duality. In particular it can be easily verified that if $c, d \in \mathbf{F}_q^{2n}$, then $\langle c|d\rangle_s = \langle \phi(c)|\phi(d)\rangle_a$. If $D \leq \mathbf{F}_{q^2}^n$, then we denote its trace-alternating dual by $D^{\perp_a} = \{v \in \mathbf{F}_{q^2}^n \mid \langle v|w\rangle_a = 0 \text{ for all } w \in D\}$. Now Theorem 10.1 can be reformulated as:

THEOREM 10.2
An $((n, K, d))_q$ stabilizer code exists if and only if there exists an additive subcode D of $\mathbf{F}_{q^2}^n$ of cardinality $|D| = q^n/K$ such that $D \leq D^{\perp_a}$ and $\mathrm{wt}(D^{\perp_a} \setminus D) = d$ if $K > 1$ (and $\mathrm{wt}(D^{\perp_a}) = d$ if $K = 1$).

PROOF From Theorem 10.1 we know that an $((n, K, d))_q$ stabilizer code exists if and only if there exists a code $C \leq \mathbf{F}_q^{2n}$ such that $|C| = q^n/K$, $C \leq C^{\perp_s}$, and $\mathrm{swt}(C^{\perp_s} \setminus C) = d$ if $K > 1$ (and $\mathrm{swt}(C^{\perp_s}) = d$ if $K = 1$). The theorem follows simply by applying the isometry ϕ. ∎

If we restrict our attention to linear codes over $\mathbf{F}_{q^2}$, then the hermitian form is more useful. The hermitian inner product of two vectors $\mathbf{x}$ and $\mathbf{y}$ in $\mathbf{F}_{q^2}^n$ is given by $\mathbf{x}^q \cdot \mathbf{y}$. From the definition of the trace-alternating form it is clear that if two vectors are orthogonal with respect to the hermitian form they are also orthogonal with respect to the trace-alternating form. Consequently, if $D \leq \mathbf{F}_{q^2}^n$, then $D^{\perp_h} \leq D^{\perp_a}$, where $D^{\perp_h} = \{v \in \mathbf{F}_{q^2}^n \mid v^q \cdot w = 0 \text{ for all } w \in D\}$.

Therefore, any self-orthogonal code with respect to the hermitian inner product is self-orthogonal with respect to the trace-alternating form. In general, the two dual spaces $D^{\perp_h}$ and $D^{\perp_a}$ are not the same. However, if D happens to be $\mathbf{F}_{q^2}$-linear, then the two dual spaces coincide.

COROLLARY 10.2

If there exists an $\mathbf{F}_{q^2}$-linear $[n,k,d]_{q^2}$ code D such that $D^{\perp_h} \leq D$, then there exists an $[[n, 2k-n, \geq d]]_q$ quantum code that is pure to d.

PROOF   Let $q = p^m$, p prime. If D is a k-dimensional subspace of $\mathbf{F}_{q^2}^n$, then $D^{\perp_h}$ is a $(n-k)$-dimensional subspace of $\mathbf{F}_{q^2}^n$. We can also view D as a $2mk$-dimensional subspace of $\mathbf{F}_p^{2mn}$, and $D^{\perp_a}$ as a $2m(n-k)$-dimensional subspace of $\mathbf{F}_p^{2mn}$. Since $D^{\perp_h} \subseteq D^{\perp_a}$ and the cardinalities of $D^{\perp_a}$ and $D^{\perp_h}$ are the same, we can conclude that $D^{\perp_a} = D^{\perp_h}$. The claim follows from Theorem 10.2.   ∎

So it is sufficient to consider the hermitian form in case of $\mathbf{F}_{q^2}$-linear codes. For additive codes (that are not linear) over $\mathbf{F}_{q^2}$ we have to use the rather inconvenient trace-alternating form.

10.4   Bounds on quantum codes

We need some bounds on the achievable minimum distance of a quantum stabilizer code. Perhaps the simplest one is the Knill–Laflamme bound, also called the quantum Singleton bound. The binary version of the quantum Singleton bound was first proved by Knill and Laflamme in [43], see also [4, 6], and later generalized by Rains using weight enumerators in [49].

THEOREM 10.3 Quantum Singleton Bound

An $((n,K,d))_q$ stabilizer code with $K > 1$ satisfies

$$K \le q^{n-2d+2}.$$

Codes which meet the quantum Singleton bound are called quantum MD-S codes. In [36] we showed that these codes cannot be indefinitely long and showed that the maximal length of a q-ary quantum MDS code is upper bounded by $2q^2 - 2$. This could probably be tightened to $q^2 + 2$. It would be interesting to find quantum MDS code of length greater than $q^2 + 2$ since it would disprove the MDS conjecture for classical codes [34]. A related open question is regarding the construction of codes with lengths between q and $q^2 - 1$. At the moment there are no analytical methods for constructing a quantum MDS code of arbitrary length in this range (see [31] for some numerical results).

Another important bound for quantum codes is the quantum Hamming bound. The quantum Hamming bound states (see [22, 20]) that:

THEOREM 10.4 Quantum Hamming Bound

Any pure $((n,K,d))_q$ stabilizer code satisfies

$$\sum_{i=0}^{\lfloor (d-1)/2 \rfloor} \binom{n}{i} (q^2 - 1)^i \le q^n/K.$$

While the quantum Singleton bound holds for all quantum codes, it is not known whether the quantum Hamming bound is of equal applicability. So far no degenerate quantum code has been found that beats this bound. Gottesman showed that impure single and double error-correcting binary quantum codes cannot beat the quantum Hamming bound [24].

Perfect Quantum Codes. A quantum code that meets the quantum Hamming bound with equality is known as a perfect quantum code. In fact the famous $[[5,1,3]]_2$ code [44] is one such. We will show that there do not exist any pure perfect quantum codes other than the ones mentioned in the following theorem. It is actually a very easy result and follows from known results on classical perfect codes, but we had not seen this result earlier in the literature.

THEOREM 10.5

There do not exist any pure perfect quantum codes with distance greater than 3.

PROOF   Assume that Q is a pure perfect quantum code with the parameters $((n,K,d))_q$. Since it meets the quantum Hamming bound we have

$$K \sum_{j=0}^{\lfloor (d-1)/2 \rfloor} \binom{n}{j} (q^2 - 1)^j = q^n.$$

By Theorem 10.2 the associated classical code C is such that $C^{\perp_a} \le C \le \mathbf{F}_{q^2}^n$ and has parameters $(n, q^n K, d)_{q^2}$. Its distance is d because the quantum code is pure. Now C obeys the classical Hamming bound (see [34, Theorem 1.12.1] or any textbook on classical codes). Hence

$$|C| = q^n K \le \frac{q^{2n}}{\sum_{j=0}^{\lfloor (d-1)/2 \rfloor} \binom{n}{j} (q^2 - 1)^j}.$$

Substituting the value of K we see that this implies that C is a perfect classical code. But the only perfect classical codes with distance greater than 3 are the Golay codes and the repetition codes [34]. The perfect Golay codes are over $\mathbf{F}_2$ and $\mathbf{F}_3$ not over a quadratic extension field as C is required to be. The repetition codes are of dimension 1 and cannot contain their duals as C is required to contain. Hence C cannot be any one of them. Therefore, there are no pure quantum codes of distance greater than 3 that meet the quantum Hamming bound.   ∎

   Since it is not known whether the quantum Hamming bound holds for degenerate quantum codes, it would be interesting to find degenerate quantum codes that either meet or beat the quantum Hamming bound.

10.5   Families of quantum codes

   We shall now restrict our attention to linear quantum codes and derive several families of quantum codes from classical linear codes. We make use of the CSS construction given in Lemma 10.4. Hence, we need to look for classical codes that are self-orthogonal with respect to the Euclidean product or for families of nested codes.

10.5.1 Quantum *m*-adic residue codes

In this section we will construct a family of quantum codes based on the *m*-adic residue codes. These codes are a generalization of the well-known quadratic residue codes and share many of their structural properties. Quantum quadratic residue codes were first constructed by Rains [49] for prime alphabet.

Let $Q_0 = \{\alpha^m | \alpha \in \mathbf{Z}_p^\times\}$ be the *m*-adic residues of $\mathbf{Z}_p^\times$, where p is a prime. And let $Q_i = b^i Q_0$, where b is a generator of $\mathbf{Z}_p^\times$ and $i \in \{0, 1, \ldots, m-1\}$. Let α be a primitive root of pth root of unity. Then we can define the following four families of *m*-adic residue codes.

Let C_i be the cyclic code with the generator polynomial

$$g_i(x) = (x^p - 1)/\prod_{z \in Q_i}(x - \alpha^z).$$

These codes C_i form the even-like codes of class I. Every code C_i has the parameters $[p, (p-1)/m]_q$. The complement of C_i is denoted by $\hat{C}_i$ and its generator polynomial is given by $\hat{g}_i(x) = \prod_{z \in Q_i}(x - \alpha^z)$. These codes constitute the family of odd-like codes of class I. These codes have the parameters $[p, p - (p-1)/m]_q$.

The code with generator polynomial $h_i(x) = (x-1)\hat{g}_i(x)$ is denoted by D_i. It has parameters $[p, p - (p-1)/m - 1]_q$. These codes form the even-like codes of class II. The complement of D_i is denoted by $\hat{D}_i$ and its generator polynomial $\hat{h}_i(x) = g_i(x)/(x-1)$. The codes $\hat{D}_i$ make up the odd-like codes of class II. Their parameters are $[p, (p-1)/m + 1]_q$.

These definitions imply that $C_i \subset \hat{D}_i$ and $D_i \subset \hat{C}_i$. Further it can be shown that $C_i^\perp = \hat{C}_i$ and $D_i^\perp = \hat{D}_i$ [35, Theorem 2,3] if -1 is a *m*-adic residue. If -1 is not a residue, then $C_i \subseteq C_i^\perp = \hat{C}_j$ and $D_i^\perp = \hat{D}_j$, where $i \neq j$. We thus have families of nested codes and the CSS construction is applicable.

THEOREM 10.6

Let q be an m-adic residue modulo of a prime p such that $\gcd(p,q) = 1$. Then there exists a quantum code with the parameters $[[p, 1, d]]_q$, where $d^m \geq p$. If -1 is a m-adic reside modulo p, then $(d^2 - d + 1)^{m/2} \geq p$.

PROOF By the CSS construction there exists a quantum code with the parameters $[[p, (p-1)/m + 1 - (p-1)/m, d]]_q$, where $d = \text{wt}\{(\hat{D}_i \setminus C_i) \cup (C_i^\perp \setminus \hat{D}_i^\perp)\}$.

If -1 is a *m*-adic residue modulo p, then we know from [35, Theorem 2,3] that $C_i^\perp = \hat{C}_i$ and $D_i^\perp = \hat{D}_i$. Since $C_i^\perp = \hat{C}_i$ and $\hat{D}_i^\perp = D_i$, this means $d = \text{wt}\{(\hat{D}_i \setminus C_i) \cup (\hat{C}_i \setminus D_i)\}$. But this is the set of odd-like vectors in $\hat{C}_i$ and $\hat{D}_i$ which is lower bounded as $d^m \geq p$ [35, Theorem 5].

If -1 is not a m-adic residue modulo p, then again from [35, Theorem 2,3] we know that $C_i \subseteq C_i^\perp = \hat{C}_j$ and $\hat{D}_i^\perp = D_j$ with $i \neq j$. Then $(d^2 - d + 1)^{m/2} \geq p$ by [35, Theorem 5]. ∎

10.5.2 Quantum projective Reed–Muller codes

We study projective Reed–Muller (PRM) codes and construct the corresponding quantum PRM codes. Let us denote by $\mathbf{F}_q[X_0, X_1, ..., X_m]$ the polynomial ring in $X_0, X_1, ..., X_m$ with coefficients in $\mathbf{F}_q$. Furthermore, let $\mathbf{F}_q[X_0, X_1, ..., X_m]_h^v \cup \{0\}$ be the vector space of homogeneous polynomials in $X_0, X_1, ..., X_m$ with coefficients in $\mathbf{F}_q$ with degree v (cf. [55]). Let $P^m(\mathbf{F}_q)$ be the m-dimensional projective space over $\mathbf{F}_q$.

Projective Reed–Muller Codes. The PRM code over $\mathbf{F}_q$ of integer order v and length $n = (q^{m+1} - 1)/(q - 1)$ is denoted by $\mathscr{P}_q(v, m)$ and defined as

$$\mathscr{P}_q(v, m) = \{(f(P_1), ..., f(P_n)) \mid f(X_0, ..., X_m) \in \mathbf{F}_q[X_0, ..., X_m]_h^v \cup \{0\}\},$$
$$\text{and } P_i \in P^m(\mathbf{F}_q) \text{ for } 1 \leq i \leq n. \tag{10.4}$$

LEMMA 10.5

The projective Reed–Muller code $\mathscr{P}_q(v, m)$, $1 \leq v \leq m(q - 1)$, is an $[n, k, d]_q$ code with length $n = (q^{m+1} - 1)/(q - 1)$, dimension

$$k(v) = \sum_{\substack{(t \equiv v \bmod (q-1)) \\ t \leq v}} \sum_{j=0}^{m+1} (-1)^j \binom{m+1}{j} \binom{t - jq + m}{t - jq} \tag{10.5}$$

and minimum distance $d(v) = (q - s)q^{m-r-1}$ where $v = r(q - 1) + s + 1$, $0 \leq s < q - 1$.

PROOF   See [55, Theorem 1]. ∎

The duals of PRM codes are also known and under some conditions they are also PRM codes. The following result gives more precise details.

LEMMA 10.6

Let $v^\perp = m(q-1) - v$, then the dual of $\mathscr{P}_q(v,m)$ is given by

$$\mathscr{P}_q(v,m)^\perp = \begin{cases} \mathscr{P}_q(v^\perp,m) & v \not\equiv 0 \bmod (q-1) \\ \mathrm{Span}_{\mathbf{F}_q}\{1, \mathscr{P}_q(v^\perp,m)\} & v \equiv 0 \bmod (q-1) \end{cases} \quad (10.6)$$

PROOF See [55, Theorem 2]. ∎

As mentioned earlier our main methods of constructing quantum codes are the CSS construction and the Hermitian construction. This requires us to identify nested families of codes and/or self-orthogonal codes. First we identify when the PRM codes are nested i.e., we find out when a PRM code contains other PRM codes as subcodes.

LEMMA 10.7

If $v_2 = v_1 + k(q-1)$, where $k > 0$, then $\mathscr{P}_q(v_1,m) \subseteq \mathscr{P}_q(v_2,m)$ and $\mathrm{wt}(\mathscr{P}_q(v_2,m) \setminus \mathscr{P}_q(v_1,m)) = \mathrm{wt}(\mathscr{P}_q(v_2,m))$.

PROOF In $\mathbf{F}_q$, we can replace any variable x_i by x_i^q, hence every function in $\mathbf{F}_q[x_0,x_1,\ldots,x_m]_v^h$ is present in $\mathbf{F}_q[x_0,x_1,\ldots,x_m]_{v+k(q-1)}^h$. Hence $\mathscr{P}_q(v_1,m) \subseteq \mathscr{P}_q(v_2,m)$. Let $v_1 = r(q-1)+s+1$, then $v_2 = (k+r)(q-1)+s+1$. By Lemma 10.5, $d(v_1) = (q-s)q^{m-r-1} > (q-s)q^{m-r-k-1} = d(v_2)$. This implies that there exists a vector of weight $d(v_2)$ in $\mathscr{P}_q(v_2,m)$ and $\mathrm{wt}(\mathscr{P}_q(v_2,m) \setminus \mathscr{P}_q(v_1,m)) = \mathrm{wt}(\mathscr{P}_q(v_2,m))$. ∎

Quantum Projective Reed–Muller Codes. We now construct stabilizer codes using the CSS construction.

THEOREM 10.7

Let $n = (q^{m+1} - 1)/(q-1)$ and $1 \le v_1 < v_2 \le m(q-1)$ such that $v_2 = v_1 + l(q-1)$ with $v_1 \not\equiv 0 \bmod (q-1)$. Then there exists an $[[n, k(v_2) - k(v_1), \min\{d(v_2), d(v_1^\perp)\}]]_q$ stabilizer code, where the parameters $k(v)$ and $d(v)$ are given in Theorem 10.5.

PROOF A direct application of the CSS construction in conjunction with Lemma 10.7. ∎

We do not need to use a pair of codes as in the previous two cases; we could use a single self-orthogonal code for constructing a quantum code. We will illustrate this idea by finding self-orthogonal PRM codes.

COROLLARY 10.3
Let $0 \leq v \leq \lfloor m(q-1)/2 \rfloor$ and $2v \equiv 0 \bmod q-1$, then $\mathscr{P}_q(v,m) \subseteq \mathscr{P}_q(v,m)^{\perp}$. If $v \not\equiv 0 \bmod q-1$ there exists an $[[n, n-2k(v), d(v^{\perp})]]_q$ quantum code where $n = (q^{m+1} - 1)/(q-1)$.

PROOF   We know that $v^{\perp} = m(q-1) - v$ and if $\mathscr{P}_q(v,m) \subseteq \mathscr{P}_q(v,m)^{\perp}$, then $v \leq v^{\perp}$ and by Lemma 10.7 $v^{\perp} = v + k(q-1)$ for some $k \geq 0$. It follows that $2v \leq \lfloor m(q-1)/2 \rfloor$ and $2v = (m-k)(q-1)$, i.e., $2v \equiv 0 \bmod q-1$. The quantum code then follows from Theorem 10.7.   ■

10.5.3   Puncturing quantum codes

Finally we will briefly touch upon another important aspect of quantum code construction, which is the topic of shortening quantum codes. In the literature on quantum codes, there is not much distinction made between puncturing and shortening of quantum codes and often the two terms are used interchangeably. Obtaining a new quantum code from an existing one is more difficult task than in the classical case, the main reason being that the code must be so modified such that the resulting code is still self-orthogonal. Fortunately, however there exists a method due to Rains [49] that can solve this problem.

From Lemma 10.4 we know that with every quantum code constructed using the CSS construction, we can associate two classical codes, C_1 and C_2. Define C to be the direct product of $C_1^{\perp}$ and $C_2^{\perp}$ viz. $C = C_1^{\perp} \times C_2^{\perp}$. Then we can associate a puncture code $P(C)$ [33, Theorem 12] which is defined as

$$P(C) = \{(a_i b_i)_{i=1}^n \mid a \in C_1^{\perp}, b \in C_2^{\perp}\}^{\perp}. \tag{10.7}$$

Surprisingly, $P(C)$ provides information about the lengths to which we can puncture the quantum codes. If there exists a vector of nonzero weight r in $P(C)$, then the corresponding quantum code can be punctured to a length r and minimum distance greater than or equal to distance of the parent code.

THEOREM 10.8
Let $0 \leq v_1 < v_2 \leq m(q-1) - 1$ where $v_2 \equiv v_1 \bmod q - 1$. Also let $0 \leq \mu \leq v_2 - v_1$ and $\mu \equiv 0 \bmod q - 1$. If $\mathscr{P}_q(\mu, m)$ has codeword of weight r, then there exists an $[[r, \geq (k(v_2) - k(v_1) - n + r), \geq d]]_q$ quantum code, where

$n = (q^m - 1)/(q-1)$, $d = \min\{d(v_2), d(v_1^{\perp})\}$. *In particular, there exists a* $[[d(\mu), \geq (k(v_2) - k(v_1) - n + d(\mu)), \geq d]]_q$ *quantum code.*

PROOF   Let $C_i = \mathscr{P}_q(v_i, m)$ with v_i as stated. Then by Theorem 10.7, an $[[n, k(v_2) - k(v_1), d]]_q$ quantum code Q exists where $d = \min\{d(v_2), d(v_1^{\perp})\}$. From equation (10.7) we find that $P(C)^{\perp} = \mathscr{P}_q(v_1 + v_2^{\perp}, m)$, so

$$P(C) = \mathscr{P}_q(m(q-1) - v_1 - v_2^{\perp}, m),$$
$$= \mathscr{P}_q(v_2 - v_1, m). \tag{10.8}$$

By [33, Theorem 11], if there exists a vector of weight r in $P(C)$, then there exists an $[[r, k', d']]_q$ quantum code, where $k' \geq (k(v_2) - k(v_1) - n + r)$ and distance $d' \geq d$. obtained by puncturing Q. Since $P(C) = \mathscr{P}_q(v_2 - v_1, m) \supseteq \mathscr{P}_q(\mu, m)$ for all $0 \leq \mu \leq v_2 - v_1$ and $\mu \equiv v_2 - v_1 \equiv 0 \mod q - 1$, the weight distributions of $\mathscr{P}_q(\mu, m)$ give all the lengths to which Q can be punctured. Moreover $P(C)$ will certainly contain vectors whose weight $r = d(\mu)$, that is the minimum weight of $PC(\mu, m)$. Thus there exist punctured quantum codes with the parameters $[[d(\mu), \geq (k(v_2) - k(v_1) - n + d(\mu)), \geq d]]_q$. ∎

10.6 Conclusion

We have given a brief introduction to the theory of nonbinary stabilizer codes. Our goal was to emphasize the key ideas so we have omitted long and cumbersome proofs. Most of these details can be found in our companion papers on stabilizer codes. After introducing the stabilizer formalism for quantum codes, we showed how these were related to classical codes. Essentially we mapped the stabilizer and its centralizer to a classical code and its dual. And from then on all properties of the quantum codes could be studied by studying the classical codes. The construction of stabilizer codes can be reduced to identifying classical codes that are self-orthogonal. Then, we discussed the question of optimal codes and some well known bounds. We showed the nonexistence of a class of perfect codes of distance greater than 3. Finally we illustrated these ideas by constructing two new families of quantum codes.

Acknowledgments

This work was supported by NSF grant CCF 0622201 and NSF CAREER award CCF 0347310.

References

[1] D. Aharonov and M. Ben-Or. Fault-tolerant quantum computation with constant error. In *Proc. of the 29th Annual ACM Symposium on Theory of Computation (STOC)*, pages 176–188, New York, 1997. ACM.

[2] V. Arvind and K.R. Parthasarathy. A family of quantum stabilizer codes based on the Weyl commutation relations over a finite field. In *A tribute to C. S. Seshadri (Chennai, 2002)*, Trends Math., pages 133–153. Birkhäuser, 2003.

[3] A. Ashikhmin and E. Knill. Nonbinary quantum stabilizer codes. *IEEE Trans. Inform. Theory*, 47(7):3065–3072, 2001.

[4] A. Ashikhmin and S. Litsyn. Upper bounds on the size of quantum codes. *IEEE Trans. Inform. Theory*, 45(4):1206–1215, 1999.

[5] A.E. Ashikhmin, A.M. Barg, E. Knill, and S.N. Litsyn. Quantum error detection I: Statement of the problem. *IEEE Trans. on Inform. Theory*, 46(3):778–788, 2000.

[6] A.E. Ashikhmin, A.M. Barg, E. Knill, and S.N. Litsyn. Quantum error detection II: Bounds. *IEEE Trans. on Inform. Theory*, 46(3):789–800, 2000.

[7] T. Beth and M. Grassl. The quantum Hamming and hexacodes. *Fortschr. Phys.*, 46(4-5):459–491, 1998.

[8] J. Bierbrauer and Y. Edel. Quantum twisted codes. *J. Comb. Designs*, 8:174–188, 2000.

[9] A.R. Calderbank, E.M. Rains, P.W. Shor, and N.J.A. Sloane. Quantum error correction and orthogonal geometry. *Phys. Rev. Lett.*, 76:405–409, 1997.

[10] A.R. Calderbank, E.M. Rains, P.W. Shor, and N.J.A. Sloane. Quantum error correction via codes over GF(4). *IEEE Trans. Inform. Theory*, 44:1369–1387, 1998.

[11] A.R. Calderbank and P. Shor. Good quantum error-correcting codes exist. *Phys. Rev. A*, 54:1098–1105, 1996.

[12] H.F. Chau. Correcting quantum errors in higher spin systems. *Phys. Rev. A*, 55:R839–R841, 1997.

[13] H.F. Chau. Five quantum register error correction code for higher spin systems. *Phys. Rev. A*, 56:R1–R4, 1997.

[14] R. Cleve. Quantum stabilizer codes and classical linear codes. *Phys. Rev. A*, 55(6):4054–4059, 1997.

[15] R. Cleve and D. Gottesman. Efficient computations of encodings for quantum error correction. *Phys. Rev. A*, 56(1):76–82, 1997.

[16] G. Cohen, S. Encheva, and S. Litsyn. On binary constructions of quantum codes. *IEEE Trans. Inform. Theory*, 45(7):2495–2498, 1999.

[17] A. Ekert and C. Macchiavello. Error correction in quantum communication. *Phys. Rev. Lett.*, 76:2585–2588, 1996.

[18] K. Feng. Quantum codes $[[6,2,3]]_p$, $[[7,3,3]]_p$ $(p \geq 3)$ exist. *IEEE Trans. Inform. Theory*, 48(8):2384–2391, 2002.

[19] K. Feng. Quantum error-correcting codes. In *Coding Theory and Cryptology*, pages 91–142. World Scientific, 2002.

[20] K. Feng and Z. Ma. A finite Gilbert-Varshamov bound for pure stabilizer quantum codes. *IEEE Trans. Inform. Theory*, 50(12):3323–3325, 2004.

[21] M.H. Freedman and D.A. Meyer. Projective plane and planar quantum codes. *Found. Comput. Math.*, 1(3):325–332, 2001.

[22] D. Gottesman. A class of quantum error-correcting codes saturating the quantum Hamming bound. *Phys. Rev. A*, 54:1862–1868, 1996.

[23] D. Gottesman. Pasting quantum codes. eprint: quant-ph/9607027, 1996.

[24] D. Gottesman. Stabilizer codes and quantum error correction. Caltech Ph. D. Thesis, eprint: quant-ph/9705052, 1997.

[25] D. Gottesman. Fault-tolerant quantum computation with higher-dimensional systems. *Chaos, Solitons, Fractals*, 10(10):1749–1758, 1999.

[26] D. Gottesman. An introduction to quantum error correction. In S. J. Lomonaco, Jr., editor, *Quantum Computation: A Grand Mathematical Challenge for the Twenty-First Century and the Millennium*, pages 221–235, Rhode Island, 2002. American Mathematical Society. eprint: quant-ph/0004072.

[27] D. Gottesman. Quantum error correction and fault-tolerance. eprint: quant-ph/0507174, 2005.

[28] M. Grassl. Algorithmic aspects of error-correcting codes. In R. Brylinski and G. Chen, editors, *The Mathematics of Quantum Computing*, pages 223–252. CRC Press, 2001.

[29] M. Grassl and T. Beth. Quantum BCH codes. In *Proc. X. Int'l. Symp. Theoretical Electrical Engineering, Magdeburg*, pages 207–212, 1999.

[30] M. Grassl and T. Beth. Cyclic quantum error-correcting codes and quantum shift registers. *Proc. Royal Soc. London Series A*, 456(2003):2689–2706, 2000.

[31] M. Grassl, T. Beth, and M. Rötteler. On optimal quantum codes. *Internat. J. Quantum Information*, 2(1):757–775, 2004.

[32] M. Grassl, W. Geiselmann, and T. Beth. Quantum Reed-Solomon codes. In *Applied algebra, algebraic algorithms and error-correcting codes (Honolulu, HI, 1999)*, volume 1719 of *Lecture Notes in Comput. Sci.*, pages 231–244. Springer, Berlin, 1999.

[33] M. Grassl, M. Rötteler, and T. Beth. Efficient quantum circuits for non-qubit quantum error-correcting codes. *Internat. J. Found. Comput. Sci.*, 14(5):757–775, 2003.

[34] W. C. Huffman and V. Pless. *Fundamentals of Error-Correcting Codes*. University Press, Cambridge, 2003.

[35] V. R. Job. *m*-adic residue codes. *IEEE Trans. Inform. Theory*, 38(2):496–501, 1992.

[36] A. Ketkar, A. Klappenecker, S. Kumar, and P. K. Sarvepalli. Nonbinary stabilizer codes over finite fields. *IEEE Trans. Inform. Theory*, 52(11):4892–4914, 2006.

[37] J.-L. Kim. New quantum-error-correcting codes from Hermitian self-orthogonal codes over GF(4). In *Proc. of the Sixth Intl. Conference on Finite Fields and Applications, Oaxaca, Mexico, May 21-25*, pages 209–213. Springer–Verlag, 2002.

[38] J.-L. Kim and V. Pless. Designs in additive codes over GF(4). *Designs, Codes and Cryptography*, 30:187–199, 2003.

[39] J.-L. Kim and J. Walker. Nonbinary quantum error-correcting codes from algebraic curves. submitted to a special issue of Com²MaC Conference on Association Schemes, Codes and Designs in Discrete Math, 2004.

[40] A.Y. Kitaev. Quantum computations: algorithms and error correction. *Russian Math. Surveys*, 52(6):1191–1249, 1997.

[41] A. Klappenecker and M. Rötteler. Beyond stabilizer codes II: Clifford codes. *IEEE Trans. Inform. Theory*, 48(8):2396–2399, 2002.

[42] E. Knill. Group representations, error bases and quantum codes. Los Alamos National Laboratory Report LAUR-96-2807, 1996.

[43] E. Knill and R. Laflamme. A theory of quantum error–correcting codes. *Physical Review A*, 55(2):900–911, 1997.

[44] R. Laflamme, C. Miquel, J. P. Paz, and W. H. Zurek. Perfect quantum error correction code. *Phys. Rev. Lett.*, 77:198–201, 1997.

[45] R. Li and X. Li. Binary construction of quantum codes of minimum distance three and four. *IEEE Trans. Inform. Theory*, 50(6):1331–1336, 2004.

[46] F.J. MacWilliams and N.J.A. Sloane. *The Theory of Error-Correcting Codes*. North-Holland, 1977.

[47] W.J. Martin. A physics-free introduction to quantum error correcting codes. *Util. Math.*, pages 133–158, 2004.

[48] R. Matsumoto and T. Uyematsu. Constructing quantum error correcting codes for p^m-state systems from classical error correcting codes. *IEICE Trans. Fundamentals*, E83-A(10):1878–1883, 2000.

[49] E.M. Rains. Nonbinary quantum codes. *IEEE Trans. Inform. Theory*, 45:1827–1832, 1999.

[50] E.M. Rains. Quantum codes of minimum distance two. *IEEE Trans. Inform. Theory*, 45(1):266–271, 1999.

[51] M. Rötteler, M. Grassl, and T. Beth. On quantum MDS codes. In *Proc. 2004 IEEE Intl. Symposium on Information Theory, Chicago, USA*, page 355, 2004.

[52] D. Schlingemann. Stabilizer codes can be realized as graph codes. *Quantum Inf. Comput.*, 2(4):307–323, 2002.

[53] D. Schlingemann and R.F. Werner. Quantum error-correcting codes associated with graphs. eprint: quant-ph/0012111, 2000.

[54] P. Shor. Scheme for reducing decoherence in quantum memory. *Phys. Rev. A*, 2:2493–2496, 1995.

[55] A. B. Sorensen. Projective Reed-Muller codes. *IEEE Trans. Inform. Theory*, 37(6):1567–1576, 1991.

[56] A. Steane. Quantum Reed-Muller codes. *IEEE Trans. Inform. Theory*, 45(5):1701–1703, 1999.

[57] A.M. Steane. Multiple-particle interference and quantum error correction. *Proc. Roy. Soc. London A*, 452:2551–2577, 1996.

[58] A.M. Steane. Simple quantum error correcting codes. *Phys. Rev. Lett.*, 77:793–797, 1996.

[59] A.M. Steane. Enlargement of Calderbank-Shor-Steane quantum codes. *IEEE Trans. Inform. Theory*, 45(7):2492–2495, 1999.

Chapter 11

Accessible information about quantum states: An open optimization problem

Jun Suzuki, Syed M. Assad, and Berthold-Georg Englert

Abstract   We give a brief summary of the current status of the problem of extracting the accessible information when a quantum system is received in one of a finite number of pre-known quantum states. We review analytical methods as well as a numerical strategy. In particular, the group-covariant positive-operator-valued measures are discussed, and several explicit examples are worked out in detail. These examples include some that occur in the security analysis of schemes for quantum cryptography.

11.1   Introduction

A sender, traditionally called Alice, sends quantum states, one by one, to a receiver, Bob. Bob then wishes to perform measurements on the quantum states he receives to find out, the best he can, what Alice has sent. Generally speaking, owing to the nature of quantum mechanics, it is impossible for Bob to obtain full knowledge about the states which he is receiving. Instead, he has to choose his measurements judiciously from all measurements permitted by quantum mechanics. A natural question one might ask is then:

> *What is the best strategy for Bob to maximize his knowledge*
> *about the states he is receiving from Alice?*   (11.1)

The answer to this question is not only of importance for our understanding of the implications of quantum mechanics, it also has great practical significance for most areas in quantum information, in particular for the capacity of quantum channels and the security analysis of schemes for quantum cryptography under powerful eavesdropping attacks. Indeed, our own interest in the matter originates in its relevance to the security of "tomographic quantum cryptography," a class of protocols for quantum key distribution developed in Singapore [1, 2, 3, 4].

The main objective of this chapter is to provide a concise introduction to this problem with a summary of ongoing research in this field. For this purpose we will not give a rigorous mathematical exposition, and we will be content with stating most of theorems without proof. We suggest that readers who are interested in the technical mathematical details consult the pertinent literature referred to in the text.

Here is a brief preview of coming attractions. In Section 11.2 we remind the reader of a few basic concepts and, at the same time, establish the terminology and the notational conventions we are using. Then, in Section 11.3, we state question (11.1) as an optimization problem, for which the mutual information between Alice and Bob is the figure of merit. Section 11.4 reports essential properties of this mutual information and important theorems about known properties of the solution. A numerical procedure for searching the optimum by a steepest-ascent method is described in Section 11.5. Examples are presented in Section 11.6, where we limit the choice to cases with a structure as one meets it in the security analysis of quantum cryptography schemes. We close with a summary and outlook.

11.2   Preliminaries

11.2.1   States and measurements

We set the stage by first providing a brief mathematical description of the physical situation that (11.1) refers to, that is: Alice sends certain physical states to Bob who measures them to find out which states she sent. For simplicity and for concreteness, we consider only finite-dimensional systems.

The quantum states prepared by Alice are denoted by $\rho_1, \rho_2, \ldots, \rho_J$ whereby $J \geq 1$ is finite, and the set $\mathscr{E} = \{\rho_j | j = 1, 2, \ldots, J\}$ is the *ensemble* of quantum states sent by Alice. Each of the ρ_js is a *density matrix*: a semi-definite positive, and therefore hermitian, matrix with finite trace.[1] One calls the jth state a *pure state* when the density matrix ρ_j is essentially a projector, otherwise it is a *mixed state*,

$$\text{state } \rho_j \text{ is } \left\{\begin{matrix} \text{pure} \\ \\ \text{mixed} \end{matrix}\right\} \text{ if } \left\{\begin{matrix} \text{Tr}(\rho_j^2) = (\text{Tr}\rho_j)^2, \\ \\ \text{Tr}(\rho_j^2) < (\text{Tr}\rho_j)^2. \end{matrix}\right. \tag{11.2}$$

By convention we normalize the ρ_js such that their traces are the probabilities a_j with which Alice is sending them. Thus, Bob knows that the probability of receiving ρ_j as the next state is $a_j = \text{Tr}\rho_j$. Since the next state is surely one of the ρ_js, these probabilities have unit sum,

$$1 = \sum_{j=1}^{J} a_j = \sum_{j=1}^{J} \text{Tr}\rho_j. \tag{11.3}$$

It follows that the total density matrix $\rho = \sum_{j=1}^{J}\rho_j$ has unit trace, $\text{Tr}\rho = 1$. The rank of ρ is the dimension d of the space under consideration, which is to say that we represent all ρ_js, and all other linear operators, by $d \times d$ matrices.[2]

It is often convenient to represent a pure-state matrix ρ_j as a product of a d-component column $|j\rangle$ and its adjoint d-component row $\langle j| = |j\rangle^\dagger$, that is $\rho_j = |j\rangle\langle j|$. In the standard terminology of quantum physics, one speaks of *kets* and *bras* when referring to the columns $|j\rangle$ and the rows $\langle j|$, respectively. The numerical row-times-column product of $\langle j_1|$ with ket $|j_2\rangle$ is denoted by $\langle j_1|j_2\rangle$ and is called their *bracket*; it is equal to the trace of their column-times-row product $|j_2\rangle\langle j_1|$,

$$\text{Tr}(|j_2\rangle\langle j_1|) = \langle j_1|j_2\rangle. \tag{11.4}$$

Bob's measurement is specified by a decomposition of the $d \times d$ identity matrix $\mathbf{1}_d$ into a set of semi-definite positive, hermitian matrices,

$$\mathbf{1}_d = \sum_{k=1}^{K} \Pi_k \quad \text{with} \quad K \geq 1 \quad \text{and} \quad \Pi_k \geq 0, \tag{11.5}$$

---

[1] More generally, a quantum state is specified by a semi-definite positive linear operator with finite trace and each of its equivalent matrix representations is a corresponding density matrix. By choosing one particular orthonormal basis in the Hilbert space, we specify one set of density matrices for the set of states under consideration.

[2] More generally, d is the dimension of the relevant subspace of a possibly much larger Hilbert space.

which is the general[3] form of a so-called *positive operator valued measure* (POVM) [5], here with K *outcomes* Π_k. Bob's *a priori* probability of getting the kth outcome is

$$b_k = \mathrm{Tr}(\rho\Pi_k), \tag{11.6}$$

which is properly normalized to unit sum as a consequence of the unit trace of ρ. Two special cases are worth mentioning: the von Neumann measurements, and the tomographically complete measurements.

We have a *von Neumann measurement* when the outcomes of the POVM are pairwise orthogonal projectors, $\Pi_k\Pi_l = \Pi_k\delta_{kl}$. When all Π_ks are rank-1 projectors, one speaks of a *maximal* von Neumann measurement, for which $K = d$, of course.

The POVM is *tomographically complete* if ρ can be inferred from the knowledge of all of Bob's probabilities b_k, which is to say that the map $\rho \mapsto \{b_k | k = 1,\ldots,K\}$ is injective. A tomographically complete POVM, has at least d^2 outcomes; in the case of $K = d^2$, one speaks of a *minimal* tomographically complete POVM.

Every outcome of a POVM can be written as a square, $\Pi_k = A_k^\dagger A_k$, but this factorization is not unique.[4] Typically, there is one such factorization for each physical implementation of the POVM. Then, given an ideal—that is, noise-free and nondestructive—implementation, the final state of the physical system after the measurement is

$$\rho^{(k)} = \frac{A_k\rho A_k^\dagger}{\mathrm{Tr}(\rho\Pi_k)} \tag{11.7}$$

if ρ is the state before the measurement and the kth outcome is obtained. Therefore, in general, the possible final states are mixed states when POVMs are performed on mixed states.

When Bob performs the POVM (11.5) on the states $\rho_1, \rho_2, \ldots, \rho_J$ sent by Alice, the joint probability that Alice sends the jth state and Bob gets the kth outcome is

$$p_{jk} = \mathrm{Tr}(\rho_j\Pi_k). \tag{11.8}$$

The respective marginal probabilities

$$a_j = \sum_{k=1}^{K} p_{jk} = \mathrm{Tr}\rho_j, \quad b_k = \sum_{j=1}^{J} p_{jk} = \mathrm{Tr}(\rho\Pi_k) \tag{11.9}$$

---

[3]Somewhat more generally, the label k could be continuous and the summation replaced by an integration. We do not need to consider such general cases.

[4]More generally, Π_k could be a sum of squares, $\Pi_k = \sum_l A_{kl}^\dagger A_{kl}$, even in the case of a von Neumann measurement, as is illustrated by $A_{kl} = V_{kl}\Pi_k^{1/2}$ with $\sum_l V_{kl}^\dagger V_{kl} = \mathbf{1}_d$ for all k. The case of $A_k = \Pi_k^{1/2}$ is sometimes referred to as an *ideal* POVM.

are the probabilities that Alice sends the jth state and the *a priori* probabilities that Bob gets the kth outcome.

The *conditional probabilities* $p(k|j) = p_{jk}/a_j$ and $p(j|k) = p_{jk}/b_k$ have the following significance, respectively: *If* Alice sends the jth state, she can predict that Bob will get the kth outcome with probability $p(k|j)$; *if* Bob receives the kth outcome, he can infer that Alice sent the jth state with probability $p(j|k)$.

It is worth noting that there is a reciprocal situation with exactly the same joint probabilities. It is specified by Alice measuring the POVM

$$1_d = \sum_{j=1}^{J} \tilde{\Pi}_j \quad \text{with} \quad \tilde{\Pi}_j = \rho^{-1/2}\rho_j\rho^{-1/2} \tag{11.10}$$

and Bob sending her the states $\tilde{\rho}_k = \rho^{1/2}\Pi_k\rho^{1/2}$.

11.2.2 Entropy and information

Next, we define several quantities that will be used for the quantification of information in the sequel [6, 7]: the von Neumann entropy, the Shannon entropy, the Kullback–Leibler relative entropy, the mutual information, and the accessible information.

von Neumann entropy: The von Neumann entropy $S(\rho)$ of a density matrix ρ is[5]

$$S(\rho) = -\text{Tr}\left(\frac{\rho}{\text{Tr}\rho}\log\frac{\rho}{\text{Tr}\rho}\right) = -\frac{\text{Tr}(\rho\log\rho)}{\text{Tr}\rho} + \log\text{Tr}\rho, \tag{11.11}$$

which has the more familiar appearance

$$S(\rho) = -\text{Tr}(\rho\log\rho) \quad \text{if} \quad \text{Tr}\rho = 1. \tag{11.12}$$

By construction, we have $S(x\rho) = S(\rho)$ for all $x > 0$. Further we note that the mapping $\rho \mapsto \text{Tr}(\rho)S(\rho)$ is concave:

$$\text{Tr}(\rho_1 + \rho_2)S(\rho_1 + \rho_2) \geq \text{Tr}(\rho_1)S(\rho_1) + \text{Tr}(\rho_2)S(\rho_2) \tag{11.13}$$

for any two density matrices ρ_1 and ρ_2.

---

[5]Historically, the von Neumann entropy involves the natural logarithm and also the Boltzmann constant to establish contact with the thermodynamical entropy, whereas the Shannon entropy uses the logarithm to base 2 and the value is usually stated in units of bits. We use the logarithm to base 2 throughout.

With the convention $\lambda \log \lambda = 0$ for $\lambda = 0$, the von Neumann entropy (11.12) is expressed in terms of the eigenvalues λ_i $(i = 1, 2, \ldots, d)$ of ρ as

$$S(\rho) = -\sum_{i=1}^{d} \lambda_i \log \lambda_i \quad \text{if} \quad \sum_{i=1}^{d} \lambda_i = 1. \qquad (11.14)$$

We remark that the von Neumann entropy is zero for pure states and only for pure states, for which a single eigenvalue is positive and all others are zero.

Shannon entropy: Given Alice's ensemble $\mathscr{E} = \{\rho_j | j = 1, 2, \ldots, J\}$, we have the set $P = \{a_j = \mathrm{Tr}\rho_j | j = 1, 2, \ldots, J\}$ that is composed of the probabilities of occurrence, which have unit sum, $\sum_j a_j = 1$. The Shannon entropy $H(P)$ of such a normalized set of probabilities P is defined by[5]

$$H(P) = -\sum_{j=1}^{J} a_j \log a_j. \qquad (11.15)$$

For any two sets of normalized probabilities $P^{(1)} = \{a_j^{(1)} | j = 1, 2, \ldots, J\}$ and $P^{(2)} = \{a_j^{(2)} | j = 1, 2, \ldots, J\}$, we can consider their convex sums $xP^{(1)} + (1 - x)P^{(2)} = \{xa_j^{(1)} + (1 - x)a_j^{(2)} | j = 1, 2, \ldots, J\}$ with $0 \le x \le 1$, for which the concavity

$$H(xP^{(1)} + (1 - x)P^{(2)}) \ge xH(P^{(1)}) + (1 - x)H(P^{(2)}) \qquad (11.16)$$

holds.

As a consequence of the concavity of the von Neumann entropy in (11.13), we have the inequalities (see, e.g., Subsection 11.3.6 in [7])

$$H(P) + \sum_{j=1}^{J} a_j S(\rho_j) \ge S(\rho) \ge \sum_{j=1}^{J} a_j S(\rho_j), \qquad (11.17)$$

where $\rho = \sum_{j=1}^{J} \rho_j$ is the total density matrix. On the left, the equal sign applies if and only if all ρ_js are pairwise orthogonal pure states. On the right, the equal sign applies if the ρ_js are essentially equal to each other in the sense that $a_j \rho_k = \rho_j a_k$ for all j and k.

Kullback–Leibler relative entropy: For any two sets of normalized probabilities $P = \{p_j | j = 1, 2, \ldots, J\}$ and $\tilde{P} = \{\tilde{p}_j | j = 1, 2, \ldots, J\}$, the Kullback–Leibler relative entropy $D(P || \tilde{P})$ is defined by

$$D(P || \tilde{P}) = \sum_{j=1}^{J} p_j \log \frac{p_j}{\tilde{p}_j} \ge 0, \qquad (11.18)$$

whereby the equal sign applies only if $p_j = \tilde{p}_j$ for all j. The Kullback–Leibler relative entropy may serve as a rough measure of difference between two probability distributions P and $\tilde{P}$. But, since it is not symmetric, $D(P||\tilde{P}) \neq D(\tilde{P}||P)$ as a rule, and does not satisfy the triangle inequality, it is not a distance or metric in the mathematical sense.

Mutual information: For any normalized set of joint probabilities $A\&B = \{p_{jk}|j = 1,2,\ldots,J; k = 1,2,\ldots,K\}$ with $\sum_{jk} p_{jk} = 1$, and its two sets of marginals $A = \{a_j = \sum_{k=1}^{K} p_{jk}|j = 1,2,\ldots,J\}$ and $B = \{b_k = \sum_{j=1}^{J} p_{jk}|k = 1,2,\ldots, K\}$, the mutual information $I(A;B)$ is the relative entropy between the joint probabilities $A\&B$ and the set $AB = \{a_j b_k|j = 1,2,\ldots,J; k = 1,2,\ldots,K\}$ of product probabilities,

$$I(A;B) = D(A\&B||AB) = \sum_{j=1}^{J} \sum_{k=1}^{K} p_{jk} \log \frac{p_{jk}}{a_j b_k}$$
$$= H(A) + H(B) - H(A\&B), \qquad (11.19)$$

where the last version expresses the mutual information in terms of the various Shannon entropies.

The mutual information is a measure of the strength of the statistical correlations in joint probabilities. If there are no correlations at all, that is, if $p_{jk} = a_j b_k$ for all j and all k, the mutual information vanishes; otherwise it is positive.

In the physical situation to which the question (11.1) refers, we have the joint probabilities of (11.8) and the marginals of (11.9). Therefore, the mutual information $I(\mathscr{E};\Pi)$ between $\mathscr{E}$, the ensemble of Alice's states, and Π, Bob's POVM, quantifies his knowledge about the quantum states she is sending. This brings us, finally, to the accessible information for Bob about Alice's quantum states.

Accessible information: The accessible information I_{acc} is the maximum of the mutual informations for all possible POVMs that Bob can perform, that is

$$I_{\text{acc}}(\mathscr{E}) = \max_{\text{all }\Pi} I(\mathscr{E};\Pi). \qquad (11.20)$$

This poses the challenge of determining the value of $I_{\text{acc}}(\mathscr{E})$ for the given set $\mathscr{E}$ of quantum states.

In addition to the accessible information, there are other numerical measures [8] that can be used for the quantification of Bob's knowledge about Alice's states, such as the Bayes cost (see, e.g., [9, 5]), which is essentially the probability for guessing wrong, or the probability that Bob can unambiguously identify the state he just received (see, e.g., chapter 11 in [10]). In the context of studying the security of quantum cryptography schemes, however, the figure of merit is the accessible information. Also, the history of the subject seems to indicate that it is substantially more difficult to determine the accessible information than the Bayes cost or the probability of unambiguous discrimination.

11.3   The optimization problem

We now state the main problem (11.1) in technical terms as a double question:

> Given an ensemble of quantum states $\mathscr{E} = \{\rho_j | j = 1, 2, \ldots, J\}$,
> (a) what is the value of the accessible information $I_{\mathrm{acc}}(\mathscr{E})$, and
> (b) what is the optimal POVM $\Pi = \{\Pi_k | k = 1, 2, \ldots, K\}$ for which the mutual information is the accessible information, $I(\mathscr{E}; \Pi) = I_{\mathrm{acc}}(\mathscr{E})$?   (11.21)

Part of the answer to query (b) is to establish the number K of outcomes in the optimal POVM.

This problem was first formulated by Holevo in 1973 [9]. After more than three decades, it remains unsolved. The major difficulty is a lack of sufficient conditions that ensure the optimality of POVMs in general. Sufficiency is known only when the ensemble of quantum states possesses certain symmetry properties; see Subsection 11.4.4 below. The obvious nonlinearity that originates in the logarithms is another hurdle.

The current situation is still rather unsatisfactory even for seemingly simple ensembles $\mathscr{E}$. For instance, we do not have analytical expressions for the optimal POVMs in the case where $\mathscr{E}$ consists of only two full-rank mixed quantum states for $d = 2$;[6] see Subsection 11.6.1 below for details.

There are, of course, very special cases for which the answer is immediate. One extreme situation is

(i) all states commute with each other, $\rho_j \rho_{j'} = \rho_{j'} \rho_j$; then the optimal POVM is a von Neumann measurement composed of the projectors to the joint eigenstates. A special case thereof is

---

[6] Two mixed single-qubit states in the jargon of quantum information.

(ii) all states are pairwise orthogonal, $\rho_j \rho_{j'} = \delta_{jj'} \rho_j^2$, so that they can be distinguished without effort and we have essentially the situation of Bob receiving a classical signal.

A related, yet different problem is the determination of the so-called *quantum channel capacity* [7, 11]. A quantum channel turns any input quantum states into an output quantum state, always preserving the positivity and usually also the trace of the input. The ensemble $\mathcal{E}$ received by Bob, for which he has to find the optimal POVM, then comes about by processing Alice's input ensemble $\mathcal{E}_{in}$ through the quantum channel. There is then a two-fold optimization problem: one needs to find both Alice's optimal input ensemble as well as Bob's optimal POVM. The quantum channel capacity problem is also an *open problem*. It is clear that any progress with the accessible-information problem (11.21) means corresponding progress with the channel-capacity problem.

11.4 Theorems

Before going to the actual computation of the accessible information, we give a brief summary of established properties of the mutual information and the accessible information [6, 7].

11.4.1 Concavity and convexity

Let us regard the joint probabilities $p_{jk} = a_j p(k|j)$ as the product of Alice's probabilities a_j and the conditional probabilities $p(k|j)$. Then, the mutual information $I(\mathcal{E}; \Pi)$ is a concave function of the a_js for given $p(k|j)$s, and a convex function of the $p(k|j)$s for given a_js. In other words, the mutual information is a convex functional on the set of all possible POVMs. Therefore, all optimal POVMs are located on the boundary of the POVM space.

Since this convexity of the mutual information is of some importance in our discussion, we give more details. Suppose we have two POVMs $\Pi^{(i)} = \{\Pi_k^{(i)} | k = 1, 2, \ldots, K\} (i = 1, 2)$, then the combined new POVM $\Pi(\lambda) = \{\lambda \Pi_k^{(1)} + (1 - \lambda) \Pi_k^{(2)} | k = 1, 2, \ldots, K\}$ with $0 < \lambda < 1$ obeys the following inequality for the mutual information:

$$I(\mathcal{E}; \Pi(\lambda)) \leq \lambda I(\mathcal{E}; \Pi^{(1)}) + (1 - \lambda) I(\mathcal{E}; \Pi^{(2)}). \qquad (11.22)$$

The equality is satisfied if and only if

$$p_{jk}^{(1)}/b_k^{(1)} = p_{jk}^{(2)}/b_k^{(2)} \quad \text{or} \quad p^{(1)}(j|k) = p^{(2)}(j|k) \tag{11.23}$$

holds for all j and k, wherein we meet the joint probabilities, $p_{jk}^{(i)} = \text{Tr}(\rho_j \Pi_k^{(i)})$ and the marginals $b_k^{(i)} = \sum_{j=1}^{J} p_{jk}^{(i)}$, as well as the resulting conditional probabilities $p^{(i)}(j|k)$.

A particular situation in which the equal sign applies in (11.22) is as follows. Let $\Pi^{(1)}$ and $\Pi^{(2)}$ be two K-outcome POVMs with null outcomes such that $\Pi_k^{(1)} = 0$ for $\bar{k} < k \leq K$ and $\Pi_k^{(2)} = 0$ for $1 \leq k \leq \bar{k}$ with $1 \leq \bar{k} < K$. Then the outcomes of $\Pi(\lambda)$ are given by $\Pi_k(\lambda) = \lambda \Pi_k^{(1)}$ for $1 \leq k \leq \bar{k}$ and $\Pi_k(\lambda) = (1-\lambda)\Pi_k^{(2)}$ for $\bar{k} < k \leq K$, and it is clear that

$$I(\mathscr{E};\Pi(\lambda)) = \lambda I(\mathscr{E};\Pi^{(1)}) + (1-\lambda)I(\mathscr{E};\Pi^{(2)}) \tag{11.24}$$

holds in this situation.

11.4.2 Necessary condition

For a POVM Π to be optimal, it is necessary that the accessible information $I(\mathscr{E};\Pi)$ is stationary with respect to infinitesimal variations of Π. These variations are, however, constrained by both the positive nature of each outcome Π_k and the unit sum of all outcomes.

The first constraint is accounted for by writing $\Pi_k = A_k^\dagger A_k$, whereby the factor A_k is rather arbitrary and may differ from the physical A_k in (11.7) by a unitary matrix multiplying A_k on the left. The second constraint, that is $\sum_{k=1}^{K} \delta \Pi_k = 0$, then requires the infinitesimal variations of the A_ks to be of the form

$$\delta A_k = i \sum_{k'=1}^{K} \varepsilon_{kk'} A_{k'} \quad \text{with} \quad \varepsilon_{kk'}^\dagger = \varepsilon_{k'k}, \tag{11.25}$$

where the $\varepsilon_{kk'}$s are otherwise arbitrary infinitesimal matrices.

We note that the mutual information is expressed as

$$I(\mathscr{E};\Pi) = \sum_{k=1}^{K} \text{Tr}(R_k \Pi_k) \tag{11.26}$$

with the hermitian matrices R_k given by

$$R_k = \sum_{j=1}^{J} \rho_j \log \frac{p_{jk}}{a_j b_k}. \tag{11.27}$$

It turns out that there is no contribution from the variation of the R_ks to

$$\delta I(\mathscr{E};\Pi) = -i \sum_{k,k'=1}^{K} \text{Tr}\big(\varepsilon_{kk'} A_{k'}(R_{k'} - R_k)A_k^\dagger\big). \tag{11.28}$$

Therefore, a necessary condition for Π to be an optimal POVM is

$$A_{k'}(R_{k'} - R_k)A_k^\dagger = 0 \quad \text{for all } k,k', \tag{11.29}$$

or

$$\Pi_{k'}(R_{k'} - R_k)\Pi_k = 0 \quad \text{for all } k,k'. \tag{11.30}$$

Upon summing over k or k' we arrive at an equivalent set of equations,

$$R_k\Pi_k = \Lambda\Pi_k \quad \text{and} \quad \Pi_k\Lambda = \Pi_k R_k \quad \text{for all } k, \tag{11.31}$$

which are adjoint statements of each other because

$$\Lambda = \sum_{k=1}^{K} R_k\Pi_k = \sum_{k=1}^{K} \Pi_k R_k \tag{11.32}$$

is hermitian. Mathematically speaking, Λ is the Lagrange multiplier of the unit-sum constraint in (11.5), and its significance is revealed by noting that $I_{\text{acc}}(\mathscr{E}) = \text{Tr}\,\Lambda$ for an optimal POVM.

Equations (11.30)–(11.32) have been investigated by Holevo [9]. These e-quations are nonlinear and there does not seem to be any efficient method for finding their solutions. Indeed, the $\frac{1}{2}K(K-1)$ equations (11.30) are not solved directly in the numerical approach described in Section 11.5. Rather, we exploit the observation that (11.28) identifies the gradient in the POVM space.

We remark that a POVM obeying (11.30) is not guaranteed to be an optimal POVM. Strictly speaking, $I(\mathscr{E};\Pi)$ is only ensured to be extremal, but it could be a local maximum rather than a global maximum, or a local minimum, or even a saddle point. Whereas local minima and saddle points tend to be unstable extrema for the numerical procedure of Section 11.5, local maxima are just as stable as global maxima.

11.4.3 Some basic theorems

We state four basic theorems about $I(\mathscr{E};\Pi)$ and $I_{\text{acc}}(\mathscr{E})$ without proof. The reader is invited to consult the respective references for proofs and further details.

Theorem 11.1: Number of outcomes
  The accessible information is always achievable by an optimal POVM whose outcomes are rank-1 operators, so that $\Pi_k^2 = \Pi_k \, \mathrm{Tr}(\Pi_k)$ for $1 \le k \le K$. The number of outcomes needed in such an optimal POVM is bounded by the rank d of the total density matrix ρ, which is also the dimension of the relevant Hilbert space,[7] in accordance with [12]

$$d \le K \le d^2. \tag{11.33}$$

When all quantum states ρ_j can be represented as matrices with real numbers, then the upper bound is reduced to $K \le d(d+1)/2$ [13].
                                           (Davies [12]; Sasaki _et al._ [13])

For the following theorems we introduce two quantities that are defined by

$$\chi(\mathscr{E}) = S(\rho) - \sum_{j=1}^{J} a_j S(\rho_j) \ge 0 \tag{11.34}$$

and

$$\chi(\mathscr{E};\Pi) = \sum_{k=1}^{K} \left(b_k S(\rho^{(k)}) - \sum_{j=1}^{J} P_{jk} S(\rho_j^{(k)}) \right) \ge 0, \tag{11.35}$$

where $\rho^{(k)}$ is the final total state conditioned on Bob's kth outcome, as in (11.7), and $\rho_j^{(k)}$ is the corresponding conditional final state when ρ_j is the initial state. That is

$$\rho^{(k)} = A_k \rho A_k^\dagger \quad \text{and} \quad \rho_j^{(k)} = A_k \rho_j A_k^\dagger, \tag{11.36}$$

where the normalizing denominators of (11.7)—respectively $\mathrm{Tr}(\rho^{(k)}) = b_k$ and $\mathrm{Tr}(\rho_j^{(k)}) = P_{jk}$—are irrelevant here because these conditional density matrices appear only as arguments of the von Neumann entropy function of (11.11).

Theorem 11.2: Upper bound on $I(\mathscr{E};\Pi)$
  The mutual information is bounded by the difference of $\chi(\mathscr{E})$ and $\chi(\mathscr{E};\Pi)$,

$$I(\mathscr{E};\Pi) \le \chi(\mathscr{E}) - \chi(\mathscr{E};\Pi). \tag{11.37}$$

                         (Schumacher, Westmoreland, and Wootters [14])

---

[7] If ρ is embedded in a larger Hilbert space, there is one more outcome in the POVM, namely, the projector on the orthogonal complement of the range of ρ.

Since the term $\chi(\mathscr{E};\Pi)$ that is subtracted on the right-hand side of (11.37) is nonnegative and vanishes if and only if all outcomes Π_k are of rank 1, we have $I(\mathscr{E};\Pi) \leq \chi(\mathscr{E})$ for all POVMs, in particular for all optimal POVMs. This implies the following theorem.

Theorem 11.3: Upper bound on the accessible information
An upper bound on the accessible information is given by

$$I_{\mathrm{acc}}(\mathscr{E}) \leq \chi(\mathscr{E}), \tag{11.38}$$

the so-called *Holevo bound*. **(Holevo [15])**

We remark that the equal sign holds in (11.38) if and only if all quantum states ρ_j commute with each other, and hence the Holevo bound is *not tight* in general.

Theorem 11.4: Lower bound on the accessible information
A lower bound of the accessible information is given by

$$I_{\mathrm{acc}}(\mathscr{E}) \geq \mathscr{Q}(\rho) - \sum_{j=1}^{J} a_j \mathscr{Q}(\rho_j/a_j), \tag{11.39}$$

wherein the so-called *subentropy* $\mathscr{Q}(\rho)$ of a unit-trace density matrix ρ with eigenvalues λ_i $(i = 1,2,\ldots,d)$ is defined by

$$\mathscr{Q}(\rho) = -\sum_{i=1}^{d} \left(\prod_{i'(\neq i)} \frac{\lambda_i}{\lambda_i - \lambda_{i'}} \right) \lambda_i \log \lambda_i. \tag{11.40}$$

If there are degenerate eigenvalues, one treats them as the limit of nondegenerate ones. **(Jozsa, Robb, and Wootters [16])**

We should also mention that one can establish substantially tighter upper and lower bounds for the accessible information by taking more specific properties of the ρ_js into account than the rather global entropies and subentropies that enter the right-hand sides of (11.38) and (11.39); see, in particular, the work of Fuchs and Caves [17, 8].

11.4.4 Group-covariant case

Following Holevo [9], an ensemble $\mathscr{E} = \{\rho_j | j = 1,2,\ldots,J\}$ of quantum states ρ_j is said to be covariant with respect to a group G if there exists a faithful projective unitary representation $\{U_g | g \in G\}$ of G such that

$$U_g \rho_j U_g^\dagger \in \mathscr{E} \quad \text{for all } \rho_j \in \mathscr{E} \text{ and all } g \in G. \tag{11.41}$$

A projective unitary representation of a group G means that for any pair g_1, g_2 of group elements $U_{g_1} U_{g_2} = U_{g_1 g_2} e^{i\phi(g_1, g_2)}$ holds with a real phase function $\phi(g_1, g_2)$. Several remarks are in order.

1. If an ensemble $\mathscr{E}$ is covariant with respect to a group G, then $\mathscr{E}$ is also covariant with respect to any subgroup of G.

2. When a group G acts transitively on an ensemble $\mathscr{E}$, then $\mathscr{E}$ constitutes a single orbit of G. In this case the order of the group $|G|$ is equal to the number of elements of the ensemble, i.e., $|G| = J$, and the group parameterizes the input states ρ_j. Furthermore, Alice's probabilities of occurrence are all equal, i.e., $a_j = 1/J$.

3. It is always possible to construct a nonprojective unitary representation of the group by a central extension of the original group. In other words, a projective unitary representation is not essential in our discussion.

In this chapter we will only consider nonprojective unitary representations.

In general, a group has a direct sum of irreducible unitary representations of the form

$$U_g = \bigoplus_{\ell=1}^{L} \mathbf{1}_{m_\ell} \otimes u_g^\ell, \qquad (11.42)$$

where m_ℓ is the multiplicity of inequivalent unitary irreducible representation of u_g^ℓ in d_ℓ dimensions, and L is the number of inequivalent irreducible representations. By construction one has $\sum_{\ell=1}^{L} m_\ell d_\ell = d$. The following theorem [18] is crucial for the discussion below.

Theorem 11.5: Optimal POVM for group-covariant ensemble

Let the ensemble of quantum states $\mathscr{E}$ be covariant with respect to the group G, which has a representation (11.42). Then there exists rank-1 projectors S_m ($m = 1, 2, \ldots, M$), the so-called *seeds*, whose orbits

$$\mathscr{C}_m = \left\{ \frac{d}{|G|} U_g S_m U_g^\dagger \,\middle|\, g \in G \right\} \qquad (11.43)$$

constitute an optimal POVM with $K = M|G|$ outcomes. The count M of the seeds is bounded by

$$M \leq \sum_{\ell=1}^{L} m_\ell^2, \qquad (11.44)$$

and the POVM is given by the weighted union of the orbits,

$$\Pi = \bigcup_{m=1}^{M} \lambda_m \mathscr{C}_m = \left\{ \frac{\lambda_m d}{|G|} U_g S_m U_g^\dagger \,\middle|\, 1 \leq m \leq M, g \in G \right\}, \qquad (11.45)$$

where the values of the nonnegative weights λ_m are determined by the identity decomposition requirement of (11.5). (**Davies [12], Decker [18]**)

We remark the following:

1. The labels k of the outcomes Π_k are here identified with the pairs (m,g) with $m = 1, 2, \ldots, M$ and $g \in G$.

2. The construction implies $\sum_{m=1}^{M} \lambda_m = 1$, which is the reason for the normalizing factor $d/|G|$ in (11.43).

3. When the group G is irreducible, we have $m_1 = d$ and $L = 1$, and theorem 11.5 reduces to the case studied by Davies and Sasaki *et al.* [12, 13].

4. Although the group-covariant POVM is an optimal POVM, it may not be the only one which maximizes the mutual information. In other words, also for group-covariant ensembles $\mathscr{E}$, the optimal POVM is not unique as a rule; there can be other POVMs that are as good as the optimal group-covariant POVM. This situation occurs typically for $|G| > d$. We will illustrate this point in several examples in Section 11.6.

5. Since $\mathscr{C}_m$ is an orbit, $U_g S_m U_g^\dagger$ and S_m are equivalent seeds. Whereas the orbits of the optimal group-covariant POVM may be unique, the seeds are not.

6. When one orbit is enough to attain the accessible information, Schur's lemma provides the following restriction on the structure of the seed:

$$S_m = \bigoplus_{\ell=1}^{L} \frac{d_\ell}{d} \mathbf{1}_{m_\ell} \otimes s^\ell, \tag{11.46}$$

where the s^ℓs are rank-1 projectors in the d_ℓ-dimensional subspaces identified by the decomposition (11.42).

7. If the group G acts transitively on the ensemble $\mathscr{E}$, we have $J = |G|$ and $U_g \rho U_g^\dagger = \rho$ for all $g \in G$, and the marginals are

$$a_j = \sum_{k=1}^{K} P_{jk} = \frac{1}{|G|}, \quad b_k = \sum_{j=1}^{J} P_{jk} = \frac{\lambda_m d}{|G|} \operatorname{Tr}(\rho S_m). \tag{11.47}$$

Bob's *a priori* probabilities b_k, with $k \equiv (m,g)$, are the same for all outcomes within one orbit $\mathscr{C}_m$; their unit sum gives

$$\sum_{m=1}^{M} \lambda_m \operatorname{Tr}(\rho S_m) = \frac{1}{d}. \tag{11.48}$$

11.5   Numerical search

Any numerical procedure that is capable of finding maxima of a function could be used in the numerical search for the optimal POVM. In particular, the method of simulated annealing performed well in practice [19]. Such general procedures, however, are unspecific; they do not take full advantage of the structural properties of the mapping $\Pi \to I(\mathcal{E}; \Pi)$ and are, therefore, not tailored to the problem at hand.

One algorithm that exploits the structure of $I(\mathcal{E}; \Pi)$ is the iterative procedure of Ref. [20]. It implements a steepest-ascent approach to the extremal points in the POVM space, locally proceeding into the direction of the gradient of $I(\mathcal{E}; \Pi)$ with respect to Π.

The gradient in steepest ascent is essentially composed of the operators that multiply the infinitesimal increments $\varepsilon_{kk'}$ in (11.28). Accordingly, if we choose the $\varepsilon_{kk'}$s proportional to the respective components of the gradient, the altered POVM will yield a larger value for $I(\mathcal{E}; \Pi)$ than the original POVM.

More specifically, we put

$$\varepsilon_{kk'} = i\alpha \left[A_{k'}(R_{k'} - R_k) A_k^\dagger \right]^\dagger , \tag{11.49}$$

where the value chosen for the "small" parameter α determines the step size. For $\alpha > 0$, the right-hand side of (11.28) is assuredly nonnegative,

$$\Delta I(\mathcal{E}; \Pi) = \alpha \sum_{k,k'=1}^{K} \mathrm{Tr}\left(\left[A_{k'}(R_{k'} - R_k) A_k^\dagger \right]^\dagger \left[A_{k'}(R_{k'} - R_k) A_k^\dagger \right] \right)$$

$$= \alpha \sum_{k,k'=1}^{K} \mathrm{Tr}\left((R_{k'} - R_k) \Pi_{k'}(R_{k'} - R_k) \Pi_k \right) \geq 0, \tag{11.50}$$

whereby the equal sign applies only if the POVM obeys the necessary condition (11.30) of an extremal point.

The increment (11.49), which is first-order in α for A_k, gives rise to a term $\propto \alpha^2$ in Π_k, so that we must ensure proper normalization of the improved POVM. This is the purpose of the $T^\dagger \cdots T$ sandwich in

$$\Pi_k \to \Pi_k^{(\mathrm{new})} = T^\dagger (\mathbf{1}_d + \alpha G_k^\dagger) \Pi_k (\mathbf{1}_d + \alpha G_k) T \tag{11.51}$$

$$\text{with} \quad G_k = R_k - \sum_{k'=1}^{K} R_{k'} \Pi_{k'} \tag{11.52}$$

$$\text{and} \quad TT^\dagger = \left(\mathbf{1}_d + \alpha^2 \sum_{k=1}^{K} G_k^\dagger \Pi_k G_k \right)^{-1} . \tag{11.53}$$

So, given the ensemble $\mathcal{E}$ of Alice's quantum states with its marginals a_j, the numerical procedure of one round of iteration is as follows. For the present nonoptimal POVM Π, we evaluate the joint probabilities p_{jk} of (11.8), the marginals b_k, and the R_ks of (11.27). Then we choose the step size $\alpha > 0$, compute the G_ks of (11.52) as well as T of (11.53), and finally determine the outcomes $\Pi_k^{(\text{new})}$ of the improved POVM in accordance with (11.51). In view of the first-order increase of (11.50), we will have $I(\mathcal{E}; \Pi^{(\text{new})}) > I(\mathcal{E}; \Pi)$ unless α is too large.

The procedure (11.51)–(11.53) is repeated until no further improvement can be achieved, which happens when the POVM obeys (11.30). Since local minima and saddle points are numerically unstable, the iteration terminates when a local maximum is reached.

Several remarks are in order.

1. If the POVM obeys (11.30), the right-hand side of (11.53) is $\mathbf{1}_d$, and then we have to choose $T = \mathbf{1}_d$ to ensure that the iteration halts. Otherwise, as long as the POVM does not obey (11.30), we have $0 < TT^\dagger < \mathbf{1}_d$ and $T = \left(\mathbf{1}_d + \alpha^2 \sum_{k=1}^K G_k^\dagger \Pi_k G_k\right)^{-1/2} U$ with U unitary and such that $U \to \mathbf{1}_d$ when $TT^\dagger \to \mathbf{1}_d$.

2. Here is an iteration that yields T in a few rounds without the need of calculating the reciprocal square root of a possibly large matrix: Starting with $T_0 = \mathbf{1}_d$ compute $T_1, T_2, \ldots$ successively with the aid of the recurrence relation

$$T_{n+1} = T_n - e^{i\pi/3} T_n \left[T_n^\dagger (TT^\dagger)^{-1} T_n - \mathbf{1}_d \right], \tag{11.54}$$

wherein $(TT^\dagger)^{-1}$ is the given inverse of the right-hand side in (11.53). As long as the step size α is so small that all eigenvalues of $(TT^\dagger)^{-1}$ are less than 2, which is typically the case in practice without particular precautions, we have $T_n \to T$ with a cubic convergence because

$$T_{n+1}^\dagger (TT^\dagger)^{-1} T_{n+1} = \mathbf{1}_d + \left[T_n^\dagger (TT^\dagger)^{-1} T_n - \mathbf{1}_d \right]^3,$$

$$\text{implying} \quad T_n^\dagger (TT^\dagger)^{-1} T_n = \mathbf{1}_d + \left(\alpha^2 \sum_{k=1}^K G_k^\dagger \Pi_k G_k \right)^{3^n}. \tag{11.55}$$

3. A quadratically convergent iteration is obtained by the replacement $e^{i\pi/3} \to \frac{1}{2}$ in (11.54); this may be preferable if $(TT^\dagger)^{-1}$ is a real matrix and one wishes to have a real matrix for T as well.

4. As mentioned earlier, the POVM resulting from the iteration procedure (11.51)–(11.53) could be a local maximum rather than a global one. Since there are no known sufficiency conditions for a global maximum, one cannot prevent convergence toward a local maximum. All numerical schemes face this generic problem. As a remedy, we run the iteration many times with different initial POVMs, and so reduce the risk of mistaking a local maximum for a global one.

5. Theorem 11.1 states that we can restrict the numerical search to POVMs with rank-1 outcomes that are no more than $K = d^2$ (or $K = \frac{1}{2}d(d+1)$ if all ρ_js are real) in number. To determine the actual value of K, we begin with optimizing for $K = d$, then for $K = d+1$, then for $K = d+2$, until an increase of K no longer gives an increase of the maximal mutual information.—Alternatively, we start with optimizing for $K = d^2$ or $K = \frac{1}{2}d(d+1)$, and then reduce the number of outcomes by identifying equivalent ones. Outcomes Π_k and $\Pi_{k'}$ are equivalent if $p_{jk}p_{j'k'} = p_{j'k}p_{jk'}$ for all j and j', for then $R_k = R_{k'}$, and the pair of outcomes $(\Pi_k + \Pi_{k'}, 0)$ is as good as the pair $(\Pi_k, \Pi_{k'})$. Incidentally, numerical experience seems to indicate [21] that by choosing the initial K value substantially larger than d^2, so that there will surely be superfluous outcomes in the POVM, one reduces substantially the risk of ending up in a local maximum.

6. The choice (11.49) is the basic steepest-ascent strategy where one proceeds in the direction of the gradient. As usual, convergence is improved markedly when one employs *conjugated gradients* instead; see Section 10.6 in [22] or Shewchuk's tutorial [23] and the references therein.

11.6   Examples

11.6.1   Two quantum states in two dimensions

We first consider the simplest example: the situation of two states, $\mathcal{E} = \{\rho_1, \rho_2\}$, in two dimensions, $d = \text{rank}(\rho_1 + \rho_2) = 2$. Since any 2×2 matrix is a linear combination of the identity matrix $\mathbf{1}_2$ and the three familiar matrices of Pauli's matrix vector $\vec{\sigma}$, we write

$$\rho_j = \frac{a_j}{2}(\mathbf{1}_2 + \vec{r}_j \cdot \vec{\sigma}), \quad j = 1, 2, \qquad (11.56)$$

for the two quantum states. The *Pauli vector* $\vec{r}_j$ is of unit length if ρ_j is a pure state, and shorter if ρ_j is a mixed state. The probabilities of occurrence are both nonzero, $0 < a_1 = 1 - a_2 < 1$.

Numerical studies by ourselves and others, such as work by Fuchs and Peres as reported by Shor [24], strongly suggest the conjecture that there is always a von Neumann measurement among the optimal POVMs if $\mathscr{E}$ is a two-state ensemble. This observation is very important in practice but, unfortunately, no proofs seem to exist in the published literature.

Bearing in mind this conjecture, we restrict the search to POVMs of the form

$$\Pi_1 = \frac{1}{2}(1_2 + \vec{n}\cdot\vec{\sigma}), \quad \Pi_2 = \frac{1}{2}(1_2 - \vec{n}\cdot\vec{\sigma}), \tag{11.57}$$

where the unit vector $\vec{n}$ specifies the POVM. Therefore the optimization of the POVM amounts to determining the direction of $\vec{n}$, which is an optimization over two angle parameters.

Then, the joint probabilities $p_{jk} = \mathrm{Tr}(\rho_j \Pi_k)$ and their marginals are

$$p_{11} = \frac{a_1}{2}(1 + x_1), \qquad\qquad p_{12} = \frac{a_1}{2}(1 - x_1),$$

$$p_{21} = \frac{a_2}{2}(1 + x_2), \qquad\qquad p_{22} = \frac{a_2}{2}(1 - x_2),$$

$$b_1 = p_{11} + p_{21} = \frac{1}{2}(1 + X), \quad b_2 = p_{12} + p_{22} = \frac{1}{2}(1 - X), \tag{11.58}$$

wherein

$$x_1 = \vec{n}\cdot\vec{r}_1, \quad x_2 = \vec{n}\cdot\vec{r}_2, \quad X = a_1 x_1 + a_2 x_2. \tag{11.59}$$

They give

$$I(\mathscr{E};\Pi) = a_1 \Phi(x_1) + a_2 \Phi(x_2) - \Phi(X) \tag{11.60}$$

with

$$\Phi(x) = \frac{1}{2}\left[(1 + x)\log_2(1 + x) + (1 - x)\log_2(1 - x)\right] \tag{11.61}$$

for the information accessed by the POVM (11.57).

When the two quantum states commute with each other, the two Pauli vectors are parallel, $\vec{r}_1 \parallel \vec{r}_2$, and then the optimal POVM is given by $\vec{n} \parallel \vec{r}_1 \parallel \vec{r}_2$. This covers as well the case that one, or both, of the Pauli vectors vanishes. Therefore, in the following we take for granted that $r_1 = |\vec{r}_1| > 0$ and $r_2 = |\vec{r}_2| > 0$, and denote by θ the angle between the two Pauli vectors, $\vec{r}_1 \cdot \vec{r}_2 = r_1 r_2 \cos\theta$ with $0 < \theta < \pi$.

An infinitesimal variation of the unit vector $\vec{n}$ is an infinitesimal rotation, $\delta\vec{n} = \vec{\varepsilon} \times \vec{n}$, where $\vec{\varepsilon}$ is an arbitrary infinitesimal vector. The resulting variation of $I(\mathscr{E};\Pi)$ is of the form $\delta I = \vec{e} \cdot [\vec{n} \times (\cdots)]$, so that $\vec{n} \parallel (\cdots)$ if the POVM (11.57) is optimal.

Since the vector $(\cdots)$ is a linear combination of $\vec{r}_1$ and $\vec{r}_2$, the POVM vector $\vec{n}$ is such a linear combination as well. In fact, then, the optimization of $\vec{n}$ is reduced to finding its orientation in the plane spanned by $\vec{r}_1$ and $\vec{r}_2$, which constitutes a one-parameter problem. Expressed in terms of the angles ϑ_1 and ϑ_2 between $\vec{n}$ and the Pauli vectors,

$$x_1 = \vec{n} \cdot \vec{r}_1 = r_1 \cos \vartheta_1, \qquad x_2 = \vec{n} \cdot \vec{r}_2 = r_2 \cos \vartheta_2 \qquad (11.62)$$

with $0 \leq \vartheta_1, \vartheta_2 \leq \pi$, we have

$$(\sin \theta)^2 \vec{n} = (\cos \vartheta_1 - \cos \vartheta_2 \cos \theta) \frac{\vec{r}_1}{r_1} + (\cos \vartheta_2 - \cos \vartheta_1 \cos \theta) \frac{\vec{r}_2}{r_2}. \quad (11.63)$$

The unit length of $\vec{n}$ implies

$$\left[\cos(\vartheta_1 + \vartheta_2) - \cos \theta \right] \left[\cos(\vartheta_1 - \vartheta_2) - \cos \theta \right] = 0. \qquad (11.64)$$

It turns out that the second, not the first, factor vanishes when $I(\mathscr{E};\Pi)$ is maximal, so that the actual constraint is $\cos(\vartheta_1 - \vartheta_2) = \cos \theta$, and since the POVM to $-\vec{n}$ is equivalent to the one to $\vec{n}$, we can insist on $\vartheta_2 - \vartheta_1 = \theta$. The optimization of $\vec{n}$ thus amounts to determining ϑ_1, say.

With $\theta = \vartheta_2 - \vartheta_1$ in (11.63), we have

$$\vec{n} = \frac{\sin \vartheta_2}{\sin \theta} \frac{\vec{r}_1}{r_1} - \frac{\sin \vartheta_1}{\sin \theta} \frac{\vec{r}_2}{r_2} \qquad (11.65)$$

and the requirement $\vec{n} \parallel (\cdots)$ reads

$$a_1 r_1 \sin \vartheta_1 \log \frac{(1+x_1)(1-X)}{(1-x_1)(1+X)} + a_2 r_2 \sin \vartheta_2 \log \frac{(1+x_2)(1-X)}{(1-x_2)(1+X)} = 0,$$
$$(11.66)$$

which we regard as the equation for ϑ_1 as the basic unknown, with $\vartheta_2 = \vartheta_1 + \theta$ and x_1, x_2, X as given in (11.62) and (11.59). The variables $a_1, a_2, r_1, r_2, \theta$ specify Alice's states, and once the value of ϑ_1 is determined, Bob's optimal POVM is known.

For arbitrary values of $a_1, a_2, r_1, r_2, \theta$, there is no known analytical solution of (11.66). But, as noted by Levitin [25] as well as Fuchs and Caves [17, 8], there is a notable special situation, for which the solution is known and simple: the case of $\det \rho_1 = \det \rho_2$ or

$$a_1^2(1 - r_1^2) = a_2^2(1 - r_2^2). \qquad (11.67)$$

When this equation is obeyed, the optimal POVM coincides with the measurement for error minimization [5], that is,

$$\vec{n} = \frac{a_1\vec{r}_1 - a_2\vec{r}_2}{|a_1\vec{r}_1 - a_2\vec{r}_2|}, \tag{11.68}$$

so that

$$x_1 = \frac{(a_1 r_1 - a_2 r_2 \cos\theta) r_1}{|a_1\vec{r}_1 - a_2\vec{r}_2|},$$

$$x_2 = \frac{(a_1 r_1 \cos\theta - a_2 r_2) r_2}{|a_1\vec{r}_1 - a_2\vec{r}_2|},$$

$$\text{and} \quad X = \frac{(a_1 r_1)^2 - (a_2 r_2)^2}{|a_1\vec{r}_1 - a_2\vec{r}_2|}. \tag{11.69}$$

To justify these remarks, we first note that, if $\vec{n}$ is of the form (11.68), (11.65) implies

$$a_1 r_1 \sin\vartheta_1 = a_2 r_2 \sin\vartheta_2, \tag{11.70}$$

and then (11.66) requires

$$\frac{(1+x_1)(1-X)}{(1-x_1)(1+X)} - 1 = \frac{(1-x_2)(1+X)}{(1+x_2)(1-X)} - 1. \tag{11.71}$$

The subtraction of 1 serves the purpose of making both sides vanish for $x_1 = x_2 = X$, which solution results in $I(\mathcal{E};\Pi) = 0$ and is, therefore, of no further interest. Upon dividing by $x_1 - x_2$, (11.71) turns into

$$a_1(1 - x_1 X) = a_2(1 - x_2 X) \quad \text{or} \quad a_1^2(1 - x_1^2) = a_2^2(1 - x_2^2). \tag{11.72}$$

The identity $(a_1 x_1)^2 - (a_2 x_2)^2 = (a_1 r_1)^2 - (a_2 r_2)^2$, which follows from (11.70), now establishes (11.67) as the condition that, indeed, must be met by Alice's states if Bob's optimal POVM is given by the unit vector in (11.68).

Two details of (11.67) are worth pointing out: It does not involve the angle θ between the two Pauli vectors; and, irrespective of the probabilities of occurrence a_1 and a_2, (11.67) is always obeyed if both states are pure ($r_1 = r_2 = 1$).

11.6.2 Trine: Z_3 symmetry in two dimensions

We next discuss the celebrated example of the "trine," where no von Neumann measurement can achieve the accessible information. This example was proposed and solved partially by Holevo in 1973 [26]. The complete solution was obtained by Sasaki *et al.* in their discussion of Z_N symmetry in the two-dimensional Hilbert space [13].

Three pure states $\rho_j = |j\rangle\langle j|$ $(j = 1, 2, 3)$ with equal probabilities of occurrence, $a_1 = a_2 = a_3 = \frac{1}{3}$, are given in $d = 2$ dimensions by their kets

$$|1\rangle = \frac{1}{2\sqrt{3}}\begin{pmatrix} -1 \\ \sqrt{3} \end{pmatrix}, \ |2\rangle = \frac{1}{2\sqrt{3}}\begin{pmatrix} -1 \\ -\sqrt{3} \end{pmatrix}, \ |3\rangle = \frac{1}{\sqrt{3}}\begin{pmatrix} 1 \\ 0 \end{pmatrix}, \quad (11.73)$$

or equivalently by their Pauli vectors,

$$\vec{r}_1 = \frac{1}{2}(-\sqrt{3}, 0, -1), \ \vec{r}_2 = \frac{1}{2}(\sqrt{3}, 0, -1), \ \vec{r}_3 = (0, 0, 1). \quad (11.74)$$

These three vectors are coplanar and point to the corners of an equilateral triangle in the *xz*-plane: they form a *trine*.

The cyclic symmetry of the trine is made explicit by noting that

$$|j\rangle = \frac{1}{\sqrt{3}}\begin{pmatrix} \cos(j\theta_0) \\ \sin(j\theta_0) \end{pmatrix} \quad \text{for} \quad j = 1, 2, 3 \quad \text{with} \quad \theta_0 = \frac{2\pi}{3} \quad (11.75)$$

and

$$U|1\rangle = |2\rangle, \ U|2\rangle = |3\rangle, \ U|3\rangle = |1\rangle \quad \text{with} \quad U = \begin{pmatrix} \cos\theta_0 & -\sin\theta_0 \\ \sin\theta_0 & \cos\theta_0 \end{pmatrix}. \quad (11.76)$$

Since $U^3 = \mathbf{1}_2$, the 2×2 matrices $\mathbf{1}_2$, U, U^2 are an irreducible unitary representation of Z_3 on a *real* field, the cyclic group of period 3, and the group acts transitively on the ensemble $\mathscr{E} = \{\rho_1, \rho_2, \rho_3\}$.

According to Subsection 11.4.4, the outcomes Π_k of the optimal POVM can be generated by these unitary matrices from a seed S:

$$\Pi_k = \frac{2}{3}U^k S U^{-k} \quad \text{for} \quad k = 1, 2, 3 \quad \text{with} \quad S = |v\rangle\langle v|. \quad (11.77)$$

The seed ket $|v\rangle$ has to be normalized to unit length, $\langle v|v\rangle = 1$, so we write

$$|v\rangle = \begin{pmatrix} \cos\theta \\ \sin\theta \end{pmatrix}, \quad (11.78)$$

where the angle parameter θ specifies the POVM.

Therefore, the problem is to maximize the mutual information $I[\theta] = I(\mathscr{E}; \Pi_\theta)$ as a function of θ, with

$$I[\theta] = \frac{1}{3} \sum_{j=1}^{3} \left(1 + \cos(2\theta + j\theta_0)\right) \log\left(1 + \cos(2\theta + j\theta_0)\right). \qquad (11.79)$$

This function is $\frac{1}{2}\theta_0$-periodic in θ, $I[\theta + \frac{1}{2}\theta_0] = I[\theta]$, because the POVM with the outcomes of (11.77) and (11.78) does not change as a whole when θ is replaced by $\theta + \frac{1}{2}\theta_0$. It is, therefore, sufficient to consider the range $0 \leq \theta < \frac{1}{2}\theta_0$, and one verifies easily that the global maximum of $I[\theta]$ is obtained for $\theta = \frac{1}{6}\pi = \frac{1}{4}\theta_0$. Accordingly, the accessible information is

$$I_{\mathrm{acc}}(\mathscr{E}; \Pi) = \log \frac{3}{2} \qquad (11.80)$$

in the case of the trine.

The optimal POVM of (11.77) with $\theta = \frac{1}{6}\pi$ consists of three rank-1 operators, $\Pi_k = \frac{1}{3}\left(1 - \vec{r}_k \cdot \vec{\sigma}\right)$, with the vectors $\vec{r}_k$ of (11.74). Thus, whereas the state ensemble $\mathscr{E}$ makes up the trine of $\vec{r}_1$, $\vec{r}_2$, and $\vec{r}_3$, the POVM makes up the "anti-trine" composed of $-\vec{r}_1$, $-\vec{r}_2$, and $-\vec{r}_3$. Since $\rho = \frac{1}{2}1_2$ here, the roles of the trine and the anti-trine are simply interchanged in the reciprocal situation of (11.10).

When we regard the three two-dimensional kets of (11.73) as spanning a plane in a three-dimensional space, we can lift them jointly out of this plane by giving each the same third component. The cyclic symmetry is maintained thereby. Such a *lifted trine* actually consists of the edges of an obtuse pyramid. As Shor established [24], one needs two seeds for the optimal six-outcome POVM of the lifted trine.

If one lifts the trine by so much that the edges of the pyramid are perpendicular to each other, then clearly a three-outcome POVM of von Neumann type is optimal. In fact, there is a large range of angles between the edges, around the perpendicular-edges geometry, for which the optimal POVM has three outcomes. But for acute pyramids with a rather small angle between the edges, one needs a four-outcome POVM [3, 27].

Instead of lifting the trine, one can distort it in the original two-dimensional space, so that the cyclic symmetry is lost. The optimal POVMs for distorted trines have been found quite recently [28].

11.6.3   Six-states protocol: symmetric group S_3

As a practical example, we now turn to an application that occurs in the security analysis in quantum cryptography. In the raw-data attack on the six-states version [29] of the BB84 protocol [30], eavesdropper Eve gains knowledge by discriminating six rank-2 states in $d = 4$ dimensions [20].

11.6.3.1   States received by Eve

We denote these states by ρ_{js} whereby $j = 1, 2, 3$ is a ternary index and $s = \pm$ is a binary index, so that we are dealing with three pairs of states. It is expedient to use the following 4×4 matrices for the six states:

$$\rho_{1\pm} = \frac{\varepsilon}{24} \begin{pmatrix} z^2 & \pm z & 0 & 0 \\ \pm z & 1 & 0 & 0 \\ 0 & 0 & 1 & \pm i \\ 0 & 0 & \mp i & 1 \end{pmatrix},$$

$$\rho_{2\pm} = \frac{\varepsilon}{24} \begin{pmatrix} z^2 & 0 & \pm z & 0 \\ 0 & 1 & 0 & \mp i \\ \pm z & 0 & 1 & 0 \\ 0 & \pm i & 0 & 1 \end{pmatrix},$$

$$\rho_{3\pm} = \frac{\varepsilon}{24} \begin{pmatrix} z^2 & 0 & 0 & \pm z \\ 0 & 1 & \pm i & 0 \\ 0 & \mp i & 1 & 0 \\ \pm z & 0 & 0 & 1 \end{pmatrix}, \qquad (11.81)$$

where the parameter ε measures the level of noise between the communicating parties that results from the eavesdropping, and $z = \sqrt{4/\varepsilon - 3}$ is a convenient abbreviation. The physically reasonable range of the noise parameter is $0 \leq \varepsilon \leq 1$ but only communications with $\varepsilon < \frac{2}{3}$ are potentially useful for the purpose of quantum cryptography. Indeed, we will see below that the optimal POVMs are structurally different for $\varepsilon < \frac{2}{3}$ and $\varepsilon \geq \frac{2}{3}$.

The two nonzero eigenvalues of each ρ_{js} are $(2 - \varepsilon)/12$ and $\varepsilon/12$, so that all six probabilities are $\frac{1}{6}$ and the six matrices of (11.81) are unitarily equivalent,

$$\rho_{js} = U_{js} \rho_{1+} U_{js}^{\dagger}. \qquad (11.82)$$

Here,

$$\rho_{1+} = |1\rangle\langle 1| + |2\rangle\langle 2| \quad \text{with} \quad \langle 1| = \sqrt{\frac{\varepsilon}{24}}(z,1,0,0)$$

$$\text{and} \quad \langle 2| = \sqrt{\frac{\varepsilon}{24}}(0,0,1,i) \qquad (11.83)$$

state the spectral decomposition of ρ_{1+} and so makes its rank-2 nature explicit, and the unitary matrices U_{js} are given by

$$
U_{1+} = \begin{pmatrix} 1 & 0 & 0 & 0 \\ 0 & 1 & 0 & 0 \\ 0 & 0 & 1 & 0 \\ 0 & 0 & 0 & 1 \end{pmatrix} = \mathbf{1}_4, \quad
U_{1-} = \begin{pmatrix} 1 & 0 & 0 & 0 \\ 0 & -1 & 0 & 0 \\ 0 & 0 & 0 & -1 \\ 0 & 0 & -1 & 0 \end{pmatrix},
$$

$$
U_{2+} = \begin{pmatrix} 1 & 0 & 0 & 0 \\ 0 & 0 & 0 & 1 \\ 0 & 1 & 0 & 0 \\ 0 & 0 & 1 & 0 \end{pmatrix}, \quad
U_{2-} = \begin{pmatrix} 1 & 0 & 0 & 0 \\ 0 & 0 & -1 & 0 \\ 0 & -1 & 0 & 0 \\ 0 & 0 & 0 & -1 \end{pmatrix},
$$

$$
U_{3+} = \begin{pmatrix} 1 & 0 & 0 & 0 \\ 0 & 0 & 1 & 0 \\ 0 & 0 & 0 & 1 \\ 0 & 1 & 0 & 0 \end{pmatrix}, \quad
U_{3-} = \begin{pmatrix} 1 & 0 & 0 & 0 \\ 0 & 0 & 0 & -1 \\ 0 & 0 & -1 & 0 \\ 0 & -1 & 0 & 0 \end{pmatrix}. \qquad (11.84)
$$

They form a multiplicative group of order 6 with this group table:

	U_{1+}	U_{1-}	U_{2+}	U_{2-}	U_{3+}	U_{3-}
U_{1+}	U_{1+}	U_{1-}	U_{2+}	U_{2-}	U_{3+}	U_{3-}
U_{1-}	U_{1-}	U_{1+}	U_{3-}	U_{3+}	U_{2-}	U_{2+}
U_{2+}	U_{2+}	U_{2-}	U_{3+}	U_{3-}	U_{1+}	U_{1-}
U_{2-}	U_{2-}	U_{2+}	U_{1-}	U_{1+}	U_{3-}	U_{3+}
U_{3+}	U_{3+}	U_{3-}	U_{1+}	U_{1-}	U_{2+}	U_{2-}
U_{3-}	U_{3-}	U_{3+}	U_{2-}	U_{2+}	U_{1-}	U_{1+}

$$(11.85)$$

which shows that it is a nonabelian group that is isomorphic to the symmetric group S_3. It is well known that the representation (11.84) is not irreducible. To get an irreducible representation, we need to carry out the similarity transformation

$$U_{js} \rightarrow \tilde{U}_{js} = T^{-1} U_{js} T \tag{11.86}$$

with the transformation matrix

$$T = \begin{pmatrix} 1 & 0 & 0 & 0 \\ 0 & 1/\sqrt{3} & -2/\sqrt{6} & 0 \\ 0 & 1/\sqrt{3} & 1/\sqrt{6} & -1/\sqrt{2} \\ 0 & 1/\sqrt{3} & 1/\sqrt{6} & 1\sqrt{2} \end{pmatrix}. \tag{11.87}$$

The transformed unitary matrices give us a direct sum of irreducible representations for the group,

$$\tilde{U}_{1\pm} = \begin{pmatrix} 1 & 0 & 0 & 0 \\ 0 & \pm 1 & 0 & 0 \\ 0 & 0 & \pm 1 & 0 \\ 0 & 0 & 0 & 1 \end{pmatrix},$$

$$\tilde{U}_{2\pm} = \begin{pmatrix} 1 & 0 & 0 & 0 \\ 0 & \pm 1 & 0 & 0 \\ 0 & 0 & \mp 1/2 & -\sqrt{3}/2 \\ 0 & 0 & \pm\sqrt{3}/2 & -1/2 \end{pmatrix},$$

$$\tilde{U}_{3\pm} = \begin{pmatrix} 1 & 0 & 0 & 0 \\ 0 & \pm 1 & 0 & 0 \\ 0 & 0 & \mp 1/2 & \sqrt{3}/2 \\ 0 & 0 & \mp\sqrt{3}/2 & -1/2 \end{pmatrix}. \tag{11.88}$$

They combine a $\phi_0 = 2\pi/3$ rotation and a reflection,

$$\tilde{U}_{js} = U_{(j-1)\phi_0} \Sigma_s \quad \text{for} \quad j = 1,2,3 \quad \text{and} \quad s = \pm, \tag{11.89}$$

where

$$U_\vartheta = \begin{pmatrix} 1 & 0 & 0 & 0 \\ 0 & 1 & 0 & 0 \\ 0 & 0 & \cos\vartheta & -\sin\vartheta \\ 0 & 0 & \sin\vartheta & \cos\vartheta \end{pmatrix}, \tag{11.90}$$

with ϑ taking on the values 0, ϕ_0, $2\phi_0$ for $j = 1, 2$, and 3, respectively, and

$$\Sigma_+ = \mathbf{1}_4 = \tilde{U}_{1+}, \quad \Sigma_- = \begin{pmatrix} 1 & 0 & 0 & 0 \\ 0 & -1 & 0 & 0 \\ 0 & 0 & -1 & 0 \\ 0 & 0 & 0 & 1 \end{pmatrix} = \tilde{U}_{1-}. \tag{11.91}$$

Eve's states ρ_{js} are transformed correspondingly, resulting in

$$\tilde{\rho}_{1\pm} = \frac{\varepsilon}{24} \begin{pmatrix} z^2 & \pm z/\sqrt{3} & \mp z\sqrt{2/3} & 0 \\ \pm z/\sqrt{3} & 1 & 0 & \pm i\sqrt{2/3} \\ \mp z\sqrt{2/3} & 0 & 1 & \pm i/\sqrt{3} \\ 0 & \mp i\sqrt{2/3} & \mp i/\sqrt{3} & 1 \end{pmatrix},$$

$$\tilde{\rho}_{2\pm} = \frac{\varepsilon}{24} \begin{pmatrix} z^2 & \pm z/\sqrt{3} & \pm z/\sqrt{6} & \mp z/\sqrt{2} \\ \pm z/\sqrt{3} & 1 & \mp i/\sqrt{2} & \mp i/\sqrt{6} \\ \pm z/\sqrt{6} & \pm i/\sqrt{2} & 1 & \pm i/\sqrt{3} \\ \mp z/\sqrt{2} & \pm i/\sqrt{6} & \mp i/\sqrt{3} & 1 \end{pmatrix},$$

$$\tilde{\rho}_{3\pm} = \frac{\varepsilon}{24} \begin{pmatrix} z^2 & \pm z/\sqrt{3} & \pm z/\sqrt{6} & \pm z/\sqrt{2} \\ \pm z/\sqrt{3} & 1 & \pm i/\sqrt{2} & \mp i/\sqrt{6} \\ \pm z/\sqrt{6} & \mp i/\sqrt{2} & 1 & \pm i/\sqrt{3} \\ \pm z/\sqrt{2} & \pm i/\sqrt{6} & \mp i/\sqrt{3} & 1 \end{pmatrix}. \tag{11.92}$$

In summary then, the inputs are generated by the group $\tilde{U}_{js}$ ($j = 1, 2, 3; s = \pm$) as

$$\tilde{\rho}_{js} = \tilde{U}_{js} \tilde{\rho}_{1+} \tilde{U}_{js}^{\dagger}, \tag{11.93}$$

where

$$\tilde{\rho}_{1+} = |\tilde{1}\rangle\langle\tilde{1}| + |\tilde{2}\rangle\langle\tilde{2}| \quad \text{with} \quad \langle\tilde{1}| = \sqrt{\frac{\varepsilon}{24}}\left(z, \sqrt{\frac{1}{3}}, -\sqrt{\frac{2}{3}}, 0\right)$$

$$\text{and} \quad \langle\tilde{2}| = \sqrt{\frac{\varepsilon}{24}}\left(0, \sqrt{\frac{2}{3}}, \sqrt{\frac{1}{3}}, i\right).$$

$$(11.94)$$

11.6.3.2   Eve's POVM

We find the optimal POVM for Eve by an application of Theorem 11.5. The group structure for the six-states protocol is given by

$$\tilde{U}_g = \bigoplus_{\ell=1}^{3}(\mathbf{1}_{m_\ell} \otimes u_g^\ell), \tag{11.95}$$

with unit multiplicity for all ℓ values,

$$m_\ell = 1 \quad \text{for} \quad \ell = 1, 2, 3, \tag{11.96}$$

and the inequivalent irreducible representations are

$$\ell = 1 \text{ or } \ell = 2: \quad u_{g\pm}^1 = 1, \quad u_{g\pm}^2 = \pm 1 \quad \text{for all } g;$$

$$\ell = 3: \quad u_{1\pm}^3 = \begin{pmatrix} \pm 1 & 0 \\ 0 & 1 \end{pmatrix},$$

$$u_{2\pm}^3 = \frac{1}{2}\begin{pmatrix} \mp 1 & -\sqrt{3} \\ \pm\sqrt{3} & -1 \end{pmatrix},$$

$$u_{3\pm}^3 = \frac{1}{2}\begin{pmatrix} \mp 1 & \sqrt{3} \\ \mp\sqrt{3} & -1 \end{pmatrix}. \tag{11.97}$$

These representations exhaust all inequivalent irreducible representations, since the sum of the squares of the dimensions of the irreducible representations is equal to the order of the group. Indeed, $1^2 + 1^2 + 2^2 = 6$ holds here.

According to Theorem 11.5, an optimal POVM can be generated by the same group by means of

$$\tilde{\Pi}_g = \frac{4}{6}\tilde{U}_g \tilde{S} \tilde{U}_g^\dagger, \tag{11.98}$$

with the seed $\widetilde{S}$ of the form

$$\widetilde{S} = \bigoplus_{\ell=1}^{3} \frac{d_\ell}{4} |\tilde{v}_\ell\rangle\langle\tilde{v}_\ell|, \tag{11.99}$$

where d_ℓ is the dimension of the respective irreducible representation, and $\langle\tilde{v}_\ell|\tilde{v}_\ell\rangle = 1$ is required for each ℓ. In general, we may need more than one rank-1 state $\widetilde{S}$, and the upper bound is $\sum_\ell m_\ell^2 = 3$. A single seed is, however, enough to reach the accessible information for the specific example under consideration.

Hence we write $\widetilde{S} = |\tilde{v}\rangle\langle\tilde{v}|$ where

$$|\tilde{v}\rangle = \begin{pmatrix} e^{i\phi_1}/2 \\ e^{i\phi_2}/2 \\ e^{i\phi_3}\cos\theta/\sqrt{2} \\ e^{i\phi_4}\sin\theta/\sqrt{2} \end{pmatrix}, \tag{11.100}$$

with real angle parameters $\phi_1, \ldots, \phi_4, \theta$. Since the global phase is irrelevant, the value of one of the ϕ_js can be chosen by a convenient convention, and we set $\phi_1 = 0$ from now on.

Upon defining f_i by

$$\langle\tilde{v}|\tilde{p}_{i\pm}|\tilde{v}\rangle = \frac{1}{24}(1 \pm f_i), \tag{11.101}$$

we find

$$f_i = \eta g_i - \frac{\varepsilon}{\sqrt{3}} h_i,$$

$$g_i = \frac{1}{2}\cos\phi_2 - \cos\phi_3\cos\varphi_i\cos\theta - \cos\phi_4\sin\varphi_i\sin\theta,$$

$$h_i = \sin\phi_{23}\sin\varphi_i\cos\theta - \sin\phi_{24}\cos\varphi_i\sin\theta - \sin\phi_{34}\cos\theta\sin\theta.$$
$$\tag{11.102}$$

Here $\eta = z\varepsilon/\sqrt{3} = \sqrt{4\varepsilon/3 - \varepsilon^2}$, $\varphi_i = 2\pi(i-1)/3$, and ϕ_{ij} denotes $\phi_{ij} = \phi_i - \phi_j$. The mutual information $I(\rho,\Pi)$ is then given by

$$I(\mathcal{E};\Pi) = \frac{1}{3}\sum_{i=1}^{3}\Phi(f_i), \tag{11.103}$$

where $\Phi(\ )$ is the function introduced in (11.61).

The accessible information is now obtained by maximizing this mutual information $I(\mathscr{E};\Pi)$ over the four parameters $\phi_2,\phi_3,\phi_4,\theta$. With the help of numerical analysis, we observe that increasing the number of seeds does not provide a larger mutual information than what we get for a single seed.

As we noted above, the cases $\varepsilon < \frac{2}{3}$ and $\varepsilon \geq \frac{2}{3}$ are physically different. This is reflected in the structural difference between the optimal POVMs in these two parameter ranges.

Case $0 \leq \varepsilon < \frac{2}{3}$: The optimal POVM is given by

$$\phi_2 = \phi_3 = \phi_4 = 0 \quad \text{and} \quad \theta = \pi. \tag{11.104}$$

In the original representation of (11.81), this is expressed as

$$|v\rangle = T|\tilde{v}\rangle = \frac{1}{2}\begin{pmatrix} 1 \\ \sqrt{3} \\ 0 \\ 0 \end{pmatrix}. \tag{11.105}$$

The accessible information is

$$I_{\text{acc}}(\mathscr{E}) = \frac{1}{3}\Phi(3\eta/2) \tag{11.106}$$

with $\Phi(\ )$ of (11.61) and η as in (11.102). We remark that this optimal POVM is independent of the noise parameter ε, and all its outcomes are real. These findings agree with those obtained in [20], which were obtained with the aid of a numerical search by the method of Section 11.5. This demonstrates the optimality of this POVM.

Case $\frac{2}{3} \leq \varepsilon \leq 1$: The optimal POVM has a more complicated structure here, namely it is specified by

$$\phi_2 = -\tan^{-1}\sqrt{\frac{2(3\varepsilon-2)}{4-3\varepsilon}},$$
$$\phi_3 = \tan^{-1}\sqrt{\frac{3\varepsilon-2}{2(4-3\varepsilon)}},$$
$$\phi_4 = 0,$$
$$\theta = \pi + \tan^{-1}\sqrt{\frac{3\varepsilon-2}{2-\varepsilon}}, \tag{11.107}$$

where $-\frac{1}{2}\pi < \phi_2, \phi_3, \theta - \pi < \frac{1}{2}\pi$. This POVM amounts to $f_1 = 1$ and $f_2 = f_3 = 0$ in (11.103), so that the accessible information is

$$I_{\mathrm{acc}}(\mathcal{E}) = \frac{1}{3}\Phi(1) = \frac{1}{3}. \tag{11.108}$$

We note in passing that there are other POVMs that also give $I_{\mathrm{acc}} = \frac{1}{3}$ for the whole range $\frac{2}{3} \le \varepsilon \le 1$.

11.6.4 Four-group in four dimensions

As a simplest nontrivial group, we study the four-group—r *Klein group*, or *vierergruppe*—which is the noncyclic group of order four. One meets this group structure in the eavesdropping analysis for the BB84 protocol [31]. Here we give a discussion based on a toy model for the four-group in a 4-dimensional Hilbert space.

Each of the four quantum states $\rho_1, \ldots, \rho_4$ is a rank-2 state, and the total state $\rho = \rho_1 + \cdots + \rho_4$ has rank 4, and we have equal probabilities of occurrence:

$$\rho_j = |\psi_j\rangle\langle\psi_j| + |\phi_j\rangle\langle\phi_j| \quad \text{with} \quad \mathrm{Tr}\rho_j = \frac{1}{4} \tag{11.109}$$

and

$$\left.\begin{array}{c}|\psi_1\rangle\\[4pt]|\psi_2\rangle\end{array}\right\} = \frac{1}{2}\begin{pmatrix} a\\ \pm b\\ 0\\ 0 \end{pmatrix}, \quad \left.\begin{array}{c}|\psi_3\rangle\\[4pt]|\psi_4\rangle\end{array}\right\} = \frac{1}{2}\begin{pmatrix} a\\ 0\\ \pm b\\ 0 \end{pmatrix},$$

$$\left.\begin{array}{c}|\phi_1\rangle\\[4pt]|\phi_2\rangle\end{array}\right\} = \frac{1}{2}\begin{pmatrix} 0\\ 0\\ \pm c\\ d \end{pmatrix}, \quad \left.\begin{array}{c}|\phi_3\rangle\\[4pt]|\phi_4\rangle\end{array}\right\} = \frac{1}{2}\begin{pmatrix} 0\\ \pm c\\ 0\\ -d \end{pmatrix}. \tag{11.110}$$

Here a, b, c, d are real constants satisfying $a^2 + b^2 + c^2 + d^2 = 1$. We express these states using unitary matrices U_j as $\rho_j = U_j \rho_1 U_j^\dagger$, whereby

$$\left.\begin{array}{c}U_1\\[4pt]U_2\end{array}\right\} = \begin{pmatrix} 1 & 0 & 0 & 0\\ 0 & \pm 1 & 0 & 0\\ 0 & 0 & \pm 1 & 0\\ 0 & 0 & 0 & 1 \end{pmatrix}, \quad \left.\begin{array}{c}U_3\\[4pt]U_4\end{array}\right\} = \begin{pmatrix} 1 & 0 & 0 & 0\\ 0 & 0 & \pm 1 & 0\\ 0 & \pm 1 & 0 & 0\\ 0 & 0 & 0 & -1 \end{pmatrix}. \tag{11.111}$$

They constitute the four-group with the familiar group table

$$
\begin{array}{c|cccc}
 & U_1 \ U_2 \ U_3 \ U_4 \\
\hline
U_1 & U_1 \ U_2 \ U_3 \ U_4 \\
U_2 & U_2 \ U_1 \ U_4 \ U_3 \\
U_3 & U_3 \ U_4 \ U_1 \ U_2 \\
U_4 & U_4 \ U_3 \ U_2 \ U_1
\end{array}
\tag{11.112}
$$

where we note that the four-group is abelian and has order-2 subgroups consisting of $U_1 = \mathbf{1}_4$ and either U_2 or U_3 or U_4.

The representation (11.111) is not irreducible. In order to obtain a direct sum of inequivalent irreducible representations $\tilde{U}_j$, we introduce the following transformation T:

$$
T = \begin{pmatrix}
1 & 0 & 0 & 0 \\
0 & 1/\sqrt{2} & -1/\sqrt{2} & 0 \\
0 & 1/\sqrt{2} & 1/\sqrt{2} & 0 \\
0 & 0 & 0 & 1
\end{pmatrix}.
\tag{11.113}
$$

As is fitting for an abelian group, the transformed unitary matrices $\tilde{U}_j = T^{-1} U_j T$ have diagonal components only:

$$
\left.\begin{array}{c} \tilde{U}_1 \\ \tilde{U}_2 \end{array}\right\} = \mathrm{diag}(1, \pm 1, \pm 1, 1), \qquad
\left.\begin{array}{c} \tilde{U}_3 \\ \tilde{U}_4 \end{array}\right\} = \mathrm{diag}(1, \pm 1, \mp 1, -1).
\tag{11.114}
$$

They are indeed the direct sum of irreducible four-dimensional representations of the four-group. These representations consist of a direct sum of four different inequivalent representations. Each of inequivalent representations is one-dimensional. We also note the unit multiplicity for all four representations.

According to Theorem 11.5, we could need as many as 4 seeds. It is important to know that if we restrict ourself to the single-orbital case, then the optimal POVM generated by this group cannot have real outcomes. This is so because the seed has to have a unit length for each component by Schur's lemma. Therefore, we encounter the perhaps unexpected situation where we need a complex seed even though all input states and group representations are expressed as real quantities. As we will see later, there also exist real seeds which provide the accessible information, but then we need more than a single orbit.

We parameterize the seed ket $|\tilde{v}_1\rangle$ by three angle parameters θ_1, θ_2 and θ_3,

$$|\tilde{v}_1\rangle = \frac{1}{2}\begin{pmatrix} 1 \\ e^{i\theta_1} \\ e^{i\theta_2} \\ e^{i\theta_3} \end{pmatrix}. \tag{11.115}$$

The group generated outcomes of the POVM are then given by

$$\tilde{\Pi}_k = \tilde{U}_k|\tilde{v}_1\rangle\langle\tilde{v}_1|\tilde{U}_k^\dagger = |\tilde{v}_k\rangle\langle\tilde{v}_k|, \tag{11.116}$$

and the optimization requires the determination of the three θ_ks.

Corresponding to the transformation on the unitary matrices, the quantum states ρ_j are transformed into $\tilde{\rho}_j = T^{-1}\rho_j T$, or $|\tilde{\psi}_j\rangle = T^{-1}|\psi_j\rangle$ and $|\tilde{\phi}_j\rangle = T^{-1}|\phi_j\rangle$. Explicitly we have

$$\left.\begin{matrix} |\tilde{\psi}_1\rangle \\ |\tilde{\psi}_2\rangle \end{matrix}\right\} = \frac{1}{2}\begin{pmatrix} a \\ \pm b/\sqrt{2} \\ \mp b/\sqrt{2} \\ 0 \end{pmatrix}, \quad \left.\begin{matrix} |\tilde{\psi}_3\rangle \\ |\tilde{\psi}_4\rangle \end{matrix}\right\} = \frac{1}{2}\begin{pmatrix} a \\ \pm b/\sqrt{2} \\ \pm b/\sqrt{2} \\ 0 \end{pmatrix},$$

$$\left.\begin{matrix} |\tilde{\phi}_1\rangle \\ |\tilde{\phi}_2\rangle \end{matrix}\right\} = \frac{1}{2}\begin{pmatrix} 0 \\ \pm c/\sqrt{2} \\ \pm c/\sqrt{2} \\ d \end{pmatrix}, \quad \left.\begin{matrix} |\tilde{\phi}_3\rangle \\ |\tilde{\phi}_4\rangle \end{matrix}\right\} = \frac{1}{2}\begin{pmatrix} 0 \\ \pm c/\sqrt{2} \\ \mp c/\sqrt{2} \\ -d \end{pmatrix}. \tag{11.117}$$

For $\bar{\rho}_j$ defined by

$$\bar{\rho}_j = \langle\tilde{v}_1|\tilde{\rho}_j|\tilde{v}_1\rangle = |\langle\tilde{v}_1|\tilde{\psi}_j\rangle|^2 + |\langle\tilde{v}_1|\tilde{\phi}_j\rangle|^2, \tag{11.118}$$

we find

$$\bar{\rho}_{1,2} = \frac{1}{16}\big[1 - (b^2 - c^2)\cos(2\theta_-) \mp 2\sqrt{2}ab\sin\theta_+\sin\theta_-$$
$$\pm 2\sqrt{2}cd\cos(\theta_+ - \theta_3)\cos\theta_-\big],$$

$$\bar{\rho}_{3,4} = \frac{1}{16}\big[1 + (b^2 - c^2)\cos(2\theta_-) \pm 2\sqrt{2}ab\cos\theta_+\cos\theta_-$$
$$\pm 2\sqrt{2}cd\sin(\theta_+ - \theta_3)\sin\theta_-\big], \tag{11.119}$$

where $\theta_\pm = (\theta_1 \pm \theta_2)/2$. Finally, the mutual information is expressed as

$$I(\mathscr{E};\Pi)[\theta_k] = \frac{1}{4} \sum_{j=1}^4 (16\bar{\rho}_j) \log(16\bar{\rho}_j), \qquad (11.120)$$

which is to be regarded as a function of the three θ_ks.

The general solution to this optimization problem is not known as yet. But if the parameters b and c are equal, we have the analytical solution at hand.

Upon setting $b = c$, the ensemble of states is characterized by two independent parameters because a, b, and d must obey $a^2 + d^2 + 2b^2 = 1$. We define A and θ_0 by

$$A = 2b\sqrt{2(a^2 + d^2)} = 2b\sqrt{2(1 - 2b^2)},$$

$$\theta_0 = \tan^{-1}\frac{d}{a}. \qquad (11.121)$$

The expression for the mutual information then simplifies to

$$I(\mathscr{E};\Pi)[\theta_k] = \frac{1}{2} \sum_{j=1}^2 \Phi(f_j), \qquad (11.122)$$

where f_1 and f_2 are

$$f_1 = A\left[\cos\theta_0 \sin\theta_+ \sin\theta_- - \sin\theta_0 \cos(\theta_+ - \theta_3)\cos\theta_-\right],$$
$$f_2 = A\left[\cos\theta_0 \cos\theta_+ \cos\theta_- + \sin\theta_0 \sin(\theta_+ - \theta_3)\sin\theta_-\right]. \quad (11.123)$$

The partial derivatives with respect to the θ_ks are

$$\frac{\partial}{\partial\theta_1}I[\theta_k] = \frac{A}{8}\left[\cos\theta_0 \sin\theta_1 \log\frac{R_1}{R_2} + \sin\theta_0 \sin(\theta_1 - \theta_3)\log(R_1 R_2)\right],$$

$$\frac{\partial}{\partial\theta_2}I[\theta_k] = \frac{A}{8}\left[-\cos\theta_0 \sin\theta_2 \log(R_1 R_2) + \sin\theta_0 \sin(\theta_2 - \theta_3)\log\frac{R_1}{R_2}\right],$$

$$\frac{\partial}{\partial\theta_3}I[\theta_k] = -\frac{\sin\theta_0 A}{8}\left[\sin(\theta_1 - \theta_3)\log(R_1 R_2) + \sin(\theta_2 - \theta_3)\log\frac{R_1}{R_2}\right],$$

with $R_j = \dfrac{1 + f_j}{1 - f_j}. \qquad (11.124)$

The right-hand sides are of the form $X_i \log R_1 + Y_i \log R_2$, and the necessary conditions for stationary points, that is: $\frac{\partial}{\partial\theta_i}I[\theta_k] = 0$ for $i = 1,2,3$, are then equivalent to

$$(X_i Y_j - X_j Y_i)\log R_l = 0 \quad \text{for} \quad l = 1,2 \quad \text{and} \quad i,j = 1,2,3. \qquad (11.125)$$

Since $\log R_1 = \log R_2 = 0$ gives zero mutual information, the coefficients must be zero, i.e.,

$$X_i Y_j - X_j Y_i = 0 \quad \text{for} \quad (i,j) = (1,2),(2,3),(3,1). \tag{11.126}$$

Explicitly, they are

$$\cos^2 \theta_0 \sin \theta_1 \sin \theta_2 + \sin^2 \theta_0 \sin(\theta_1 - \theta_3)\sin(\theta_2 - \theta_3) = 0,$$
$$\left[\cos \theta_0 \sin \theta_2 + \sin \theta_0 \sin(\theta_1 - \theta_3)\right]\sin(\theta_2 - \theta_3) = 0,$$
$$\left[\cos \theta_0 \sin \theta_1 - \sin \theta_0 \sin(\theta_2 - \theta_3)\right]\sin(\theta_1 - \theta_3) = 0, \tag{11.127}$$

two of which imply the third. One verifies immediately that

$$\theta_1 = \theta_0, \quad \theta_2 = -\theta_0, \quad \theta_3 = -\frac{\pi}{2} \tag{11.128}$$

solve these equations, and this solution gives the accessible information.

The optimal POVM thus found consists of a single orbit with the seed ket given by

$$|v_1\rangle = T|\tilde{v}_1\rangle = \frac{1}{2}\begin{pmatrix} 1 \\ \sqrt{2}i\sin\theta_0 \\ \sqrt{2}\cos\theta_0 \\ -i \end{pmatrix} \tag{11.129}$$

in the original representation of (11.109). This corresponds to $f_1 = A$ and $f_2 = 0$ in (11.123). The resulting accessible information is

$$I_{\text{acc}}(\mathscr{E}) = \frac{1}{2}\Phi(A) = \frac{1}{2}\Phi\left(2b\sqrt{2(1-2b^2)}\right). \tag{11.130}$$

Rather intriguingly, the accessible information depends only on one of the parameters.

We next show how to construct a real optimal POVM out of this complex solution. Split the optimal seed $|\tilde{v}_1\rangle$ into real and imaginary parts,

$$|\tilde{v}_1\rangle = |\tilde{v}_{1r}\rangle + i|\tilde{v}_{1i}\rangle, \tag{11.131}$$

and consider another set of outcomes generated by the complex conjugate seed

$$|\tilde{v}_1^*\rangle = |\tilde{v}_{1r}\rangle - i|\tilde{v}_{1i}\rangle. \tag{11.132}$$

These two POVMs give the same joint probabilities and, therefore, the same amount of mutual information. It then follows from the convexity of the mutual information (11.22) that a real rank-2 seed

$$\widetilde{S}_{\text{real}} = \frac{1}{2}(|\tilde{v}_1\rangle\langle\tilde{v}_1| + |\tilde{v}_1^*\rangle\langle\tilde{v}_1^*|) = |\tilde{v}_{1r}\rangle\langle\tilde{v}_{1r}| + |\tilde{v}_{1i}\rangle\langle\tilde{v}_{1i}| \qquad (11.133)$$

gives the accessible information as well.

As we mentioned before, the optimal POVM is not unique as a rule. Here we have already a choice between a POVM with four complex rank-1 outcomes, its complex conjugate POVM, or a POVM with four real rank-2 outcomes. These three POVMs can be regarded as equivalent in the sense that they give rise to the same joint probabilities.

In addition, there is a one-parameter family of inequivalent POVMs, each having four real rank-2 outcomes. In the original representation of (11.109) the outcomes are of the form

$$\Pi_k = |u_k\rangle\langle u_k| + |v_k\rangle\langle v_k| \quad \text{for} \quad k = 1,\ldots,4 \qquad (11.134)$$

with the kets $|u_k\rangle$ and $|v_k\rangle$ depending on the real parameter r in the following way:

$$\left.\begin{array}{c}|u_1\rangle\\[2mm]|u_2\rangle\end{array}\right\} = \frac{\sqrt{\cos^2\theta_0 + r}}{2}\begin{pmatrix}1/\cos\theta_0\\ \pm\sqrt{2}\\ 0\\ 0\end{pmatrix},$$

$$\left.\begin{array}{c}|u_3\rangle\\[2mm]|u_4\rangle\end{array}\right\} = \frac{\sqrt{\cos^2\theta_0 - r}}{2}\begin{pmatrix}1/\cos\theta_0\\ 0\\ \pm\sqrt{2}\\ 0\end{pmatrix},$$

$$\left.\begin{array}{c}|v_1\rangle\\[2mm]|v_2\rangle\end{array}\right\} = \frac{\sqrt{\sin^2\theta_0 + r}}{2}\begin{pmatrix}0\\ 0\\ \pm\sqrt{2}\\ 1/\sin\theta_0\end{pmatrix},$$

$$\left.\begin{array}{c} |v_3\rangle \\ |v_4\rangle \end{array}\right\} = \frac{\sqrt{\sin^2\theta_0 - r}}{2}\begin{pmatrix} 0 \\ \mp\sqrt{2} \\ 0 \\ 1/\sin\theta_0 \end{pmatrix}. \tag{11.135}$$

The parameter r is restricted to the range

$$|r| \leq \min\left(\cos^2\theta_0, \sin^2\theta_0\right) = \frac{\min(a^2, d^2)}{a^2 + d^2} \tag{11.136}$$

but is otherwise arbitrary. We note that, when $|r|$ is maximal, the corresponding optimal POVM has two rank-2 outcomes and two rank-1 outcomes, rather than four rank-2 outcomes. We note further that the POVM is not group-covariant when $r \neq 0$.

11.7   Summary and outlook

We have given a brief introduction to, and summary of, the problem of determining the accessible information about a given set of quantum states. At present, the problem (11.21) remains open because there is no generally applicable method by which we can determine the optimal POVM and the accessible information. We note in particular the lack of sufficient conditions by which one could judge whether a candidate POVM is optimal. Until such conditions are established, the strategy of choice is a combination of a numerical search—possibly by the method described in Section 11.5—with an analytical check of the necessary conditions (11.30).

We recall further that the seemingly simple conjecture mentioned after (11.56) has not been proven as yet. A proof would surely constitute a major step forward because in practice one often encounters the situation of the conjecture, namely the task of distinguishing optimally between two quantum states.

We also emphasize that obtaining analytical expressions for the optimal POVM usually requires solving a set of nonlinear equations, and we would not expect that they can be solved routinely, with closed-form solutions. This point is well illustrated by the example in Subsection 11.6.1, arguably the simplest nontrivial situation.

In practice, however, we are rarely looking for the accessible information about random quantum states. Rather, the quantum states of interest tend to possess certain symmetries among them. We can then apply the group-covariant POVM method of Subsection 11.4.4 for solving the problem as demonstrated by the examples of Subsections 11.6.2–11.6.4. Nevertheless, the numerical strategy explained in detail in Section 11.5 lends us significant help in the search for optimal POVMs. A major problem thereby is, of course, to discriminate between local and global maxima. Further studies are clearly necessary.

We remark that in general the optimal POVM is not unique for a given set of quantum states. We have demonstrated this nonuniqueness by the example of Subsection 11.6.4, where we report an optimal group-covariant von Neumann measurement, an optimal group-covariant POVM that is not of von Neumann type, and a family of inequivalent optimal POVMs that are not group-covariant. From the purely theoretical point of view, these POVMs are equally good in the sense of providing the accessible information. On the other hand, however, there are great differences between them when a physical implementation of the POVM is required. Generally speaking, von Neumann projection measurements and nonprojection measurements belong to different classes of measurement schemes.

This suggests that one should examine thoroughly under which conditions a von Neumann measurement can extract the accessible information about the given quantum states. The conjecture mentioned above is particularly relevant in this context.

Acknowledgments

We are grateful for numerous discussions with Janet Anders, Wee Kang Chua, Thomas Decker, Dagomir Kaszlikowski, Shang Yong Looi, and Jaroslav Řeháček. J. S. and B.-G. E. wish to thank Hans Briegel for the generous hospitality extended to them at the Institute for Quantum Optics and Quantum Information in Innsbruck, where part of this work was done. This work is supported by A*STAR Temasek Grant 012-104-0040 and NUS Grant WBS R144-000-116-101.

References

[1] D. Bruß, M. Christandl, A. Ekert, B.-G. Englert, D. Kaszlikowski, and C. Macchiavello, Phys. Rev. Lett. **91** (2003) art. 097901 (4 pages).

[2] Y. C. Liang, D. Kaszlikowski, B.-G. Englert, L. C. Kwek, and C. H. Oh, Phys. Rev. A **68** (2003) art. 022324 (9 pages).

[3] D. Kaszlikowski, A. Gopinathan, Y. C. Liang, L. C. Kwek, and B.-G. Englert, Phys. Rev. A **70** (2004) art. 032306 (5 pages).

[4] B.-G. Englert, D. Kaszlikowski, H. K. Ng, W. K. Chua, J. Řeháček, and J. Anders, *Highly Efficient Quantum Key Distribution With Minimal State Tomography*, eprint arXiv:quant-ph/0412089 (2004).

[5] C. W. Helstrom, *Quantum Detection and Estimation Theory*, (Academic Press, New York 1976).

[6] T. Cover and J. Thomas, *Elements of Information Theory*, (John Wiley & Sons, New York 1991).

[7] M. A. Nielsen and I. L. Chuang, *Quantum Computation and Quantum Information*, (Cambridge University Press, Cambridge 2000).

[8] C. A. Fuchs, *Distinguishability and Accessible Information in Quantum Theory*, Ph.D. Thesis, University of New Mexico (1995), eprint arXiv:quant-ph/9601020 (1996).

[9] A. S. Holevo, J. Multivariate Anal. **3** (1973) 337–394.

[10] *Quantum State Estimation*, edited by M. Paris and J. Řeháček, Lecture Notes in Physics, Vol. 649 (Springer, Berlin 2004).

[11] A. S. Holevo, *Coding Theorems for Quantum Channels*, eprint arXiv:quant-ph/9809023 (1998).

[12] E. B. Davies, IEEE Trans. Inf. Theory **24** (1978) 596–599.

[13] M. Sasaki, S. M. Barnett, R. Jozsa, M. Osaki, and O. Hirota, Phys. Rev. A **59** (1999) 3325–3335.

[14] B. Schumacher, M. Westmoreland, and W. K. Wootters, Phys. Rev. Lett. **76** (1996) 3452–3455.

[15] A. S. Holevo, Probl. Peredachi Inf. **9** (1973) 3–11; English translation: Probl. Inf. Transm. (USSR) **9** (1973) 177–183.

[16] R. Jozsa, D. Robb, and W. K. Wootters, Phys. Rev. A **49** (1994) 668–677.

[17] C. A. Fuchs and C. M. Caves, Phys. Rev. Lett. **73** (1994) 3047–3050.

[18] T. Decker, *Symmetric measurements attaining the accessible informa-tion*, eprint arXiv:quant-ph/0509122 (2005).

[19] F. H. Willeboordse, A. Gopinathan, and D. Kaszlikowski, Phys. Rev. A **71** (2005) art. 042310 (4 pages).

[20] J. Řeháček, B. -G. Englert, and D. Kaszlikowski, Phys. Rev. A **71** (2005) art. 054303 (4 pages).

[21] J. Řeháček, private communication (2005).

[22] W. H. Press, B. P. Flannery, S. A. Teukolsky, W. T. Vetterling, *Numerical Recipes in C: The Art of Scientific Computing* (2nd edition, Cambridge University Press 1992).

[23] J. R. Shewchuk, *An Introduction to the Conjugate Gradien-t Method Without the Agonizing Pain (Edition* $1\frac{1}{4}$*)*, available at www.cs.cmu.edu/~quake-papers/painless-conjugate-gra-dient.pdf.

[24] P. W. Shor, *On the Number of Elements in a POVM Attaining the Accessible Information*, eprint arXiv:quant-ph/0009077 (2000).

[25] L. B. Levitin, *Optimal quantum measurements for two pure and mixed states*, in: *Quantum Communications and Measurement*, edited by V. P. Belavkin, O. Hirota, and R. L. Hudson, (Plenum Press, New York 1995) 439–448.

[26] A. S. Holevo, Probl. Peredachi Inf. **9** (1973) 31–42; English translation: Probl. Inf. Transm. (USSR) **9** (1973) 110–118.

[27] D. Kaszlikowski, A. Gopinathan, Y. C. Liang, L. C. Kwek, and B.-G. Englert, *How well can you know the edge of a quantum pyramid?* eprint arXiv:quant-ph/0307086 (2003).

[28] M. R. Frey, Phys. Rev. A **73** (2006) art. 032309 (7 pages).

[29] D. Bruß, Phys. Rev. Lett. **81** (1998) 3018–3021.

[30] C. H. Bennett and G. Brassard, in: *Proceedings of the IEEE Conference on Computers, Systems, and Signal Processing Bangalore, India, December 1984* (IEEE, New York 1984) 175–179.

[31] S. M. Assad, J. Suzuki, and B.-G. Englert, Int. J. Quant. Inf. **4** (2006) 1003–1012.

Chapter 12

Quantum Entanglement: Concepts and Criteria

Fu-li Li and M. Suhail Zubairy

Abstract   Entanglement is one of the key properties of the quantum mechanical systems that is fundamentally different from a classical system. Quantum entanglement plays an important role in the debate concerning the foundations of physics but also is an important resource in various quantum information processes. In this article, we introduce basic concepts on EPR correlations and quantum entanglement, and review established entanglement criteria, and show how to use these criteria in the generation of entanglement by considering coherence-induced entanglement and correlated spontaneous emission laser.

12.1   Introduction

The concept of locality lies at the foundation of classical physics. The locality implies that measurement performed on one of two systems which are spatially separated and without interaction cannot disturb one another. Based on the locality concept, Einstein, Podolsky and Rosen [1] (EPR) proposed a gedanken experiment that showed that both position and momentum variables could be simultaneously assigned to a single localized particle with certainty if two particles are initially prepared in the ideal position and momentum correlated state. This obviously violates the principle of quantum mechanics that two physical quantities represented by noncommutable operators can not have simultaneous reality. Therefore, assuming that the locality concept is one of the universal physical principles, EPR argued that quantum mechanics is incomplete. In order to verify the EPR argument, Bell [2] derived inequalities of measurable physical quantities for all theories based on reality and locality. Standard quantum mechanics violates these inequalities. Over the past forty years, a variety of experimental and theoretical efforts have been devoted to investigating the violation of Bell inequalities in quantum mechanics.

In the original gedanken experiment proposed by EPR [1], continuous spectra of physical quantities were involved. Bohm [3] established a version of the EPR argument in discrete variables. Following Bohm's suggestion, Bell-like inequalities have been tested in optical experiments, trapped ions experiments, and single-neutron interferometry [4, 5, 6, 7, 8, 9, 10, 11, 12].

Besides Bell inequalities, the EPR argument implies that the Heisenberg uncertainty relation could be violated, which is termed the EPR paradox. A direct demonstration of the EPR paradox can also prove the violation of locality by standard quantum mechanics. Two quadrature-phase amplitudes of an optical field satisfy the same commutator as the position and the momentum of a massive particle do, and are the conjugate "position" and "momentum"-like operators. These quadrature-phase amplitudes of the electromagnetic field are measurable by use of homodyne detection schemes. In [13], a scheme was proposed that used two correlated quadrature-phase amplitudes of an optical field to demonstrate a violation of an inferred Heisenberg uncertainty relation, i.e., the continuous variable EPR paradox. Following this proposal, a subthreshold nondegenerate optical parametric oscillator was employed to generate correlated amplitudes for signal and idler beams of light, and then inferred the amplitudes of the signal beam (X_s, Y_s) from measurements of the spatially separated amplitudes (X_i, Y_i) of the idler beam [14]. The errors of these inferences are quantified by the variances $\Delta_{inf}^2 X$ and $\Delta_{inf}^2 Y$, with the EPR paradox requiring that $\Delta_{inf}^2 X \Delta_{inf}^2 Y < 1$. In the experiment, it was shown that $\Delta_{inf}^2 X \Delta_{inf}^2 Y = 0.7 \pm 0.01$, thus demonstrating the paradox. Using the Kerr nonlinearity of an optical fiber, Silberhorn et al. [15] generated the EPR correlations with the product of the inferred uncertainties 0.64 ± 0.08 well below the EPR limit of unity. Bowen et al. [16] generated a pair of entangled beams from the interference of two amplitude squeezed beams and demonstrated the EPR paradox with the inferred uncertainty product 0.58 ± 0.02. In the recent experiment, Howell et al. [17] investigated correlations of transverse position and momentum of photons produced from parametric down conversion and achieved a measured two-photon momentum-position variance product of $0.01\hbar^{-1}$, which dramatically violates the bounds for the EPR correlations. Up to now, nearly all the experiments on the EPR correlations show that quantum mechanics is not a local-realistic theory.

In contrast to classical theories, quantum mechanics promises the existence of nonlocal correlations between two systems spatially separated and without any direct interactions, which Schrödinger termed as entanglement [18]. It has been shown that entanglement is a key ingredient in various quantum information processes such as quantum teleportation [19], quantum dense coding [20], quantum cryptography [21] and quantum computing [22]. Because of the crucial role of entanglement in quantum information processes, the study of entanglement has attracted a lot of interest in recent years. Among various studies on entanglement, the first question which may be asked is how to know that a quantum state is entangled. For a pure bipartite state, the Schmidt decomposition [23] can be used to judge whether the state is entangled and the degree of entanglement can be quantified by the partial von Neumann entropy [24]. Thus, in principle, the problem of entanglement for pure states of a bipartite system has been completely solved.

In the real world, quantum systems inevitably undergo decoherence processes and quantum systems are mostly in mixed states. For density matrix of a quantum system consisting of two subsystems, several criteria on entanglement have been established [25]-[33]. Peres [25] found that a sufficient condition for the density matrix of a bipartite system being inseparable or entangled is that a matrix, obtained by partial transpose of the density matrix, becomes negative. Horodecki [26] showed that this condition is necessary and sufficient for entanglement only in the case of 2×2 and 2×3 systems. Based on the partial transposition criterion, Simon [30] found a necessary and sufficient condition for entanglement of two-mode Gaussian states. Using the Heisenberg uncertainty relation and the Cauchy–Schwarz inequality, Duan et al. [31] derived the same necessary and sufficient condition. Giedke et al. [32] found a necessary and sufficient condition for entanglement of Gaussian states of bipartite systems of arbitrarily many modes. Hillery and Zubairy [33] derived a class of inequalities for various moments of bosonic creation and annihilation operators, whose violation shows the presence of entanglement in two-mode systems. Shchukin and Vogel [34] found necessary and sufficient conditions for the partial transposition of bipartite harmonic quantum states to be nonnegative. The conditions are expressed as an infinite series of inequalities for the moments of creation and annihilation operators of bosons. The violation of any one of the inequalities provides a sufficient condition for entanglement.

The generation of entangled states has been investigated in various systems from atoms or ions, photons and quadrature-phase amplitudes of the electromagnetic field. In 1997, Hagley et al. [35] produced the atomic entangled state in which two atoms are in two different circular Rydberg states and separated by a distance of the order of 1 cm. Turchette et al. [36] showed that the internal states of two trapped ions can be prepared in both the Bell-like singlet and triplet entangled states in a deterministic fashion. For applications, the more particles are entangled, the more useful the states can be. Along the lines of the proposal suggested by Molmer and Sorensen [37], Sackett et al. [38] realized experimentally entangled states of four trapped ions. Based on Duan et al.'s proposal [39], Julsgaard et al. [40] demonstrated experimentally at the level of macroscopy the entanglement between two separate samples of atoms each of which contains 10^{12} atoms. As for the generation of entangled states of photons, a great progress has been made in recent years. Using a single circular Rydberg atom, Rauschenbeutel et al. [41] prepared two modes of a superconducting cavity in a maximally entangled state in which the two modes share a single photon. In most EPR optical experiments, pairs of polarization photons flying apart can be created in an entangled state by either spontaneous emission in a cascade atomic system [42, 43, 44], or down-conversion in a nonlinear medium [45, 46]. Entangled states of three [47], four [48] and five photons [49] have been realized. For the purpose of application, an entangled state containing more photons is an interesting problem. Tsujino et al. [50] showed the experimental generation of two-photon-polarization states by parametric down-conversion. Eisenberg et al. [51] created a bipartite multiphoton entangled state through stimulated parametric down-conversion of strong laser pulses in a nonlinear crystal. Quantum information processes based on the entangled quadrature-phase amplitudes of the electromagnetic field show some advantages and the generation of the entangled quadrature-phase amplitudes of an optical field has also attracted much attention [52]. In usual experiments, the entangled quadrature-phase optical beams are generated by two vacuum squeezed states via a beam splitter and are in a two-mode quadrature squeezed vacuum. In a recent experiment, Zhang et al. [53] showed that the bright entangled signal and idler beams can be generated by a nondegenerate optical parametric amplifier.

It is evident that the experimental and theoretical studies of bipartite systems have made a great progress in recent years. In this article, we introduce basic concepts on EPR correlations and entanglement, and review established entanglement criteria. We also discuss coherence-induced entanglement and correlated spontaneous emission laser as examples for explaining how to use these criteria in the generation of entanglement. The paper is organized as follows. In Section 12.2, basic concepts on EPR correlations and quantum entanglement are introduced. In Section 12.3, entanglement criterion for pure states is discussed on the basis of Schmidt decomposition theory. In Section

12.4, various established entanglement criteria for mixed states are reviewed. In Sections 12.5 and 12.5, we discuss two examples as applications of entanglement criteria by considering the interaction of a V-type three-level atom with two modes of the cavity field and correlated spontaneous emission laser, respectively. In Section 12.6, several remarks on entanglement criterion and measurement are given.

12.2   EPR correlations and quantum entanglement

In 1935, Einstein, Podolsky, and Rosen (EPR) [1] published one of the most controversial papers of the 20th century which argued the completeness of quantum theory. The paper involved three important concepts: completeness, reality and locality. About completeness of a physical theory, they stated: "Every element of the physical reality must have a counterpart in the physical theory". Regarding reality, they suggested the criterion: "If, without in any way disturbing a system, we can predict with certainty (i.e., with probability equal to unity) the value of a physical quantity, then there exists an element of physical reality corresponding to this physical quantity". The statement "without in any way disturbing a system" implied a measurement only on one system that can not disturb another one in any way if the two systems are spatially separated and do not interact. This implied locality. With these concepts in mind, we go into the details of EPR's argument.

Consider two particles that are spatially separated and without interaction but have been prepared in a state described by wave function $|\Psi(1,2)\rangle$ after the foregoing interaction between them ended. Let $\hat{A}$ and $\hat{B}$ be two operators which represent two physical quantities of the first particle. The operators $\hat{A}$ and $\hat{B}$ have eigenvalues $(a_1, a_2, a_3, \cdots)$ and $(b_1, b_2, b_3, \cdots)$, and the corresponding eigenfunctions $(|a_1\rangle, |a_2\rangle, |a_3\rangle, \cdots)$ and $(|b_1\rangle, |b_2\rangle, |b_3\rangle, \cdots)$, respectively. At first, we consider a measurement of the eigenvalues of the operator $\hat{A}$ on the first particle. For describing the measurement, we expand the wave function in terms of the eigenfunctions of the operator $\hat{A}$ as

$$|\Psi(1,2)\rangle = \sum_n |\varphi_n(2)\rangle |a_n(1)\rangle, \qquad (12.1)$$

where $|\varphi_n(2)\rangle$ are state vectors resulting from the expansion and depend only on coordinates of the second particle. According to the quantum theory, when the result a_n is obtained in the measurement, the whole system simultaneously collapses into the state $|\varphi_n(2)\rangle |a_n(1)\rangle$. Suppose that $|\varphi_n(2)\rangle$ is one of the eigenfunctions of operator $\hat{P}$ of the second particle with eigenvalue p_n. We can then assign the eigenfunction $|\varphi_n(2)\rangle$ to the second particle and predict the result p_n for a measurement of eigenvalues of the operator $\hat{P}$ on the second particle with certainty even if the real measurement was not carried out on the second particle.

Instead of measuring the eigenvalue of the operator $\hat{A}$, we can also perform a measurement of the eigenvalue of the operator $\hat{B}$ on the first particle. We now expand the wave function in terms of the eigenfunctions of the operator $\hat{B}$

$$|\Psi(1,2)\rangle = \sum_n |\phi_n(2)\rangle |b_n(1)\rangle, \qquad (12.2)$$

where $|\phi_n(2)\rangle$ are the state vectors in the expansion and depend only on the coordinates of the second system. As noted above, once the result b_n is obtained as a result of measurement, the whole system collapses into the state $|\phi_n(2)\rangle |b_n(1)\rangle$. Suppose that $|\phi_n(2)\rangle$ is one of the eigenfunctions of operator $\hat{Q}$ of the second particle with eigenvalue q_n. We can then assign the eigenfunction $|\phi_n(2)\rangle$ to the second particle and predict with certainty the result q_n for a measurement of eigenvalues of the operator $\hat{Q}$ although the measurement is in fact not performed on the second particle.

Now the following argument was raised by EPR. According to the locality condition, either of the above two measurements which were performed on the first particle could not disturb the second particle in any way. The state of the second particle after the measurements must be the same as the state before the measurements. However, based on the measurements, we can assign with certainty the eigenstate $|\varphi_n(2)\rangle$ of the operator $\hat{P}$ as well as the eigenstate $|\phi_n(2)\rangle$ of the operator $\hat{Q}$ to the same reality! Moreover, if $\hat{P}$ and $\hat{Q}$ are non-commutable, the above gedanken experiment means that the same reality could simultaneously have the eigenvalues p_n and q_n which are eigenvalues of the two non-commutable operators. This is in contradiction with the standard quantum mechanics. This is often referred to as the EPR paradox.

In order to make the problem more clear, we consider two spin $1/2$ particles. Suppose that the two particles are initially prepared in the state [3]

$$|\Psi(1,2)\rangle = \frac{1}{\sqrt{2}}(|\downarrow_2\rangle|\uparrow_1\rangle - |\uparrow_2\rangle|\downarrow_1\rangle), \qquad (12.3)$$

where $|\uparrow\rangle$ and $|\downarrow\rangle$ are eigenfunctions of the spin operator $\hat{\sigma}_z$ with eigenvalues $+1$ and -1, respectively. The state vector (12.3) has the expansion form in terms of eigenstates of the spin operator $\hat{\sigma}_z$ of the first particle. The expansion state vectors are the eigenfunction of the spin operator $\hat{\sigma}_z$ of the second particle. The state vector (12.3) can also be expanded in terms of eigenfunction of the spin operator $\hat{\sigma}_x$ of the first particle. In this way, one has [54]

$$|\Psi(1,2)\rangle = \frac{1}{\sqrt{2}}(|-_2\rangle|+_1\rangle - |+_2\rangle|-_1\rangle), \qquad (12.4)$$

where $|+\rangle$ and $|-\rangle$ are eigenfunctions of the spin operator $\hat{\sigma}_x$ with eigenvalues $+1$ and -1, respectively. Therefore, according to the above EPR argument, one could assign eigenfunctions of the spin operators $\hat{\sigma}_z$ and $\hat{\sigma}_x$ to the second particle with certainty. In other words, the second particle may simultaneously have eigenvalues of $\hat{\sigma}_z$ and $\hat{\sigma}_x$.

Now we face the following problem. If the EPR argument is correct, the standard quantum mechanics is an incomplete theory. If the EPR argument was incorrect, we may have to give up the either the notion of reality or locality (12.3).

For a long time, the debate on the EPR argument had rested on theoretical or philosophical aspects until 1964 when Bell established inequalities to measure physical quantities so that a direct testing of the consequences of the EPR premises becomes possible [2]. Many experiments of testing Bell inequalities have been performed [4, 8, 12]. The results obtained from these experiments are all favorable to quantum mechanics. These experimental results show that the locality concept on correlations between two particles, which is ingrained in classical theories, may not be a universal concept and nonlocal correlations between particles separated spatially and without interaction may exist.

To have a concrete idea on quantum nonlocal correlations, we reconsider the state (12.3). Let **n** be a unit vector pointing at direction (θ, φ). The state (12.3) can be written in term of eigenvectors $|+1\rangle$ and $|-1\rangle$ of $\hat{\sigma} \cdot \mathbf{n}$ with eigenvalues $+1$ and -1, respectively, as

$$|\Psi(1,2)\rangle = \frac{e^{i\varphi}}{\sqrt{2}}(|-1\rangle_2|+1\rangle_1 - |+1\rangle_2|-1\rangle_1). \qquad (12.5)$$

From this state, we see that the spin states of the two particles along any direction are maximally correlated. If spin state of one of the particles along an arbitrary direction (θ, φ) is measured up or down, we immediately know spin state of another particle along the direction (θ, φ).

We should emphasize that nonclassical correlations revealed by the EPR paradox are not related to any interactions and result from the description of the wave function for states of physical systems and the superposition principle of states. Schrödinger termed this correlation as entanglement [18].

12.3   Entanglement of pure states

A pure state of a bipartite system is called separable if and only if it can be written as a product of two states of the subsystems. A state of a bipartite system is entangled if it is not separable. The separability of pure two-party states can be recognized by the Schmidt decomposition [23].

Suppose that the quantum system under consideration consists of two subsystems A and B whose states are defined in the Hilbert space $\mathcal{H}_{\mathscr{A}}$ with the dimension N_A and $\mathcal{H}_{\mathscr{B}}$ with the dimension N_B. Without any loss of generality, we assume $N_B \leq N_A$ in the following discussion. In general, an arbitrary pure state of the bipartite system can be expressed as

$$|\Psi_{AB}\rangle = \sum_{i,j} a_{ij}|u_i\rangle \otimes |v_j\rangle, \tag{12.6}$$

where $\{|u_i\rangle \in \mathcal{H}_{\mathscr{A}}, i = 1, 2, \ldots N_A\}$ and $\{|v_j\rangle \in \mathcal{H}_{\mathscr{B}}, j = 1, 2, \ldots N_B\}$ are two complete sets of orthonormal basis vectors in their respective Hilbert spaces.

The density matrix $\hat{\rho}_{AB}$ for the whole system is given by $|\Psi_{AB}\rangle\langle\Psi_{AB}|$. Tracing out the variables of the subsystem A in $\hat{\rho}_{AB}$, we obtain the reduced density matrix for the subsystem B

$$\hat{\rho}_B = tr_A(\hat{\rho}_{AB}) = \sum_{j,j'}(\sum_i a_{ij}a_{ij'}^*)|v_j\rangle\langle v_{j'}|. \tag{12.7}$$

Next we consider the eigenvalue problem

$$\hat{\rho}_B|b_v\rangle = p_v|b_v\rangle, \tag{12.8}$$

where $v = 1, 2, \ldots, N_B$. From Eq. (12.7), the eigenvalues $p_v(\geq 0)$ can be expressed as

$$p_v = \sum_{j,j'}(\sum_i a_{ij}a_{ij'}^*)\langle b_v|v_j\rangle\langle v_{j'}|b_v\rangle. \tag{12.9}$$

The eigenstates $\{|b_v\rangle\}$ compose a new complete set of orthonormal basis vectors in the Hilbert space $\mathcal{H}_{\mathscr{B}}$. From Eq. (12.9), we have $\sum_v p_v = 1$.

The basis states $\{|v_j\rangle\}$ can be expanded in terms of the eigenstates $\{|b_v\rangle\}$ as

$$|v_j\rangle = \sum_v \langle b_v|v_j\rangle|b_v\rangle. \tag{12.10}$$

It follows, on substituting from Eq. (12.10) into Eq. (12.6), that

$$|\Psi_{AB}\rangle = \sum_v \sum_{i,j} a_{ij}\langle b_v|v_j\rangle|u_i\rangle \otimes |b_v\rangle. \tag{12.11}$$

Next we define a new set of basis vectors for the subsystem A

$$|a_v\rangle = \sum_{i,j} a_{ij}\langle b_v|v_j\rangle|u_i\rangle = \langle b_v|\Psi_{AB}\rangle. \tag{12.12}$$

It is easily shown that $\langle a_{v'}|a_v\rangle = p_v\delta_{v,v'}$. Therefore, for $p_v = 0$, we must have $|a_v\rangle = 0$. On redefining the basis vectors

$$|a_v\rangle = \frac{1}{\sqrt{p_v}}\langle b_v|\Psi_{AB}\rangle, \tag{12.13}$$

we can rewrite the state (12.11) as

$$|\Psi_{AB}\rangle = \sum_v^{r_B} \sqrt{p_v}|a_v\rangle \otimes |b_v\rangle, \tag{12.14}$$

where $r_B \leq N_B$. The expansion (12.14) is called the Schmidt decomposion of the quantum states of a bipartite quantum system. The state (12.14) has the form of the EPR state (12.3) if the Schmidt rank r_B is larger than one. Therefore, a pure two-party state is the EPR state or entangled state if and only if number of nonzero Schmidt coefficients $\sqrt{p_n}$ is larger than one. We note that the state (12.3) has been written as the Schmidt decomposition form with the Schmidt number 2.

For any bipartite pure state, we can transform it to the Schmidt decomposition form via unitary transformations. According to different forms of the Schmidt decomposition, we may classify the states. For entangled states of a bipartite system whose Hilbert space is 2×2, there are four basic maximally entangled states, i.e. the four Bell states,

$$|\Phi^{\pm}\rangle = \frac{1}{\sqrt{2}}(|00\rangle \pm |11\rangle), \tag{12.15}$$

$$|\Psi^{\pm}\rangle = \frac{1}{\sqrt{2}}(|01\rangle \pm |10\rangle), \tag{12.16}$$

where $|0\rangle$ and $|1\rangle$ are two independent basis vectors of the Hilbert space for either of the subsystems. For tripartite systems which are defined in the Hilbert space $2 \times 2 \times 2$, there are two basic Schmidt decomposition entangled states, namely the Greenberger–Horne–Zeilinger (GHZ) state [55] and the W state [56],

$$|GHZ\rangle = \frac{1}{\sqrt{2}}(|000\rangle + |111\rangle), \tag{12.17}$$

$$|W\rangle = \frac{1}{\sqrt{3}}(|100\rangle + |010\rangle + |001\rangle). \tag{12.18}$$

It has been shown that, for the case of three qubits, any pure and fully entangled state can be transformed to either the GHZ state or the W state via stochastic local operations and classical communication [55]. Some efforts have been also been devoted to the Schmidt decomposition for tripartite states in the higher dimensional Hilbert space [56, 57] and multi-party states [58].

12.4 Criteria on entanglement of mixed states

12.4.1 Peres–Horodecki criterion

Because of the inevitable interaction with the environment, a quantum system in the real world is always in a mixed state. An important problem therefore relates to verifying whether a given mixed state is entangled. As the mixed states may simultaneously possess both classical and quantum correlations, the problem becomes more complicated than in the case of pure states as discussed in Section 12.3. Among various criteria on entanglement of mixed states, the partial transposition criterion which was established by Peres [25] and Horodecki [26] is more basic since several useful criteria that can be derived are based on it.

Consider a bipartite system whose states are defined in the Hilbert space $\mathcal{H} = \mathcal{H}_1 \otimes \mathcal{H}_2$. In general, a mixed state of the system can be described by the density matrix

$$\hat{\rho} = \sum_i p_i |\psi_i\rangle\langle\psi_i|, \tag{12.19}$$

where the state vectors $|\psi_i\rangle \in \mathcal{H}$, p_i is the probability of finding the system in the pure state $|\psi_i\rangle$, $p_i \geq 0$ and $\sum_i p_i = 1$. Let us suppose that $|\phi_\lambda\rangle$ are eigenstates of $\hat{\rho}$ with eigenvalues ρ_λ. It can be shown that

$$\rho_\lambda = \sum_i p_i |\langle\psi_i|\phi_\lambda\rangle|^2. \tag{12.20}$$

Therefore, all the eigenvalues ρ_λ are always positive. In this sense, the density matrix is called a positive operator. Any density matrices for describing a physical system must be positive. State (12.20) also shows that ρ_λ is the probability for finding the system in the eigenstate $|\phi_\lambda\rangle$. So, we have $\sum_\lambda \rho_\lambda = 1$.

As for pure states, we can divide mixed states also into two kinds: separable and inseparable or entangled. A separable state can always be written as a statistically mixed form of products of states of each subsystem [59]

$$\hat{\rho} = \sum_j p_j \hat{\rho}_j^{(1)} \otimes \hat{\rho}_j^{(2)}, \tag{12.21}$$

where $\hat{\rho}_j^{(i)} \in \mathcal{H}_i$, $p_j \geq 0$ and $\sum_j p_j = 1$. A state $\hat{\rho}$ is called inseparable or entangled if it is not separable. State (12.21) shows that the two subsystems are still correlated with each other even if they are in a separable state. This correlation is called the classical correlation, for example, which may result from simultaneously preparing the system 1 in the state $\hat{\rho}_j^{(1)}$ and the system 2 in the state $\hat{\rho}_j^{(2)}$ with the probability p_j.

Let $\hat{A}$ and $\hat{B}$ represent operators acting on states in the Hilbert spaces $\mathcal{H}_1$ and $\mathcal{H}_2$, respectively. For any states $\hat{\rho}^{(1)} \in \mathcal{H}_1$, $\hat{A}(\hat{\rho}^{(1)}) = \hat{A}\hat{\rho}^{(1)}\hat{A}^\dagger$. If $\hat{A}\hat{\rho}^{(1)}\hat{A}^\dagger$ are positive, i.e, all of eigenvalues of $\hat{A}\hat{\rho}^{(1)}\hat{A}^\dagger$ are positive, the operator $\hat{A}$ is called a positive operator. There is the same definition for the operator $\hat{B}$. Acting the direct product of $\hat{A}$ and $\hat{B}$ on the state (12.21), we have

$$\hat{A} \otimes \hat{B}(\hat{\rho}) = \sum_j p_j (\hat{A}\hat{\rho}_j^{(1)}\hat{A}^\dagger) \otimes (\hat{B}\hat{\rho}_j^{(2)}\hat{B}^\dagger). \tag{12.22}$$

Therefore, the resulting density matrix must be separable if the original density matrix is separable. This means that any local operations such as $\hat{A}$ and $\hat{B}$ can not transfer a separable state to an inseparable one. In other words, any local operations can not create entanglement.

For a separable state, state (12.22) also shows that the resulting state is still positive if the operators $\hat{A}$ and $\hat{B}$ are positive. In an entangled state, two subsystems are quantum mechanically correlated. When we perform an operation on one subsystem, we inevitably disturb another even if we did not really perform any operations on it. Therefore, intuitively, the above conclusion may not be true for entangled states. In order to detect entanglement sensibly and efficiently, the key problem is how to choose the local positive operators $\hat{A}$ and $\hat{B}$.

Peres [25] and Horodecki [26] considered two special operations: $\hat{A} = \hat{I}_1$ and $\hat{B} = \hat{T}_2$, where $\hat{I}_1$ is the unity operator for subsystem 1 and $\hat{T}_2$ is the transpose operator for subsystem 2. Obviously, both $\hat{I}_1$ and $\hat{T}_2$ are positive. The direct product operator $\hat{I}_1 \otimes \hat{T}_2$ is called a partial transpose operator. In the representation spanned by a set of orthonormal basis states $|\varphi_n\rangle \cdot |\phi_\nu\rangle$ where $|\varphi_n\rangle \in \mathcal{H}_1$ and $|\phi_\nu\rangle \in \mathcal{H}_2$, a density matrix $\hat{\rho}$ is represented by the matrix with matrix elements $\rho_{n\nu,m\mu} = \langle\varphi_n|\langle\phi_\nu|\hat{\rho}|\phi_\mu\rangle|\varphi_m\rangle$. After performing the partial transpose operation on the density matrix, we obtain the resulting matrix σ. If the original density matrix ρ is separable, state (12.22) shows that the partial transposed matrix σ must be positive. However, if the original density matrix ρ is inseparable, the partial transposed matrix σ may become negative. If it is true, the partial transposed matrix σ does not still represent a physical state. Therefore, the product of the local operators may change an entangled physical state to an unphysical state. The positivity of the partial transposed matrix provides a sufficiency condition for inseparability of a two-party state. This condition is named as the Peres–Horodecki criterion for entanglement. This is a very useful and powerful criterion in judging whether a bipartite state is entangled or not. Horodecki finally showed that this condition is a necessary and sufficient condition of entanglement only for 2×2 and 2×3 systems [26, 60]. Based on the positivity of density operator with various choices of the local operators, Horodecki also established some other criteria for entanglement of bipartite states [26, 60].

Now let us see how to use the Peres–Horodecki criterion through an example. Consider a pair of particles each of which may be in either the state $|0\rangle$ or the state $|1\rangle$. Here, the state $|0\rangle(|1\rangle)$ may represent either the spin-$\frac{1}{2}$ down(up) state or zero(one) photon-number state. The state $|n_1, n_2\rangle$ is a state in which the first particle is in $|n_1\rangle$ and the second one in $|n_2\rangle$ with $n_i = 0, 1$. Suppose that the two particle are prepared in a Werner state [59]

$$\hat{\rho} = x|\Phi_1\rangle\langle\Phi_1| + \frac{1}{4}(1-x)\sum_{i=1}^{4}|\Phi_i\rangle\langle\Phi_i|, \tag{12.23}$$

where

$$|\Phi_1\rangle = \frac{1}{\sqrt{2}}[|1,0\rangle - |0,1\rangle], \tag{12.24}$$

$$|\Phi_2\rangle = \frac{1}{\sqrt{2}}[|1,0\rangle + |0,1\rangle], \tag{12.25}$$

$$|\Phi_3\rangle = |1,1\rangle\langle1,1|, \tag{12.26}$$

$$|\Phi_4\rangle = |0,0\rangle\langle0,0|. \tag{12.27}$$

On performing the partial transpose operation $\hat{I}_1 \otimes \hat{T}_2$ on the state (12.23), we have the resulting matrix

$$\hat{\sigma} = \frac{x}{2}[|1,0\rangle\langle1,0| + |0,1\rangle\langle0,1| - |0,0\rangle\langle1,1| - |1,1\rangle\langle0,0|]$$
$$+ \frac{1}{4}(1-x)[|1,0\rangle\langle1,0| + |0,1\rangle\langle0,1| + |0,0\rangle\langle0,0|$$
$$+ |1,1\rangle\langle1,1|]. \tag{12.28}$$

In the representation spanned by the orthonormal basis states $\{|n_1, n_2\rangle, n_i = 0, 1\}$, we have the matrix representation for $\hat{\sigma}$

$$
\sigma = \frac{1}{4}
\begin{pmatrix}
1-x & 0 & 0 & -2x \\
0 & 1+x & 0 & 0 \\
0 & 0 & 1+x & 0 \\
-2x & 0 & 0 & 1-x
\end{pmatrix}.
\tag{12.29}
$$

A straightforward calculation shows that the matrix σ has the tri-degenerate eigenvalue $(1+x)/4$ and one single eigenvalue $(1-3x)/4$. It is seen that the eigenvalue $(1-3x)/4$ becomes negative if the parameter $x > 1/3$. According to the Peres–Horodecki criterion, therefore, the state (12.23) must be entangled when $x > 1/3$. By use of Bell's inequality and the α–entropic inequality to detect entanglement of the state (12.23), one finds that the state (12.23) becomes entangled only when $x < 1/\sqrt{2}$ and $x < 1/\sqrt{3}$ [61]. Compared to these criteria, the Peres–Horodecki criterion is stronger for detecting entanglement.

The Peres–Horodecki criterion would intuitively be considered to be just a mathematical condition. In the very recent experiment for generating and detecting entanglement in a bipartite multiphoton entangled state [67], however, Eisenberg et al. directly detected the positivity of the partial-transposed density matrix. So, the Peres–Horodecki criterion is also one of entanglement criteria which can be tested in experiments.

12.4.2 Simon criterion

Using the Peres–Horodecki criterion, Simon found that the separability condition of bipartite continuous-variable states can be expressed in term of variances of the phase space variables [30].

Consider a bipartite system of two modes which are described by annihilation operators $\hat{a}_1 = (\hat{q}_1 + i\hat{p}_1)/\sqrt{2}$ and $\hat{a}_2 = (\hat{q}_2 + i\hat{p}_2)/\sqrt{2}$. One can arrange the phase space variables and the phase-quadrature Hermitian operators into four-dimensional column vectors

$$
\xi = (q_1, p_1, q_2, p_2), \quad \hat{\xi} = (\hat{q}_1, \hat{p}_1, \hat{q}_2, \hat{p}_2).
\tag{12.30}
$$

The commutation relations between the phase-quadrature operators take the compact form [63]

$$
[\hat{\xi}_\alpha, \hat{\xi}_\beta] = i\Omega_{\alpha\beta},
\tag{12.31}
$$

where $\alpha, \beta = 1, 2, 3, 4$, and

$$\Omega = \begin{pmatrix} J & 0 \\ 0 & J \end{pmatrix}, \quad J = \begin{pmatrix} 0 & 1 \\ -1 & 0 \end{pmatrix}. \tag{12.32}$$

Given a bipartite density operator $\hat{\rho}$, let us define $\Delta \hat{\xi}_\alpha = \hat{\xi}_\alpha - <\hat{\xi}_\alpha>$, where $<\hat{\xi}_\alpha> = tr(\hat{\xi}_\alpha \hat{\rho})$, and introduce the Hermitian operators $\{\Delta \hat{\xi}_\alpha, \Delta \hat{\xi}_\beta\} = (\Delta \hat{\xi}_\alpha \Delta \hat{\xi}_\beta + \Delta \hat{\xi}_\beta \Delta \hat{\xi}_\alpha)/2$. The expectations $<\{\Delta \hat{\xi}_\alpha, \Delta \hat{\xi}_\beta\}> = tr(\{\Delta \hat{\xi}_\alpha, \Delta \hat{\xi}_\beta\} \hat{\rho})$ can be arranged into a 4×4 real variance matrix V, defined through $V_{\alpha\beta} = <\{\Delta \hat{\xi}_\alpha, \Delta \hat{\xi}_\beta\}>$. Using the variance matrix V, we can write the uncertainty relations for the phase-quadrature operators in the following compact form

$$V + \frac{i}{2}\Omega \geq 0. \tag{12.33}$$

Simon showed that under the Peres–Horodecki partial transpose the variance matrix is changed to $V \longrightarrow \tilde{V} = \Lambda V \Lambda$, where $\Lambda = diag(1, 1, 1, -1)$. The partial transposed density matrix still describes a physical state if the state under consideration is separable. Therefore, we have

$$\tilde{V} + \frac{i}{2}\Omega \geq 0 \tag{12.34}$$

as a necessary condition for separability. The variance matrix V can be written in the block form

$$V = \begin{pmatrix} A & C \\ C^T & B \end{pmatrix}, \tag{12.35}$$

where A, B, C are 2×2 real matrices and C^T is the transpose of C. The physical condition (12.33) implies $A \geq 1/4, B \geq 1/4$. Simon found that the condition (12.34) can be simplified in terms of the submatrices as the form

$$detAdetB + (\frac{1}{4} - |detC|)^2 - tr(AJCJBJC^T J) \geq \frac{1}{4}(detA + detB). \tag{12.36}$$

This condition should necessarily be satisfied by every separable state of any bipartite systems. A violation of the inequality (12.36) represents a sufficient condition for entangled states. Simon also showed that the condition (12.36) becomes necessary and sufficient for separability of all bipartite Gaussian states.

12.4.3   Duan–Giedke–Cirac–Zoller criterion

In quantum mechanics, physical observables of a system are represented by a set of Hermitian operators $\{\hat{A}_i\}$. According to the Heisenberg uncertainty principle, it is never possible to simultaneously predict measurement results in arbitrary precision for all observables. The uncertainty of a Hermitian operator in a given quantum state is described by the variance $\langle(\Delta\hat{A}_i)^2\rangle = \langle\hat{A}_i^2\rangle - \langle\hat{A}_i\rangle^2$. Only when the quantum state is one of eigenstates of $\hat{A}_i$, the variance $\langle(\Delta\hat{A}_i)^2\rangle$ can be zero. For any two operators $\hat{A}_i$ and $\hat{A}_j$ which satisfy the commutator $[\hat{A}_i, \hat{A}_j] = i\hat{C}_{ij}$, the Heisenberg uncertainty principle sets an inequality to the variances of $\hat{A}_i$ and $\hat{A}_j$

$$\langle(\Delta\hat{A}_i)^2\rangle\langle(\Delta\hat{A}_j)^2\rangle \geq \frac{1}{4}|\langle\hat{C}_{ij}\rangle|^2. \tag{12.37}$$

It follows that

$$\langle(\Delta\hat{A}_i)^2\rangle + \langle(\Delta\hat{A}_j)^2\rangle \geq |\langle\hat{C}_{ij}\rangle|. \tag{12.38}$$

As an example, for the coordinate and momentum operators $\hat{X}$ and $\hat{P}$, we have

$$\langle(\Delta\hat{X})^2\rangle + \langle(\Delta\hat{P})^2\rangle \geq 1. \tag{12.39}$$

The result can be generalized to the case of multi operators. Suppose that $\{\hat{A}_i\}$ are a set of Hermitian operators some of which are not commutable. Therefore, they have no common eigenstates. According to the Heisenberg uncertainty principle, there must be a nontrivial constant limit $U_A > 0$ for the sum of the uncertainties [64]

$$\sum_i \langle(\Delta\hat{A}_i)^2\rangle \geq U_A. \tag{12.40}$$

It should be pointed out that the limit U_A is to be determined and is defined as the global minimum of the uncertainty sum for all quantum states of the Hilbert space under consideration.

Let us consider two subsystems A and B which are characterized by two sets of operators $\{\hat{A}_i\}$ and $\{\hat{B}_i\}$, respectively. According to Eq. (12.40), we may have the sum of uncertainty relations

$$\sum_i \langle(\Delta\hat{A}_i)^2\rangle \geq U_A, \tag{12.41}$$

$$\sum_i \langle(\Delta\hat{B}_i)^2\rangle \geq U_B. \tag{12.42}$$

Consider the sum operators $\{\hat{A}_i + \hat{B}_i\}$. For a direct product state $\hat{\rho}^{(AB)} = \hat{\rho}^{(A)} \otimes \hat{\rho}^{(B)}$, it is obvious that $\langle[\Delta(\hat{A}_i + \hat{B}_i)]^2\rangle = \langle(\Delta\hat{A}_i)^2\rangle + \langle(\Delta\hat{B}_i)^2\rangle$. Therefore, the sum of the variances of the operators $\{\hat{A}_i + \hat{B}_i\}$ is given by

$$\sum_i \langle [\Delta(\hat{A}_i + \hat{B}_i)]^2 \rangle \geq U_A + U_B. \tag{12.43}$$

For any separable state with the form (12.21), one has

$$\sum_i \langle [\Delta(\hat{A}_i + \hat{B}_i)]^2 \rangle = \sum_j p_j \sum_i tr\{\rho_j^{(A)} \otimes \rho_j^{(B)} [\Delta(\hat{A}_i + \hat{B}_i)]^2\}$$

$$= \sum_j p_j \sum_i [\langle (\Delta\hat{A}_i)^2 \rangle_j + \langle (\Delta\hat{B}_i)^2 \rangle_j]$$

$$\geq \sum_j p_j (U_A + U_B) = U_A + U_B. \tag{12.44}$$

In any separable states of a bipartite system, the sum of the variances of the operators $\{\hat{A}_i + \hat{B}_i\}$ must satisfy inequality (12.44). Any violation of this uncertainty relation shows that the quantum state must be an entangled state. The variances involved in inequality (12.44) can be determined by local measurements of the operators $\hat{A}_i$ and $\hat{B}_i$. The local uncertainty relation (12.44) provides a sufficient condition for the existence of entanglement [64].

As an example of applications of inequality (12.44), we consider a two-mode electromagnetic field. We define the phase-quadrature operators as

$$\hat{x}_j = \frac{1}{\sqrt{2}}(\hat{a}_j + \hat{a}_j^{\dagger}), \quad \hat{p}_j = \frac{1}{\sqrt{2}i}(\hat{a}_j - \hat{a}_j^{\dagger}), \tag{12.45}$$

where $\hat{a}_j$ and $\hat{a}_j^{\dagger}$ are the annihilation and creation operators of the j-th mode (j=1,2). The phase-quadrature operators satisfy the commutators $[\hat{x}_j, \hat{p}_{j'}] = i\delta_{jj'}$. According to (12.39), the sum of the variances of the operators $|a|\hat{x}_1$ and $|b|\hat{p}_1$ for any quantum state obeys the inequality

$$\langle (\Delta|a|\hat{x}_1)^2 \rangle + \langle (\Delta|b|\hat{p}_1)^2 \rangle \geq |ab|. \tag{12.46}$$

In the same way, for the operators $\hat{x}_2/c$ and $\hat{p}_2/d$, we have

$$\langle (\Delta\frac{1}{c}\hat{x}_2)^2 \rangle + \langle (\Delta\frac{1}{d}\hat{p}_2)^2 \rangle \geq \frac{1}{|cd|}. \tag{12.47}$$

In Eqs. (12.46) and (12.47), a, b, c, d are arbitrary (nonzero) real numbers. We note that Eqs. (12.46) and (12.47) have the same forms as Eqs. (12.41) and (12.42). We now consider the sum operators

$$\hat{u} = |a|\hat{x}_1 - \frac{1}{c}\hat{x}_2, \tag{12.48}$$

$$\hat{v} = |b|\hat{p}_1 - \frac{1}{d}\hat{p}_2. \tag{12.49}$$

From Eq. (12.44), we obtain the following sufficient condition for entanglement

$$\langle(\Delta\hat{u})^2\rangle + \langle(\Delta\hat{v})^2\rangle < |ab| + \frac{1}{|cd|}. \tag{12.50}$$

Duan et al. [39] derived the criterion (12.50) by using the commutators of the operators $\hat{x}_j$ and $\hat{p}_j$ and the Cauchy–Schwarz inequality. Compared to their method, the present derivation is much simpler. The inequality (12.50) contains four arbitrary real numbers. With arbitrary choices of these numbers, the inequality (12.50) is a sufficient condition for entanglement of the two-mode continuous-variable quantum systems. Duan et al. [39] showed that the criterion (12.50) becomes necessary and sufficient for two-mode Gaussian states if the numbers are specifically chosen.

The Wigner characteristic function of a two-mode Gaussian state takes the form [65]

$$\chi^{(w)}(\lambda_1,\lambda_2) = \exp\left[-\frac{1}{2}(\lambda_1^I,\lambda_1^R,\lambda_2^I,\lambda_2^R)M(\lambda_1^I,\lambda_1^R,\lambda_2^I,\lambda_2^R)^T\right], \tag{12.51}$$

where λ_j^R and λ_j^I are real variables and M is the 4×4 real symmetric correlation matrix. Duan et al. [39] showed that by use of local linear unitary Bogoliubov operators (LLUBO) the correlation matrix M can be transformed into the standard form

$$M_s = \begin{pmatrix} n_1 & c_1 & & \\ & n_2 & & c_2 \\ c_1 & & m_1 & \\ & c_2 & & m_2 \end{pmatrix}, \tag{12.52}$$

where the n_i, m_i, and c_i satisfy

$$\frac{n_1-1}{m_1-1} = \frac{n_2-1}{m_2-1}, \tag{12.53}$$

$$|c_1| - |c_2| = \sqrt{(n_1-1)(m_1-1)} - \sqrt{(n_2-1)(m_2-1)}. \tag{12.54}$$

They further showed that the inequality (12.50) becomes necessary and sufficient for entanglement of a two-mode Gaussian state with the standard form (12.52) of the correlation matrix M if the four numbers in (12.50) are chosen as $a = b = a_0, c = c_1/|c_1|a_0$ and $d = c_2/|c_2|a_0$, where $a_0^2 = \sqrt{(m_1-1)/(n_1-1)} = \sqrt{(m_2-1)/(n_2-1)}$.

12.4.4 Hillery–Zubairy criterion

Consider a two-mode continuous-variable quantum system. Let a and $a^\dagger$ be the annihilation and creation operators of the first mode, and b and $b^\dagger$ are the annihilation and creation operators of the second. The set of the quadratic forms: $ab^\dagger, a^\dagger b, a^\dagger a, b^\dagger b$ constitutes a $u(2)$ Lie algebra. The set of the linear combinations of these quadratic operators: $L_1 = ab^\dagger + a^\dagger b, L_2 = i(ab^\dagger - a^\dagger b), L_3 = a^\dagger a + b^\dagger b$, forms a $su(2)$ Lie algebra. Based on the Schwarz inequality, Hillery and Zubairy [33] showed that the inequality

$$\langle (\Delta L_1)^2 \rangle + \langle (\Delta L_2)^2 \rangle < 2\langle L_3 \rangle \qquad (12.55)$$

holds for entangled states of the system. This criterion can be used to witness entanglement of the mixed state

$$\rho = s|\psi_{01}\rangle\langle\psi_{01}| + \frac{1-s}{4}P_{01}, \qquad (12.56)$$

where $|\psi_{01}\rangle = (|0\rangle_a|1\rangle_b + |1\rangle_a|0\rangle_b)/\sqrt{2}$, $0 \le s \le 1$, and P_{01} is the projection operator onto the space spanned by the vectors $\{|0\rangle_a|0\rangle_b, |0\rangle_a|1\rangle_b, |1\rangle_a|0\rangle_b, |1\rangle_a|1\rangle_b\}$. Direct calculation finds that in the state (12.56) $\langle (\Delta L_1)^2 \rangle + \langle (\Delta L_2)^2 \rangle = 3 - s - s^2$ and $\langle L_3 \rangle = 1$. The inequality (12.55) shows that the state is entangled if $1 \ge s > (\sqrt{5} - 1)/2$.

For any state of the system, we can show that

$$\langle (\Delta L_1)^2 \rangle + \langle (\Delta L_2)^2 \rangle = 2(\langle (N_a + 1)N_b \rangle + \langle (N_b + 1)N_a \rangle - 2|\langle ab^\dagger|^2), \quad (12.57)$$

where $N_a = a^\dagger a$ and $N_b = b^\dagger b$. On replacing the right-hand-side of (12.55) by the right-hand-side of (12.57), it can be seen that the inequality (12.55) is equivalent to

$$|\langle ab^\dagger|^2 > \langle N_a N_b \rangle. \qquad (12.58)$$

This condition motivates a family of inequalities similar to (12.58) for detecting entanglement by considering the higher-order operators $a^m(b^\dagger)^n$ without the restriction $n = m = 1$. In fact, Hillery and Zubairy [33] showed that a state is entangled if

$$|\langle a^m(b^\dagger)^n \rangle|^2 > \langle (a^\dagger)^m a^m (b^\dagger)^n b^n \rangle, \qquad (12.59)$$

where n, m are any positive integer numbers.

Instead of considering the quadratic forms : $ab^\dagger, a^\dagger b, a^\dagger a, b^\dagger b$, we may consider another kind of the quadratic forms: $ab, b^\dagger a^\dagger, a^\dagger a, b^\dagger b$, which constitutes a $u(1,1)$ Lie algebra. Similar to the inequality (12.59), Hillery and Zubairy [33] found that for any positive integers m and n, a state is entangled if

$$|\langle a^m b^n \rangle| > [\langle (a^\dagger)^m a^m \rangle \langle (b^\dagger)^n b^n \rangle]^{1/2}. \qquad (12.60)$$

As an example of applications of the inequality (12.60), we consider the two-mode squeezed vacuum state

$$|\psi\rangle = (1-x^2)^{1/2} \sum_{n=0}^{\infty} x^n |n\rangle_a |n\rangle_b, \qquad (12.61)$$

where $0 \le x \le 1$. For this state, it is easily shown that $[\langle N_a N_b \rangle]^{1/2} = x^2/(1-x^2)$ and $|\langle ab \rangle| = x/(1-x^2)$. It is clear that the inequality (12.60) with $n = m = 1$ holds, and hence this state is entangled as long as $x \ne 0$.

12.4.5 Shchukin–Vogel criterion

As mentioned earlier, the Simon criterion involves only second-order moments of position and momentum operators. Shchukin and Vogel [34] found necessary and sufficient conditions for the partial transpose of bipartite harmonic quantum states to be nonnegative. The conditions can be formulated as an infinite series of inequalities for the various order moments of annihilation and creation operators. The violation of any inequality of this series is a sufficient condition for entanglement.

Let $\hat{A}$ be a Hermitian and nonnegative operator. For all states $|\psi\rangle$, we have the inequality

$$\langle \psi | \hat{A} | \psi \rangle = tr(\hat{A} | \psi \rangle \langle \psi |) \ge 0. \qquad (12.62)$$

Any pure bipartite state $|\psi\rangle$ can be expressed as

$$|\psi\rangle = \hat{g}^\dagger |00\rangle, \qquad (12.63)$$

with an appropriate operator function $\hat{g} = \hat{g}(\hat{a}, \hat{b})$, where $\hat{a}$ and $\hat{b}$ are the annihilation operators of the first and the second mode. In this way, the nonnegative operator $|\psi\rangle\langle\psi|$ can be represented in the form $\hat{f}^\dagger \hat{f}$, where

$$\hat{f} = |00\rangle\langle 00|\hat{g}. \qquad (12.64)$$

Since

$$\langle n| : e^{-\hat{a}^\dagger \hat{a}} : |n\rangle = \begin{cases} 0, \text{ if } n \ne 0 \\ 1, \text{ if } n = 0, \end{cases} \qquad (12.65)$$

with $: \cdots :$ denoting the normal ordering of the annihilation and creation operators, the two-mode vacuum density operator may be expressed in the normally ordered form

$$|00\rangle\langle 00| \equiv\; : \exp(-\hat{a}^\dagger \hat{a} - \hat{b}^\dagger \hat{b}) : . \qquad (12.66)$$

Upon substituting from Eq. (12.66) into Eq. (12.64), we see that the normal ordered form of the operator $\hat{f}$ always exists. Hence, we conclude that a Hermitian operator is nonnegative if and only if, for any operator $\hat{f}$ whose normally ordered form exists, the inequality

$$tr(\hat{A}\hat{f}^{\dagger}\hat{f}) \geq 0 \qquad (12.67)$$

is satisfied.

For any separable state $\hat{\rho}$, the Peres–Horodecki condition states that the partial transpose $\hat{\rho}^{PT}$ of $\hat{\rho}$ must be positive. According to (12.67), this statement can be formulated as the inequality

$$tr(\hat{\rho}^{PT}\hat{f}^{\dagger}\hat{f}) \geq 0 \qquad (12.68)$$

for any operator $\hat{f}$ whose normally ordered form exists. If the operator $\hat{f}$ is expanded as

$$\hat{f} = \sum_{n,m,k,l=0}^{+\infty} c_{nmkl}\hat{a}^{\dagger n}\hat{a}^{m}\hat{b}^{\dagger k}\hat{b}^{l}, \qquad (12.69)$$

the condition (12.68) becomes

$$\sum_{n,m,k,l,p,q,r,s=0}^{+\infty} c_{pqrs}^{*}c_{nmkl}M_{pqrs,nmkl} \geq 0, \qquad (12.70)$$

where

$$M_{pqrs,nmkl} = tr(\hat{\rho}^{PT}\hat{a}^{\dagger q}\hat{a}^{p}\hat{a}^{\dagger n}\hat{a}^{m}\hat{b}^{\dagger s}\hat{b}^{r}\hat{b}^{\dagger k}\hat{b}^{l}) = tr(\hat{\rho}\hat{a}^{\dagger q}\hat{a}^{p}\hat{a}^{\dagger n}\hat{a}^{m}\hat{b}^{\dagger l}\hat{b}^{k}\hat{b}^{\dagger r}\hat{b}^{s}).$$
$$(12.71)$$

The left-hand side of the inequality (12.70) is a quadratic form with respect to the coefficients c_{nmkl} of the expansion (12.69). The inequality (12.70) holds for all c_{nmkl} if and only if all the main minors of the form (12.70) are nonnegative, i.e., for all

$$D_N = \begin{vmatrix} M_{11} & M_{12} & \cdots & M_{1N} \\ M_{21} & M_{22} & \cdots & M_{2N} \\ \cdots & \cdots & \cdots & \cdots \\ M_{N1} & M_{N2} & \cdots & M_{NN} \end{vmatrix} \geq 0. \qquad (12.72)$$

In order to explicitly write out the condition (12.72), we have to relate multi-indices (n,m,k,l) and (p,q,r,s) of the moments defined in (12.71) to row index i and column index j of the determinant in (12.72). In general, one can at will prescribe the relation between multi-indices (n,m,k,l) and single row or column index. Once the relation is defined well, we have an infinite series of the inequalities (12.72) with $N = 1, 2, 3, \cdots$. According to the Peres–Horodecki criterion, we obtain a sufficient condition for inseparability or entanglement: A bipartite quantum state is entangled if there exits a negative determinant such that there exists an N, for which

$$D_N < 0. \tag{12.73}$$

Let us illustrate the condition (12.73) for the example of an entangled quantum composed of two coherent states

$$|\psi_-\rangle = N_-(\alpha, \beta)(|\alpha, \beta\rangle - |-\alpha, -\beta\rangle), \tag{12.74}$$

where the normalization is

$$N_-(\alpha, \beta) = [2(1 - e^{-2(|\alpha|^2 + |\beta|^2)})]^{-1/2}. \tag{12.75}$$

It is easily shown that the Simon criterion (12.36) fails to demonstrate the entanglement of this state. Properly relating row index and column index of D_N to multi-indices (n, m, k, l) of the moments (12.71), from (12.72), one can establish the sub-determinant

$$D_3 = \begin{vmatrix} 1 & \langle \hat{b}^\dagger \rangle & \langle \hat{a}\hat{b}^\dagger \rangle \\ \langle \hat{b} \rangle & \langle \hat{b}^\dagger \hat{b} \rangle & \langle \hat{a}\hat{b}^\dagger \hat{b} \rangle \\ \langle \hat{a}^\dagger \hat{b} \rangle & \langle \hat{a}\hat{b}^\dagger \hat{b} \rangle & \langle \hat{a}^\dagger \hat{a}\hat{b}^\dagger \hat{b} \rangle \end{vmatrix}. \tag{12.76}$$

Explicitly, it reads as

$$D_3 = -|\alpha|^2 |\beta|^4 \frac{\coth(|\alpha|^2 + |\beta|^2)}{\sinh(|\alpha|^2 + |\beta|^2)}, \tag{12.77}$$

which is clearly negative for all nonzero coherent amplitudes. Hence, according to the criterion (12.73), the state (12.74) is always entangled if both coherent amplitudes are nonzero.

12.5 Coherence-induced entanglement

As mentioned in Subsection 12.4.1, in order to successfully detect entanglement of a state $\hat{\rho}$ according to the positivity of the resulting state $\hat{A} \otimes \hat{B}(\hat{\rho})$, one has to properly choose the local positive operations $\hat{A} \otimes \hat{B}$. If choosing $\hat{A} = \hat{I}_1$ and $\hat{B} = \hat{T}_2$, one has to diagonalize the resulting partial transposed matrix for checking the positivity of eigenvalues. If dimension of the Hilbert space under consideration is high, this method becomes difficult. In this section, we first discuss a modification version of the Peres–Horodecki criterion and then apply the modified criterion to atomic coherence induced entanglement between two thermal fields [66].

Let $\hat{P}$ and $\hat{Q}$ represent projection operators onto subspaces of the Hilbert spaces $\mathscr{H}_1$ and $\mathscr{H}_2$, respectively. Instead of taking $\hat{A} = \hat{I}_1$ and $\hat{B} = \hat{T}_2$ as Peres and Horodecki did, we choose the local positive operators $\hat{A} = \hat{P}$ and $\hat{B} = \hat{T}_2 \hat{Q}$. From (12.22), we see that the resulting matrix through the projection and partial transposition operations must still be positive if the original state is separable. Therefore, we can definitely claim that the original state must be entangled as long as the projected and partial transposed matrix is negative. Here, by use of the Peres–Horodecki criterion, we detect entanglement of the original state in the subspace which is projected by operators $\hat{P} \otimes \hat{Q}$ out of the whole Hilbert space. The projection operation may make the criterion weaker. However, this method becomes useful when we have to deal with entanglement of states in the high dimensional Hilbert space. Bose et al. [67] used this method in studying subsystem-purity-induced atom-field entanglement in the Jaynes–Cummings model. Here, we detect entanglement between two thermal fields which interact with a single three-level atom of V-configuration. We are originally involved in an infinite dimensional Hilbert space.

The model under consideration is shown in Fig. 12.1. The transitions be-

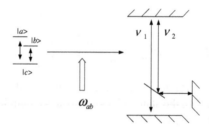

FIGURE 12.1: Atomic level scheme, atomic coherence preparation via a classical field, and doubly resonant cavity.

tween the upper levels $|a\rangle$ and $|b\rangle$ to the ground state $|c\rangle$ are dipole allowed and these transitions are coupled resonantly with two modes of the electromagnetic field inside a cavity at temperature T. The transition between the upper levels $|a\rangle$ and $|b\rangle$ is dipole forbidden. The interaction picture Hamiltonian of the system is given by

$$\hat{H} = \hbar g_1(|a\rangle\langle c|\hat{a}_1 + |c\rangle\langle a|\hat{a}_1^\dagger) + \hbar g_2(|b\rangle\langle c|\hat{a}_2 + |c\rangle\langle b|\hat{a}_2^\dagger), \qquad (12.78)$$

where $\hat{a}_1(\hat{a}_1^\dagger)$ and $\hat{a}_2(\hat{a}_2^\dagger)$ are the annihilation (creation) operators for the two cavity modes and $g_{1,2}$ are coupling constants of the atom with the fields.

We consider the initial states of the cavity fields to be diagonal in the Fock-state representation and the atom to be prepared in a coherent superposition of the upper levels by a classical field of frequency ω_{ab} as shown in Fig. 12.1. The initial state of the atom-field system is written as

$$\hat{\rho}_{af}(0) = \sum_{n_1=0}^{\infty} P_{n_1}|n_1\rangle\langle n_1| \otimes \sum_{n_2=0}^{\infty} P_{n_2}|n_2\rangle\langle n_2| \qquad (12.79)$$

$$\otimes (\rho_{aa}|a\rangle\langle a| + \rho_{bb}|b\rangle\langle b| + \rho_{cc}|c\rangle\langle c| + \rho_{ab}|a\rangle\langle b| + \rho_{ba}|b\rangle\langle a|),$$

where $P_{n_{1,2}}$ are the probabilities for having photon number states $|n_{1,2}\rangle$. An example of fields with vanishing off-diagonal matrix elements in the Fock-state representation is a thermal state, which has

$$P_{n_{1,2}} = \frac{\langle n_{1,2}\rangle^{n_{1,2}}}{(1 + \langle n_{1,2}\rangle)^{n_{1,2}+1}}. \qquad (12.80)$$

In (12.80), $\langle n_{1,2}\rangle = (e^{\hbar\nu_{1,2}\beta} - 1)^{-1}$ are the mean photon number of the fields at temperature T with $\nu_{1,2}$ being the field frequencies, and $\beta^{-1} = k_B T$ with k_B being the Boltzmann constant.

The density matrix operator at time t is given by $\hat{\rho}_{af}(t) = \hat{U}(t)\hat{\rho}_{af}(0)\hat{U}^\dagger(t)$ where $\hat{U}(t) = \exp(-i\hat{H}t/\hbar)$ is the time evolution operator. It follows, on taking a trace over the atomic variables, that the reduced density matrix operator for the fields is given by

$$\hat{\rho}_f(t) = \sum_{n_1=0}^{\infty}\sum_{n_2=0}^{\infty} \rho_{n_1,n_2;n_1,n_2}|n_1,n_2\rangle\langle n_1,n_2|$$

$$+ \rho_{ab}\sum_{n_1=0}^{\infty}\sum_{n_2=0}^{\infty} \rho_{n_1+1,n_2;n_1,n_2+1}|n_1+1,n_2\rangle\langle n_1,n_2+1|$$

$$+ \rho_{ba}\sum_{n_1=0}^{\infty}\sum_{n_2=0}^{\infty} \rho_{n_1,n_2+1;n_1+1,n_2}|n_1,n_2+1\rangle\langle n_1+1,n_2|, \qquad (12.81)$$

where the matrix elements are given by

$$\rho_{n_1,n_2;n_1,n_2} = P_{n_1}P_{n_2}\{\rho_{aa}[1 - g_1^2(n_1+1)A_{n_1+1,n_2}(1 - C_{n_1+1,n_2})]^2$$

$$+ \rho_{bb}[1 - g_2^2(n_2+1)A_{n_1,n_2+1}(1-C_{n_1,n_2+1})]^2\}$$
$$+ g_1^2 g_2^2 \{\rho_{aa} P_{n_1-1} P_{n_2+1} n_1(n_2+1)A_{n_1,n_2+1}^2(1-C_{n_1,n_2+1})^2$$
$$+ \rho_{bb} P_{n_1+1} P_{n_2-1} n_2(n_1+1)A_{n_1+1,n_2}^2(1-C_{n_1+1,n_2})^2\}$$
$$+ \{\rho_{aa} P_{n_1-1} P_{n_2} g_1^2 n_1 + \rho_{bb} P_{n_1} P_{n_2-1} g_2^2 n_2\} A_{n_1,n_2} S_{n_1,n_2}^2$$
$$+ \rho_{cc} \{P_{n_1} P_{n_2} C_{n_1,n_2}^2 + P_{n_1+1} P_{n_2} g_1^2(n_1+1)A_{n_1+1,n_2} S_{n_1+1,n_2}^2$$
$$+ P_{n_1} P_{n_2+1} g_2^2(n_2+1)A_{n_1,n_2+2} S_{n_1,n_2+1}^2\}, \tag{12.82}$$

$$\rho_{n_1+1,n_2;n_1,n_2+1} = -g_1 g_2 \sqrt{(n_1+1)(n_2+1)} \{A_{n_1+1,n_2+1}(1-C_{n_1+1,n_2+1})$$
$$\times (P_{n_1+1} P_{n_2}[1 - g_1^2(n_1+2)A_{n_1+2,n_2}(1-C_{n_1+2,n_2})]$$
$$+ P_{n_1} P_{n_2+1}[1 - g_2^2(n_2+2)A_{n_1,n_2+2}(1-C_{n_1,n_2+2})])$$
$$- P_{n_1} P_{n_2} \sqrt{A_{n_1+1,n_2} A_{n_1,n_2+1}} S_{n_1+1,n_2} S_{n_1,n_2+1}\}, \tag{12.83}$$

$$\rho_{n_1,n_2+1;n_1+1,n_2} = (\rho_{n_1+1,n_2;n_1,n_2+1})^*, \tag{12.84}$$

with $A_{n_1,n_2} = (g_1^2 n_1 + g_2^2 n_2)^{-1}$, $S_{n_1,n_2} = \sin(\sqrt{g_1^2 n_1 + g_2^2 n_2} t)$ and $C_{n_1,n_2} = \cos(\sqrt{g_1^2 n_1 + g_2^2 n_2} t)$.

The density matrix (12.81) is defined in an infinite dimensional Hilbert space. To estimate entanglement of (12.81), we consider the local operators $\hat{A}_{n_1} = |n_1\rangle\langle n_1| + |n_1+1\rangle\langle n_1+1|$ and $\hat{B}_{n_2} = \hat{T}_2(|n_2\rangle\langle n_2| + |n_2+1\rangle\langle n_2+1|)$ with $n_{1,2} = 0,2,4,\cdots$. The projection of (12.81) on the subspace spanned by basis vectors $(|n_1\rangle, |n_1+1\rangle) \otimes (|n_2\rangle, |n_2+1\rangle)$ with fixed photon numbers $n_{1,2}(= 0,2,4,...)$ leads to the state

$$[\hat{\sigma}_f(t)]_{n_1,n_2} = \hat{A}_{n_1}\hat{B}_{n_2}\hat{\rho}_f(t)\hat{B}_{n_2}^\dagger\hat{A}_{n_1}^\dagger. \tag{12.85}$$

In the subspace under consideration, the projected and partial transposed density matrix (12.85) becomes a 4×4 Hermitian matrix. Then, the Peres–Horodecki sufficient condition can be directly applied for the inseparability of matrix (12.85). It is easily found that the matrix (12.85) has a negative eigenvalue if the condition

$$|\rho_{ab}|^2 > R_{n_1,n_2} = \frac{\rho_{n_1,n_2;n_1,n_2}\rho_{n_1+1,n_2+1;n_1+1,n_2+1}}{|\rho_{n_1+1,n_2;n_1,n_2+1}|^2} \tag{12.86}$$

is satisfied. According to the Peres–Horodecki condition, one can claim that the state (12.81) is an entangled state if the condition (12.86) is satisfied.

In order to conveniently control the populations and atomic coherence at the same time, we consider the atom whose level populations initially are $\rho_{ii}(0)$ $(i = a,b,c)$ and off diagonal matrix elements $\rho_{ij} = 0$ for $i \neq j$. A coherence between the excited states a and b is created when the atom interacts resonantly with a classical field of frequency ω_{ab} for a time τ. After the interaction with the classical field, the populations and the atomic coherence are given by [68]

$$\rho_{aa} = \rho_{aa}(0)\cos^2(\Omega\tau) + \rho_{bb}(0)\sin^2(\Omega\tau), \qquad (12.87)$$

$$\rho_{bb} = \rho_{aa}(0)\sin^2(\Omega\tau) + \rho_{bb}(0)\cos^2(\Omega\tau), \qquad (12.88)$$

$$\rho_{ab} = (\rho_{ba})^* = ie^{i\theta}(\rho_{aa}(0) - \rho_{bb}(0))\sin(\Omega\tau)\cos(\Omega\tau), \qquad (12.89)$$

where Ω is the Rabi frequency and θ is the phase of the driving field. All the other density matrix elements remain unchanged. In this way, we can unitarily and continuously control the level populations and atomic coherence by use of the single parameter $\Omega\tau$. After passing through the classical field, the atom acquires a coherence. When this atom passes through the cavity with two thermal fields, the state of the fields is described by the density matrix (12.81). The entanglement of the resulting states of the field is determined by the condition (12.86).

In Fig. 12.2, the right side of (12.86) with $n_1 = n_2 = 0$ and the squared modulus of the atomic coherence (12.89) are shown as a function of $\Omega\tau$ when the atom and the fields are initially in thermal equilibrium. In the calculation,

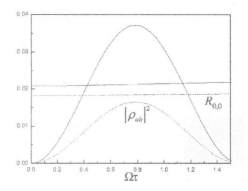

FIGURE 12.2: The solid lines are for the case with $\langle n_1 \rangle = 0.1$ and $\langle n_2 \rangle = 5.0$. The dashed lines are for the case with $\langle n_1 \rangle = 0.1$ and $\langle n_2 \rangle = 1.0$. In the two cases, $gt = 11.0$.

we take $g_1 = g_2 = g$. We also find that, as a function of n_1 and n_2, the right side of the inequality (12.86) takes the minimal value with $n_1 = n_2 = 0$. From Fig. 12.2, it can be noticed that the entanglement condition (12.86) can be satisfied if the difference between the mean thermal photon numbers of the two fields is sufficiently large. This situation may not be easily realizable because it requires a large frequency difference between the two upper levels.

Equation (12.89) shows that the atomic coherence is proportional to the population inversion of the upper levels. On the other hand, the numerator of the right side of the condition (12.86) decreases if the level populations ρ_{aa}, ρ_{bb} or ρ_{cc} are small. Therefore, the best initial condition of the atom for satisfying the condition (12.86) is that the atom is in one of the upper levels. For this case, Fig. 12.3 shows that the entanglement condition (12.86) can be satisfied even if the temperature becomes arbitrarily high.

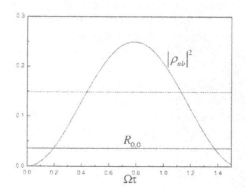

FIGURE 12.3: The solid lines are for the case with $\langle n_1 \rangle = \langle n_2 \rangle = 1.0$. The dashed lines are for the case with $T \to \infty$. In the two cases, $gt = 5.0$ and $\rho_{aa} = 1.0$.

As discussed earlier, the Hilbert space for the complete system is infinite dimensional, i.e., the dimension of the density matrix (12.81) is infinite. We can therefore obtain an infinite number of the projected 4×4 Hermitian matrices (12.85) with different photon numbers n_1 and n_2 through the projection and partial transpose operations. We then use the quantity [69]

$$\langle \mathscr{E} \rangle = -2 \sum_{n_1, n_2 = 0, 2, 4, \cdots}^{\infty} p_{n_1, n_2} \lambda_{n_1, n_2} \tag{12.90}$$

to measure the entanglement of (12.81), where λ_{n_1, n_2} is the negative eigenvalue of the density matrix (12.85) and $p_{n_1, n_2} = \rho_{n_1, n_2; n_1, n_2} + \rho_{n_1+1, n_2; n_1, n_2} +$

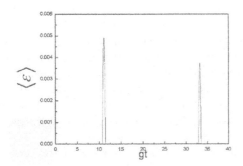

FIGURE 12.4: The time evolution of the entanglement measurement with $\langle n_1 \rangle = 0.1$ and $\langle n_2 \rangle = 5.0$, and $\Omega\tau = \pi/4$.

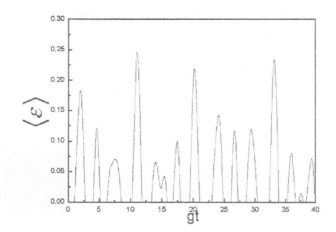

FIGURE 12.5: The time evolution of the entanglement measurement with $\langle n_1 \rangle = \langle n_2 \rangle = 1.0$. $\Omega\tau = \pi/4$, and $\rho_{aa} = 1.0$.

$P_{n_1,n_2+1;n_1,n_2} + P_{n_1+1,n_2+1;n_1+1,n_2+1}$ is the probability of taking the 4×4 matrix (12.85) out of the matrix (12.81). If $\langle \mathscr{E} \rangle = 0$, it does not mean nonentanglement. If $\langle \mathscr{E} \rangle \neq 0$, however, we can ensure that the infinite dimensional density matrix (12.81) must be an entangled state. In Fig. 12.4, the time evolution of the entanglement (12.90) is shown when the atom and the fields are initially in thermal equilibrium. It is seen that for this case the weak entanglement is detected at several time points. As pointed out earlier, the atomic coherence will become stronger when the atom is initially in one of the upper levels. Therefore, we may expect that in this case the stronger entanglement will be detected. Fig. 12.5 shows the time evolution of the entanglement (12.90) when the atom is initially in the level $|a\rangle$.

From this example, we see that the Peres and Horodecki criterion can also apply to the case of a high dimensional Hilbert space provided proper projections are performed prior to the partial transpose operation.

12.6   Correlated spontaneous emission laser as an entanglement amplifier

In the present section, as an example of applications of the criterion (12.50), we consider an entanglement amplifier based on the correlated spontaneous emission [70].

A system under consideration is composed of three-level atoms in a cascade configuration which interact with two modes of the field inside a doubly resonant cavity as shown in Fig. 12.6. The dipole allowed transitions $|a\rangle - |b\rangle$

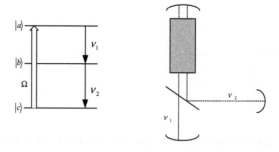

FIGURE 12.6: Atomic level scheme and doubly resonant cavity.

and $|b\rangle - |c\rangle$ are resonantly coupled with the two non-degenerate modes v_1 and

v_2 of the cavity, while the dipole forbidden transition $|a\rangle - |c\rangle$ is driven by a semiclassical field. The interaction Hamiltonian of this system in the rotating-wave-approximation is

$$H_I = \hbar g_1 (a_1|a\rangle\langle b| + a_1^\dagger|b\rangle\langle a|) + \hbar g_2 (a_2|b\rangle\langle c| + a_2^\dagger|c\rangle\langle b|)$$
$$- \frac{1}{2}\hbar\Omega(e^{-i\phi}|a\rangle\langle c| + e^{i\phi}|c\rangle\langle a|), \tag{12.91}$$

where $a_1(a_1^\dagger)$ and $a_2(a_2^\dagger)$ are the annihilation (creation) operators of the two nondegenerate modes of the cavities, g_1 and g_2 are the associated vacuum Rabi frequencies and Ω is the Rabi frequency of the classical driving field.

With the assumption that the atoms are injected in the cavity in the lower level $|c\rangle$ at a rate r_a, the equation of motion for the reduced density matrix of the cavity field modes is [72, 71].

$$\dot\rho = -[\beta_{11}^* a_1 a_1^\dagger \rho + \beta_{11}\rho a_1 a_1^\dagger - (\beta_{11} + \beta_{11}^*)a_1^\dagger \rho a_1 + \beta_{22}^* a_2^\dagger a_2 \rho + \beta_{22}\rho a_2^\dagger a_2$$
$$-(\beta_{22} + \beta_{22}^*)a_2\rho a_2^\dagger]$$
$$- [\beta_{12}^* a_1 a_2\rho + \beta_{21}\rho a_1 a_2 - (\beta_{12}^* + \beta_{21})a_2\rho a_1]e^{i\phi}$$
$$- [\beta_{21}^* a_1^\dagger a_2^\dagger\rho + \beta_{12}\rho a_1^\dagger a_2^\dagger - (\beta_{12} + \beta_{21}^*)a_1^\dagger\rho a_2^\dagger]e^{-i\phi}$$
$$- \kappa_1(a_1^\dagger a_1\rho - 2a_1\rho a_1^\dagger + \rho a_1^\dagger a_1) - \kappa_2(a_2^\dagger a_2\rho - 2a_2\rho a_2^\dagger + \rho a_2^\dagger a_2), \tag{12.92}$$

where the cavity damping terms are included in the usual way and κ_1 and κ_2 are decay rates of mode 1 and mode 2, respectively. The coefficients $\beta_{11}, \beta_{22}, \beta_{12}$ and β_{21} are given by

$$\beta_{11} = \frac{g_1^2 r_a}{4} \frac{3\Omega^2}{(\gamma^2 + \Omega^2)(\gamma^2 + \frac{\Omega^2}{4})}, \tag{12.93}$$

$$\beta_{22} = g_2^2 r_a \frac{1}{\gamma^2 + \Omega^2}, \tag{12.94}$$

$$\beta_{12} = g_1 g_2 r_a \frac{i\Omega}{\gamma(\gamma^2 + \Omega^2)}, \tag{12.95}$$

$$\beta_{21} = \frac{g_1 g_2 r_a}{4} \frac{i\Omega(\Omega^2 - 2\gamma^2)}{\gamma(\gamma^2 + \Omega^2)(\gamma^2 + \frac{\Omega^2}{4})}, \tag{12.96}$$

where γ is the atomic decay rate for all the three atomic levels.

In the limit when $\Omega \gg \gamma$, from (12.93)–(12.96), we have that

$$\beta_{11} \sim 0, \beta_{22} \sim 0, \beta_{12} \approx \beta_{21} \sim i g_1 g_2 r_a \frac{1}{\gamma\Omega}. \tag{12.97}$$

Under these conditions, Eq. (12.92) is simplified considerably as

$$
\begin{aligned}
\dot{\rho} = &-i\alpha(\rho a_1 a_2 - a_2\rho a_1)e^{i\phi} - i\alpha(\rho a_1^\dagger a_2^\dagger - a_1^\dagger \rho a_2^\dagger)e^{-i\phi} \\
&+i\alpha(a_1 a_2\rho - a_2\rho a_1)e^{i\phi} + i\alpha(a_1^\dagger a_2^\dagger\rho - a_1^\dagger\rho a_2^\dagger)e^{-i\phi} \\
&-\kappa_1(a_1^\dagger a_1\rho - 2a_1\rho a_1^\dagger + \rho a_1^\dagger a_1) \\
&-\kappa_2(a_2^\dagger a_2\rho - 2a_2\rho a_2^\dagger + \rho a_2^\dagger a_2),
\end{aligned}
\tag{12.98}
$$

with $i\alpha = \beta_{12} = \beta_{21}$. This equation describes a parametric oscillator in the parametric approximation. Based on (12.97), we can calculate the time evolution of the quantum fluctuations of the operators (12.48) and (12.49) with the choice $a = b = 1$ and $d = -c = 1$. The resulting expressions are

$$
\left[(\Delta\hat{u})^2 + (\Delta\hat{v})^2\right](t) = \left\{[(\Delta\hat{u})^2 + (\Delta\hat{v})^2](0) - \frac{2\kappa}{\alpha+\kappa}\right\}e^{-2(\alpha+\kappa)t} + \frac{2\kappa}{\alpha+\kappa}.
\tag{12.99}
$$

When deriving out (12.99), we have taken the phase of the driven field to be $\phi = -\pi/2$, since only under this special phase, the positive exponential terms in $(\Delta\hat{u})^2 + (\Delta\hat{v})^2$ can be canceled out and ensure that this quantity does not grow with time.

It is clear that, for any initial state of the field, the quantity $(\Delta\hat{u})^2 + (\Delta\hat{v})^2$ decreases as time evolves. When $(\alpha + \kappa)t >> 1$, we have $(\Delta\hat{u})^2 + (\Delta\hat{v})^2 = 2\frac{\kappa}{\alpha+\kappa} < 2$, i.e., the entanglement criterion (12.50) is satisfied. Thus the system evolves into an entangled state.

For the general case, we have to numerically calculate the field moments required in the inequality (12.50) according to (12.92). In Figs. 12.7 and 12.8 we show the time development of $(\Delta\hat{u})^2 + (\Delta\hat{v})^2$ for different Ω/γ and fixed κ/g. In Fig. 12.8, we plot these quantities for an initial coherent state with 10^4 photons in each mode. The choice of the phase for the coherent amplitude is such that the condition $\alpha_1\alpha_2 = -|\alpha_1\alpha_2|$ is satisfied. The parameter values are such that they correspond to the micromaser experiments in Garching [73]. We find that the two states remain entangled for a long time. The parametric results are valid only for $gt < 10$. In Fig. 12.8, we plot $(\Delta\hat{u})^2 + (\Delta\hat{v})^2$ for initial vacuum states for the two modes. The time scale for the two modes to remain entangled increases as the Rabi frequency of the driving field is increased.

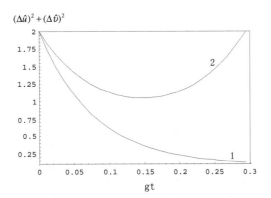

FIGURE 12.7: Time evolution of $(\Delta\hat{u})^2 + (\Delta\hat{v})^2$ for initial coherent states $|100, -100\rangle$ in terms of the normalized time gt. Various parameters are $r_a = 22kHz$, $g = g_1 = g_2 = 43kHz$, $\kappa = \kappa_1 = \kappa_2 = 3.85kHz$, $\gamma = 20kHz$, $\Omega = 400kHz$. The curves labeled by 1 and 2 represent the results for the parametric case and the general case, respectively.

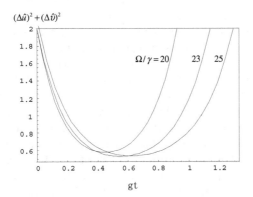

FIGURE 12.8: Time evolution of $(\Delta\hat{u})^2 + (\Delta\hat{v})^2$ for initial vacuum state of the two modes with $\Omega/\gamma = 20, 23, 25$; $r_a = g = \gamma$ and $\kappa/g = 0.001$.

12.6   Remarks

Since entanglement is one of the key properties of quantum systems that is fundamentally different from classical mechanics and it plays a key role in quantum information processing, a lot of experimental and theoretical effort has been devoted to investigating entanglement of subsystems in recent years. Some entanglement criteria and measures have been established. However, several important questions still remain open.

The existing entanglement criteria may be divided into three different classes. The first one is based on the locality concept of correlations between subsystems. Various Bell inequalities and inferred uncertainty relations resulting from the EPR correlations belong to this class. These criteria are often used in various optical experiments for testing the EPR correlations and creating entanglement between photons such as in frequency down-conversion. The second one is based on variances of variables and the definition of separable states which are statistically mixed products of subsystem states. Variances of all physical variables for any quantum states must obey the Heisenberg uncertainty relation. As a specific class of quantum states, however, separable states set a loose relation between either a product or a sum of variances of physical variables. Since these criteria are expressed in terms of variances of measurable quantities, they may be easy to use in experimental tests even if not all matrix elements of a density matrix under study are known. If testing operators are properly chosen, entanglement criteria for very weakly entangled states such as bound entangled states can also be established in this way. But the problem is how to choose the testing operators. Until now, we have had no general rule for this choice. Moreover, in this case, a necessary and sufficient condition for entanglement of Gaussian states has been established. Closely related to these criteria, one can also establish some criteria which may be more useful and stronger for states in the vicinity of extreme entangled states. The third one is based on the general properties of a density matrix such as the positivity of density matrix operators. The Peres–Horodecki criterion of the negativity of partial transpose matrices belongs to this kind. In many cases, this criterion is more powerful than others. This criterion also provides a necessary and sufficient condition for 2×2 and 2×3 systems. Compared to the other criteria, this criterion has a drawback that one has to know all the matrix elements of a density matrix when using it. On the other hand, implementing the criterion becomes difficult when dimension of states involved in investigation is high. We may use a projection method to reduce the dimension of matrices which are dealt with, but in this way the resulting criterion may become weak. Although the above criteria of entanglement have been established, more powerful and generic criteria of entanglement are still needed. The reason is that the above criteria work for certain states but fail for others. On the other hand, most of the entanglement criteria provide only a sufficient condition and a necessary condition for entanglement of generic mixed states is still lacking. For entanglement of multi-subsystems, effective criteria have not been established yet.

Another important point relates to the quantification of entanglement. According to general physical considerations, several entanglement measures such as entanglement of formation, the relative entropy and the negativity of partial transpose matrices have been established. For generic states of two qubits, or some specific states such as Werner states, some explicit expressions for entanglement measure have been found. Since all the entanglement measures involve the minimum problem over a whole set of either entangled or separable states, this makes it extremely difficult to work out these entanglement measures for given mixed states. The quantification of entanglement is still an open question.

References

[1] A. Einstein, B. Podolsky, and N. Rosen, Phys. Rev. **47**, 777 (1935).

[2] J. S. Bell, Physics **1**, 195(1964); Rev. Mod. Phys. **38**, 447 (1966).

[3] D. Bohm, *Quantum Theory* (Prentice-Hall, Englewood Cliffs, NJ, 1951).

[4] S. J. Freedman and J. Clauser, Phys. Rev. Lett. **28**, 938 (1972).

[5] E. S. Fry and R. C. Thompson, Phys. Rev. Lett. **37**, 465 (1976).

[6] A. Aspect, P. Grangier, and G. Roger, Phys. Rev. Lett. **49**, 91 (1982).

[7] W. Perrie, A. J. Duncan, H. J. Beyer, and H. Kleinpoppen, Phys. Rev. Lett. **54**, 1790 (1985).

[8] Y. H. Shih and C. O. Alley, Phys. Rev. Lett. **61**, 2921 (1988).

[9] Z. Y. Ou and L. Mandel, Phys. Rev. Lett. **61**, 50 (1988).

[10] J. G. Rarity and P. R. Tapster, Phys. Rev. Lett. **64**, 2495 (1990).

[11] P. G. Kwiat, P. H. Eberhard, A. M. Steinberg, and R. Y. Chiao, Phys. Rev A **52**, 4381 (1995).

[12] Y. Hasegawa, R. Loidl, G. Badurek, M. Baron, and H. Rauch, Nature **425**, 45 (2003).

[13] M. D. Reid, Phys. Rev. A **40**, 913 (1989).

[14] Z. Y. Ou, S. F. Pereira, H. J. Kimble, and K. C. Peng, Phys. Rev. Lett. **68**, 3663 (1992).

[15] Ch. Silberhorn, P. K. Lam, O. Weiß, F. König, N. Korolkova1, and G. Leuchs, Phys. Rev. Lett. **86**, 4267 (2001).

[16] W. P. Bowen, R. Schnabel, and P. K. Lam, Phys. Rev. Lett. **90**, 043601 (2003).

[17] J. C. Howell, R. S. Bennink, S. J. Bentley, and R. W. Boyd, Phys. Rev. Lett. **92**, 210403 (2004).

[18] E. Schrödinger, Naturwissenschaften **23**, 807(1935); D. Brus, J. Math. Phys. **43**, 4237 (2002).

[19] C. H. Bennett, G. Brassard, C. Crepeau, R. Jozsa, A. Peres, and W. K. Wootters, Phys. Rev. Lett. **70**, 1895 (1993).

[20] C. H. Bennett and S. J. Wiesner, Phys. Rev. Lett. **69**, 2881 (1992).

[21] A. K. Ekert, Phys. Rev. Lett. **67**, 6961 (1991).

[22] A. Barenco, D. Deutsch, A. Ekert, and R. Jozsa, Phys. Rev. Lett. **74**, 4083 (1995).

[23] E. Schmidt, Math. Annalen **63**, 433 (1906).

[24] C. H. Bennett, H. J. Bernstein, S. Popescu, and B. Schumacher, Phys. Rev. A **53**, 2046 (1996).

[25] A. Peres, Phys. Rev. Lett. **77**, 1413 (1996).

[26] M. Horodecki, P. Horodecki, and R. Horodecki, Phys. Lett. A **223**, 1 (1996).

[27] M. Horodecki, P. Horodecki, Phys. Rev. A **59**, 4206 (1996).

[28] M. Nielsen and J. Kempe, quant-ph/0011117.

[29] B. Terhal, Phys. Lett. A **271**, 319 (2000).

[30] R. Simon, Phys. Rev. Lett. **84**, 2726 (2000).

[31] L. M. Duan, G. Giedke, J. I. Cirac, and P. Zoller, Phys. Rev. Lett. **84**, 2722 (2000).

[32] G. Giedke, B. Kraus, M. Lewenstein, and J. I. Cirac, Phys. Rev. Lett. **87**, 167904 (2001).

[33] M. Hillery and M. S. Zubairy, Phys. Rev. Lett. **96**, 050503 (2006); Phys. Rev. A **74**, 032333 (2006).

[34] E. Shchukin and W. Vogel, Phys. Rev. Lett. **95**, 230502 (2005).

[35] E. Hagley, X. Maitre, G. Nogues, C. Wunderlich, M. Brune, J. M. Raimond, and S. Haroche , Phys. Rev. Lett. **79**, 1 (1997).

[36] Q. A. Turchette, C. S. Wood, B. E. King, C. J. Myatt, D. Leibfried W. M. Itano, C. Monroe, and D. J. Wineland , Phys. Rev. Lett. **81**, 3631 (1998).

[37] K. Molmer and A. Sorensen, Phys. Rev. Lett. **82**, 1835 (1999).

[38] C. A. Sackett, D. Kielpinski, B. E. King, C. Langer, V. Meyer, C. J. Myatt, M. Rowe, Q. A. Turchette, W. M. Itano, D. J. Wineland, and C. Monroe, Nature **404**, 256 (2000).

[39] L. M. Duan, J. I. Cirac, P. Zoller, and E. S. Polzik, Phys. Rev. Lett. **85**, 5643 (2000).

[40] B. Julsgaard, A. Kozhekln, and E. S. Polzlk, Nature **413**, 400 (2001).

[41] A. Rauschenbeutel, P. Bertet, S. Osnaghi, G. Nogues, M. Brune, J. M. Raimond, and S. Haroche, Phys. Rev. A **64**, 050301 (2001).

[42] S. J. Freedman and J. F. Clauser, Phys. Rev. Lett. **28**, 938 (1972).

[43] J. F. Clauser, Phys. Rev. Lett. **36**, 1223 (1975).

[44] E. S. Fry and R. C. Thompson, Phys. Rev. Lett. **37**, 465 (1976).

[45] Z. Y. Ou and L. Mandel, Phys. Rev. Lett. **61**, 50 (1988).

[46] P. G. Kwiat, K. Mattle, H. Weinfurter, and A. Zeilinger Phys. Rev. Lett. **75**, 4337 (1995).

[47] D. Bouwmeester, J.-W Pan, M. Daniell, H. Weinfurter, and A. Zeilinger Phys. Rev. Lett. **82**, 1345 (1998).

[48] J.-W. Pan, M. Daniell, S. Gasparoni, G. Weihs, and A. Zeilinger, Phys. Rev. Lett. **86**, 4435 (2001).

[49] Z. Zhao, Y.-A. Chen, A.-N. Zhang, T. Yang, H. J. Briegel, and J.-W. Pan, Nature **430**, 54 (2004).

[50] K. Tsujino, H. F. Hofmann, S. Takeuchi, and K. Sasaki, Phys. Rev. Lett. **92**, 153602 (2004).

[51] H. S. Eisenberg, G. Khoury, G. A. Durkin, C. Simon, and D. Bouwmeester, Phys. Rev. Lett. **93**, 193901 (2004).

[52] S. L. Braunstein and P. van Loock, quant-ph/0410100 and references therein.

[53] Y. Zhang, H. Wang, X. Y. Li, J. T. Jing, C. D. Xie, and K. C. Peng, Phys. Rev. A **62**, 023813 (2000).

[54] M. O. Scully and M. S. Zubairy, Quantum Optics (Cambridge University Press, 1997).

[55] D. M. Greenberger, M. A. Horne, A. Shimony, and A. Zeilinger, Am. J. Phys. **58**, 1131 (1990).

[56] W. Dur, G. Vidal, and J. I. Cirac, Phys. Rev. A **62**, 062314 (2000).

[57] A. Peres, Phys. Lett. A **202**, 16 (1995); A. K. Pati, quant-ph/9911073.

[58] A. V. Thapliyal, Phys. Rev. A **59**, 3336 (1999).

[59] R. F. Werner, Phys. Rev. A **40**, 4277 (1989).

[60] P. Horodecki, Phys. Lett. A**232**, 333 (1997).

[61] R. Horodecki, P. Horodecki, and M. Horodecki, Phys. Lett. A **210**, 377 (1996).

[62] H. S. Eisenberg, G. Khoury, G. A. Durkin, C. Simon, and D. Bouwmeester, Phys. Rev. Lett. **93**, 193901 (2004).

[63] R. Simon, E. C. G. Sudarshan, and N. Mukunda, Phys. Rev. A **36**, 3868 (1987).

[64] H. F. Hofmann and S. Takeuchi, Phys. Rev. A**68**, 032103 (2003).

[65] C. W. Gardiner and P. Zoller, *Quantum Noise* (Springer–Verlag, Berlin, 1999), 2nd ed.

[66] Fu-li Li, H. Xiong, and M. S. Zubairy, Phys. Rev. A**72**, 010303(R) (2005).

[67] S. Bose, I. Fuentes-Guridi, P. L. Knight, and V. Vedral, Phys. Rev. Lett. **87**, 050401 (2001).

[68] M. S. Zubairy,in *Quantum Limits to the Second Law*, edited by D. P. Sheehan, 2002, pp. 92-97.

[69] J. Lee and M. S. Kim, Phys. Rev. Lett. **84**, 4236 (2000).

[70] H. Xiong, M. O. Scully and M. S. Zubairy, Phys. Rev. Lett. **94**, 023902(2005).

[71] N. A. Ansari, J. Gea-Banacloche, and M. S. Zubairy, Phys. Rev. A **41**, 5179 (1990).

[72] C. A. Blockley and D. F. Walls, Phys. Rev. A **43**, 5049 (1991).

[73] D. Meschede, H. Walther, and G. Muller, Phys. Rev. Lett. **54**, 551 (1985); G. Raqithel, C. Wagner, H. Walther, L. M. Narducci, and M. O. Scully, in *Advances in Atomic, Molecular, and Optical Physics*, edited by P. Berman (Academic, New York 1994), Supp. 2, p. 57.

Chapter 13

Parametrizations of Positive Matrices With Applications

M. Tseng, H. Zhou, and V. Ramakrishna

Abstract   The purpose of this work is twofold. The first is to survey some parametrizations of positive matrices which have found applications in quantum information theory. The second is to provide some more applications of a parametrization of quantum states and channels introduced by T. Constantinescu and the last author, and thereby to provide further evidence of the utility of this parametrization. This work is dedicated to the memory of our colleague and teacher, the late Professor T. Constantinescu.

13.1   Introduction

Positive matrices play a vital role in quantum mechanics and its applications (in particular, quantum information processing). Indeed the two basic ingredients in the theory of quantum information, viz., quantum states and quantum channels involve positive matrices. See, for instance, [17, 21]. Thus, a study of parametrizations of positive matrices seems very much warranted. In particular, the very useful Bloch sphere picture, [17, 21], for the quantum state of a qubit has prompted several attempts at the extension of this picture to higher dimensions. In the process, several groups of researchers have looked into the question of finding tractable characterizations of positive matrices, which could lead to useful parametrizations of positive matrices, [3, 5, 15, 23, 27].

This paper is organized as follows. In the next section we set up basic notation and also point out some sources for positive matrices in quantum mechanics and its applications. The third section introduces six (perhaps well-known) characterizations of positive matrices, and reviews some putative parametrizations of states of qudits. In the next section, we review a parametrization proposed in [5], reiterating its utility. The final section offers two more applications of the parametrization in [5]. The first concerns Toeplitz states, i.e., density matrices which are also Toeplitz. The second investigates constraints imposed on relaxation rates of an open quantum system by the requirement of complete positivity.

13.2   Sources of positive matrices in quantum theory

Let us recall that a matrix is positive semidefinite (positive, for short) if $z^*Pz \geq 0$ for all $z \in C^n$. One can easily extend this definition to infinite positive matrices. In effect such a matrix is what is called a *positive kernel*, [6], viz., a map $K : N_0 \times N_0 \to C$, where N_0 is the set of non-negative integers, with the property that for each $n > 0$, and each choice $p_1, \ldots, p_n$ in N_0 and each choice $z_1, \ldots, z_n$ of elements of C we have

$$\sum_{i,j=1}^{n} K(p_i, p_j) \bar{z}_i z_j \geq 0.$$

Positive matrices intervene in at least two of the basic ingredients of quantum mechanics and quantum information theory, viz., quantum states and quantum channels. There are, of course, more sources for positive matrices, but, for reasons of brevity, we will confine ourselves to discussing states and channels.

The state of a d-dimensional quantum system is described by a $d \times d$ positive density matrix of trace 1, that is, a positive element of trace 1 in the algebra $\mathcal{M}_d$ of complex $d \times d$ matrices. States described by rank-one density matrices are called pure states.

A quantum channel is a completely positive map $\Phi : \mathcal{A} \to \mathcal{L}(\mathcal{H})$ from a C^*-algebra $\mathcal{A}$ into the set $\mathcal{L}(\mathcal{H})$ of all bounded linear operators on the Hilbert space $\mathcal{H}$ (in the situations most frequently met in quantum information processing, $\mathcal{A} = \mathcal{M}_d$ and $\mathcal{L}(\mathcal{H}) = \mathcal{M}_{d'}$). By the Stinespring theorem, (Theorem 4.1 of [19]), such a map is the compression of a $*$-homomorphism.

For $\mathscr{A} = \mathscr{M}_d$, there is a somewhat more explicit representation, given in [4] (see also [13]). Thus, $\Phi : \mathscr{M}_d \to \mathscr{L}(\mathscr{H})$ is completely positive if and only if the matrix

$$S = S_\Phi = \left[\Phi(E_{k,j}) \right]_{k,j=1}^d \tag{13.1}$$

is positive, where $E_{k,j}$, $k, j = 1, \ldots, d$, are the standard matrix units of $\mathscr{M}_d$. Each $E_{k,j}$ is a $d \times d$ matrix consisting of 1 in the $(k, j) - th$ entry and zeros elsewhere.

REMARK 13.1   Usually one requires a quantum channel to satisfy two additional requirements: i) Φ be trace preserving, and/or ii) Φ be unital (see below).   ∎

A Kraus operator representation of a completely positive map is a (non-unique) choice of operators V_i such that one can express the effect of Φ via

$$\Phi(\rho) = \sum_{i=1}^r V_i \rho V_i^*.$$

Usually only the non-zero V_i are taken into account in the above equation (though sometimes it is convenient to ignore this convention).

Then Φ is trace-preserving iff $\sum_{i=1}^r V_i^* V_i = \mathrm{Id}$, while Φ is *unital* iff $\sum_{i=1}^r V_i V_i^* = \mathrm{Id}$. These properties can also be verified (without any reference to Kraus representations) by computing the partial traces of S_Φ viewed as an unnormalized state (see [26]).

All choices of Kraus operator representations for Φ come from square roots of S_Φ, i.e., matrices T such that $S_\Phi = T T^*$. One then obtains the V_i from the ith column of T by reversing the *vec* operation, [10, 26]. Recall that the *vec* operator associates to a $d \times e$ matrix, V, a vector in C^{de} obtained by stacking the columns of V. It is precisely because of lack of uniqueness in the square roots of S_Φ that the Kraus operator representation of Φ is non-unique.

We should point out that some of the definitions for quantum channel notions used in [5], though equivalent to the standard ones (i.e., the ones used here), are different.

13.3   Characterizations of positive matrices

All positive matrices are Hermitian (unlike the real case, the definition of a positive matrix automatically forces Hermiticity). There are several characterizations of positive matrices as a subclass of Hermitian matrices. Some of these yield useful parametrizations of positive matrices.

The following theorem, which for the most part is standard textbook material (see, for instance, the classic [11]), reviews some of these characterizations.

THEOREM 13.1
Let P be a Hermitian matrix. Then the following are equivalent:

- **P1** *P is positive.*

- **P2** *All the eigenvalues of P are non-negative.*

- **P3** *There is an upper-triangular matrix T such that $P = T^*T$ (Cholesky decomposition)*

- **P4** *All principal minors of P are non-negative.*

- **P5** *Let $p(t) = t^n + \sum_{i=0}^{n-1}(-1)^i b_i t^{n-i}$ be the characteristic polynomial of P. Then $b_i \geq 0$, for all i.*

- **P6** *There is a Hermitian matrix H such that $P = H^2$.$\diamondsuit$*

REMARK 13.2

- **P3** is normally mentioned only for positive definite matrices in the bulk of the literature. However, a limiting argument shows that it is valid for positive semidefinite matrices as well.

- That **P5** is equivalent to **P2** is just a consequence of Descartes' rule of signs.

- **P4** should be folklore. Quite surprisingly, we were unable to find any source where **P4** is stated explicitly (even in a venerable text such as [11]). Since a similar statement for positive definite matrices (viz., positive definiteness is equivalent to the positivity of the *leading* principal minors) is well documented and we have seen this statement occasionally incorrectly applied to positive semidefinite matrices, we will include a brief proof here. Clearly if P is positive, all principal submatrices of P are positive, and hence all principal minors are non-negative. Conversely suppose all principal minors of P are non-negative. Since the coefficients b_i of the characteristic polynomial of any matrix are just the sum of all the $i \times i$ principal minors of P, it follows that $b_i \geq 0$. Hence P is positive. $\diamond$

$\blacksquare$

Whilst the above conditions are equivalent to positivity, they typically do not lead to useful parametrizations of positive matrices. For instance, parametrizing P by its eigenvalues only describes the $U(n)$ orbit to which P belongs. For the same reason one cannot parametrize P by the coefficients b_i of the characteristic polynomial $p(t)$.

However, one can turn these characterizations into *potential* parametrizations. To illustrate this consider the problem of parametrizing quantum states in dimension d, i.e., $d \times d$ positive matrices with unit trace. The standard starting point is to represent a state ρ via

$$\rho = \frac{1}{d}\left(I_d + \sum_{i=1}^{d^2-1} \beta_i \lambda_i\right). \tag{13.2}$$

Here $\beta_i \in R$ and the λ_i form an orthogonal basis for the space of traceless Hermitian matrices, specifically $Tr(\lambda_i \lambda_j) = 2\delta_{ij}$. One typical choice is the so-called generalized Gell–Mann matrices, [15, 27]. This basis is obtained from the matrices $E_{kj}, k, j = 1, \ldots, d$ ($E_{kj} = e_k e_j^*$) via the following construction:

$$f_{k,j}^d = E_{k,j} + E_{j,k}, \quad k < j,$$

$$f_{k,j}^d = \frac{1}{i}\left(E_{j,k} - E_{k,j}\right), \quad k > j,$$

$$h_1^d = I_d, \quad h_k^d = h_k^{d-1} \oplus 0, \quad 1 < k < d,$$

$$h_d^d = \sqrt{\frac{2}{d(d-1)}}\left(h_1^{d-1} \oplus (1-d)\right).$$

These matrices, $f_{k,j}^d, h_1^d, h_k^d, h_d^d$ together form one choice of the $\{\lambda_i, I_d\}$ basis for the space of $d \times d$ Hermitian matrices. When $d = 2$ this is precisely the Pauli matrix basis. When $d = 3$ one gets the usual Gell–Mann matrices.

With Equation (13.2) as the starting point one can restrict the vector $\beta = (\beta_1, \ldots, \beta_{d^2-1}) \in R^{d^2-1}$ to satisfy any of the characterizations **P1–P6**. In principle, this provides a bijection from a subset of R^{d^2-1}, say D_β, to the space of

$d \times d$ density matrices. This is precisely what is proposed simultaneously in [3, 15] for the characterization **P5**. However, now by conservation of difficulty, the burden of the analysis of quantum states in dimension $d > 2$ is shifted to obtaining a concrete analysis of the subset D_β. In particular, these do not lead to easily computed parametrization of quantum states (cf. the conclusions section of [15]). Interestingly enough each of these characterizations leads precisely to the Bloch sphere picture when $d = 2$, as we encourage the reader to verify. However, this approach has some utility in higher dimensions as well. For instance, depending on which characterization one uses, it is at least possible to be more precise about the set of pure states (i.e., rank-one states). We shall explain this via the characterization **P6** because pure states are precisely those states, ρ for which $\rho^2 = \rho$, and this fits in nicely with **P6**.

In order to state a precise result, let us introduce the tensor d_{kli} obtained from considering the Jordan structure of the λ_i. Specifically, if $\{\lambda_k, \lambda_l\}$ denotes the Jordan commutator (i.e., the anti-commutator) of λ_k, λ_l, then

$$\{\lambda_k, \lambda_l\} = \lambda_k\lambda_l + \lambda_l\lambda_k = \frac{4}{d}I_d\delta_{kl} + \sum_{i=1}^{d^2-1} d_{kli}\lambda_i.$$

We use the d_{kli} to introduce an operation amongst vectors $x, y \in R^{d^2-1}$, via

$$x \cup y = (\sum_{j,k=1}^{d^2-1} d_{1jk}x_jy_k, \sum_{j,k=1}^{d^2-1} d_{2jk}x_jy_k, \ldots, \sum_{j,k=1}^{n^2-1} d_{ijk}x_jy_k, \ldots).$$

$x \cup y$ is thus a vector in R^{d^2-1}. We can now state the following.

PROPOSITION 13.1

Every density matrix can be represented in the form in Eqn. (13.2) with $\beta = \frac{2\kappa}{d}\beta_0 + \frac{\beta_0 \cup \beta_0}{d}$, where β_0 is any vector in R^{d^2-1} with $\|\beta_0\|^2 \leq \frac{d^2}{2}$ and $\kappa = +\sqrt{\frac{d^2-2\|\beta_0\|^2}{d}}$. Conversely any Hermitian matrix admitting such a representation is necessarily a density matrix. A state ρ is pure precisely if it can be represented in the form in Eqn. 13.2) with $< \beta, \beta >= \frac{d^2-d}{2}$ and $(d-2)\beta = \beta \cup \beta$.

The proof is straightforward. Since $\rho = H^2$ and H itself can be expanded as a linear combination of I_d and the λ_i (albeit with the coefficient of I_d different from $\frac{1}{d}$), the first part of the result follows from the linear independence of $\{I_d, \lambda_i\}$. For the second part, we represent ρ as in Eqn. (13.2) and equate it to its square.

Once again, the difficulty is in the analysis of states which are not pure. It is worth mentioning that the pure state condition in the above proposition is essentially the same as that obtained from the characterization **P5** (for a pure state the characteristic polynomial is $p(t) = t^d + (-1)^d t^{d-1}$, i.e., $b_1 = 1, b_i = 0, i \geq 2$).

It should be pointed out that even an analysis of the pure state conditions is far from trivial. The condition $(d-2)\beta = \beta \cup \beta$ is vacuously true when $d = 2$ (since the Pauli matrices anti-commute). For $d \geq 3$, this condition imposes genuine restrictions. It is an interesting problem to find an orthogonal basis for the space of Hermitian matrices (the generalized Gell–Mann matrices form just one amongst many) which is close to "abelian", i.e., one for which many of the d_{kli} vanish, to facilitate the analysis of the condition $(d-2)\beta = \beta \cup \beta$.

In contrast, the parametrization discussed in the next section yields a very simple characterization of pure states.

13.4 A different parametrization of positive matrices

In this section we recall informally the main result of [5] on the parametrization of positive matrices. In order to do that a few preliminary definitions and notions are needed.

To any contraction T, one defines its defect operator via

$$D_T = (I - T^*T)^{1/2}.$$

Here M^* is the adjoint of an operator (when M is a scalar, this is merely complex conjugation).

To such a contraction one can also associate a certain unitary operator, called the *Julia operator* of T via

$$U(T) = \begin{bmatrix} T & D_{T^*} \\ D_T & -T^* \end{bmatrix}. \tag{13.3}$$

Thus, $U(T)$ is a unitary dilation of T.

If we are given a family of contractions $\Gamma_{kj}, j \geq k$ with $\Gamma_{kk} = 0$ for all k, then we associate it to a family of unitary operations via the Julia operator construction as follows. We first let $U_{k,k} = \text{Id}$, while for $j > k$ we set

$$U_{k,j} = U_{j-k}(\Gamma_{k,k+1})U_{j-k}(\Gamma_{k,k+2})\ldots U_{j-k}(\Gamma_{k,j})(U_{k+1,j} \oplus I_{\mathscr{D}_{\Gamma_{k,j}^*}}),$$

where

$$U_{j-k}(\Gamma_{k,k+l}) = I \oplus U(\Gamma_{k,k+l}) \oplus I.$$

To a family of contractions, $\Gamma_{k,j}$ one can associate a row contraction via

$$R_{k,j} = \left[\Gamma_{k,k+1}, D_{\Gamma^*_{k,k+1}} \Gamma_{k,k+2}, \ldots, D_{\Gamma^*_{k,k+1}} \cdots D_{\Gamma^*_{k,j-1}} \Gamma_{k,j} \right],$$

and a column contraction via

$$C_{k,j} = \left[\Gamma_{j-1,j}, \Gamma_{j-2,j} D_{\Gamma_{j-1,j}}, \ldots, \Gamma_{k,j} D_{\Gamma_{k+1,j}} \cdots D_{\Gamma_{j-1,j}} \right]^t,$$

where "t" stands for matrix transpose. For more details on the ranges and domains of these operators see [5].

Then the main theorem regarding positive matrices can be stated informally as follows (for a precise statement, especially concerning the ranges and domains of all operators involved, see [5]).

THEOREM 13.2

The matrix $S = \left[S_{k,j} \right]_{k,j=1}^{d}$ as above, satisfying $S^*_{jk} = S_{kj}$, is positive if and only if i) $S_{kk} \geq 0, k = 1,\ldots,d$ and ii) there exists a family $\{\Gamma_{k,j} \mid k,j = 1,\ldots,d, k \leq j\}$ of contractions such that $\Gamma_{k,k} = 0$ for $k = 1,\ldots,d$ valid, and

$$S_{k,j} = L^*_{k,k}(R_{k,j-1}U_{k+1,j-1}C_{k+1,j} + D_{\Gamma^*_{k,k+1}} \cdots D_{\Gamma^*_{k,j-1}} \Gamma_{k,j} D_{\Gamma_{k+1,j}}$$
$$\cdots D_{\Gamma_{j-1,j}})L_{j,j} \tag{13.4}$$

where $L_{k,k}$ is any square root of S_{kk} $\Diamond$.

DEFINITION 13.1   The contractions $\Gamma_{k,j}$, with $j > k$, will be called the Schur–Constantinescu parameters of S.

These parameters were first discovered for Toeplitz matrices by Schur, [24], albeit in the guise of a problem about power series which are bounded in the unit circle. In our humble opinion, it was our late colleague and teacher, T. Constantinescu, who championed the study of these parameters to cover all positive matrices (more generally to matrices with displacement structure, [8]) and most adroitly brought to fore many of their interesting features. Therefore, we have chosen to call these parameters, the Schur–Constantinescu parameters, in his honor.

We will illustrate Theorem (13.2) via the case of 3×3 positive matrices.

Thus, let $S = \begin{bmatrix} S_{11} & S_{12} & S_{13} \\ S_{12}^* & S_{22} & S_{23} \\ S_{13}^* & S_{23}^* & S_{33} \end{bmatrix}$ be a positive matrix. Then $S_{ii} > 0$ and let us

pick L_{ii} as the positive square roots of S_{ii}. In this case $L_{ii}^* = L_{ii}$. Then per Theorem 13.2, there are complex numbers $\Gamma_{12}, \Gamma_{13}, \Gamma_{23}$ in the unit disc such that:

$$S_{12} = L_{11}^* \Gamma_{12} L_{22},$$

$$S_{23} = L_{22}^* \Gamma_{23} L_{33},$$

$$S_{13} = L_{11}^* \left(\Gamma_{12} \Gamma_{23} + D_{\Gamma_{12}^*} \Gamma_{13} D_{\Gamma_{23}} \right) L_{33}.$$

Note that there is a recursive procedure to determine the Γ_{kj}. The first and the second equations yield Γ_{12}, Γ_{23} from quantities already known, while the last equation yields Γ_{13} from quantities already determined from the first two equations.

Whilst the Schur–Constantinescu parameters are defined directly in terms of the entries of S, one could also seek expressions for them in terms of the vector β of Equation (13.2) (i.e., when S is a density matrix). See [5, 7] for such expressions. In particular, for $d = 2$ the analog of the Bloch sphere is now a cylinder.

It is appropriate to make several comments about these parameters at this point:

- **C1** As can be expected from the form of Eqn. (13.4), Theorem 13.2 is valid for operator matrices, i.e., matrices whose entries are matrices or even operators in infinite-dimensional spaces, i.e., for elements of $\mathcal{M}_d \otimes \mathcal{L}(\mathcal{H})$, with $\mathcal{H}$ allowed to be infinite-dimensional. In fact, one can easily extend the result to infinite matrices with (possibly infinite-dimensional) operator entries.

- **C2** Though we only called the Γ_{kj} as the Schur–Constantinescu parameters, a full parametrization is provided by the $\frac{d(d-1)}{2}$ contractions $\Gamma_{kj}, k < j$ and the $L_{ii}, i = 1, \ldots, d$. In the case of scalar valued matrices, i.e., when $\mathcal{H} = C$, we thus get the right count of d^2 real parameters. Note the $\Gamma_{kk} = 0$ are just some fake parameters, included in the statement of the theorem to avoid an artificial separation of the $j = k+1$ case from that for other values of j.

- **C3** Since the L_{ii} are allowed to be any choice of square root of S_{ii} (i.e., $S_{ii} = L_{ii}L_{ii}^*$), the parametrization will be different for different choices of the L_{ii}. A most natural choice would be the Cholesky factorization of S_{ii}. In fact, as described in [5], there is an algorithmic proof of Theorem (13.2) which automatically yields the Cholesky factorization of S. In the infinite-dimensional case, some of the algorithmic flavor of the proof is lost.

- **C4** While Eqn. (13.4) in Theorem 13.2 is nonlinear and looks quite complicated, there is an iterative feature to it (as mentioned in the 3×3 example given before), inasmuch as in each equation there is just one of the Γ_{kj} being solved for. It is precisely because of this that the Schur–Constantinescu parameters have an inheritance property, namely that the parameters of any leading principal submatrix (recall that these will be positive themselves) are the same as that obtained from the original matrix.

- **C5** Since the proof of Theorem 13.2 supplies the Cholesky factorization of S, we get an algorithmic recipe for finding one Kraus operator representation of a quantum channel Φ. Since the Cholesky factor, V is lower triangular, the Kraus operators, V_i, thereby obtained from V (as described in Section 13.2), tend to be sparse. This can be useful in determining sufficient conditions for a channel to be entanglement breaking, or for computing quantities associated to channels such as the entanglement fidelity, for instance. The utility of using the Cholesky factorization lies not just in the avoidance of spectral calculations (as would be the case if T was found from the spectral factorization of S), but that most of the Kraus operators V_i are then sparse.

- **C6** Returning to a positive matrix, S, whose entries are scalar, it is known that if $S_{ii} = 0$ for some i, then the entire row and column to which S_{ii} belongs has to be zero. Therefore, a reasonable convention to assume is that $\Gamma_{kj} = 0$, whenever $S_{jj}S_{kk} = 0$. With this convention, the Γ_{kj}, L_{ii} provide a one-one parametrization of positive matrices.

- **C7** In the previous section we saw that even the problem of characterizing pure states via the proposed parametrizations of that section was not fully resolved. However, the Schur–Constantinescu *parametrization provides a very simple and effective characterization* of rank one states, viz., S is rank one iff all $\Gamma_{kj} = 0$, except for those cases in which $S_{jj}S_{kk} \neq 0$, in which case Γ_{kj} should be on the unit circle.

- **C8** Let S be a positive matrix. Then there is a very simple formula for its determinant in terms of the Γ_{kj}, viz.,

$$\det(S) = \left(\prod_{k=1}^{d} S_{k,k}\right) \prod_{k<j}(1 - |\Gamma_{kj}|^2).$$

  This is useful since some entropic quantities can often be expressed in terms of determinants [18].

- **C9** While Eqn. (13.4) is intricate, there is a useful diagram (called a transmission line diagram) which keeps track of all the matrix products in it $\Diamond$.

13.5 Two further applications

In this section, two additional applications of the parametrization of the previous section are provided. The first is to show that block Toeplitz states have positive partial transpose. The second is to examine the restrictions on the relaxation rates for an open quantum N-level system imposed by the requirements of complete positivity (cf. [22]).

13.5.1 Toeplitz states

The positive partial trace condition of [12, 20] has been found to be a very useful operational condition for entanglement. While for general states, it is known to be necessary and sufficient only for 2×2 and 2×3 states, there have been several arguments in favor of the notion that states which satisfy this positive partial trace condition (PPT states) are "close" to being unentangled, at least inasmuch as they are not useful for tasks such as dense coding. Similarly, there have been several attempts at studying the PPT property for positive matrices which satisfy additional conditions, see [2]. In this section we provide a contribution along the same vein. We show that positive Toeplitz matrices are PPT states.

The proof of this result was first found by considering the Schur–Constantinescu parameters for 3×3 block Toeplitz positive matrices. This proof can be extended in a simple but tedious manner for $d_i \times d_2$ states. But there is, in fact, a second proof which works for all dimensions. We provide this first and then discuss the parameter-based proof.

PROPOSITION 13.2

A Toeplitz mixed state is PPT. ◇

Sketch of the proof: Let $A \in C^{N \times N}$ be a Toeplitz matrix given by

$$
\begin{bmatrix}
a_0 & a_{-1} & \cdots & a_{-n} \\
a_1 & a_0 & \cdots & a_{-n+1} \\
\vdots & \vdots & \ddots & \vdots \\
a_n & a_{n-1} & \cdots & a_0
\end{bmatrix}.
$$

We will first, for illustration purposes, show that A^T is also positive. This is, of course, true for arbitrary positive matrices, but it will serve to illustrate the proof in the partial transpose case. Then the ij-th entry of A is given by $A_{ij} = a_{i-j}$. The transpose of A, denoted by A^T, is

$$
\begin{bmatrix}
a_0 & a_1 & \cdots & a_n \\
a_{-1} & a_0 & \cdots & a_{n-1} \\
\vdots & \vdots & \ddots & \vdots \\
a_{-n} & a_{-n+1} & \cdots & a_0
\end{bmatrix},
$$

with $A_{ij}^T = a_{j-i}$. Next, in the cycle notation, let σ_0 be the element of the symmetric group S_N on N letters, $\{1, 2, \cdots, N\}$, defined by

$$
\sigma_0 = \prod_{1 \leq k \leq N} (k \, (N-k)).
$$

The cycle σ_0 induces two simple operations on $N \times N$ matrices. If $M \in C^{N \times N}$ takes the form $M = \begin{bmatrix} w_1 \\ w_2 \\ \vdots \\ w_n \end{bmatrix}$, where w_k's are rows of M, we define the operation

R_{σ_0} by

$$
M \xrightarrow{R_{\sigma_0}} \begin{bmatrix} w_{\sigma_0(1)} \\ w_{\sigma_0(2)} \\ \vdots \\ w_{\sigma_0(n)} \end{bmatrix},
$$

i.e., R_{σ_0} simply permutes the rows of M as specified by σ_0. Another operation on columns, C_{σ_0}, is defined in the same way. Now, we notice that if A is Toeplitz as given above, then

$$
\left[R_{\sigma_0}(C_{\sigma_0}(A)) \right]_{i,j} = A_{N-i,N-j} = a_{j-i} = A_{i,j}^T.
$$

Since R_{σ_0} and C_{σ_0} preserve the characteristic polynomial, we have shown that if a Hermitian Toeplitz matrix is positive then so is its transpose.

The above fact can be extended to the partial transpose of an $NM \times NM$ Toeplitz matrix A in the following way: Let σ_m be the same permutation as σ_0 on the letters $\{mn, mn+1, \cdots, (m+1)n-1\}$. If $\sigma \in S_{N^2}$ is defined to be the disjoint product $\sigma_0\sigma_1 \cdots \sigma_{M-1}$, and R_σ and C_σ are the induced operators, then by the same argument as above, we have $R_\sigma(C_\sigma(A)) = A^{PT}$, where A^{PT} denotes the partial transpose of A. Once again these operations preserve the characteristic polynomial for Toeplitz matrices and hence if A is positive, in addition, we find that so is A^{PT}. Thus a positive Toeplitz matrix is PPT.

The Schur parametrization of positive matrices gives another proof of proposition 1 that is immediate. If B is a Toeplitz matrix, then B is also block Toeplitz. So let B be, for instance, a 3×3 block Toeplitz matrix. Using the Schur–Constantinescu parameters and the block Toeplitz property of B, we can write B explicitly as

$$
\begin{bmatrix} A & A^{\frac{1}{2}}\Gamma_1 A^{\frac{1}{2}} & A^{\frac{1}{2}}(\Gamma_1^2 + D_{\Gamma_1^*}\Gamma_2 D_{\Gamma_1})A^{\frac{1}{2}} \\ A^{\frac{1}{2}}\Gamma_1^* A^{\frac{1}{2}} & A & A^{\frac{1}{2}}\Gamma_1 A^{\frac{1}{2}} \\ A^{\frac{1}{2}}((\Gamma_1^*)^2 + D_{\Gamma_1}\Gamma_2^* D_{\Gamma_1^*})A^{\frac{1}{2}} & A^{\frac{1}{2}}\Gamma_1^* A^{\frac{1}{2}} & A \end{bmatrix},
$$

where each entry is an $N \times N$ matrix. Note that due to the block-Toeplitz nature of B its Schur–Constantinescu parameters Γ_{ij} can be indexed by only one subscript. Transpose block-wise gives us A^{PT}. By the spectral theorem, $D_{\Gamma_1^T} = (D_{\Gamma_1^*}^*)^T$. So, simply by inspection, we see that A^{PT} has Schur parameters $\{(A^{\frac{1}{2}})^T, \Gamma_1^T, \Gamma_2^T\}$. Therefore $A^{PT} \geq 0$. This is in fact true in general, according to the following:

PROPOSITION 13.3
If $A \in C^{MN \times MN}$ is block Toeplitz, then A is PPT.

The basic idea is to show that If A is parametrized by $\{\Gamma_i\}$, then A^{PT} is parametrized by $\{\Gamma_i^T\}$. Note that the block-Toeplitz property means that Schur–Constantinescu parameters of A depend only on one index (cf. the 3×3 block case). We will omit the proof, which is straightforward but tedious. Via the combinatorial structure of the Schur parameters, one can see how the parametrization of A gives rise to that of A^{PT}. The so-called lattice structure of the Schur parameters for the 4×4 case is shown in Fig. 13.1 below. Each transfer box in Fig. 13.1 describes the action of the Julia operator $U(\Gamma_i)$. Let $U^T(\Gamma)$ denote the transpose of the Julia operator of Γ, i.e.

$$U^T(\Gamma) = \begin{bmatrix} \Gamma^T & (D_\Gamma)^T \\ (D_{\Gamma^*})^T & -(\Gamma^*)^T \end{bmatrix} = \begin{bmatrix} \Gamma^T & D_{\Gamma^{T*}} \\ (D_{\Gamma^T}) & -(\Gamma^T)^* \end{bmatrix} = U(\Gamma^T).$$

Each entry of the positive semidefinite kernel $\{A_{ij}\}$ corresponds to those paths in the diagram that start from L_{jj} and end at $L_{ii}{}^*$. For example, each path from L_{33} to $L_{11}{}^*$ describes a summand in the expression for A_{13}. So we can see that the transmission line diagram of A^{PT} is then obtained by replacing each $U(\Gamma_i)$ transfer box by that of $U^T(\Gamma_i)$.

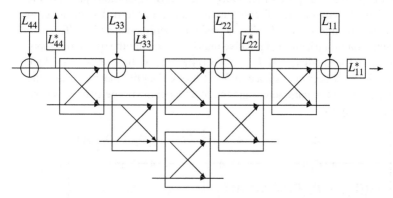

FIGURE 13.1: Lattice structure for 4×4 positive matrices.

13.5.2 Constraints on relaxation rates

In this subsection we revisit the very interesting work of [22] on the constraints imposed on the relaxation rates of an open N-level quantum system by the requirement that its evolution be completely positive. *In order to keep the notation the same as in [22], we will, in this subsection only denote the Schur–Constantinescu parameters by g_{ij} (but not Γ_{ij}).*

Let us first briefly review the contents of [22]. Let $\rho(t)$ be the state of an open N-level quantum system and let $\tilde{\rho}$ be the vector in C^{N^2} which represents $vec(\rho)$. Then its evolution can be expressed via

$$\dot{\tilde{\rho}} = (-\frac{i}{\hbar}L_H + L_D)\tilde{\rho}. \tag{13.5}$$

where L_H and L_D are $N^2 \times N^2$ matrices representing the Hamiltonian and dissipative parts, respectively, in the evolution of $\tilde{\rho}$. Let the index (m,n) denote the number $m + (n-1)N$. Then the non-zero entries of L_D are given by

$$(L_D)_{(m,n)(m,n)} = -\Gamma_{mn}, m \neq n,$$
$$(L_D)_{(m,m)(l,l)} = \gamma_{ml}, m \neq l,$$
$$(L_D)_{(m,m)(m,m)} = -\sum_{k=1,k\neq m}^{N} \gamma_{km}.$$

Here γ_{kn} is the population relaxation rate from level $\mid n >$ to $\mid k >$. The γ_{kn} are real and non-negative. Γ_{kn} (for $k \neq n$) is the dephasing rate for the transition from $\mid k >$ to $\mid n >$. Since $\Gamma_{kn} = \Gamma_{nk}$, it is easily seen that $(L_D)_{(m,n)(m,n)} = (L_D)_{(n,m)(n,m)}$. A key step in the work of [22] is to express Γ_{kn} as a sum of two summands, in recognition of the fact that dephasing is also enhanced by population relaxation, to wit,

$$\Gamma_{kn} = \Gamma_{kn}^p + \Gamma_{kn}^d,$$

with Γ_{kn}^p, the decoherence rate due to population relaxation and Γ_{kn}^d the decoherence rate due to pure phase relaxation. The requirement that the open quantum system's evolution be completely positive, [1, 9], imposes restrictions on γ_{kn} and Γ_{kn}. These restrictions can be expressed as the requirement that a certain $(N^2 - 1) \times (N^2 - 1)$ matrix concocted out of the γ_{kn} and Γ_{kn} be positive [22]. However, per [22], this requirement can be reduced to verifying that a related $(N-1) \times (N-1)$ matrix be positive. The form of this $(N-1) \times (N-1)$ matrix will depend on a choice of an orthogonal basis for the space of traceless, Hermitian $(N^2 - 1) \times (N^2 - 1)$ matrices. However, positivity of this matrix itself is independent of the choice of basis. The excellent analysis in [22] is unfortunately marred for the $N = 4$ case by an incorrect criterion for positivity. Indeed, Eqn. (28) of [22] is only necessary for positivity, while Eqns. (31)–(32)

are (as correctly claimed in [22]) also just necessary (though they come closer to sufficiency than Eqn. (28) of [22]).

In the sequel, we will use the Schur–Constantinescu parameters to analyse the $N = 4$ case of [22]. As in [22] the evolution Equation (13.5) is completely positive iff the 3×3 real symmetric matrix $B = (b_{ij})$ is positive. To specify the entries of the b_{ij}, we denote by Γ_{tot}^d the quantity $\frac{1}{2} \sum_{n=2}^{4} \sum_{m=1}^{n-1} \Gamma_{mn}^d$. Then the entries of B are given by

$$b_{11} = \Gamma_{tot}^d - (\Gamma_{13}^d + \Gamma_{24}^d,$$
$$b_{22} = \Gamma_{tot}^d - (\Gamma_{13}^d + \Gamma_{24}^d,$$
$$b_{33} = \Gamma_{tot}^d - (\Gamma_{12}^d + \Gamma_{34}^d,$$
$$b_{12} = \frac{(\Gamma_{12}^d - \Gamma_{34}^d)}{2},$$
$$b_{13} = \frac{(\Gamma_{14}^d - \Gamma_{23}^d)}{2},$$
$$b_{23} = \frac{(\Gamma_{13}^d - \Gamma_{24}^d)}{2}.$$

Now B is positive iff $b_{ii} \geq 0, i = 1, \ldots, 3$ and the Schur–Constantinescu parameters g_{12}, g_{13}, g_{23} are in the closed unit disc. Since B is real this is equivalent to demanding that the g_{ij} belong to the interval $[-1, 1]$.

The conditions $b_{ii} \geq 0$ become

$$\Gamma_{12}^d + \Gamma_{14}^d + \Gamma_{23}^d + \Gamma_{34}^d \geq \Gamma_{13}^d + \Gamma_{24}^d,$$
$$\Gamma_{12}^d + \Gamma_{13}^d + \Gamma_{24}^d + \Gamma_{34}^d \geq \Gamma_{14}^d + \Gamma_{23}^d,$$
$$\Gamma_{13}^d + \Gamma_{14}^d + \Gamma_{23}^d + \Gamma_{24}^d \geq \Gamma_{12}^d + \Gamma_{34}^d.$$

Next $b_{12} = \sqrt{b_{11}} g_{12} \sqrt{b_{22}}$. So $g_{12} \in [-1, 1]$ becomes

$$4\Gamma_{12}^d \Gamma_{34}^d - (\Gamma_{13}^d - \Gamma_{14}^d)^2 - (\Gamma_{13}^d - \Gamma_{23}^d)^2 + (\Gamma_{13}^d - \Gamma_{24}^d)^2 + (\Gamma_{14}^d - \Gamma_{23}^d)^2$$
$$- (\Gamma_{14}^d - \Gamma_{24}^d)^2 - (\Gamma_{23}^d - \Gamma_{24}^d)^2 \geq 0.$$

Likewise, the condition $g_{23} \in [-1, 1]$ becomes

$$4\Gamma_{13}^d \Gamma_{24}^d - (\Gamma_{12}^d - \Gamma_{14}^d)^2 - (\Gamma_{12}^d - \Gamma_{23}^d)^2 + (\Gamma_{12}^d - \Gamma_{34}^d)^2 + (\Gamma_{14}^d - \Gamma_{23}^d)^2$$
$$- (\Gamma_{14}^d - \Gamma_{34}^d)^2 - (\Gamma_{23}^d - \Gamma_{34}^d)^2 \geq 0.$$

Finally, $g_{13} \in [-1, 1]$ becomes

$$b_{11} b_{22} b_{33} + 2 b_{12} b_{13} b_{23} \geq b_{11} b_{23}^2 + b_{22} b_{13}^2 + b_{33} b_{12}^2.$$

Note that the condition $|g_{13}| \leq 1$ is not similar to the condition for the other g_{ij} to be in $[-1, 1]$. This is to be expected since the formula for g_{jk} for $k > j + 1$ is more intricate than those for the case $g_{jk}, k = j + 1$. Furthermore, this last condition is precisely one of those obtained in [22]. However, the conditions obtained here are necessary and sufficient.

13.6 Conclusions

Since positive matrices play a vital role in many applications, it is of importance to obtain computable parametrizations of them. In this paper we discussed several such potential parametrizations. Which one of them is most useful is, of course, a matter dictated by the application. We argued, hopefully persuasively, in favor of the versatility of the parametrization proposed in [5]. There are several other applications besides the ones discussed here, to which one could apply this parametrization. This will be the subject of future work.

References

[1] R.Alicki and K. Lendi, *Quantum Dynamical Semigroups and Applications*, Springer–Verlag, Berlin (1987).

[2] S. Braunstein, S. Ghosh and S. Severini, The Laplacian of a graph as a density matrix, arXiv, quant-ph 0406165, (2004).

[3] M. Byrd and N. Khaneja, Characterization of the positivity of the density matrix in terms of the coherence vector representation, *Phys. Rev A*, **68**, 062322 (2003).

[4] M. D. Choi, Completely positive linear maps on complex matrices, *Lin. Alg. Appl.*, **10** (1975), 285–290.

[5] T. Constantinescu and V. Ramakrishna, Parametrizing quantum states and channels, *Quantum Information Processing*, **2**, 221–248, 2003.

[6] T. Constantinescu, *Schur Parameters, Factorization and Dilation Problems*, Birkhäuser, 1996.

[7]   T. Constantinescu and V. Ramakrishna, On a parametrization of purifi-
      cations of a qubit, *Quantum Information Processing*, **1**, No 5, 109–124,
      2003.

[8]   T. Constantinescu, A. H. Sayed and T. Kailath, Displacement structure
      and completion problems, *SIAM J. Matrix Anal. Appl.*, **16** (1995), 58–78.

[9]   V. Gorini, A. Kossakowski and E. Sudarshan, *J. Math Physics*, **17**, 821
      (1976).

[10] T. F Havel, Procedures for converting among Lindblad, Kraus and ma-
      trix representations of quantum dynamical semigroups, *J. Math Physics*,
      44 (2003), 534–557.

[11] R. Horn and C. Johnson, *Matrix Analysis*, Cambridge University Press,
      Cambridge (1986).

[12] M. Horodecki, P. Horodecki and R. Horodecki, Separability of mixed
      states: necessary and sufficient conditions, *Phys. Lett*, **A223** (1996), 1.

[13] A. Jamiolkowski, Linear transformations which preserve and positive
      semidefiniteness of operators, *Rep Math Phys.*, **3** (1972), 275-278.

[14] T. Kailath and A. H. Sayed, Displacement structure: theory and appli-
      cations, *SIAM Rev.*, **37** (1995), 297–386.

[15] G. Kimura, The Bloch vector for N-level systems, arXiv:quant-
      ph/0301152, 2003.

[16] K. Kraus, General state changes in quantum theory, *Ann. Physics*, **64**
      (1971), 311–335.

[17] M. Nielsen, I. Chuang, *Quantum Computation and Quantum Informa-
      tion*, Cambridge University Press, 1999.

[18] M. Ohya and D. Petz, *Quantum entropy and its use*, Springer, Berlin,
      1993.

[19] V. Paulsen, *Completely bounded maps and dilation*, Pitman Research
      Notes in Math. 146, Longman, Wiley, New York, 1986.

[20] A. Peres, Separability criterion for density matrices, *Phys. Review Lett.*,
      77 (1996), 1413–1415.

[21] J. Preskill, website for Physics 219,
      www.theory.caltech.edu/people/preskill/ph229/

[22] S. G. Schirmer and A. Solomon, Constraints on the relaxation rates for
      N-level systems, *Phys. Rev A*, **70**, 022107 (2004).

[23] S. G. Schirmer, T. Zhang and J. Leahy, Orbits of quantum states and geometry of Bloch vectors for N-Level systems, *J. Phys A*, **37** (4), 1389–1402, (2004).

[24] I. Schur, *Über* potenzreihen die im Inneren des Einheitskreises beschränkt sind, *J. Reine und Angewandte Mathematik*, **147**, 205–232 (1917) (english translation in *Operator Theory: Advances and Applications*, **18**, 31–88, I. Gohberg ed, Birkhauser, Boston, 1986).

[25] W. F. Stinespring, Positive functions on C^* algebras, *Proc. Amer. Math. Soc.*, **6** (1955), 211–216.

[26] F. Verstraete and H. Verschelde, On quantum channels, arXiv, quant-ph 0202124 (2002).

[27] P. Zanardi, A note on quantum cloning in d dimensions, arXiv:quantum-ph/9804011 v3, 1998.

Quantum Topology, Categorical Algebra, and Logic

Chapter 14

Quantum Computing and Quantum Topology

Louis H. Kauffman and Samuel J. Lomonaco

Abstract This paper is an introduction to relationships between quantum topology and quantum computing. We discuss unitary solutions to the Yang–Baxter equation that are universal quantum gates, quantum entanglement and topological entanglement, and we give an exposition of knot-theoretic recoupling theory, its relationship with topological quantum field theory and apply these methods to produce unitary representations of the braid groups that are dense in the unitary groups. Our methods are rooted in the bracket state sum model for the Jones polynomial. We give our results for a large class of representations based on values for the bracket polynomial that are roots of unity. We make a separate and self-contained study of the quantum universal Fibonacci model in this framework. We apply our results to give quantum algorithms for the computation of the colored Jones polynomials for knots and links, and the Witten–Reshetikhin–Turaev invariant of three manifolds.

14.1 Introduction

This paper describes relationships between quantum topology and quantum computing. Quantum topology is, roughly speaking, that part of low-dimensional topology that interacts with statistical and quantum physics. Many invariants of knots, links and three-dimensional manifolds have been born of this interaction, and the form of the invariants is closely related to the form of the computation of amplitudes in quantum mechanics. Consequently, it is fruitful to move back and forth between quantum topological methods and the techniques of quantum information theory.

This paper is an expanded version of [57] that includes more expository and background material. We hope that enough background material has been included here to make the paper useful to both topologists and quantum information specialists.

We sketch the background topology, discuss analogies (such as topological entanglement and quantum entanglement), show direct correspondences between certain topological operators (solutions to the Yang–Baxter equation) and universal quantum gates. We then describe the background for topological quantum computing in terms of Temperley–Lieb (we will sometimes abbreviate this to TL) recoupling theory. This is a recoupling theory that generalizes standard angular momentum recoupling theory, generalizes the Penrose theory of spin networks and is inherently topological. Temperley–Lieb recoupling theory is based on the bracket polynomial model [36, 43] for the Jones polynomial. It is built in terms of diagrammatic combinatorial topology. The same structure can be explained in terms of the $SU(2)_q$ quantum group, and has relationships with functional integration and Witten's approach to topological quantum field theory. Nevertheless, the approach given here will be unrelentingly elementary. Elementary, does not necessarily mean simple. In this case an architecture is built from simple beginnings and this architecture and its recoupling language can be applied to many things including, e.g., colored Jones polynomials, Witten–Reshetikhin–Turaev invariants of three manifolds, topological quantum field theory and quantum computing.

In quantum computing, the application of topology is most interesting because the simplest non-trivial example of the Temperley–Lieb recoupling theory gives the so-called Fibonacci model. The recoupling theory yields representations of the Artin braid group into unitary groups $U(n)$ where n is a Fibonacci number. These representations are *dense* in the unitary group, and can be used to model quantum computation universally in terms of representations of the braid group. Hence the term: topological quantum computation.

In this paper, we outline the basics of the Temperely–Lieb recoupling theory, and show explicitly how the Fibonacci model arises from it. The diagrammatic computations in the Sections 14.11 and 14.12 are completely self-contained and can be used by a reader who has just learned the bracket polynomial, and wants to see how these dense unitary braid group representations arise from it. The subjects covered in this paper are listed below.

Knots and braids

Quantum mechanics and quantum computation

Braiding operators and universal quantum gates

A remark about *EPR*, entanglement, and Bell's inequality

The Aravind hypothesis

$SU(2)$ representations of Artin braid group

Bracket polynomial and Jones polynomial

Quantum topology, cobordism categories, Temperley–Lieb algebra and topological quantum field theory

Braiding and topological quantum field theory

Spin networks and Temperley–Lieb recoupling theory

Fibonacci particles

Fibonacci recoupling model

Quantum computation of colored Jones polynomials and Witten–Reshetikhin–Turaev invariant

We should point out that while this paper attempts to be self-contained, and hence has some expository material, most of the results are either new, or are new points of view on known results. The material on $SU(2)$ representations of the Artin braid group is new, and the relationship of this material to the recoupling theory is new. The treatment of elementary cobordism categories is well-known, but new in the context of quantum information theory. The reformulation of Temperley–Lieb recoupling theory for the purpose of producing unitary braid group representations is new for quantum information theory, and directly related to much of the recent work of Freedman and his collaborators. The treatment of the Fibonacci model in terms of two-strand recoupling theory is new and at the same time, the most elementary non-trivial example of the recoupling theory. The models for quantum computation of colored Jones polynomials and for quantum computation of the Witten–Reshetikhin–Turaev invariant are new in this form of the recoupling theory. They take a particularly simple aspect in this context.

Here is a very condensed presentation of how unitary representations of the braid group are constructed via topological quantum field theoretic methods. One has a mathematical particle with label P that can interact with itself to produce either itself labeled P or itself with the null label $*$. We shall denote the interaction of two particles P and Q by the expression PQ, but it is understood that the "value" of PQ is the result of the interaction, and this may partake of a number of possibilities. Thus for our particle P, we have that PP may be equal to P or to $*$ in a given situation. When $*$ interacts with P the result is always P. When $*$ interacts with $*$ the result is always $*$. One considers process spaces where a row of particles labeled P can successively interact, subject to the restriction that the end result is P. For example the space $V[(ab)c]$ denotes the space of interactions of three particles labeled P. The particles are placed in the positions a, b, c. Thus we begin with $(PP)P$. In a typical sequence of interactions, the first two P's interact to produce a $*$, and the $*$ interacts with P to produce P.

$$(PP)P \longrightarrow (*)P \longrightarrow P.$$

In another possibility, the first two P's interact to produce a P, and the P interacts with P to produce P.

$$(PP)P \longrightarrow (P)P \longrightarrow P.$$

It follows from this analysis that the space of linear combinations of processes $V[(ab)c]$ is two-dimensional. The two processes we have just described can be taken to be the qubit basis for this space. One obtains a representation of the three strand Artin braid group on $V[(ab)c]$ by assigning appropriate phase changes to each of the generating processes. One can think of these phases as corresponding to the interchange of the particles labeled a and b in the association $(ab)c$. The other operator for this representation corresponds to the interchange of b and c. This interchange is accomplished by a *unitary change of basis mapping*

$$F : V[(ab)c] \longrightarrow V[a(bc)].$$

If

$$A : V[(ab)c] \longrightarrow V[(ba)c]$$

is the first braiding operator (corresponding to an interchange of the first two particles in the association) then the second operator

$$B : V[(ab)c] \longrightarrow V[(ac)b]$$

is accomplished via the formula $B = F^{-1}RF$ where the R in this formula acts in the second vector space $V[a(bc)]$ to apply the phases for the interchange of b and c. These issues are illustrated in Fig. 14.1, where the parenthesization of the particles is indicated by circles and by also by trees. The trees can be taken to indicate patterns of particle interaction, where two particles interact at the branch of a binary tree to produce the particle product at the root. See also Fig. 14.28 for an illustration of the braiding $B = F^{-1}RF$.

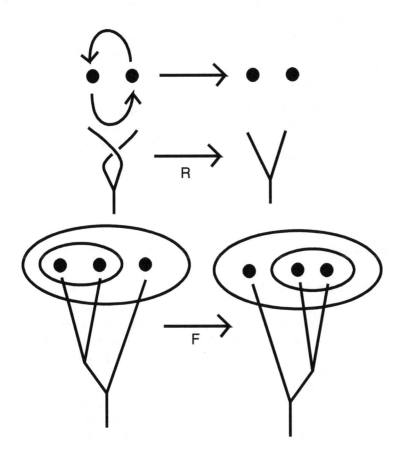

FIGURE 14.1: Braiding anyons.

In this scheme, vector spaces corresponding to associated strings of particle interactions are interrelated by *recoupling transformations* that generalize the mapping F indicated above. A full representation of the Artin braid group on each space is defined in terms of the local interchange phase gates and the recoupling transformations. These gates and transformations have to satisfy a number of identities in order to produce a well-defined representation of the braid group. These identities were discovered originally in relation to topological quantum field theory. In our approach the structure of phase gates and recoupling transformations arise naturally from the structure of the bracket model for the Jones polynomial. Thus we obtain a knot-theoretic basis for topological quantum computing.

In modeling the quantum Hall effect [15, 16, 25, 84], the braiding of quasiparticles (collective excitations) leads to non-trivial representations of the Artin braid group. Such particles are called *Anyons*. The braiding in these models is related to topological quantum field theory. It is hoped that the mathematics we explain here will form a bridge between theoretical models of anyons and their applications to quantum computing.

14.2   Knots and braids

The purpose of this section is to give a quick introduction to the diagrammatic theory of knots, links and braids. A *knot* is an embedding of a circle in three-dimensional space, taken up to ambient isotopy. The problem of deciding whether two knots are isotopic is an example of a *placement problem*, a problem of studying the topological forms that can be made by placing one space inside another. In the case of knot theory we consider the placements of a circle inside three-dimensional space. There are many applications of the theory of knots. Topology is a background for the physical structure of real knots made from rope of cable. As a result, the field of practical knot tying is a field of applied topology that existed well before the mathematical discipline of topology arose. Then again long molecules such as rubber molecules and DNA molecules can be knotted and linked. There have been a number of intense applications of knot theory to the study of DNA [79] and to polymer physics [59]. Knot theory is closely related to theoretical physics as well with applications in quantum gravity [52, 76, 83] and many applications of ideas in physics to the topological structure of knots themselves [43].

Quantum topology is the study and invention of topological invariants via the use of analogies and techniques from mathematical physics. Many invariants such as the Jones polynomial are constructed via partition functions and generalized quantum amplitudes. As a result, one expects to see relationships between knot theory and physics. In this paper we will study how knot theory can be used to produce unitary representations of the braid group. Such representations can play a fundamental role in quantum computing.

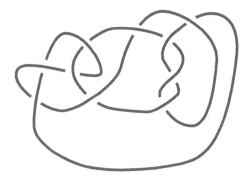

FIGURE 14.2: A knot diagram.

That is, two knots are regarded as equivalent if one embedding can be obtained from the other through a continuous family of embeddings of circles in three-space. A *link* is an embedding of a disjoint collection of circles, taken up to ambient isotopy. Fig. 14.2 illustrates a diagram for a knot. The diagram is regarded both as a schematic picture of the knot, and as a plane graph with extra structure at the nodes (indicating how the curve of the knot passes over or under itself by standard pictorial conventions).

Ambient isotopy is mathematically the same as the equivalence relation generated on diagrams by the *Reidemeister moves*. These moves are illustrated in Fig. 14.3. Each move is performed on a local part of the diagram that is topologically identical to the part of the diagram illustrated in this figure (these figures are representative examples of the types of Reidemeister moves) without changing the rest of the diagram. The Reidemeister moves are useful in doing combinatorial topology with knots and links, notably in working out the behaviour of knot invariants. A *knot invariant* is a function defined from knots and links to some other mathematical object (such as groups or polynomials or

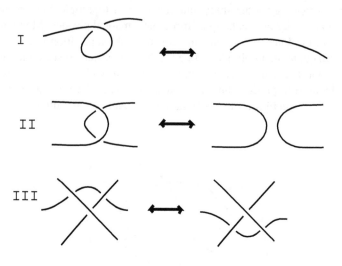

FIGURE 14.3: The Reidemeister moves.

numbers) such that equivalent diagrams are mapped to equivalent objects (iso-morphic groups, identical polynomials, identical numbers). The Reidemeister moves are of great use for analyzing the structure of knot invariants and they are closely related to the *Artin braid group*, which we discuss below.

A *braid* is an embedding of a collection of strands that have their ends in two rows of points that are set one above the other with respect to a choice of vertical. The strands are not individually knotted and they are disjoint from one another. See Figs. 14.4, 14.5 and 14.6 for illustrations of braids and moves on braids. Braids can be multiplied by attaching the bottom row of one braid to the top row of the other braid. Taken up to ambient isotopy, fixing the endpoints, the braids form a group under this notion of multiplication. In Fig. 14.4 we illustrate the form of the basic generators of the braid group, and the form of the relations among these generators. Fig. 14.5 illustrates how to close a braid by attaching the top strands to the bottom strands by a collection of parallel arcs. A key theorem of Alexander states that every knot or link can be represented as a closed braid. Thus the theory of braids is critical to the theory of knots and links. Fig. 14.6 illustrates the famous Borromean rings (a link of three unknotted loops such that any two of the loops are unlinked) as the closure of a braid.

Let B_n denote the Artin braid group on n strands. We recall here that B_n is

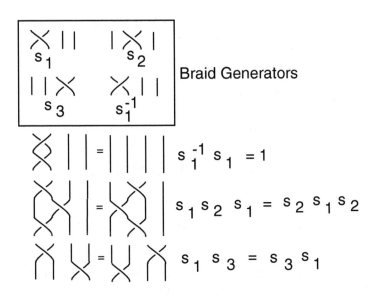

Braid Generators

$s_1^{-1} \, s_1 \, = 1$

$s_1 \, s_2 \, s_1 \, = \, s_2 \, s_1 \, s_2$

$s_1 \, s_3 \, = \, s_3 \, s_1$

FIGURE 14.4: Braid generators.

generated by elementary braids $\{s_1, \cdots, s_{n-1}\}$ with relations

1. $s_i s_j = s_j s_i$ for $|i - j| > 1$,

2. $s_i s_{i+1} s_i = s_{i+1} s_i s_{i+1}$ for $i = 1, \cdots n - 2$.

See Fig. 14.4 for an illustration of the elementary braids and their relations. Note that the braid group has a diagrammatic topological interpretation, where a braid is an intertwining of strands that lead from one set of n points to another set of n points. The braid generators s_i are represented by diagrams where the i-th and $(i+1)$-th strands wind around one another by a single half-twist (the sense of this turn is shown in Fig. 14.4) and all other strands drop straight to

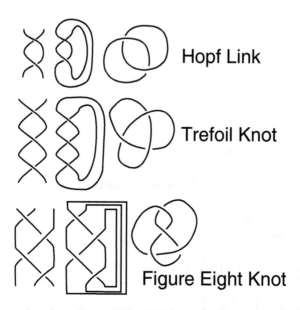

Hopf Link

Trefoil Knot

Figure Eight Knot

FIGURE 14.5: Closing braids to form knots and links.

the bottom. Braids are diagrammed vertically as in Fig. 14.4, and the products are taken in order from top to bottom. The product of two braid diagrams is accomplished by adjoining the top strands of one braid to the bottom strands of the other braid.

In Fig. 14.4 we have restricted the illustration to the four-stranded braid group B_4. In that figure the three braid generators of B_4 are shown, and then the inverse of the first generator is drawn. Following this, one sees the identities $s_1 s_1^{-1} = 1$ (where the identity element in B_4 consists in four vertical strands), $s_1 s_2 s_1 = s_2 s_1 s_2$, and finally $s_1 s_3 = s_3 s_1$.

Braids are a key structure in mathematics. It is not just that they are a collection of groups with a vivid topological interpretation. From the algebraic point of view the braid groups B_n are important extensions of the symmetric groups S_n. Recall that the symmetric group S_n of all permutations of n distinct objects has presentation as shown below.

1. $s_i^2 = 1$ for $i = 1, \cdots n - 1$,

2. $s_i s_j = s_j s_i$ for $|i - j| > 1$,

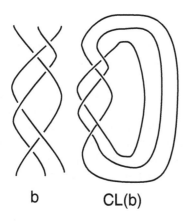

b   CL(b)

FIGURE 14.6: Borromean rings as a braid closure.

3. $s_i s_{i+1} s_i = s_{i+1} s_i s_{i+1}$ for $i = 1, \cdots n-2$.

Thus S_n is obtained from B_n by setting the square of each braiding generator equal to one. We have an exact sequence of groups

$$1 \longrightarrow B_n \longrightarrow S_n \longrightarrow 1$$

exhibiting the Artin braid group as an extension of the symmetric group.

In the next sections we shall show how representations of the Artin braid group are rich enough to provide a dense set of transformations in the unitary groups. Thus the braid groups are *in principle* fundamental to quantum computation and quantum information theory.

14.3 Quantum mechanics and quantum computation

We shall quickly indicate the basic principles of quantum mechanics. The quantum information context encapsulates a concise model of quantum theory:

The initial state of a quantum process is a vector $|v\rangle$ in a complex vector space H. Measurement returns basis elements β of H with probability

$$|\langle \beta \,|v\rangle|^2 / \langle v\,|v\rangle$$

where $\langle v\,|w\rangle = v^\dagger w$ with $v^\dagger$ the conjugate transpose of v. A physical process occurs in steps $|v\rangle \longrightarrow U\,|v\rangle = |Uv\rangle$ where U is a unitary linear transformation.

Note that since $\langle Uv\,|Uw\rangle = \langle v\,|U^\dagger U\,|w\rangle = \langle v\,|w\rangle =$ when U is unitary, it follows that probability is preserved in the course of a quantum process.

One of the details required for any specific quantum problem is the nature of the unitary evolution. This is specified by knowing appropriate information about the classical physics that supports the phenomena. This information is used to choose an appropriate Hamiltonian through which the unitary operator is constructed via a correspondence principle that replaces classical variables with appropriate quantum operators. (In the path integral approach one needs a Langrangian to construct the action on which the path integral is based.) One needs to know certain aspects of classical physics to solve any specific quantum problem.

A key concept in the quantum information viewpoint is the notion of the superposition of states. If a quantum system has two distinct states $|v\rangle$ and $|w\rangle$, then it has infinitely many states of the form $a|v\rangle + b|w\rangle$ where a and b are complex numbers taken up to a common multiple. States are "really" in the projective space associated with H. There is only one superposition of a single state $|v\rangle$ with itself. On the other hand, it is most convenient to regard the states $|v\rangle$ and $|w\rangle$ as vectors in a vector space. We than take it as part of the procedure of dealing with states to normalize them to unit length. Once again, the superposition of a state with itself is again itself.

Dirac [22] introduced the "bra -(c)-ket" notation $\langle A\,|B\rangle = A^\dagger B$ for the inner product of complex vectors $A, B \in H$. He also separated the parts of the bracket into the *bra* $\langle A\,|$ and the *ket* $|B\rangle$. Thus

$$\langle A\,|B\rangle = \langle A\,|\,|B\rangle$$

In this interpretation, the ket $|B\rangle$ is identified with the vector $B \in H$, while the bra $\langle A\,|$ is regarded as the element dual to A in the dual space H^*. The dual element to A corresponds to the conjugate transpose $A^\dagger$ of the vector A, and the inner product is expressed in conventional language by the matrix product $A^\dagger B$ (which is a scalar since B is a column vector). Having separated the bra and the ket, Dirac can write the "ket-bra" $|A\rangle\langle B| = AB^\dagger$. In conventional notation, the ket-bra is a matrix, not a scalar, and we have the following formula for the square of $P = |A\rangle\langle B|$:

$$P^2 = |A\rangle\langle B||A\rangle\langle B| = A(B^\dagger A)B^\dagger = (B^\dagger A)AB^\dagger = \langle B|A\rangle P.$$

The standard example is a ket-bra $P = |A\rangle\langle A|$ where $\langle A|A\rangle = 1$ so that $P^2 = P$. Then P is a projection matrix, projecting to the subspace of H that is spanned by the vector $|A\rangle$. In fact, for any vector $|B\rangle$ we have

$$P|B\rangle = |A\rangle\langle A||B\rangle = |A\rangle\langle A|B\rangle = \langle A|B\rangle|A\rangle.$$

If $\{|C_1\rangle, |C_2\rangle, \cdots |C_n\rangle\}$ is an orthonormal basis for H, and

$$P_i = |C_i\rangle\langle C_i|,$$

then for any vector $|A\rangle$ we have

$$|A\rangle = \langle C_1|A\rangle|C_1\rangle + \cdots + \langle C_n|A\rangle|C_n\rangle.$$

Hence

$$\langle B|A\rangle = \langle B|C_1\rangle\langle C_1|A\rangle + \cdots + \langle B|C_n\rangle\langle C_n|A\rangle.$$

One wants the probability of starting in state $|A\rangle$ and ending in state $|B\rangle$. The probability for this event is equal to $|\langle B|A\rangle|^2$. This can be refined if we have more knowledge. If the intermediate states $|C_i\rangle$ are a complete set of orthonormal alternatives then we can assume that $\langle C_i|C_i\rangle = 1$ for each i and that $\Sigma_i|C_i\rangle\langle C_i| = 1$. This identity now corresponds to the fact that 1 is the sum of the probabilities of an arbitrary state being projected into one of these intermediate states.

If there are intermediate states between the intermediate states this formulation can be continued until one is summing over all possible paths from A to B. This becomes the path integral expression for the amplitude $\langle B|A\rangle$.

14.3.1 What is a quantum computer?

A *quantum computer* is, abstractly, a composition U of unitary transformations, together with an initial state and a choice of measurement basis. One runs the computer by repeatedly initializing it, and then measuring the result of applying the unitary transformation U to the initial state. The results of these measurements are then analyzed for the desired information that the computer was set to determine. The key to using the computer is the design of the initial state and the design of the composition of unitary transformations. The reader should consult [69] for more specific examples of quantum algorithms.

Let H be a given finite dimensional vector space over the complex numbers C. Let $\{W_0, W_1, ..., W_n\}$ be an orthonormal basis for H so that with $|i\rangle := |W_i\rangle$ denoting W_i and $\langle i|$ denoting the conjugate transpose of $|i\rangle$, we have

$$\langle i|j\rangle = \delta_{ij}$$

where δ_{ij} denotes the Kronecker delta (equal to one when its indices are equal to one another, and equal to zero otherwise). Given a vector v in H let $|v|^2 := \langle v|v\rangle$. Note that $\langle i|v$ is the i-th coordinate of v.

A *measurement of v* returns one of the coordinates $|i\rangle$ of v with probability $|\langle i|v|^2$. This model of measurement is a simple instance of the situation with a quantum mechanical system that is in a mixed state until it is observed. The result of observation is to put the system into one of the basis states.

When the dimension of the space H is two ($n = 1$), a vector in the space is called a *qubit*. A qubit represents one quantum of binary information. On measurement, one obtains either the ket $|0\rangle$ or the ket $|1\rangle$. This constitutes the binary distinction that is inherent in a qubit. Note however that the information obtained is probabilistic. If the qubit is

$$|\psi\rangle = \alpha|0\rangle + \beta|1\rangle,$$

then the ket $|0\rangle$ is observed with probability $|\alpha|^2$, and the ket $|1\rangle$ is observed with probability $|\beta|^2$. In speaking of an idealized quantum computer, we do not specify the nature of measurement process beyond these probability postulates.

In the case of general dimension n of the space H, we will call the vectors in H *qunits*. It is quite common to use spaces H that are tensor products of two-dimensional spaces (so that all computations are expressed in terms of qubits) but this is not necessary in principle. One can start with a given space, and later work out factorizations into qubit transformations.

A *quantum computation* consists in the application of a unitary transformation U to an initial qunit $\psi = a_0|0\rangle + ... + a_n|n\rangle$ with $|\psi|^2 = 1$, plus a measurement of $U\psi$. A measurement of $U\psi$ returns the ket $|i\rangle$ with probability $|\langle i|U\psi|^2$. In particular, if we start the computer in the state $|i\rangle$, then the probability that it will return the state $|j\rangle$ is $|\langle j|U|i\rangle|^2$.

It is the necessity for writing a given computation in terms of unitary transformations, and the probabilistic nature of the result that characterize quantum computation. Such computation could be carried out by an idealized quantum mechanical system. It is hoped that such systems can be physically realized.

14.4 Braiding operators and universal quantum gates

A class of invariants of knots and links called quantum invariants can be constructed by using representations of the Artin braid group, and more specifically by using solutions to the Yang–Baxter equation [10], first discovered in relation to $1 + 1$ dimensional quantum field theory, and two-dimensional statistical mechanics. Braiding operators feature in constructing representations of the Artin braid group, and in the construction of invariants of knots and links.

A key concept in the construction of quantum link invariants is the association of a Yang–Baxter operator R to each elementary crossing in a link diagram. The operator R is a linear mapping

$$R: V \otimes V \longrightarrow V \otimes V$$

defined on the two-fold tensor product of a vector space V, generalizing the permutation of the factors (i.e., generalizing a swap gate when V represents one qubit). Such transformations are not necessarily unitary in topological applications. It is useful to understand when they can be replaced by unitary transformations for the purpose of quantum computing. Such unitary R-matrices can be used to make unitary representations of the Artin braid group.

A solution to the Yang–Baxter equation, as described in the last paragraph is a matrix R, regarded as a mapping of a two-fold tensor product of a vector space $V \otimes V$ to itself that satisfies the equation

$$(R \otimes I)(I \otimes R)(R \otimes I) = (I \otimes R)(R \otimes I)(I \otimes R).$$

From the point of view of topology, the matrix R is regarded as representing an elementary bit of braiding represented by one string crossing over another. In Fig. 14.7 we have illustrated the braiding identity that corresponds to the Yang–Baxter equation. Each braiding picture with its three input lines (below) and output lines (above) corresponds to a mapping of the three-fold tensor product of the vector space V to itself, as required by the algebraic equation quoted above. The pattern of placement of the crossings in the diagram corresponds to the factors $R \otimes I$ and $I \otimes R$. This crucial topological move has an algebraic expression in terms of such a matrix R. Our approach in this section to relate topology, quantum computing, and quantum entanglement is through the use of the Yang–Baxter equation. In order to accomplish this aim, *we need to study solutions of the Yang–Baxter equation that are unitary.* Then the R matrix can be seen *either* as a braiding matrix *or* as a quantum gate in a quantum computer.

The problem of finding solutions to the Yang–Baxter equation that are unitary turns out to be surprisingly difficult. Dye [24] has classified all such

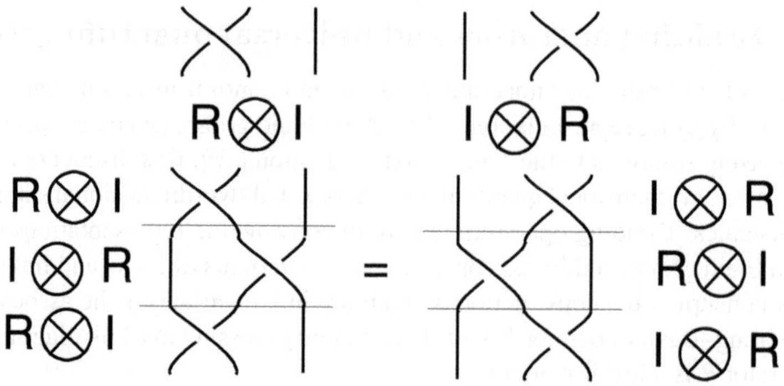

FIGURE 14.7: The Yang–Baxter equation—$(R \otimes I)(I \otimes R)(R \otimes I) = (I \otimes R)(R \otimes I)(I \otimes R)$.

matrices of size 4×4. A rough summary of her classification is that all 4×4 unitary solutions to the Yang–Baxter equation are similar to one of the following types of matrix:

$$R = \begin{pmatrix} 1/\sqrt{2} & 0 & 0 & 1/\sqrt{2} \\ 0 & 1/\sqrt{2} & -1/\sqrt{2} & 0 \\ 0 & 1/\sqrt{2} & 1/\sqrt{2} & 0 \\ -1/\sqrt{2} & 0 & 0 & 1/\sqrt{2} \end{pmatrix}$$

$$R' = \begin{pmatrix} a & 0 & 0 & 0 \\ 0 & 0 & b & 0 \\ 0 & c & 0 & 0 \\ 0 & 0 & 0 & d \end{pmatrix}$$

$$R'' = \begin{pmatrix} 0 & 0 & 0 & a \\ 0 & b & 0 & 0 \\ 0 & 0 & c & 0 \\ d & 0 & 0 & 0 \end{pmatrix}$$

where a,b,c,d are unit complex numbers.

For the purpose of quantum computing, one should regard each matrix as acting on the standard basis $\{|00\rangle, |01\rangle, |10\rangle, |11\rangle\}$ of $H = V \otimes V$, where V is a two-dimensional complex vector space. Then, for example we have

$$
\begin{aligned}
R|00\rangle &= (1/\sqrt{2})|00\rangle - (1/\sqrt{2})|11\rangle, \\
R|01\rangle &= (1/\sqrt{2})|01\rangle + (1/\sqrt{2})|10\rangle, \\
R|10\rangle &= -(1/\sqrt{2})|01\rangle + (1/\sqrt{2})|10\rangle, \\
R|11\rangle &= (1/\sqrt{2})|00\rangle + (1/\sqrt{2})|11\rangle.
\end{aligned}
$$

The reader should note that R is the familiar change-of-basis matrix from the standard basis to the Bell basis of entangled states.

In the case of R', we have

$$
\begin{aligned}
R'|00\rangle &= a|00\rangle, R'|01\rangle = c|10\rangle, \\
R'|10\rangle &= b|01\rangle, R'|11\rangle = d|11\rangle.
\end{aligned}
$$

Note that R' can be regarded as a diagonal phase gate P, composed with a swap gate S.

$$
P = \begin{pmatrix} a & 0 & 0 & 0 \\ 0 & b & 0 & 0 \\ 0 & 0 & c & 0 \\ 0 & 0 & 0 & d \end{pmatrix}
$$

$$
S = \begin{pmatrix} 1 & 0 & 0 & 0 \\ 0 & 0 & 1 & 0 \\ 0 & 1 & 0 & 0 \\ 0 & 0 & 0 & 1 \end{pmatrix}.
$$

Compositions of solutions of the (braiding) Yang–Baxter equation with the swap gate S are called *solutions to the algebraic Yang–Baxter equation*. Thus the diagonal matrix P is a solution to the algebraic Yang–Baxter equation.

REMARK 14.1   Another avenue related to unitary solutions to the Yang–Baxter equation as quantum gates comes from using extra physical parameters in this equation (the rapidity parameter) that are related to statistical physics. In [88] we discovered that solutions to the Yang–Baxter equation with the rapidity parameter allow many new unitary solutions. The significance of these gates for quantum computing is still under investigation.   ∎

14.4.1   Universal gates

A *two-qubit gate* G is a unitary linear mapping $G : V \otimes V \longrightarrow V$ where V is a two complex dimensional vector space. We say that the gate G is *universal for quantum computation* (or just *universal*) if G together with local unitary transformations (unitary transformations from V to V) generates all unitary transformations of the complex vector space of dimension 2^n to itself. It is well-known [69] that *CNOT* is a universal gate. (On the standard basis, *CNOT* is the identity when the first qubit is 0, and it flips the second qubit, leaving the first alone, when the first qubit is 1.)

A gate G, as above, is said to be *entangling* if there is a vector

$$|\alpha\beta\rangle = |\alpha\rangle \otimes |\beta\rangle \in V \otimes V$$

such that $G|\alpha\beta\rangle$ is not decomposable as a tensor product of two qubits. Under these circumstances, one says that $G|\alpha\beta\rangle$ is *entangled*.

In [17], the Brylinskis give a general criterion of G to be universal. They prove that *a two-qubit gate G is universal if and only if it is entangling.*

REMARK 14.2   A two-qubit pure state

$$|\phi\rangle = a|00\rangle + b|01\rangle + c|10\rangle + d|11\rangle$$

is entangled exactly when $(ad - bc) \neq 0$. It is easy to use this fact to check when a specific matrix is, or is not, entangling.   ∎

REMARK 14.3   There are many gates other than *CNOT* that can be used as universal gates in the presence of local unitary transformations. Some of these are themselves topological (unitary solutions to the Yang–Baxter equation, see [55]) and themselves generate representations of the Artin braid group. Replacing *CNOT* by a solution to the Yang–Baxter equation does not place the local unitary transformations as part of the corresponding representation of the braid group. Thus such substitutions give only a partial solution to creating topological quantum computation. In this paper we are concerned with braid group representations that include all aspects of the unitary group. Accordingly, in the next section we shall first examine how the braid group on three strands can be represented as local unitary transformations. ∎

THEOREM 14.1

Let D denote the phase gate shown below. D is a solution to the algebraic Yang–Baxter equation (see the earlier discussion in this section). Then D is a universal gate.

$$D = \begin{pmatrix} 1 & 0 & 0 & 0 \\ 0 & 1 & 0 & 0 \\ 0 & 0 & 1 & 0 \\ 0 & 0 & 0 & -1 \end{pmatrix}.$$

PROOF   It follows at once from the Brylinski theorem that D is universal. For a more specific proof, note that $CNOT = QDQ^{-1}$, where $Q = H \otimes I$, H is the 2×2 Hadamard matrix. The conclusion then follows at once from this identity and the discussion above. We illustrate the matrices involved in this proof below:

$$H = (1/\sqrt{2}) \begin{pmatrix} 1 & 1 \\ 1 & -1 \end{pmatrix}$$

$$Q = (1/\sqrt{2}) \begin{pmatrix} 1 & 1 & 0 & 0 \\ 1 & -1 & 0 & 0 \\ 0 & 0 & 1 & 1 \\ 0 & 0 & 1 & -1 \end{pmatrix}$$

$$D = \begin{pmatrix} 1 & 0 & 0 & 0 \\ 0 & 1 & 0 & 0 \\ 0 & 0 & 1 & 0 \\ 0 & 0 & 0 & -1 \end{pmatrix}$$

$$QDQ^{-1} = QDQ = \begin{pmatrix} 1\,0\,0\,0 \\ 0\,1\,0\,0 \\ 0\,0\,0\,1 \\ 0\,0\,1\,0 \end{pmatrix} = CNOT.$$

This completes the proof of the theorem. ∎

REMARK 14.4   We thank Martin Roetteles [75] for pointing out the specific factorization of *CNOT* used in this proof. ∎

THEOREM 14.2
The matrix solutions R' and R'' to the Yang–Baxter equation, described above, are universal gates exactly when $ad - bc \neq 0$ for their internal parameters a, b, c, d. In particular, let R_0 denote the solution R' (above) to the Yang–Baxter equation with $a = b = c = 1, d = -1$.

$$R' = \begin{pmatrix} a\,0\,0\,0 \\ 0\,0\,b\,0 \\ 0\,c\,0\,0 \\ 0\,0\,0\,d \end{pmatrix}$$

$$R_0 = \begin{pmatrix} 1\,0\,0\,\ 0 \\ 0\,0\,1\,\ 0 \\ 0\,1\,0\,\ 0 \\ 0\,0\,0\,{-1} \end{pmatrix}.$$

Then R_0 is a universal gate.

PROOF   The first part follows at once from the Brylinski theorem. In fact, letting H be the Hadamard matrix as before, and

$$\sigma = \begin{pmatrix} 1/\sqrt{2} & i/\sqrt{2} \\ i/\sqrt{2} & 1/\sqrt{2} \end{pmatrix}, \lambda = \begin{pmatrix} 1/\sqrt{2} & 1/\sqrt{2} \\ i/\sqrt{2} & -i/\sqrt{2} \end{pmatrix}$$

$$\mu = \begin{pmatrix} (1-i)/2 & (1+i)/2 \\ (1-i)/2 & (-1-i)/2 \end{pmatrix}.$$

Then

$$CNOT = (\lambda \otimes \mu)(R_0(I \otimes \sigma)R_0)(H \otimes H).$$

This gives an explicit expression for *CNOT* in terms of R_0 and local unitary transformations (for which we thank Ben Reichardt). ∎

REMARK 14.5   Let *SWAP* denote the Yang–Baxter Solution R' with $a = b = c = d = 1$.

$$SWAP = \begin{pmatrix} 1 & 0 & 0 & 0 \\ 0 & 0 & 1 & 0 \\ 0 & 1 & 0 & 0 \\ 0 & 0 & 0 & 1 \end{pmatrix}.$$

SWAP is the standard swap gate. Note that *SWAP* is not a universal gate. This also follows from the Brylinski theorem, since *SWAP* is not entangling. Note also that R_0 is the composition of the phase gate D with this swap gate. ∎

THEOREM 14.3
   Let

$$R = \begin{pmatrix} 1/\sqrt{2} & 0 & 0 & 1/\sqrt{2} \\ 0 & 1/\sqrt{2} & -1/\sqrt{2} & 0 \\ 0 & 1/\sqrt{2} & 1/\sqrt{2} & 0 \\ -1/\sqrt{2} & 0 & 0 & 1/\sqrt{2} \end{pmatrix}$$

be the unitary solution to the Yang–Baxter equation discussed above. Then R is a universal gate. The proof below gives a specific expression for CNOT in terms of R.

PROOF   This result follows at once from the Brylinski theorem, since R is highly entangling. For a direct computational proof, it suffices to show that *CNOT* can be generated from R and local unitary transformations. Let

$$\alpha = \begin{pmatrix} 1/\sqrt{2} & 1/\sqrt{2} \\ 1/\sqrt{2} & -1/\sqrt{2} \end{pmatrix}$$

$$\beta = \begin{pmatrix} -1/\sqrt{2} & 1/\sqrt{2} \\ i/\sqrt{2} & i/\sqrt{2} \end{pmatrix}$$

$$\gamma = \begin{pmatrix} 1/\sqrt{2} & i/\sqrt{2} \\ 1/\sqrt{2} & -i/\sqrt{2} \end{pmatrix}$$

$$\delta = \begin{pmatrix} -1 & 0 \\ 0 & -i \end{pmatrix}.$$

Let $M = \alpha \otimes \beta$ and $N = \gamma \otimes \delta$. Then it is straightforward to verify that

$$CNOT = MRN.$$

This completes the proof.   ∎

REMARK 14.6   See [55] for more information about these calculations.   ∎

14.5   A remark about EPR, entanglement and Bell's inequality

A state $|\psi\rangle \in H^{\otimes n}$, where H is the qubit space, is said to be *entangled* if it cannot be written as a tensor product of vectors from non-trivial factors of $H^{\otimes n}$. Such states turn out to be related to subtle nonlocality in quantum physics. It helps to place this algebraic structure in the context of a gedanken experiment to see where the physics comes in. Thought experiments of the sort we are about to describe were first devised by Einstein, Podolosky and Rosen, referred henceforth as *EPR*.

Consider the entangled state

$$S = (|0\rangle|1\rangle + |1\rangle|0\rangle)/\sqrt{2}.$$

In an EPR thought experiment, we think of two "parts" of this state that are separated in space. We want a notation for these parts and suggest the following:

$$L = (\{|0\rangle\}|1\rangle + \{|1\rangle\}|0\rangle)/\sqrt{2},$$
$$R = (|0\rangle\{|1\rangle\} + |1\rangle\{|0\rangle\})/\sqrt{2}.$$

In the left state L, an observer can only observe the left hand factor. In the right state R, an observer can only observe the right hand factor. These "states" L and R together comprise the EPR state S, but they are accessible individually just as are the two photons in the usual thought experiment. One can transport L and R individually and we shall write

$$S = L * R$$

to denote that they are the "parts" (but not tensor factors) of S.

The curious thing about this formalism is that it includes a little bit of macroscopic physics implicitly, and so it makes it a bit more apparent what EPR were concerned about. After all, lots of things that we can do to L or R do not affect S. For example, transporting L from one place to another, as in the original experiment where the photons separate. On the other hand, if Alice has L and Bob has R and Alice performs a local unitary transformation on "her" tensor factor, this applies to both L and R since the transformation is actually being applied to the state S. This is also a "spooky action at a distance" whose consequence does not appear until a measurement is made.

To go a bit deeper it is worthwhile seeing what entanglement, in the sense of tensor indecomposability, has to do with the structure of the EPR thought experiment. To this end, we look at the structure of the Bell inequalities using the Clauser, Horne, Shimony, Holt formalism ($CHSH$) as explained in the book by Nielsen and Chuang [69]. For this we use the following observables with eigenvalues ± 1.

$$Q = \begin{pmatrix} 1 & 0 \\ 0 & -1 \end{pmatrix}_1,$$

$$R = \begin{pmatrix} 0 & 1 \\ 1 & 0 \end{pmatrix}_1,$$

$$S = \begin{pmatrix} -1 & -1 \\ -1 & 1 \end{pmatrix}_2 /\sqrt{2},$$

$$T = \begin{pmatrix} 1 & -1 \\ -1 & -1 \end{pmatrix}_2 /\sqrt{2}.$$

The subscripts 1 and 2 on these matrices indicate that they are to operate on the first and second tensor factors, repsectively, of a quantum state of the form

$$\phi = a|00\rangle + b|01\rangle + c|10\rangle + d|11\rangle.$$

To simplify the results of this calculation we shall here assume that the coefficients a, b, c, d are real numbers. We calculate the quantity

$$\Delta = \langle\phi|QS|\phi\rangle + \langle\phi|RS|\phi\rangle + \langle\phi|RT|\phi\rangle - \langle\phi|QT|\phi\rangle,$$

finding that

$$\Delta = (2 - 4(a+d)^2 + 4(ad - bc))/\sqrt{2}.$$

Classical probability calculation with random variables of value ± 1 gives the value of $QS + RS + RT - QT = \pm 2$ (with each of Q, R, S and T equal to ± 1). Hence the classical expectation satisfies the Bell inequality

$$E(QS) + E(RS) + E(RT) - E(QT) \le 2.$$

That quantum expectation is not classical is embodied in the fact that Δ can be greater than 2. The classic case is that of the Bell state

$$\phi = (|01\rangle - |10\rangle)/\sqrt{2}.$$

Here

$$\Delta = 6/\sqrt{2} > 2.$$

In general we see that the following inequality is needed in order to violate the Bell inequality

$$(2 - 4(a+d)^2 + 4(ad - bc))/\sqrt{2} > 2.$$

This is equivalent to

$$(\sqrt{2} - 1)/2 < (ad - bc) - (a+d)^2.$$

Since we know that ϕ is entangled exactly when $ad - bc$ is non-zero, this shows that an unentangled state cannot violate the Bell inequality. This formula *also* shows that it is possible for a state to be entangled and yet not violate the Bell inequality. For example, if

$$\phi = (|00\rangle - |01\rangle + |10\rangle + |11\rangle)/2,$$

then $\Delta(\phi)$ satisfies Bell's inequality, but ϕ is an entangled state. We see from this calculation that entanglement in the sense of tensor indecomposability, and entanglement in the sense of Bell inequality violation for a given choice of Bell operators are not equivalent concepts. On the other hand, Benjamin Schumacher has pointed out [77] that any entangled two-qubit state will violate Bell inequalities for an appropriate choice of operators. This deepens the context for our question of the relationship between topological entanglement and quantum entanglement. The Bell inequality violation is an indication of quantum mechanical entanglement. One's intuition suggests that it is *this* sort of entanglement that should have a topological context.

14.6 The Aravind hypothesis

Link diagrams can be used as graphical devices and holders of information. In this vein Aravind [5] proposed that the entanglement of a link should correspond to the entanglement of a state. *Measurement of a link would be modeled by deleting one component of the link.* A key example is the Borromean rings. See Fig. 14.8.

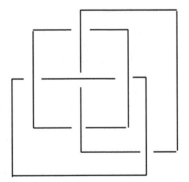

FIGURE 14.8: Borromean rings.

Deleting any component of the Boromean rings yields a remaining pair of unlinked rings. The Borromean rings are entangled, but any two of them are unentangled. In this sense the Borromean rings are analogous to the *GHZ* state $|GHZ\rangle = (1/\sqrt{2})(|000\rangle + |111\rangle)$. Measurement in any factor of the *GHZ* yields an unentangled state. Aravind points out that this property is basis dependent. *We point out that there are states whose entanglement after a measurement is a matter of probability (via quantum amplitudes).* Consider for example the state

$$|\psi\rangle = (1/2)(|000\rangle + |001\rangle + |101\rangle + |110\rangle).$$

Measurement in any coordinate yields an entangled or an unentangled state with equal probability. For example

$$|\psi\rangle = (1/2)(|0\rangle(|00\rangle + |01\rangle) + |1\rangle(|01\rangle + |10\rangle))$$

so that projecting to $|0\rangle$ in the first coordinate yields an unentangled state, while projecting to $|1\rangle$ yields an entangled state, each with equal probability.

New ways to use link diagrams must be invented to map the properties of such states. See [54].

14.7 SU(2) representations of the Artin braid group

The purpose of this section is to determine all the representations of the three-strand Artin braid group B_3 to the special unitary group $SU(2)$ and concomitantly to the unitary group $U(2)$. One regards the groups $SU(2)$ and $U(2)$ as acting on a single qubit, and so $U(2)$ is usually regarded as the group of local unitary transformations in a quantum information setting. If one is looking for a coherent way to represent all unitary transformations by way of braids, then $U(2)$ is the place to start. Here we will show that there are many representations of the three-strand braid group that generate a dense subset of $U(2)$. Thus it is a fact that local unitary transformations can be "generated by braids" in many ways.

We begin with the structure of $SU(2)$. A matrix in $SU(2)$ has the form

$$M = \begin{pmatrix} z & w \\ -\bar{w} & \bar{z} \end{pmatrix},$$

where z and w are complex numbers, and $\bar{z}$ denotes the complex conjugate of z. To be in $SU(2)$ it is required that $Det(M) = 1$ and that $M^\dagger = M^{-1}$ where Det denotes determinant, and $M^\dagger$ is the conjugate transpose of M. Thus if $z = a + bi$

and $w = c + di$ where a, b, c, d are real numbers, and $i^2 = -1$, then

$$M = \begin{pmatrix} a+bi & c+di \\ -c+di & a-bi \end{pmatrix}$$

with $a^2 + b^2 + c^2 + d^2 = 1$. It is convenient to write

$$M = a \begin{pmatrix} 1 & 0 \\ 0 & 1 \end{pmatrix} + b \begin{pmatrix} i & 0 \\ 0 & -i \end{pmatrix} + c \begin{pmatrix} 0 & 1 \\ -1 & 0 \end{pmatrix} + d \begin{pmatrix} 0 & i \\ i & 0 \end{pmatrix},$$

and to abbreviate this decomposition as

$$M = a + bi + cj + dk$$

where

$$1 \equiv \begin{pmatrix} 1 & 0 \\ 0 & 1 \end{pmatrix}, i \equiv \begin{pmatrix} i & 0 \\ 0 & -i \end{pmatrix}, j \equiv \begin{pmatrix} 0 & 1 \\ -1 & 0 \end{pmatrix}, k \equiv \begin{pmatrix} 0 & i \\ i & 0 \end{pmatrix}$$

so that

$$i^2 = j^2 = k^2 = ijk = -1$$

and

$$ij = k, jk = i, ki = j$$
$$ji = -k, kj = -i, ik = -j.$$

The algebra of $1, i, j, k$ is called the *quaternions* after William Rowan Hamilton who discovered this algebra prior to the discovery of matrix algebra. Thus the unit quaternions are identified with $SU(2)$ in this way. We shall use this identification, and some facts about the quaternions to find the $SU(2)$ representations of braiding. First we recall some facts about the quaternions.

1. Note that if $q = a + bi + cj + dk$ (as above), then $q^\dagger = a - bi - cj - dk$ so that $qq^\dagger = a^2 + b^2 + c^2 + d^2 = 1$.

2. A general quaternion has the form $q = a + bi + cj + dk$ where the value of $qq^\dagger = a^2 + b^2 + c^2 + d^2$, is not fixed to unity. The *length* of q is by definition $\sqrt{qq^\dagger}$.

3. A quaternion of the form $ri + sj + tk$ for real numbers r, s, t is said to be a *pure* quaternion. We identify the set of pure quaternions with the vector space of triples (r, s, t) of real numbers R^3.

4. Thus a general quaternion has the form $q = a + bu$ where u is a pure quaternion of unit length and a and b are arbitrary real numbers. A unit quaternion (element of $SU(2)$) has the addition property that $a^2 + b^2 = 1$.

5. If u is a pure unit length quaternion, then $u^2 = -1$. Note that the set of pure unit quaternions forms the two-dimensional sphere $S^2 = \{(r,s,t)|r^2 + s^2 + t^2 = 1\}$ in R^3.

6. If u, v are pure quaternions, then

$$uv = -u \cdot v + u \times v$$

where $u \cdot v$ is the dot product of the vectors u and v, and $u \times v$ is the vector cross product of u and v. In fact, one can take the definition of quaternion multiplication as

$$(a + bu)(c + dv) = ac + bc(u) + ad(v) + bd(-u \cdot v + u \times v),$$

and all the above properties are consequences of this definition. Note that quaternion multiplication is associative.

7. Let $g = a + bu$ be a unit length quaternion so that $u^2 = -1$ and $a = cos(\theta/2), b = sin(\theta/2)$ for a chosen angle θ. Define $\phi_g : R^3 \longrightarrow R^3$ by the equation $\phi_g(P) = gPg^\dagger$, for P any point in R^3, regarded as a pure quaternion. Then ϕ_g is an orientation preserving rotation of R^3 (hence an element of the rotation group $SO(3)$). Specifically, ϕ_g is a rotation about the axis u by the angle θ. The mapping

$$\phi : SU(2) \longrightarrow SO(3)$$

is a two-to-one surjective map from the special unitary group to the rotation group. In quaternionic form, this result was proved by Hamilton and by Rodrigues in the middle of the nineteeth century. The specific formula for $\phi_g(P)$ is shown below:

$$\phi_g(P) = gPg^{-1} = (a^2 - b^2)P + 2ab(P \times u) + 2(P \cdot u)b^2 u.$$

We want a representation of the three-strand braid group in $SU(2)$. This means that we want a homomorphism $\rho : B_3 \longrightarrow SU(2)$, and hence we want elements $g = \rho(s_1)$ and $h = \rho(s_2)$ in $SU(2)$ representing the braid group generators s_1 and s_2. Since $s_1 s_2 s_1 = s_2 s_1 s_2$ is the generating relation for B_3, the only requirement on g and h is that $ghg = hgh$. We rewrite this relation as $h^{-1}gh = ghg^{-1}$, and analyze its meaning in the unit quaternions.

Suppose that $g = a + bu$ and $h = c + dv$ where u and v are unit pure quaternions so that $a^2 + b^2 = 1$ and $c^2 + d^2 = 1$. then $ghg^{-1} = c + d\phi_g(v)$ and $h^{-1}gh =$ $a + b\phi_{h^{-1}}(u)$. Thus it follows from the braiding relation that $a = c$, $b = \pm d$, and that $\phi_g(v) = \pm\phi_{h^{-1}}(u)$. However, in the case where there is a minus sign we have $g = a + bu$ and $h = a - bv = a + b(-v)$. Thus we can now prove the following theorem.

THEOREM 14.4

If $g = a + bu$ and $h = c + dv$ are pure unit quaternions, then, without loss of generality, the braid relation $ghg = hgh$ is true if and only if $h = a + bv$, and $\phi_g(v) = \phi_{h^{-1}}(u)$. Furthermore, given that $g = a + bu$ and $h = a + bv$, the condition $\phi_g(v) = \phi_{h^{-1}}(u)$ is satisfied if and only if $u \cdot v = \frac{a^2-b^2}{2b^2}$ when $u \neq v$. If $u = v$ then then $g = h$ and the braid relation is trivially satisfied.

PROOF   We have proved the first sentence of the theorem in the discussion prior to its statement. Therefore assume that $g = a + bu, h = a + bv$, and $\phi_g(v) = \phi_{h^{-1}}(u)$. We have already stated the formula for $\phi_g(v)$ in the discussion about quaternions:

$$\phi_g(v) = gvg^{-1} = (a^2 - b^2)v + 2ab(v \times u) + 2(v \cdot u)b^2u.$$

By the same token, we have

$$\phi_{h^{-1}}(u) = h^{-1}uh = (a^2 - b^2)u + 2ab(u \times -v) + 2(u \cdot (-v))b^2(-v)$$
$$= (a^2 - b^2)u + 2ab(v \times u) + 2(v \cdot u)b^2(v).$$

Hence we require that

$$(a^2 - b^2)v + 2(v \cdot u)b^2u = (a^2 - b^2)u + 2(v \cdot u)b^2(v).$$

This equation is equivalent to

$$2(u \cdot v)b^2(u - v) = (a^2 - b^2)(u - v).$$

If $u \neq v$, then this implies that

$$u \cdot v = \frac{a^2 - b^2}{2b^2}.$$

This completes the proof of the theorem.   ∎

An Example. Let

$$g = e^{i\theta} = a + bi$$

where $a = cos(\theta)$ and $b = sin(\theta)$. Let

$$h = a + b[(c^2 - s^2)i + 2csk]$$

where $c^2 + s^2 = 1$ and $c^2 - s^2 = \frac{a^2 - b^2}{2b^2}$. Then we can rewrite g and h in matrix form as the matrices G and H. Instead of writing the explicit form of H, we write $H = FGF^\dagger$ where F is an element of $SU(2)$ as shown below.

$$G = \begin{pmatrix} e^{i\theta} & 0 \\ 0 & e^{-i\theta} \end{pmatrix}$$

$$F = \begin{pmatrix} ic & is \\ is & -ic \end{pmatrix}.$$

This representation of braiding where one generator G is a simple matrix of phases, while the other generator $H = FGF^\dagger$ is derived from G by conjugation by a unitary matrix, has the possibility for generalization to representations of braid groups (on greater than three strands) to $SU(n)$ or $U(n)$ for n greater than 2. In fact we shall see just such representations constructed later in this paper, by using a version of topological quantum field theory. The simplest example is given by

$$g = e^{7\pi i/10}$$
$$f = i\tau + k\sqrt{\tau}$$
$$h = frf^{-1}$$

where $\tau^2 + \tau = 1$. Then g and h satisfy $ghg = hgh$ and generate a representation of the three-strand braid group that is dense in $SU(2)$. We shall call this the *Fibonacci* representation of B_3 to $SU(2)$.

Density. Consider representations of B_3 into $SU(2)$ produced by the method of this section. That is consider the subgroup $SU[G,H]$ of $SU(2)$ generated by a pair of elements $\{g,h\}$ such that $ghg = hgh$. We wish to understand when such a representation will be dense in $SU(2)$. We need the following lemma.

LEMMA 14.1
$e^{ai}e^{bj}e^{ci} = cos(b)e^{i(a+c)} + sin(b)e^{i(a-c)}j$. Hence any element of $SU(2)$ can be written in the form $e^{ai}e^{bj}e^{ci}$ for appropriate choices of angles a, b, c.

In fact, if u and v are linearly independent unit vectors in R^3, then any element of $SU(2)$ can be written in the form

$$e^{au}e^{bv}e^{cu}$$

for appropriate choices of the real numbers a,b,c.

PROOF   It is easy to check that

$$e^{ai}e^{bj}e^{ci} = cos(b)e^{i(a+c)} + sin(b)e^{i(a-c)}j.$$

This completes the verification of the identity in the statement of the lemma.

Let v be any unit direction in R^3 and λ an arbitrary angle. We have

$$e^{v\lambda} = cos(\lambda) + sin(\lambda)v,$$

and

$$v = r + si + (p + qi)j$$

where $r^2 + s^2 + p^2 + q^2 = 1$. So

$$\begin{aligned} e^{v\lambda} &= cos(\lambda) + sin(\lambda)[r + si] + sin(\lambda)[p + qi]j \\ &= [(cos(\lambda) + sin(\lambda)r) + sin(\lambda)si] + [sin(\lambda)p + sin(\lambda)qi]j. \end{aligned}$$

By the identity just proved, we can choose angles a,b,c so that

$$e^{v\lambda} = e^{ia}e^{jb}e^{ic}.$$

Hence

$$cos(b)e^{i(a+c)} = (cos(\lambda) + sin(\lambda)r) + sin(\lambda)si$$

and

$$sin(b)e^{i(a-c)} = sin(\lambda)p + sin(\lambda)qi.$$

Suppose we keep v fixed and vary λ. Then the last equations show that this will result in a full variation of b.

Now consider

$$e^{ia'}e^{v\lambda}e^{ic'} = e^{ia'}e^{ia}e^{jb}e^{ic}e^{ib'} = e^{i(a'+a)}e^{jb}e^{i(c+c')}.$$

By the basic identity, this shows that any element of $SU(2)$ can be written in the form

$$e^{ia'}e^{v\lambda}e^{ic'}.$$

Then, by applying a rotation, we finally conclude that if u and v are linearly independent unit vectors in R^3, then any element of $SU(2)$ can be written in the form

$$e^{au}e^{bv}e^{cu}$$

for appropriate choices of the real numbers a, b, c.   ∎

This lemma can be used to verify the density of a representation, by finding two elements A and B in the representation such that the powers of A are dense in the rotations about its axis, and the powers of B are dense in the rotations about its axis, and such that the axes of A and B are linearly independent in R^3. Then by the lemma the set of elements $A^{a+c}B^bA^{a-c}$ is dense in $SU(2)$. It follows, for example, that the Fibonacci representation described above is dense in $SU(2)$, and indeed the generic representation of B_3 into $SU(2)$ will be dense in $SU(2)$. Our next task is to describe representations of the higher braid groups that will extend some of these unitary repressentations of the three-strand braid group. For this we need more topology.

14.8 The bracket polynomial and the Jones polynomial

We now discuss the Jones polynomial. We shall construct the Jones polynomial by using the bracket state summation model [36]. The bracket polynomial, invariant under Reidmeister moves II and III, can be normalized to give an invariant of all three Reidemeister moves. This normalized invariant, with a change of variable, is the Jones polynomial [34, 35]. The Jones polynomial was originally discovered by a different method than the one given here.

The *bracket polynomial*, $\langle K \rangle = \langle K \rangle (A)$, assigns to each unoriented link diagram K a Laurent polynomial in the variable A, such that

1. If K and K' are regularly isotopic diagrams, then $\langle K \rangle = \langle K' \rangle$.

2. If $K \sqcup O$ denotes the disjoint union of K with an extra unknotted and unlinked component O (also called 'loop' or 'simple closed curve' or 'Jordan curve'), then

$$\langle K \sqcup O \rangle = \delta \langle K \rangle,$$

where

$$\delta = -A^2 - A^{-2}.$$

3. $\langle K \rangle$ satisfies the following formulas

$$\langle \chi \rangle = A \langle \asymp \rangle + A^{-1} \langle \rangle \langle \rangle$$

$$\langle \overline{\chi} \rangle = A^{-1} \langle \asymp \rangle + A \langle \rangle \langle \rangle ,$$

where the small diagrams represent parts of larger diagrams that are identical except at the site indicated in the bracket. We take the convention that the letter chi, χ, denotes a crossing where *the curved line is crossing over the straight segment*. The barred letter denotes the switch of this crossing, where *the curved line is undercrossing the straight segment*. See Fig. 14.9 for a graphic illustration of this relation, and an indication of the convention for choosing the labels A and A^{-1} at a given crossing.

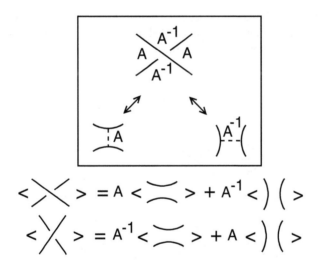

FIGURE 14.9: Bracket smoothings.

It is easy to see that Properties 2 and 3 define the calculation of the bracket on arbitrary link diagrams. The choices of coefficients (A and A^{-1}) and the value of δ make the bracket invariant under the Reidemeister moves II and III. Thus Property 1 is a consequence of the other two properties.

In computing the bracket, one finds the following behavior under Reidemeister move I:

$$\langle \gamma \rangle = -A^3 \langle \smile \rangle$$

and

$$\langle \overline{\gamma} \rangle = -A^{-3} \langle \smile \rangle$$

where γ denotes a curl of positive type as indicated in Fig. 14.10, and $\overline{\gamma}$ indicates a curl of negative type, as also seen in this figure. The type of a curl is the sign of the crossing when we orient it locally. Our convention of signs is also given in Fig. 14.10. Note that the type of a curl does not depend on the orientation we choose. The small arcs on the right hand side of these formulas indicate the removal of the curl from the corresponding diagram.

The bracket is invariant under regular isotopy and can be normalized to an invariant of ambient isotopy by the definition

$$f_K(A) = (-A^3)^{-w(K)} \langle K \rangle (A),$$

where we chose an orientation for K, and where $w(K)$ is the sum of the crossing signs of the oriented link K. $w(K)$ is called the *writhe* of K. The convention for crossing signs is shown in Fig. 14.10.

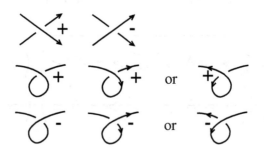

FIGURE 14.10: Crossing signs and curls.

One useful consequence of these formulas is the following *switching formula*

$$A \langle \chi \rangle - A^{-1} \langle \overline{\chi} \rangle = (A^2 - A^{-2}) \langle \asymp \rangle.$$

Note that in these conventions the A-smoothing of χ is $\asymp$, while the A-smoothing of $\overline{\chi}$ is $)($. Properly interpreted, the switching formula above says that you can switch a crossing and smooth it either way and obtain a three diagram relation. This is useful since some computations will simplify quite quickly with the proper choices of switching and smoothing. Remember that it is necessary to keep track of the diagrams up to regular isotopy (the equivalence relation generated by the second and third Reidemeister moves). Here is an example.

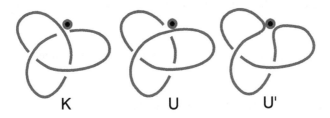

FIGURE 14.11: Trefoil and two relatives.

Fig. 14.11 shows a trefoil diagram K, an unknot diagram U and another unknot diagram U'. Applying the switching formula, we have

$$A^{-1}\langle K\rangle - A\langle U\rangle = (A^{-2} - A^2)\langle U'\rangle$$

and $\langle U\rangle = -A^3$ and $\langle U'\rangle = (-A^{-3})^2 = A^{-6}$. Thus

$$A^{-1}\langle K\rangle - A(-A^3) = (A^{-2} - A^2)A^{-6}.$$

Hence

$$A^{-1}\langle K\rangle = -A^4 + A^{-8} - A^{-4}.$$

Thus

$$\langle K\rangle = -A^5 - A^{-3} + A^{-7}.$$

This is the bracket polynomial of the trefoil diagram K.
Since the trefoil diagram K has writhe $w(K) = 3$, we have the normalized polynomial

$$f_K(A) = (-A^3)^{-3}\langle K\rangle = -A^{-9}(-A^5 - A^{-3} + A^{-7}) = A^{-4} + A^{-12} - A^{-16}.$$

The bracket model for the Jones polynomial is quite useful both theoretically and in terms of practical computations. One of the neatest applications is to simply compute, as we have done, $f_K(A)$ for the trefoil knot K and determine that $f_K(A)$ is not equal to $f_K(A^{-1}) = f_{-K}(A)$. This shows that the trefoil is not ambient isotopic to its mirror image, a fact that is much harder to prove by classical methods.

The State Summation. In order to obtain a closed formula for the bracket, we now describe it as a state summation. Let K be any unoriented link diagram. Define a *state*, S, of K to be a choice of smoothing for each crossing of K. There are two choices for smoothing a given crossing, and thus there are 2^N states of a diagram with N crossings. In a state we label each smoothing with A or A^{-1} according to the left-right convention discussed in Property 3 (see Fig. 14.9). The label is called a *vertex weight* of the state. There are two evaluations related to a state. The first one is the product of the vertex weights, denoted

$$\langle K|S \rangle.$$

The second evaluation is the number of loops in the state S, denoted

$$||S||.$$

Define the *state summation*, $\langle K \rangle$, by the formula

$$\langle K \rangle = \sum_S \langle K|S \rangle \delta^{||S||-1}.$$

It follows from this definition that $\langle K \rangle$ satisfies the equations

$$\langle \chi \rangle = A \langle \asymp \rangle + A^{-1} \langle \rangle \langle \rangle,$$
$$\langle K \sqcup O \rangle = \delta \langle K \rangle,$$
$$\langle O \rangle = 1.$$

The first equation expresses the fact that the entire set of states of a given diagram is the union, with respect to a given crossing, of those states with an A-type smoothing and those with an A^{-1}-type smoothing at that crossing. The second and the third equation are clear from the formula defining the state summation. Hence this state summation produces the bracket polynomial as we have described it at the beginning of the section.

REMARK 14.7   By a change of variables one obtains the original Jones polynomial, $V_K(t)$, for oriented knots and links from the normalized bracket:

$$V_K(t) = f_K(t^{-\frac{1}{4}}).$$

I

REMARK 14.8 The bracket polynomial provides a connection between knot theory and physics, in that the state summation expression for it exhibits it as a generalized partition function defined on the knot diagram. Partition functions are ubiquitous in statistical mechanics, where they express the summation over all states of the physical system of probability weighting functions for the individual states. Such physical partition functions contain large amounts of information about the corresponding physical system. Some of this information is directly present in the properties of the function, such as the location of critical points and phase transition. Some of the information can be obtained by differentiating the partition function, or performing other mathematical operations on it.

There is much more in this connection with statistical mechanics in that the local weights in a partition function are often expressed in terms of solutions to a matrix equation called the Yang–Baxter equation, that turns out to fit perfectly invariance under the third Reidemeister move. As a result, there are many ways to define partition functions of knot diagrams that give rise to invariants of knots and links. The subject is intertwined with the algebraic structure of Hopf algebras and quantum groups, useful for producing systematic solutions to the Yang–Baxter equation. In fact Hopf algebras are deeply connected with the problem of constructing invariants of three-dimensional manifolds in relation to invariants of knots. We have chosen, in this survey paper, to not discuss the details of these approaches, but rather to proceed to Vassiliev invariants and the relationships with Witten's functional integral. The reader is referred to [3, 34, 35, 36, 37, 39, 42, 43, 44, 73, 74, 81, 82] for more information about relationships of knot theory with statistical mechanics, Hopf algebras and quantum groups. For topology, the key point is that Lie algebras can be used to construct invariants of knots and links. I

14.8.1 Quantum computation of the Jones polynomial

Can the invariants of knots and links such as the Jones polynomial be configured as quantum computers? This is an important question because the algorithms to compute the Jones polynomial are known to be *NP*-hard, and so corresponding quantum algorithms may shed light on the relationship of this level of computational complexity with quantum computing (See [28]). Such models can be formulated in terms of the Yang–Baxter equation [36, 37, 43, 48]. The next paragraph explains how this comes about.

In Fig. 14.12, we indicate how topological braiding plus maxima (caps) and minima (cups) can be used to configure the diagram of a knot or link. This also can be translated into algebra by the association of a Yang–Baxter matrix R (not necessarily the R of the previous sections) to each crossing and other matrices to the maxima and minima. There are models of very effective invariants of knots and links such as the Jones polynomial that can be put into this form [48]. In this way of looking at things, the knot diagram can be viewed as a picture, with time as the vertical dimension, of particles arising from the vacuum, interacting (in a two-dimensional space) and finally annihilating one another. The invariant takes the form of an amplitude for this process that is computed through the association of the Yang–Baxter solution R as the scattering matrix at the crossings and the minima and maxima as creation and annihilation operators. Thus we can write the amplitude in the form

$$Z_K = \langle CUP|M|CAP\rangle$$

where $\langle CUP|$ denotes the composition of cups, M is the composition of elementary braiding matrices, and $|CAP\rangle$ is the composition of caps. We regard $\langle CUP|$ as the preparation of this state, and $|CAP\rangle$ as the measurement of this state. In order to view Z_K as a quantum computation, M must be a unitary operator. This is the case when the R-matrices (the solutions to the Yang–Baxter equation used in the model) are unitary. Each R-matrix is viewed as a a quantum gate (or possibly a composition of quantum gates), and the vacuum-vacuum diagram for the knot is interpreted as a quantum computer. This quantum computer will probabilistically (via quantum amplitudes) compute the values of the states in the state sum for Z_K.

We should remark, however, that it is not necessary that the invariant be modeled via solutions to the Yang–Baxter equation. One can use unitary representations of the braid group that are constructed in other ways. In fact, the presently successful quantum algorithms for computing knot invariants indeed use such representations of the braid group, and we shall see this below. Nevertheless, it is useful to point out this analogy between the structure of the knot invariants and quantum computation.

Quantum algorithms for computing the Jones polynomial have been discussed elsewhere. See [48, 55, 1, 58, 2, 86]. Here, as an example, we give a local unitary representation that can be used to compute the Jones polynomial for closures of 3-braids. We analyze this representation by making explicit how the bracket polynomial is computed from it, and showing how the quantum computation devolves to finding the trace of a unitary transformation.

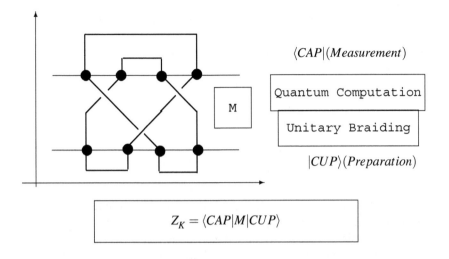

$$Z_K = \langle CAP|M|CUP\rangle$$

FIGURE 14.12: A knot quantum computer.

The idea behind the construction of this representation depends upon the algebra generated by two single qubit density matrices (ket-bras). Let $|v\rangle$ and $|w\rangle$ be two qubits in V, a complex vector space of dimension two over the complex numbers. Let $P = |v\rangle\langle v|$ and $Q = |w\rangle\langle w|$ be the corresponding ket-bras. Note that

$$P^2 = |v|^2 P,$$
$$Q^2 = |w|^2 Q,$$
$$PQP = |\langle v|w\rangle|^2 P,$$
$$QPQ = |\langle v|w\rangle|^2 Q.$$

P and Q generate a representation of the Temperley–Lieb algebra (See Section 14.5 of the present paper). One can adjust parameters to make a representation of the three-strand braid group in the form

$$s_1 \longmapsto rP + sI,$$
$$s_2 \longmapsto tQ + uI,$$

where I is the identity mapping on V and r, s, t, u are suitably chosen scalars. In the following we use this method to adjust such a representation so that it is unitary. Note also that this is a local unitary representation of B_3 to $U(2)$. We leave it as an exercise for the reader to verify that it fits into our general classification of such representations as given in Section 14.3 of the present paper.

The representation depends on two symmetric but non-unitary matrices U_1 and U_2 with

$$U_1 = \begin{bmatrix} d & 0 \\ 0 & 0 \end{bmatrix} = d|w\rangle\langle w|$$

and

$$U_2 = \begin{bmatrix} d^{-1} & \sqrt{1-d^{-2}} \\ \sqrt{1-d^{-2}} & d-d^{-1} \end{bmatrix} = d|v\rangle\langle v|$$

where $w = (1,0)$, and $v = (d^{-1}, \sqrt{1-d^{-2}})$, assuming the entries of v are real. Note that $U_1^2 = dU_1$ and $U_2^2 = dU_2$. Moreover, $U_1U_2U_1 = U_1$ and $U_2U_1U_2 = U_1$. This is an example of a specific representation of the Temperley–Lieb algebra [36, 48]. The desired representation of the Artin braid group is given on the two braid generators for the three-strand braid group by the equations:

$$\Phi(s_1) = AI + A^{-1}U_1,$$
$$\Phi(s_2) = AI + A^{-1}U_2.$$

Here I denotes the 2×2 identity matrix.

For any A with $d = -A^2 - A^{-2}$ these formulas define a representation of the braid group. With $A = e^{i\theta}$, we have $d = -2cos(2\theta)$. We find a specific range of angles θ in the following disjoint union of angular intervals

$$\theta \in [0, \pi/6] \sqcup [\pi/3, 2\pi/3] \sqcup [5\pi/6, 7\pi/6] \sqcup [4\pi/3, 5\pi/3] \sqcup [11\pi/6, 2\pi]$$

that give unitary representations of the three-strand braid group. Thus a specialization of a more general represention of the braid group gives rise to a continuous family of unitary representations of the braid group.

LEMMA 14.2

Note that the traces of these matrices are given by the formulas $tr(U_1) = tr(U_2) = d$ while $tr(U_1U_2) = tr(U_2U_1) = 1$. If b is any braid, let $I(b)$ denote the sum of the exponents in the braid word that expresses b. For b a three-strand braid, it follows that

$$\Phi(b) = A^{I(b)}I + \Pi(b)$$

where I is the 2×2 identity matrix and $\Pi(b)$ is a sum of products in the Temperley–Lieb algebra involving U_1 and U_2.

We omit the proof of this lemma. It is a calculation. To see it, consider an example. Suppose that $b = s_1 s_2^{-1} s_1$. Then

$$\Phi(b) = \Phi(s_1 s_2^{-1} s_1) = \Phi(s_1)\Phi(s_2^{-1})\Phi(s_1)$$
$$= (AI + A^{-1}U_1)(A^{-1}I + AU_2)(AI + A^{-1}U_1).$$

The sum of products over the generators U_1 and U_2 of the Temperley–Lieb algebra comes from expanding this expression.

Since the Temperley–Lieb algebra in this dimension is generated by I, U_1, U_2, $U_1 U_2$ and $U_2 U_1$, it follows that the value of the bracket polynomial of the closure of the braid b, denoted $\langle \bar{b} \rangle$, can be calculated directly from the trace of this representation, except for the part involving the identity matrix. The result is the equation

$$\langle \bar{b} \rangle = A^{I(b)}d^2 + tr(\Pi(b))$$

where $\bar{b}$ denotes the standard braid closure of b, and the sharp brackets denote the bracket polynomial. From this we see at once that

$$\langle \bar{b} \rangle = tr(\Phi(b)) + A^{I(b)}(d^2 - 2).$$

It follows from this calculation that the question of computing the bracket polynomial for the closure of the three-strand braid b is mathematically equivalent to the problem of computing the trace of the unitary matrix $\Phi(b)$.

The Hadamard Test

In order to (quantum) compute the trace of a unitary matrix U, one can use the *Hadamard test* to obtain the diagonal matrix elements $\langle \psi | U | \psi \rangle$ of U. The trace is then the sum of these matrix elements as $|\psi\rangle$ runs over an orthonormal basis for the vector space. We first obtain

$$\frac{1}{2} + \frac{1}{2} Re \langle \psi | U | \psi \rangle$$

as an expectation by applying the Hadamard gate H

$$H|0\rangle = \frac{1}{\sqrt{2}}(|0\rangle + |1\rangle)$$

$$H|1\rangle = \frac{1}{\sqrt{2}}(|0\rangle - |1\rangle)$$

to the first qubit of

$$C_U \circ (H \otimes 1)|0\rangle|\psi\rangle = \frac{1}{\sqrt{2}}(|0\rangle \otimes |\psi\rangle + |1\rangle \otimes U|\psi\rangle).$$

Here C_U denotes controlled U, acting as U when the control bit is $|1\rangle$ and the identity mapping when the control bit is $|0\rangle$. We measure the expectation for the first qubit $|0\rangle$ of the resulting state

$$\frac{1}{2}(H|0\rangle \otimes |\psi\rangle + H|1\rangle \otimes U|\psi\rangle) = \frac{1}{2}((|0\rangle + |1\rangle) \otimes |\psi\rangle + (|0\rangle - |1\rangle) \otimes U|\psi\rangle)$$

$$= \frac{1}{2}(|0\rangle \otimes (|\psi\rangle + U|\psi\rangle) + |1\rangle \otimes (|\psi\rangle - U|\psi\rangle)).$$

This expectation is

$$\frac{1}{2}((\langle\psi| + \langle\psi|U^\dagger)(|\psi\rangle + U|\psi\rangle)) = \frac{1}{2} + \frac{1}{2}Re\langle\psi|U|\psi\rangle.$$

The imaginary part is obtained by applying the same procedure to

$$\frac{1}{\sqrt{2}}(|0\rangle \otimes |\psi\rangle - i|1\rangle \otimes U|\psi\rangle.$$

This is the method used in [1], and the reader may wish to contemplate its efficiency in the context of this simple model. Note that the Hadamard test enables this quantum computation to estimate the trace of any unitary matrix U by repeated trials that estimate individual matrix entries $\langle\psi|U|\psi\rangle$.

14.9   Quantum topology, cobordism categories, Temperley–Lieb algebra, and topological quantum field theory

The purpose of this section is to discuss the general idea behind topological quantum field theory, and to illustrate its application to basic quantum mechanics and quantum mechanical formalism. It is useful in this regard to have available the concept of *category*, and we shall begin the section by discussing this far-reaching mathematical concept.

DEFINITION 14.1   *A category* Cat *consists in two related collections:*

1. *Obj(Cat), the* objects *of Cat, and*

2. *Morph(Cat), the* morphisms *of Cat.*

satisfying the following axioms:

1. *Each morphism f is associated to two objects of Cat, the* domain *of f and the* codomain *of f. Letting A denote the domain of f and B denote the codomain of f, it is customary to denote the morphism f by the arrow notation $f : A \longrightarrow B$.*

2. *Given $f : A \longrightarrow B$ and $g : B \longrightarrow C$ where A, B and C are objects of Cat, then there exists an associated morphism $g \circ f : A \longrightarrow C$ called the* composition *of f and g.*

3. *To each object A of Cat there is a* unique *identity morphism $1_A : A \longrightarrow A$ such that $1_A \circ f = f$ for any morphism f with codomain A, and $g \circ 1_A = g$ for any morphism g with domain A.*

4. *Given three morphisms $f : A \longrightarrow B$, $g : B \longrightarrow C$ and $h : C \longrightarrow D$, then composition is associative. That is*

$$(h \circ g) \circ f = h \circ (g \circ f).$$

If Cat_1 and Cat_2 are two categories, then a functor *$F : Cat_1 \longrightarrow Cat_2$ consists in functions $F_O : Obj(Cat_1) \longrightarrow Obj(Cat_2)$ and $F_M : Morph(Cat_1) \longrightarrow Morph(Cat_2)$ such that identity morphisms and composition of morphisms are preserved under these mappings. That is (writing just F for F_O and F_M),*

1. *$F(1_A) = 1_{F(A)}$,*

2. *$F(f : A \longrightarrow B) = F(f) : F(A) \longrightarrow F(B)$,*

3. *$F(g \circ f) = F(g) \circ F(f)$.*

A functor $F : Cat_1 \longrightarrow Cat_2$ is a structure preserving mapping from one category to another. It is often convenient to think of the image of the functor F as an interpretation *of the first category in terms of the second. We shall use this terminology below and sometimes refer to an interpretation without specifying all the details of the functor that describes it.*

The notion of category is a broad mathematical concept, encompassing many fields of mathematics. Thus one has the category of sets where the objects are sets (collections) and the morphisms are mappings between sets. One has the category of topological spaces where the objects are spaces and the morphisms are continuous mappings of topological spaces. One has the category of groups where the objects are groups and the morphisms are homomorphisms of groups. Functors are structure preserving mappings from one category to another. For example, the fundamental group is a functor from the category of topological spaces with base point to the category of groups. In all the examples mentioned so far, the morphisms in the category are restrictions of mappings in the category of sets, but this is not necessarily the case. For example, any group G can be regarded as a category, $Cat(G)$, with one object $*$. The morphisms from $*$ to itself are the elements of the group and composition is group multiplication. In this example, the object has no internal structure and all the complexity of the category is in the morphisms.

The Artin braid group B_n can be regarded as a category whose single object is an ordered row of points $[n] = \{1, 2, 3, ..., n\}$. The morphisms are the braids themselves and composition is the multiplication of the braids. A given ordered row of points is interpreted as the starting or ending row of points at the bottom or the top of the braid. In the case of the braid category, the morphisms have both external and internal structure. Each morphism produces a permutation of the ordered row of points (corresponding to the beginning and ending points of the individual braid strands), and weaving of the braid is extra structure beyond the object that is its domain and codomain. Finally, for this example, we can take all the braid groups B_n (n a positive integer) under the wing of a single category, $Cat(B)$, whose objects are all ordered rows of points $[n]$, and whose morphisms are of the form $b : [n] \longrightarrow [n]$ where b is a braid in B_n. The reader may wish to have morphisms between objects with different n. We will have this shortly in the Temperley–Lieb category and in the category of tangles.

The *n-cobordism category*, $Cob[n]$, has as its objects smooth manifolds of dimension n, and as its morphisms, smooth manifolds M^{n+1} of dimension $n+1$ with a partition of the boundary, ∂M^{n+1}, into two collections of n-manifolds that we denote by $L(M^{n+1})$ and $R(M^{n+1})$. We regard M^{n+1} as a morphism from $L(M^{n+1})$ to $R(M^{n+1})$

$$M^{n+1} : L(M^{n+1}) \longrightarrow R(M^{n+1}).$$

As we shall see, these cobordism categories are highly significant for quantum mechanics, and the simplest one, $Cob[0]$ is directly related to the Dirac notation of bras and kets and to the Temperley–Lieb algebra. We shall concentrate in this section on these cobordism categories, and their relationships with quantum mechanics.

One can choose to consider either oriented or non-oriented manifolds, and within unoriented manifolds there are those that are orientable and those that are not orientable. In this section we will implicitly discuss only orientable manifolds, but we shall not specify an orientation. In the next section, with the standard definition of topological quantum field theory, the manifolds will be oriented. The definitions of the cobordism categories for oriented manifolds go over mutatis mutandis.

Lets begin with $Cob[0]$. Zero dimensional manifolds are just collections of points. The simplest zero dimensional manifold is a single point p. We take p to be an object of this category and also $*$, where $*$ denotes the empty manifold (i.e. the empty set in the category of manifolds). The object $*$ occurs in $Cob[n]$ for every n, since it is possible that either the left set or the right set of a morphism is empty. A line segment S with boundary points p and q is a morphism from p to q.

$$S : p \longrightarrow q.$$

See Fig. 14.13. In this figure we have illustrated the morphism from p to p. The simplest convention for this category is to take this morphism to be the identity. Thus if we look at the subcategory of $Cob[0]$ whose only object is p, then the only morphism is the identity morphism. Two points occur as the boundary of an interval. The reader will note that $Cob[0]$ and the usual arrow notation for morphisms are very closely related. This is a place where notation and mathematical structure share common elements. In general the objects of $Cob[0]$ consist in the empty object $*$ and non-empty rows of points, symbolized by

$$p \otimes p \otimes \cdots \otimes p \otimes p.$$

Fig. 14.13 also contains a morphism

$$p \otimes p \longrightarrow *$$

and the morphism

$$* \longrightarrow p \otimes p.$$

The first represents a cobordism of two points to the empty set (via the bounding curved interval). The second represents a cobordism from the empty set to two points.

In Fig. 14.14, we have indicated more morphisms in $Cob[0]$, and we have named the morphisms just discussed as

$$|\Omega\rangle : p \otimes p \longrightarrow *,$$

$$\langle\Theta| : * \longrightarrow p \otimes p.$$

The point to notice is that the usual conventions for handling Dirac bra-kets are essentially the same as the compostion rules in this topological category.

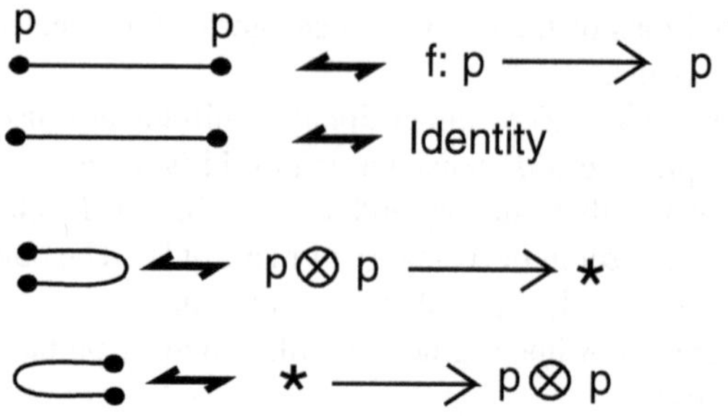

FIGURE 14.13: Elementary cobordisms.

Thus in Fig. 14.14

$$\langle\Theta|\circ|\Omega\rangle = \langle\Theta|\Omega\rangle : * \longrightarrow *$$

represents a cobordism from the empty manifold to itself. This cobordism is topologically a circle and, in the Dirac formalism is interpreted as a scalar. In order to interpret the notion of scalar we would have to map the cobordism category to the category of vector spaces and linear mappings. We shall discuss this after describing the similarities with quantum mechanical formalism. Nevertheless, the reader should note that if V is a vector space over the complex numbers $\mathscr{C}$, then a linear mapping from $\mathscr{C}$ to $\mathscr{C}$ is determined by the image of 1, and hence is characterized by the scalar that is the image of 1. In this sense a mapping $\mathscr{C} \longrightarrow \mathscr{C}$ can be regarded as a possible image in vector spaces of the abstract structure $\langle\Theta|\Omega\rangle : * \longrightarrow *$. It is therefore assumed that in $Cob[0]$ the composition with the morphism $\langle\Theta|\Omega\rangle$ commutes with any other morphism. In that way $\langle\Theta|\Omega\rangle$ behaves like a scalar in the cobordism category. In general, an $n+1$ manifold without boundary behaves as a scalar in $Cob[n]$, and if a manifold M^{n+1} can be written as a union of two submanifolds L^{n+1} and R^{n+1} so that that an n-manifold W^n is their common boundary:

$$M^{n+1} = L^{n+1} \cup R^{n+1}$$

with

$$L^{n+1} \cap R^{n+1} = W^n$$

then, we can write

$$\langle M^{n+1} \rangle = \langle L^{n+1} \cup R^{n+1} \rangle = \langle L^{n+1} | R^{n+1} \rangle,$$

and $\langle M^{n+1} \rangle$ will be a scalar (morphism that commutes with all other morphisms) in the category $Cob[n]$.

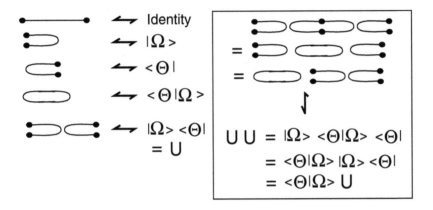

FIGURE 14.14: Bras, kets and projectors.

Getting back to the contents of Fig. 14.14, note how the zero dimensional cobordism category has structural parallels to the Dirac ket–bra formalism

$$U = |\Omega\rangle\langle\Theta|$$
$$UU = |\Omega\rangle\langle\Theta|\Omega\rangle\langle\Theta| = \langle\Theta|\Omega\rangle|\Omega\rangle\langle\Theta| = \langle\Theta|\Omega\rangle U.$$

In the cobordism category, the bra–ket and ket–bra formalism is seen as patterns of connection of the one-manifolds that realize the cobordisms.

Now view Fig. 14.15. This figure illustrates a morphism S in $Cob[0]$ that requires two crossed line segments for its planar representation. Thus S can be regarded as a non-trivial permutation, and $S^2 = I$ where I denotes the identity morphisms for a two-point row. From this example, it is clear that $Cob[0]$ contains the structure of all the symmetric groups and more. In fact, if we take the subcategory of $Cob[0]$ consisting of all morphisms from $[n]$ to $[n]$ for a fixed positive integer n, then this gives the well-known *Brauer algebra* (see [13]) extending the symmetric group by allowing any connections among the points in the two rows. In this sense, one could call $Cob[0]$ the *Brauer category*. We shall return to this point of view later.

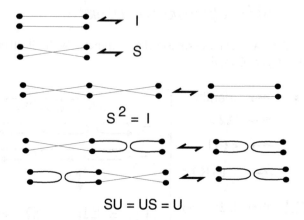

$$S^2 = I$$

$$SU = US = U$$

FIGURE 14.15: Permutations.

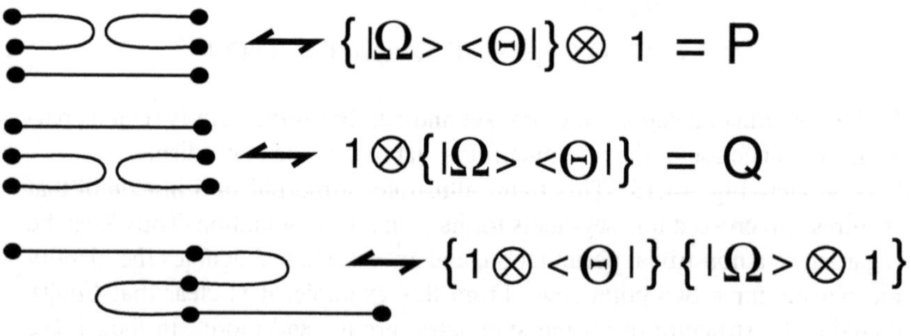

$$\{|\Omega> <\Theta|\} \otimes 1 = P$$

$$1 \otimes \{|\Omega> <\Theta|\} = Q$$

$$\{1 \otimes <\Theta|\}\{|\Omega> \otimes 1\}$$

FIGURE 14.16: Projectors in tensor lines and elementary topology.

In this section, we shall be concentrating on the part of $Cob[0]$ that does not involve permutations. This part can be characterized by those morphisms that can be represented by planar diagrams without crossings between any of the line segments (the one-manifolds). We shall call this crossingless subcategory of $Cob[0]$ the *Temperley–Lieb Category* and denote it by $CatTL$. In $CatTL$ we have the subcategory $TL[n]$ whose only objects are the row of n points and the empty object $*$, and whose morphisms can all be represented by configurations that embed in the plane as in the morphisms P and Q in Fig. 14.16. Note that with the empty object $*$, the morphism whose diagram is a single loop appears in $TL[n]$ and is taken to commute with all other morphisms.

The *Temperley–Lieb Algebra*, $AlgTL[n]$ is generated by the morphisms in $TL[n]$ that go from $[n]$ to itself. Up to multiplication by the loop, the product (composition) of two such morphisms is another flat morphism from $[n]$ to itself. For algebraic purposes the loop $* \longrightarrow *$ is taken to be a scalar algebraic variable δ that commutes with all elements in the algebra. Thus the equation

$$UU = \langle \Theta | \Omega \rangle U$$

becomes

$$UU = \delta U$$

in the algebra. In the algebra we are allowed to add morphisms formally and this addition is taken to be commutative. Initially the algebra is taken with coefficients in the integers, but a different commutative ring of coefficients can be chosen and the value of the loop may be taken in this ring. For example, for quantum mechanical applications it is natural to work over the complex numbers. The multiplicative structure of $AlgTL[n]$ can be described by generators and relations as follows: Let I_n denote the identity morphism from $[n]$ to $[n]$. Let U_i denote the morphism from $[n]$ to $[n]$ that connects k with k for $k < i$ and $k > i+1$ from one row to the other, and connects i to $i+1$ in each row. Then the algebra $AlgTL[n]$ is generated by $\{I_n, U_1, U_2, \cdots, U_{n-1}\}$ with relations

$$U_i^2 = \delta U_i$$
$$U_i U_{i+1} U_i = U_i$$
$$U_i U_j = U_j U_i : |i - j| > 1.$$

These relations are illustrated for three strands in Fig. 14.16. We leave the commuting relation for the reader to draw in the case where n is four or greater. For a proof that these are indeed all the relations, see [51].

Figs. 14.16 and 14.17 indicate how the zero dimensional cobordism category contains structure that goes well beyond the usual Dirac formalism. By tensoring the ket–bra on one side or another by identity morphisms, we obtain the beginnings of the Temperley–Lieb algebra and the Temperley–Lieb category. Thus Fig. 14.17 illustrates the morphisms P and Q obtained by such tensoring, and the relation $PQP = P$ which is the same as $U_1 U_2 U_1 = U_1$

Note the composition at the bottom of the Fig. 14.17. Here we see a composition of the identity tensored with a ket, followed by a bra tensored with the identity. The diagrammatic for this association involves "straightening" the curved structure of the morphism to a straight line. In Fig. 14.18 we have elaborated this situation even further, pointing out that in this category each of the morphisms $\langle \Theta |$ and $| \Omega \rangle$ can be seen, by straightening, as mappings from the generating object to itself. We have denoted these corresponding morphisms by Θ and Ω respectively. In this way there is a correspondence between morphisms $p \otimes p \longrightarrow *$ and morphisms $p \longrightarrow p$.

In Fig. 14.18 we have illustrated the generalization of the straightening procedure of Fig. 14.17. In Fig. 14.17 the straightening occurs because the connection structure in the morphism of $Cob[0]$ does not depend on the wandering of curves in diagrams for the morphisms in that category. Nevertheless, one can envisage a more complex interpretation of the morphisms where each one-manifold (line segment) has a label, and a multiplicity of morphisms can correspond to a single line segment. This is exactly what we expect in interpretations. For example, we can interpret the line segment $[1] \longrightarrow [1]$ as a mapping from a vector space V to itself. Then $[1] \longrightarrow [1]$ is the diagrammatic abstraction for $V \longrightarrow V$, and there are many instances of linear mappings from V to V.

At the vector space level there is a duality between mappings $V \otimes V \longrightarrow \mathscr{C}$ and linear maps $V \longrightarrow V$. Specifically, let

$$\{|0\rangle, \cdots, |m\rangle\}$$

be a basis for V. Then $\Theta : V \longrightarrow V$ is determined by

$$\Theta |i\rangle = \Theta_{ij} |j\rangle$$

(where we have used the Einstein summation convention on the repeated index j) corresponds to the bra

$$\langle \Theta | : V \otimes V \longrightarrow \mathscr{C}$$

defined by

$$\langle \Theta | ij \rangle = \Theta_{ij}.$$

Given $\langle \Theta | : V \otimes V \longrightarrow \mathscr{C}$, we associate $\Theta : V \longrightarrow V$ in this way.

Comparing with the diagrammatic for the category $Cob[0]$, we say that $\Theta : V \longrightarrow V$ is obtained by *straightening* the mapping

$$\langle \Theta | : V \otimes V \longrightarrow \mathscr{C}.$$

Note that in this interpretation, the bras and kets are defined relative to the tensor product of V with itself and $[2]$ is interpreted as $V \otimes V$. If we interpret $[2]$ as a single vector space W, then the usual formalisms of bras and kets still pass over from the cobordism category.

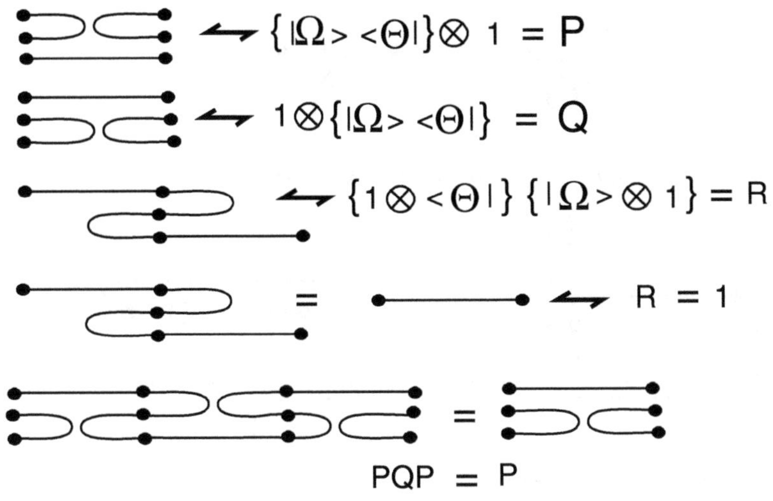

$\{|\Omega\rangle \langle\Theta|\} \otimes 1 = P$

$1 \otimes \{|\Omega\rangle \langle\Theta|\} = Q$

$\{1 \otimes \langle\Theta|\}\{|\Omega\rangle \otimes 1\} = R$

$= \qquad R = 1$

$PQP = P$

FIGURE 14.17: The basic Temperley–Lieb relation.

Fig. 14.18 illustrates the straightening of $|\Theta\rangle$ and $\langle\Omega|$, and the straightening of a composition of these applied to $|\psi\rangle$, resulting in $|\phi\rangle$. In the left-hand part of the bottom of Fig. 14.18 we illustrate the preparation of the tensor product $|\Theta\rangle \otimes |\psi\rangle$ followed by a successful measurement by $\langle\Omega|$ in the second two tensor factors. The resulting single qubit state, as seen by straightening, is $|\phi\rangle = \Theta \circ \Omega |\psi\rangle$.

From this, we see that it is possible to reversibly, indeed unitarily, transform a state $|\psi\rangle$ via a combination of preparation and measurement just so long as the straightenings of the preparation and measurement (Θ and Ω) are each invertible (unitary). This is the key to teleportation [50, 19, 20]. In the standard teleportation procedure one chooses the preparation Θ to be (up to normalization) the two-dimensional identity matrix so that $|\theta\rangle = |00\rangle + |11\rangle$. If the successful measurement Ω is also the identity, then the transmitted state $|\phi\rangle$ will be equal to $|\psi\rangle$. In general we will have $|\phi\rangle = \Omega |\psi\rangle$. One can then choose a basis of measurements $|\Omega\rangle$, each corresponding to a unitary transformation Ω so that the recipient of the transmission can rotate the result by the inverse of

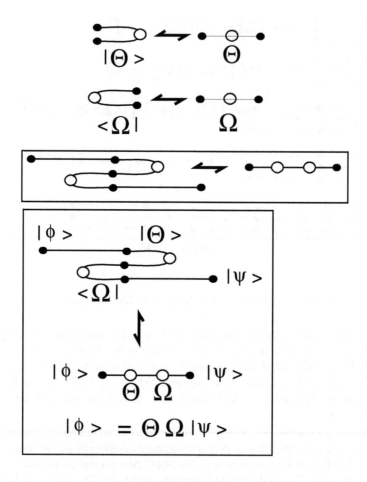

FIGURE 14.18: The key to teleportation.

Ω to reconstitute $|\psi\rangle$ if he is given the requisite information. This is the basic design of the teleportation procedure.

There is much more to say about the category $Cob[0]$ and its relationship with quantum mechanics. We will stop here, and invite the reader to explore further. Later in this paper, we shall use these ideas in formulating our representations of the braid group. For now, we point out how things look as we move upward to $Cob[n]$ for $n > 0$. In Fig. 14.19 we show typical cobordisms (morphisms) in $Cob[1]$ from two circles to one circle and from one circle to two circles. These are often called "pairs of pants". Their composition is a surface of genus one seen as a morphism from two circles to two circles. The bottom of the figure indicates a ket-bra in this dimension in the form of a mapping from one circle to one circle as a composition of a cobordism of a circle to the empty set and a cobordism from the empty set to a circle (circles bounding disks). As we go to higher dimensions the structure of cobordisms becomes more interesting and more complicated. It is remarkable that there is so much structure in the lowest dimensions of these categories.

14.10   Braiding and topological quantum field theory

The purpose of this section is to discuss in a very general way how braiding is related to topological quantum field theory. In the section to follow, we will use the Temperley–Lieb recoupling theory to produce specific unitary representations of the Artin braid group.

The ideas in the subject of topological quantum field theory (TQFT) are well expressed in the book [6] by Michael Atiyah and the paper [85] by Edward Witten. Here is Atiyah's definition:

DEFINITION 14.2   *A TQFT in dimension d is a functor* $Z(\Sigma)$ *from the cobordism category* $Cob[d]$ *to the category Vect of vector spaces and linear mappings which assigns*

1. *a finite dimensional vector space* $Z(\Sigma)$ *to each compact, oriented d-dimensional manifold* Σ.

2. *a vector* $Z(Y) \in Z(\Sigma)$ *for each compact, oriented* $(d+1)$-*dimensional manifold* Y *with boundary* Σ.

3. *a linear mapping* $Z(Y) : Z(\Sigma_1) \longrightarrow Z(\Sigma_2)$ *when* Y *is a* $(d+1)$-*manifold*

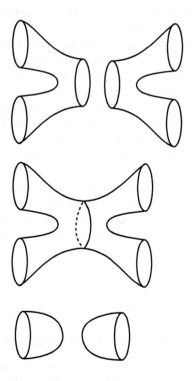

FIGURE 14.19: Corbordisms of one-manifolds are surfaces.

*that is a cobordism between Σ_1 and Σ_2 (whence the boundary of Y
is the union of Σ_1 and $-\Sigma_2$.*

The functor satisfies the following axioms.

1. *$Z(\Sigma^\dagger) = Z(\Sigma)^\dagger$ where $\Sigma^\dagger$ denotes the manifold Σ with the opposite
   orientation and $Z(\Sigma)^\dagger$ is the dual vector space.*

2. *$Z(\Sigma_1 \cup \Sigma_2) = Z(\Sigma_1) \otimes Z(\Sigma_2)$ where $\cup$ denotes disjoint union.*

3. *If Y_1 is a cobordism from Σ_1 to Σ_2, Y_2 is a cobordism from Σ_2 to Σ_3
   and Y is the composite cobordism $Y = Y_1 \cup_{\Sigma_2} Y_2$, then*

$$Z(Y) = Z(Y_2) \circ Z(Y_1) : Z(\Sigma_1) \longrightarrow Z(\Sigma_2)$$

is the composite of the corresponding linear mappings.

4. $Z(\phi) = \mathscr{C}$ ($\mathscr{C}$ denotes the complex numbers) for the empty manifold ϕ.

5. With $\Sigma \times I$ (where I denotes the unit interval) denoting the identity cobordism from Σ to Σ, $Z(\Sigma \times I)$ is the identity mapping on $Z(\Sigma)$.

Note that, in this view a TQFT is basically a functor from the cobordism categories defined in the last section to vector spaces over the complex numbers. We have already seen that in the lowest dimensional case of cobordisms of zero-dimensional manifolds, this gives rise to a rich structure related to quantum mechanics and quantum information theory. The remarkable fact is that the case of three-dimensions is also related to quantum theory, and to the lower-dimensional versions of the TQFT. This gives a significant way to think about three-manifold invariants in terms of lower dimensional patterns of interaction. Here follows a brief description.

Regard the three-manifold as a union of two handlebodies with boundary an orientable surface S_g of genus g. The surface is divided up into trinions as illustrated in Fig. 14.20. A *trinion* is a surface with boundary that is topologically equivalent to a sphere with three punctures. The trinion constitutes in itself a cobordism in $Cob[1]$ from two circles to a single circle, or from a single circle to two circles, or from three circles to the empty set. The *pattern* of a trinion is a trivalent graphical vertex, as illustrated in Fig. 14.21. In that figure we show the trivalent vertex graphical pattern drawn on the surface of the trinion, forming a graphical pattern for this combordism. It should be clear from this figure that any cobordism in $Cob[1]$ can be diagrammed by a trivalent graph, so that the category of trivalent graphs (as morphisms from ordered sets of points to ordered sets of points) has an image in the category of cobordisms of compact one-dimensional manifolds. Given a surface S (possibly with boundary) and a decomposition of that surface into triions, we associate to it a trivalent graph $G(S,t)$ where t denotes the particular trinion decomposition.

In this correspondence, distinct graphs can correspond to topologically i-dentical cobordisms of circles, as illustrated in Fig. 14.22. It turns out that the graphical structure is important, and that it is extraordinarily useful to articulate transformations between the graphs that correspond to the homeomorphisms of the corresponding surfaces. The beginning of this structure is indicated in the bottom part of Fig. 14.22.

In Fig. 14.23 we illustrate another feature of the relationship between surfaces and graphs. At the top of the figure we indicate a homeomorphism between a twisted trinion and a standard trinion. The homeomorphism leaves the ends of the trinion (denoted A,B and C) fixed while undoing the internal twist. This can be accomplished as an ambient isotopy of the embeddings in three-dimensional space that are indicated by this figure. Below this isotopy we indicate the corresponding graphs. In the graph category there will have to be a transformation between a braided and an unbraided trivalent vertex that corresponds to this homeomorphism.

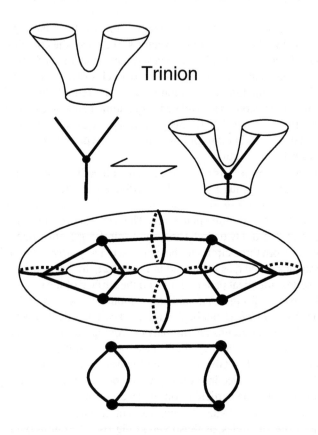

FIGURE 14.20: Decomposition of a surface into trinions.

From the point of view that we shall take in this paper, the key to the mathematical structure of three-dimensional TQFT lies in the trivalent graphs, including the braiding of graphical arcs. We can think of these braided graphs as representing idealized Feynman diagrams, with the trivalent vertex as the basic particle interaction vertex, and the braiding of lines representing an interaction resulting from an exchange of particles. In this view one thinks of the particles as moving in a two-dimensional medium, and the diagrams of braid-

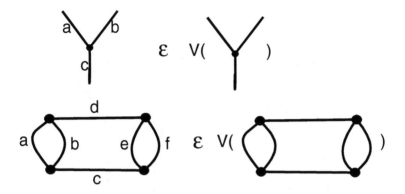

FIGURE 14.21: Trivalent vectors.

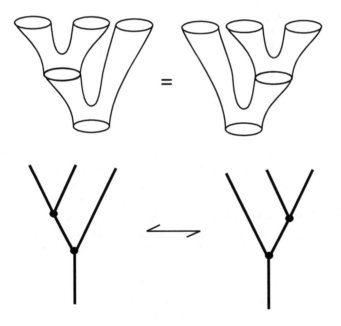

FIGURE 14.22: Trinion associativity.

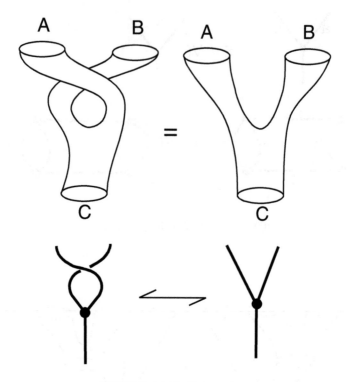

FIGURE 14.23: Tube twist.

ing and trivalent vertex interactions as indications of the temporal events in the system, with time indicated in the direction of the morphisms in the category. Adding such graphs to the category of knots and links is an extension of the *tangle category* where one has already extended braids to allow any embedding of strands and circles that start in n ordered points and end in m ordered points. The tangle category includes the braid category and the Temperley–Lieb category. These are both included in the category of braided trivalent graphs.

Thinking of the basic trivalent vertex as the form of a particle interaction there will be a set of particle states that can label each arc incident to the vertex. In Fig. 14.21 we illustrate the labeling of the trivalent graphs by such particle states. In the next two sections we will see specific rules for labeling such states. Here it suffices to note that there will be some restrictions on these labels, so that a trivalent vertex has a set of possible labelings. Similarly, any trivalent graph will have a set of admissible labelings. These are the possible particle processes that this graph can support. We take the set of admissible labelings of a given graph G as a basis for a vector space $V(G)$ over the complex numbers. This vector space is the space of *processes* associated with the graph G. Given a surface S and a decomposition t of the surface into trinions, we have the associated graph $G(S,t)$ and hence a vector space of processes $V(G(S,t))$. It is desirable to have this vector space independent of the particular decomposition into trinions. If this can be accomplished, then the set of vector spaces and linear mappings associated to the surfaces can consitute a functor from the category of cobordisms of one-manifolds to vector spaces, and hence gives rise to a one-dimensional topological quantum field theory. To this end we need some properties of the particle interactions that will be described below.

A *spin network* is, by definition a labelled trivalent graph in a category of graphs that satisfy the properties outlined in the previous paragraph. We shall detail the requirements below.

The simplest case of this idea is C. N. Yang's original interpretation of the Yang–Baxter equation [87]. Yang articulated a quantum field theory in one dimension of space and one dimension of time in which the R-matrix giving the scattering amplitudes for an interaction of two particles whose (let us say) spins corresponded to the matrix indices so that R_{ab}^{cd} is the amplitude for particles of spin a and spin b to interact and produce particles of spin c and d. Since these interactions are between particles in a line, one takes the convention that the particle with spin a is to the left of the particle with spin b, and the particle with spin c is to the left of the particle with spin d. If one follows the concatenation of such interactions, then there is an underlying permutation that is obtained by following strands from the bottom to the top of the diagram (thinking of time as moving up the page). Yang designed the Yang–Baxter equation for R so that *the amplitudes for a composite process depend only on the underlying permutation corresponding to the process and not on the individual sequences of interactions.*

In taking over the Yang–Baxter equation for topological purposes, we can use the same interpretation, but think of the diagrams with their under- and over-crossings as modeling events in a spacetime with two dimensions of space and one dimension of time. The extra spatial dimension is taken in displacing the woven strands perpendicular to the page, and allows us to use braiding operators R and R^{-1} as scattering matrices. Taking this picture to heart, one can add other particle properties to the idealized theory. In particular one can add fusion and creation vertices where in fusion two particles interact to become a single particle and in creation one particle changes (decays) into two particles. These are the trivalent vertices discussed above. Matrix elements corresponding to trivalent vertices can represent these interactions. See Fig. 14.24.

FIGURE 14.24: Creation and fusion.

Once one introduces trivalent vertices for fusion and creation, there is the question how these interactions will behave in respect to the braiding operators. There will be a matrix expression for the compositions of braiding and fusion or creation as indicated in Fig. 14.25. Here we will restrict ourselves to showing the diagrammatics with the intent of giving the reader a flavor of these structures. It is natural to assume that braiding intertwines with creation as shown in Fig. 14.27 (similarly with fusion). This intertwining identity is clearly the sort of thing that a topologist will love, since it indicates that the diagrams can be interpreted as embeddings of graphs in three-dimensional s-pace, and it fits with our interpretation of the vertices in terms of trinions. Fig. 14.25 illustrates the Yang–Baxter equation. The intertwining identity is an assumption like the Yang–Baxter equation itself, that simplifies the mathematical structure of the model.

It is to be expected that there will be an operator that expresses the recoupling of vertex interactions as shown in Fig. 14.28 and labeled by Q. This corresponds to the associativity at the level of trinion combinations shown in Fig. 14.22. The actual formalism of such an operator will parallel the mathematics of recoupling for angular momentum. See for example [38]. If one just considers the abstract structure of recoupling then one sees that for trees with four branches (each with a single root) there is a cycle of length five as shown in Fig. 14.29. One can start with any pattern of three-vertex interactions and go through a sequence of five recouplings that bring one back to the same tree from which one started. *It is a natural simplifying axiom to assume that this composition is the identity mapping.* This axiom is called the *pentagon identity*.

Finally there is a hexagonal cycle of interactions between braiding, recou-

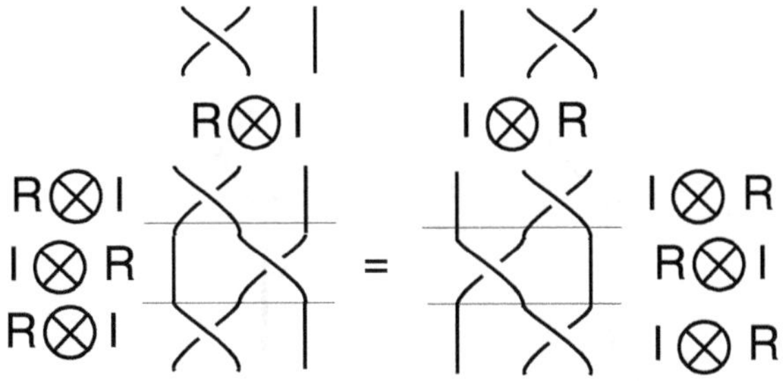

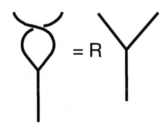

FIGURE 14.25: Yang–Baxter equation.

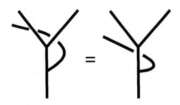

FIGURE 14.26: Braiding.

FIGURE 14.27: Intertwining.

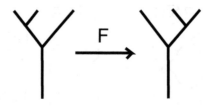

FIGURE 14.28: Recoupling.

pling and the intertwining identity as shown in Fig. 14.30. One says that the interactions satisfy the *hexagon identity* if this composition is the identity.

A *graphical three-dimensional topological quantum field theory* is an algebra of interactions that satisfies the Yang–Baxter equation, the intertwining identity, the pentagon identity and the hexagon identity. There is not room in this summary to detail the way that these properties fit into the topology of knots and three-dimensional manifolds, but a sketch is in order. For the case of topological quantum field theory related to the group $SU(2)$ there is a construction based entirely on the combinatorial topology of the bracket polynomial (See Sections 14.7, 14.9 and 14.10 of this article.). See [43, 38] for more information on this approach.

Now return to Fig. 14.20 where we illustrate trinions, shown in relation to a trivalent vertex, and a surface of genus three that is decomposed into four trinions. It turns out that the vector space $V(S_g) = V(G(S_g,t))$ to a surface with a trinion decomposition as t described above, and defined in terms of the graphical topological quantum field theory, does not depend upon the choice of trinion decomposition. This independence is guaranteed by the braiding, hexagon and pentagon identities. One can then associate a well-defined vector $|M\rangle$ in $V(S_g)$ whenenver M is a three manifold whose boundary is S_g. Furthermore, if a closed three manifold M^3 is decomposed along a surface S_g into the union of M_- and M_+ where these parts are otherwise disjoint three manifolds with boundary S_g, then the inner product $I(M) = \langle M_-|M_+\rangle$ is, up to normalization, an invariant of the three manifold M_3. With the definition of graphical topological quantum field theory given above, knots and links can be incorporated as well, so that one obtains a source of invariants $I(M^3,K)$ of knots and links in orientable three manifolds. Here we see the uses of the relationships

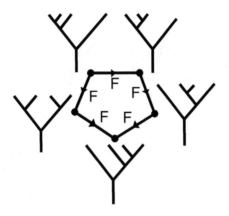

FIGURE 14.29: Pentagon identity.

that occur in the higher dimensional cobordism categories, as described in the previous section.

The invariant $I(M^3, K)$ can be formally compared with the Witten [85] integral

$$Z(M^3, K) = \int DA e^{(ik/4\pi)S(M,A)} W_K(A).$$

It can be shown that up to limits of the heuristics, $Z(M, K)$ and $I(M^3, K)$ are essentially equivalent for appropriate choice of gauge group and corresponding spin networks.

By these graphical reformulations, a three-dimensional $TQFT$ is, at base, a highly simplified theory of point particle interactions in $2 + 1$ dimensional spacetime. It can be used to articulate invariants of knots and links and invariants of three manifolds. The reader interested in the $SU(2)$ case of this structure and its implications for invariants of knots and three manifolds can consult [18, 34, 43, 63, 68]. One expects that physical situations involving $2 + 1$ spacetime will be approximated by such an idealized theory. There are also applications to $3 + 1$ quantum gravity [4, 7, 52]. Aspects of the quan-

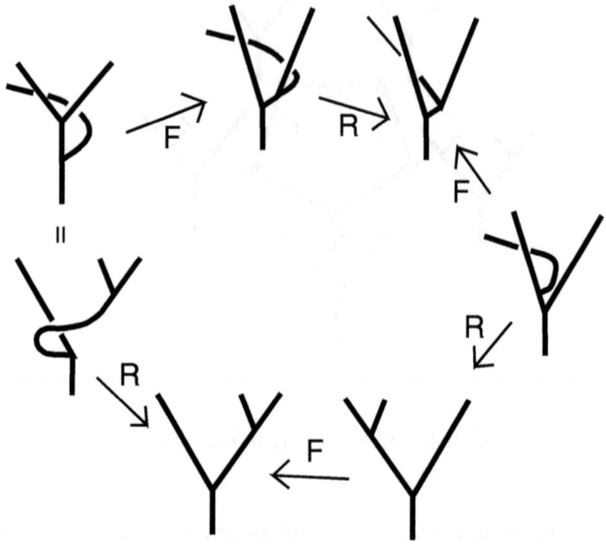

FIGURE 14.30: Hexagon identity.

tum Hall effect may be related to topological quantum field theory [84]. One can study a physics in two-dimensional space where the braiding of particles or collective excitations leads to non-trivial representations of the Artin braid group. Such particles are called *Anyons*. Such $TQFT$ models would describe applicable physics. One can think about applications of anyons to quantum computing along the lines of the topological models described here.

A key point in the application of $TQFT$ to quantum information theory is contained in the structure illustrated in Fig. 14.31. There we show a more complex braiding operator, based on the composition of recoupling with the elementary braiding at a vertex. (This structure is implicit in the Hexagon identity of Fig. 14.30.) The new braiding operator is a source of unitary rep-

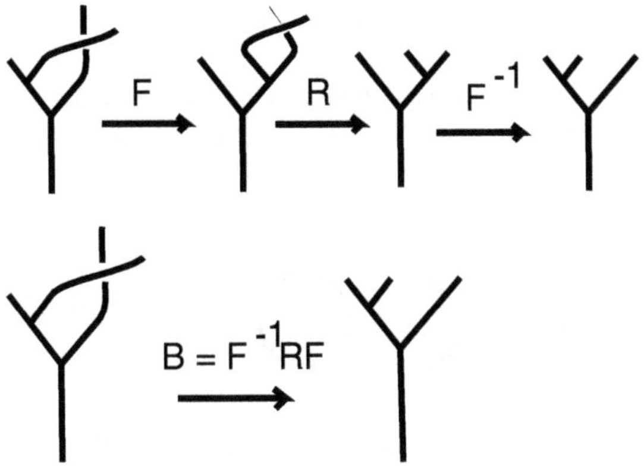

FIGURE 14.31: A more complex braiding operator.

resentations of braid group in situations (which exist mathematically) where the recoupling transformations are themselves unitary. This kind of pattern is utilized in the work of Freedman and collaborators [26, 27, 28, 29, 30] and in the case of classical angular momentum formalism has been dubbed a "spin-network quantum simlator" by Rasetti and collaborators [65, 66]. In the next section we show how certain natural deformations [38] of Penrose spin networks [32] can be used to produce these unitary representations of the Artin braid group and the corresponding models for anyonic topological quantum computation.

14.11 Spin networks and Temperley–Lieb recoupling theory

In this section we discuss a combinatorial construction for spin networks that generalizes the original construction of Roger Penrose. The result of this generalization is a structure that satisfies all the properties of a graphical $TQFT$ as described in the previous section, and specializes to classical angular momentum recoupling theory in the limit of its basic variable. The construction is based on the properties of the bracket polynomial (as already described in Section 14.4). A complete description of this theory can be found in the book "Temperley–Lieb Recoupling Theory and Invariants of Three-Manifolds" by Kauffman and Lins [38].

The "q-deformed" spin networks that we construct here are based on the bracket polynomial relation. View Fig. 14.32 and Fig. 14.33.

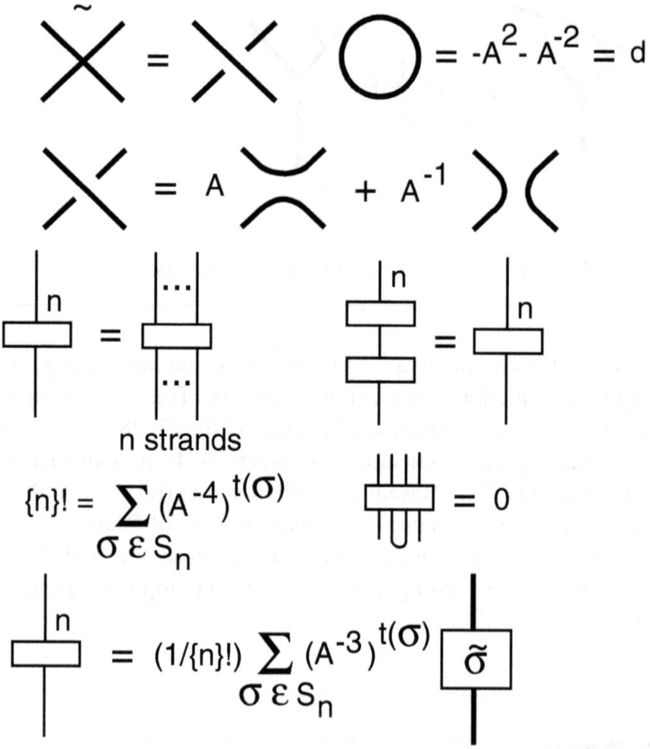

FIGURE 14.32: Basic projectors.

In Fig. 14.32 we indicate how the basic projector 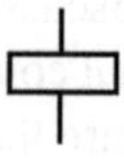 (symmetrizer, Jones–Wenzl projector) is constructed on the basis of the bracket polynomial expansion. In this tech-

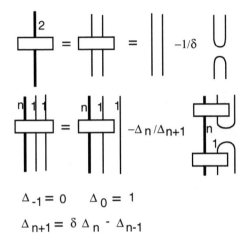

$$\Delta_{-1} = 0 \qquad \Delta_0 = 1$$
$$\Delta_{n+1} = \delta \Delta_n - \Delta_{n-1}$$

FIGURE 14.33: Two-strand projector.

nology a symmetrizer is a sum of tangles on n strands (for a chosen integer n). The tangles are made by summing over braid lifts of permutations in the symmetric group on n letters, as indicated in Fig. 14.32. Each elementary braid is then expanded by the bracket polynomial relation as indicated in Fig. 14.32 so that the resulting sum consists of flat tangles without any crossings (these can be viewed as elements in the Temperley–Lieb algebra). The projectors have the property that the concatenation of a projector with itself is just that projector, and if you tie two lines on the top or the bottom of a projector together, then the evaluation is zero. This general definition of projectors is very useful for this theory. The two-strand projector is shown in Fig. 14.33. Here the formula for that projector is particularly simple. It is the sum of two parallel arcs and two turn-around arcs (with coefficient $-1/d$, with $d = -A^2 - A^{-2}$ as the loop value for the bracket polynomial). Fig. 14.33 also shows the recursion formula for the general projector. This recursion formula is due to Jones and Wenzl and the projector in this form, developed as a sum in the Temperley–Lieb algebra (see Section 14.5 of this paper), is usually known as the *Jones–Wenzl projector*.

The projectors are combinatorial analogs of irreducible representations of a

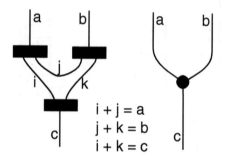

$$i + j = a$$
$$j + k = b$$
$$i + k = c$$

FIGURE 14.34: Vertex.

group (the original spin nets were based on $SU(2)$ and these deformed nets are based on the corresponding quantum group to SU(2)). As such the reader can think of them as "particles". The interactions of these particles are governed by how they can be tied together into three vertices. See Fig. 14.34. In Fig. 14.34 we show how to tie three projectors, of a, b, c strands respectively, together to form a three-vertex. In order to accomplish this interaction, we must share lines between them as shown in that figure so that there are non-negative integers i, j, k so that $a = i + j, b = j + k, c = i + k$. This is equivalent to the condition that $a + b + c$ is even and that the sum of any two of a, b, c is greater than or equal to the third. For example $a + b \geq c$. One can think of the vertex as a possible particle interaction where $[a]$ and $[b]$ interact to produce $[c]$. That is, any two of the legs of the vertex can be regarded as interacting to produce the third leg.

There is a basic orthogonality of three vertices as shown in Fig. 14.35. Here if we tie two of the three vertices together so that they form a "bubble" in the middle, then the resulting network with labels a and b on its free ends is a multiple of an a-line (meaning a line with an a-projector on it) or zero (if *a* is not equal to *b*). The multiple is compatible with the results of closing the diagram in the equation of Fig. 14.35 so the two free ends are identified with one another. On closure, as shown in the figure, the left hand side of the equation becomes a theta graph and the right hand side becomes a multiple of a delta where Δ_a denotes the bracket polynomial evaluation of the *a*-strand loop with a projector on it. The $\Theta(a,b,c)$ denotes the bracket evaluation of a theta graph made from three trivalent vertices and labeled with a,b,c on its edges.

There is a recoupling formula in this theory in the form shown in Fig. 14.36. Here there are "6-j symbols", recoupling coefficients that can be expressed, as shown in Fig. 14.36, in terms of tetrahedral graph evaluations and theta graph evaluations. The tetrahedral graph is shown in Fig. 14.37. One derives the formulas for these coefficients directly from the orthogonality relations for the trivalent vertices by closing the left hand side of the recoupling formula and using orthogonality to evaluate the right hand side. This is illustrated in Fig. 14.38. The reader should be advised that there are specific calculational formulas for the theta and tetrahedral nets. These can be found in [38]. Here we are indicating only the relationships and external logic of these objects.

Finally, there is the braiding relation, as illustrated in Fig. 14.36.

With the braiding relation in place, this *q*-deformed spin network theory satisfies the pentagon, hexagon and braiding naturality identities needed for a topological quantum field theory. All these identities follow naturally from the basic underlying topological construction of the bracket polynomial. One can apply the theory to many different situations.

14.11.1   Evaluations

In this section we discuss the structure of the evaluations for Δ_n and the theta and tetrahedral networks. We refer to [38] for the details behind these formulas. Recall that Δ_n is the bracket evaluation of the closure of the *n*-strand projector, as illustrated in Fig. 14.35. For the bracket variable A, one finds that

$$\Delta_n = (-1)^n \frac{A^{2n+2} - A^{-2n-2}}{A^2 - A^{-2}}.$$

One sometimes writes the *quantum integer*

$$[n] = (-1)^{n-1} \Delta_{n-1} = \frac{A^{2n} - A^{-2n}}{A^2 - A^{-2}}.$$

If

$$A = e^{i\pi/2r}$$

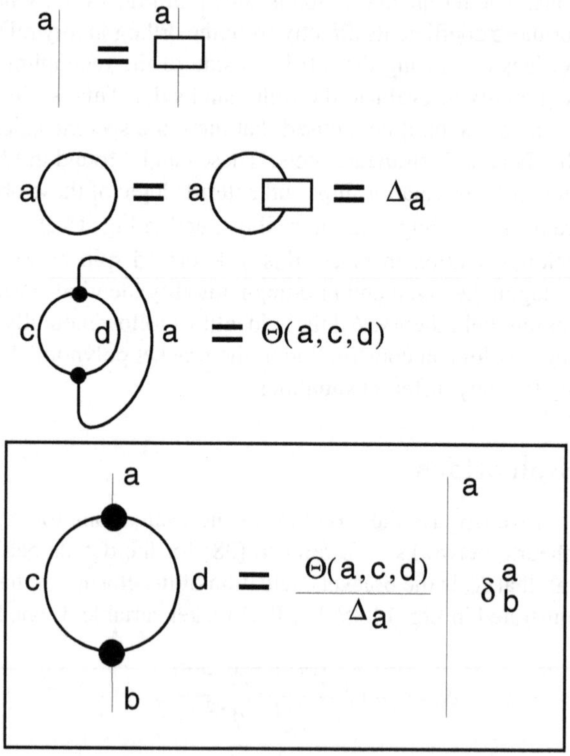

FIGURE 14.35: Orthogonality of trivalent vertices.

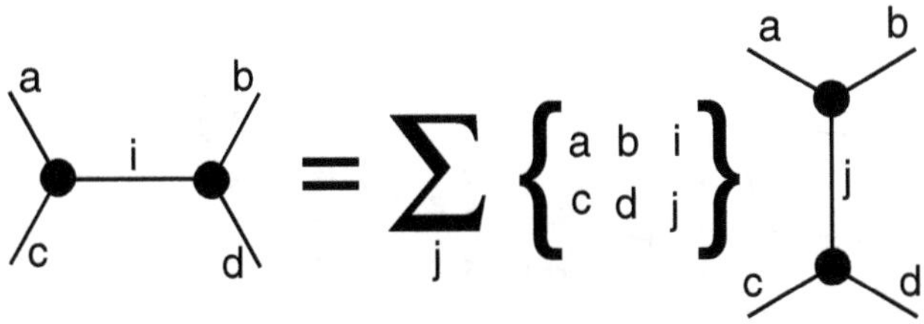

FIGURE 14.36: Recoupling formula.

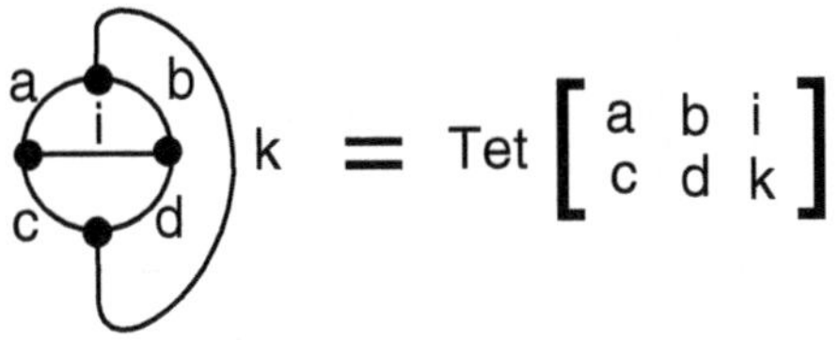

FIGURE 14.37: Tetrahedron network.

where r is a positive integer, then

$$\Delta_n = (-1)^n \frac{sin((n+1)\pi/r)}{sin(\pi/r)}.$$

Here the corresponding quantum integer is

$$[n] = \frac{sin(n\pi/r)}{sin(\pi/r)}.$$

Note that $[n+1]$ is a positive real number for $n = 0,1,2,...r-2$ and that $[r-1] = 0$.

The evaluation of the theta net is expressed in terms of quantum integers by the formula

$$\Theta(a,b,c) = (-1)^{m+n+p} \frac{[m+n+p+1]![n]![m]![p]!}{[m+n]![n+p]![p+m]!}$$

where

$$a = m+p, b = m+n, c = n+p.$$

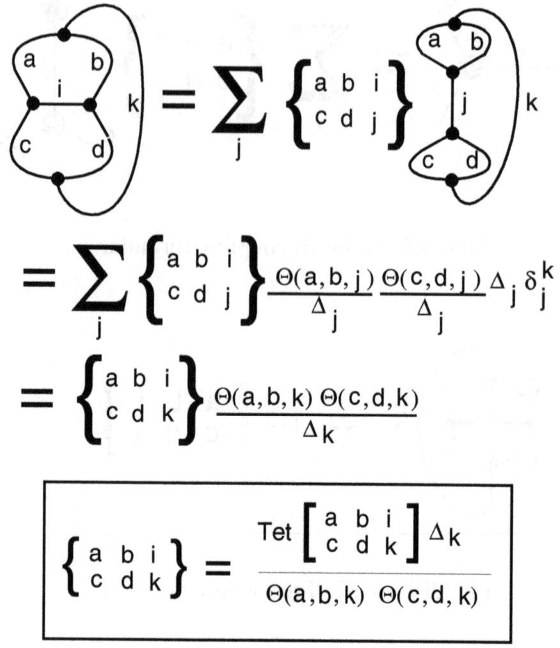

$$= \sum_j \left\{ \begin{matrix} a & b & i \\ c & d & j \end{matrix} \right\} \frac{\Theta(a,b,j)}{\Delta_j} \frac{\Theta(c,d,j)}{\Delta_j} \Delta_j \delta_j^k$$

$$= \left\{ \begin{matrix} a & b & i \\ c & d & k \end{matrix} \right\} \frac{\Theta(a,b,k)\, \Theta(c,d,k)}{\Delta_k}$$

$$\left\{ \begin{matrix} a & b & i \\ c & d & k \end{matrix} \right\} = \frac{\mathrm{Tet}\left[\begin{matrix} a & b & i \\ c & d & k \end{matrix} \right] \Delta_k}{\Theta(a,b,k)\ \Theta(c,d,k)}$$

FIGURE 14.38: Tetrahedron formula for recoupling coefficients.

Note that

$$(a+b+c)/2 = m+n+p.$$

When $A = e^{i\pi/2r}$, the recoupling theory becomes finite with the restriction that only three-vertices (labeled with a, b, c) are *admissible* when $a+b+c \leq 2r-4$. All the summations in the formulas for recoupling are restricted to admissible triples of this form.

14.11.2   Symmetry and unitarity

The formula for the recoupling coefficients given in Fig. 14.38 has less symmetry than is actually inherent in the structure of the situation. By multiplying all the vertices by an appropriate factor, we can reconfigure the formulas in this theory so that the revised recoupling transformation is orthogonal, in the sense that its transpose is equal to its inverse. This is a very useful fact. It means

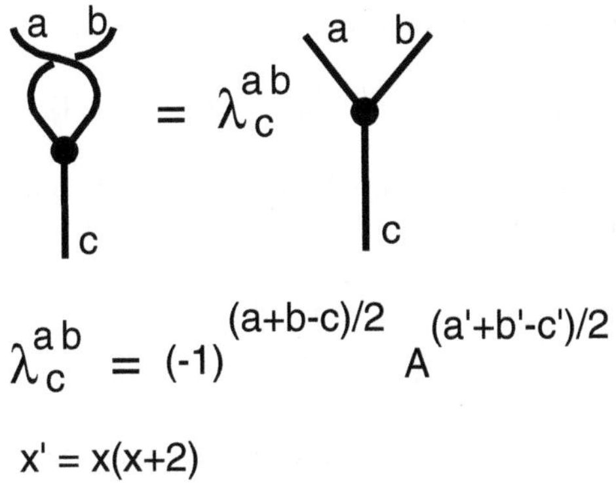

$$\lambda_c^{ab} = (-1)^{(a+b-c)/2} A^{(a'+b'-c')/2}$$

$$x' = x(x+2)$$

FIGURE 14.39: Local braiding formula.

that when the resulting matrices are real, then the recoupling transformations are unitary. We shall see particular applications of this viewpoint later in the paper.

Fig. 14.40 illustrates this modification of the three-vertex. Let $Vert[a,b,c]$ denote the original three-vertex of the Temperley–Lieb recoupling theory. Let $ModVert[a,b,c]$ denote the modified vertex. Then we have the formula

$$ModVert[a,b,c] = \frac{\sqrt{\sqrt{\Delta_a \Delta_b \Delta_c}}}{\sqrt{\Theta(a,b,c)}} \, Vert[a,b,c].$$

LEMMA 14.3

For the bracket evaluation at the root of unity $A = e^{i\pi/2r}$ the factor

$$f(a,b,c) = \frac{\sqrt{\sqrt{\Delta_a \Delta_b \Delta_c}}}{\sqrt{\Theta(a,b,c)}}$$

is real, and can be taken to be a positive real number for (a,b,c) admissible (i.e. $a+b+c \leq 2r-4$).

PROOF   By the results from the previous subsection,

$$\Theta(a,b,c) = (-1)^{(a+b+c)/2}\hat{\Theta}(a,b,c)$$

where $\hat{\Theta}(a,b,c)$ is positive real, and

$$\Delta_a\Delta_b\Delta_c = (-1)^{(a+b+c)}[a+1][b+1][c+1]$$

where the quantum integers in this formula can be taken to be positive real. It follows from this that

$$f(a,b,c) = \sqrt{\frac{\sqrt{[a+1][b+1][c+1]}}{\hat{\Theta}(a,b,c)}},$$

showing that this factor can be taken to be positive real.   ∎

In Fig. 14.41 we show how this modification of the vertex affects the non-zero term of the orthogonality of trivalent vertices (compare with Fig. 14.35). We refer to this as the "modified bubble identity." The coefficient in the modified bubble identity is

$$\sqrt{\frac{\Delta_b\Delta_c}{\Delta_a}} = (-1)^{(b+c-a)/2}\sqrt{\frac{[b+1][c+1]}{[a+1]}}$$

where (a,b,c) form an admissible triple. In particular $b+c-a$ is even and hence this factor can be taken to be real.

We rewrite the recoupling formula in this new basis and emphasize that the recoupling coefficients can be seen (for fixed external labels a,b,c,d) as a matrix transforming the horizontal "double-Y" basis to a vertically disposed double-Y basis. In Figs. 14.42, 14.43 and 14.44 we have shown the form of this transformation, using the matrix notation

$$M[a,b,c,d]_{ij}$$

for the modified recoupling coefficients. In Fig. 14.42 we derive an explicit formula for these matrix elements. The proof of this formula follows directly from trivalent vertex orthogonality (see Figs. 14.35 and 14.38), and is given in Fig. 14.42. The result shown in Fig. 14.42 and Fig. 14.43 is the following formula for the recoupling matrix elements.

$$M[a,b,c,d]_{ij} = ModTet \begin{pmatrix} a & b & i \\ c & d & j \end{pmatrix} / \sqrt{\Delta_a \Delta_b \Delta_c \Delta_d}$$

where $\sqrt{\Delta_a \Delta_b \Delta_c \Delta_d}$ is short-hand for the product

$$\sqrt{\frac{\Delta_a \Delta_b}{\Delta_j}} \sqrt{\frac{\Delta_c \Delta_d}{\Delta_j}} \Delta_j$$

$$= (-1)^{(a+b-j)/2}(-1)^{(c+d-j)/2}(-1)^j \sqrt{\frac{[a+1][b+1]}{[j+1]}} \sqrt{\frac{[c+1][d+1]}{[j+1]}} [j+1]$$

$$= (-1)^{(a+b+c+d)/2} \sqrt{[a+1][b+1][c+1][d+1]}.$$

In this form, since (a,b,j) and (c,d,j) are admissible triples, we see that this coefficient can be taken to be real, and its value is independent of the choice of i and j. The matrix $M[a,b,c,d]$ is real-valued.
It follows from Fig. 14.36 (turn the diagrams by ninety degrees) that

$$M[a,b,c,d]^{-1} = M[b,d,a,c].$$

In Fig. 14.45 we illustrate the formula

$$M[a,b,c,d]^T = M[b,d,a,c].$$

It follows from this formula that

$$M[a,b,c,d]^T = M[a,b,c,d]^{-1}.$$

Hence $M[a,b,c,d]$ is an orthogonal, real-valued matrix.

THEOREM 14.5
In the Temperley–Lieb theory we obtain unitary (in fact real orthogonal) recoupling transformations when the bracket variable A has the form $A = e^{i\pi/2r}$ for r a positive integer. Thus we obtain families of unitary representations of the Artin braid group from the recoupling theory at these roots of unity.

PROOF   The proof is given the discussion above.   ∎

In Section 14.9 we shall show explictly how these methods work in the case of the Fibonacci model where $A = e^{3i\pi/5}$.

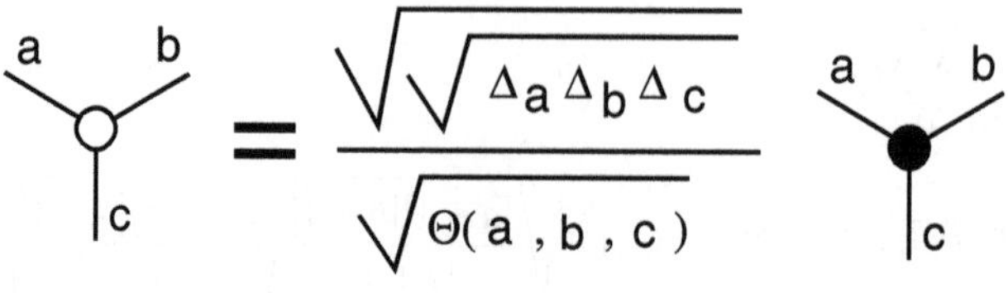

$$
\overset{a \quad b}{\underset{c}{\Y}} \; = \; \frac{\sqrt{\sqrt{\Delta_a \Delta_b \Delta_c}}}{\sqrt{\Theta(a,b,c)}} \quad \overset{a \qquad b}{\underset{c}{\Y}}
$$

FIGURE 14.40: Modified three-vertex.

14.12   Fibonacci particles

In this section and the next we detail how the Fibonacci model for anyonic quantum computing [60, 71] can be constructed by using a version of the two-stranded bracket polynomial and a generalization of Penrose spin networks. This is a fragment of the Temperly–Lieb recoupling theory [38]. We already gave in the preceding sections a general discussion of the theory of spin networks and their relationship with quantum computing.

The Fibonacci model is a $TQFT$ that is based on a single "particle" with two states that we shall call the *marked state* and the *unmarked state*. The particle in the marked state can interact with itself either to produce a single particle in the marked state, or to produce a single particle in the unmarked state. The particle in the unmarked state has no influence in interactions (an unmarked state interacting with any state S yields that state S). One way to indicate these two interactions symbolically is to use a box for the marked state and a blank space for the unmarked state. Then one has two modes of interaction of a box with itself:

1. Adjacency: □ □

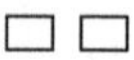

   and

2. Nesting: [□].

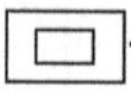

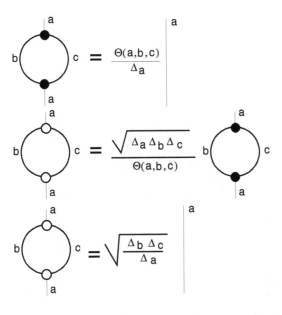

FIGURE 14.41: Modified bubble identity.

With this convention we take the adjacency interaction to yield a single box, and the nesting interaction to produce nothing:

$$\square\,\square = \square$$
$$\boxed{\square} =$$

We take the notational opportunity to denote nothing by an asterisk (*). The asterisk is a stand-in for no mark at all and it can be erased or placed wherever it is convenient to do so. Thus

$$\boxed{\square} = *.$$

We shall make a recoupling theory based on this particle, but it is worth noting some of its purely combinatorial properties first. The arithmetic of combining boxes (standing for acts of distinction) according to these rules has been studied and formalized in [80] and correlated with Boolean algebra and classical logic. Here *within* and *next to* are ways to refer to the two sides delineated by the given distinction. From this point of view, there are two modes

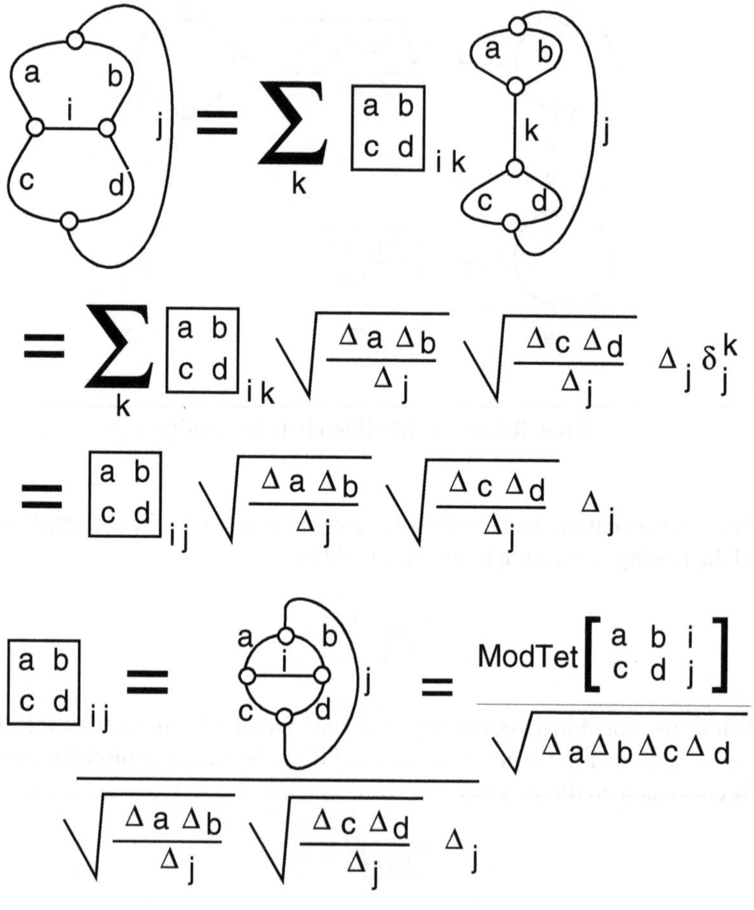

FIGURE 14.42: Derivation of modified recoupling coefficients.

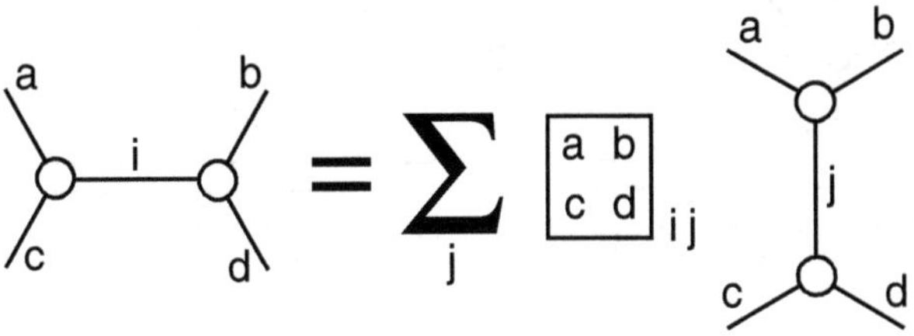

FIGURE 14.43: Modified recoupling formula.

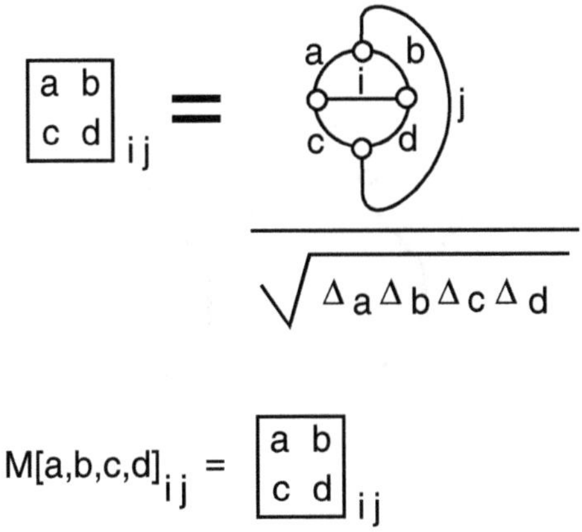

$$M[a,b,c,d]_{ij} = \begin{bmatrix} a & b \\ c & d \end{bmatrix}_{ij}$$

FIGURE 14.44: Modified recoupling matrix.

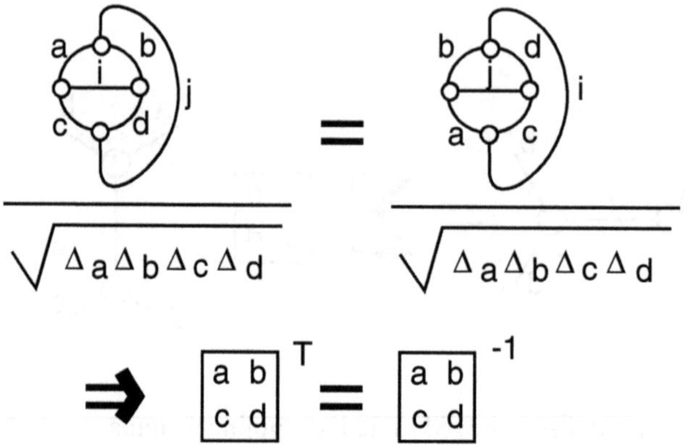

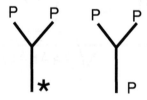

$$\frac{\begin{array}{c} a \; i \; b \\ c \; d \end{array}}{\sqrt{\Delta_a \Delta_b \Delta_c \Delta_d}} = \frac{\begin{array}{c} b \; d \\ a \; c \end{array}}{\sqrt{\Delta_a \Delta_b \Delta_c \Delta_d}}$$

$$\Rightarrow \begin{bmatrix} a & b \\ c & d \end{bmatrix}^T = \begin{bmatrix} a & b \\ c & d \end{bmatrix}^{-1}$$

FIGURE 14.45: Modified matrix transpose.

FIGURE 14.46: Fibonacci particle interaction.

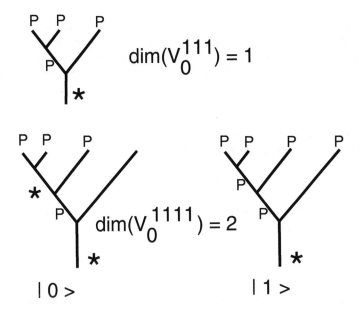

FIGURE 14.47: Fibonacci trees.

of relationship (adjacency and nesting) that arise at once in the presence of a distinction.

From here on we shall denote the Fibonacci particle by the letter P. Thus the two possible interactions of P with itself are as follows.

1. $P,P \longrightarrow *$

2. $P,P \longrightarrow P$.

In Fig. 14.47 we indicate in small tree diagrams the two possible interactions of the particle P with itself. In the first interaction the particle vanishes, producing the asterisk. In the second interaction a single copy of P is produced. These are the two basic actions of a single distinction relative to itself, and they constitute our formalism for this very elementary particle.

In Fig. 14.47, we have indicated the different results of particle processes where we begin with a left-associated tree structure with three branches, all marked and then four branches all marked. In each case we demand that the particles interact successively to produce an unmarked particle in the end, at the root of the tree. More generally one can consider a left-associated tree with n upward branches and one root. Let $T(a_1, a_2, \cdots, a_n : b)$ denote such a tree with particle labels $a_1, \cdots, a_n$ on the top and root label b at the bottom of the tree. We consider all possible processes (sequences of particle interactions) that start with the labels at the top of the tree, and end with the labels at the bottom of the tree. Each such sequence is regarded as a basis vector in a complex vector space

$$V_b^{a_1, a_2, \cdots, a_n}$$

associated with the tree. In the case where all the labels are marked at the top and the bottom label is unmarked, we shall denote this tree by

$$V_0^{111\cdots11} = V_0^{(n)}$$

where n denotes the number of upward branches in the tree. We see from Fig. 14.47 that the dimension of $V_0^{(3)}$ is 1, and that

$$dim(V_0^{(4)}) = 2.$$

This means that $V_0^{(4)}$ is a natural candidate in this context for the two-qubit space.

Given the tree $T(1,1,1,\cdots,1:0)$ (n marked states at the top, an unmarked state at the bottom), a process basis vector in $V_0^{(n)}$ is in direct correspondence with a string of boxes and asterisks (1's and 0's) of length $n-2$ with no repeated asterisks and ending in a marked state. See Fig. 14.47 for an illustration of the simplest cases. It follows from this that

$$dim(V_0^{(n)}) = f_{n-2}$$

where f_k denotes the k-th Fibonacci number:

$$f_0 = 1, f_1 = 1, f_2 = 2, f_3 = 3, f_4 = 5, f_5 = 8, \cdots$$

where

$$f_{n+2} = f_{n+1} + f_n.$$

The dimension formula for these spaces follows from the fact that there are f_n sequences of length $n-1$ of marked and unmarked states with no repetition of an unmarked state. This fact is illustrated in Fig. 14.48.

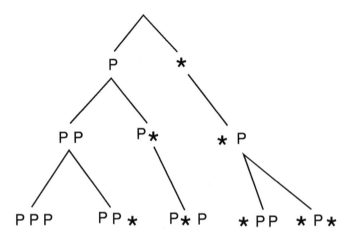

Tree of squences with no occurence of ✶ ✶

FIGURE 14.48: Fibonacci sequence.

14.13 The Fibonacci recoupling model

We now show how to make a model for recoupling the Fibonacci particle by using the Temperley–Lieb recoupling theory and the bracket polynomial. Everything we do in this section will be based on the 2-projector, its properties and evaluations based on the bracket polynomial model for the Jones polynomial. While we have outlined the general recoupling theory based on the bracket polynomial in earlier sections of this paper, the present section is self-contained, using only basic information about the bracket polynomial, and the essential properties of the 2-projector as shown in Fig. 14.49. In this figure we state the definition of the 2-projector, list its two main properties (the operator is idempotent and a self-attached strand yields a zero evaluation) and give diagrammatic proofs of these properties.

In Fig. 14.50, we show the essence of the Temperley–Lieb recoupling model for the Fibonacci particle. The Fibonacci particle is, in this mathematical model, identified with the 2-projector itself. As the reader can see from Fig. 14.50, there are two basic interactions of the 2-projector with itself, one giving a 2-projector, the other giving nothing. This is the pattern of self-interaction

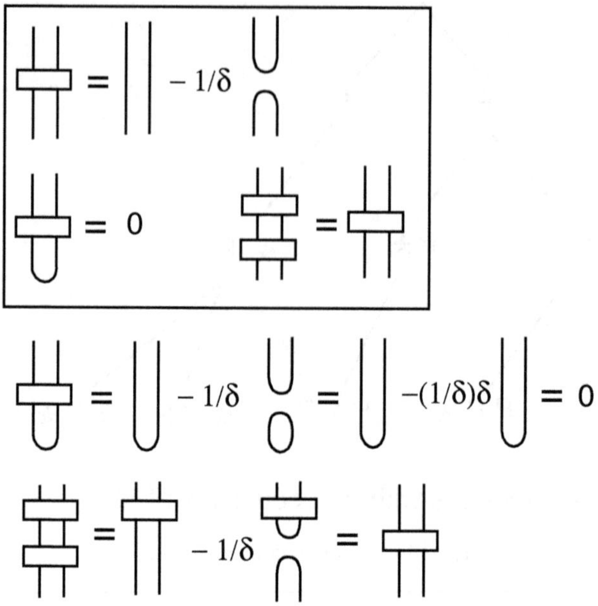

FIGURE 14.49: The 2-projector.

of the Fibonacci particle. There is a third possibility, depicted in Fig. 14.50, where two 2-projectors interact to produce a 4-projector. We could remark at the outset, that the 4-projector will be zero if we choose the bracket polynomial variable $A = e^{3\pi/5}$. Rather than start there, we will assume that the 4-projector is forbidden and deduce (below) that the theory has to be at this root of unity. Note that in Fig. 14.50 we have adopted a single strand notation for the particle interactions, with a solid strand corresponding to the marked particle, a dotted strand (or nothing) corresponding to the unmarked particle. A dark vertex indicates either an interaction point, or it may be used to indicate the single strand is shorthand for two ordinary strands. Remember that these are all shorthand expressions for underlying bracket polynomial calculations.

In Figs. 14.51, 14.52, 14.53, 14.54, 14.55 and 14.56 we have provided complete diagrammatic calculations of all of the relevant small nets and evaluations that are useful in the two-strand theory that is being used here. The reader may wish to skip directly to Fig. 14.57 where we determine the form of the recoupling coefficients for this theory. We will discuss the resulting algebra below.

For the reader who does not want to skip the next collection of figures, here

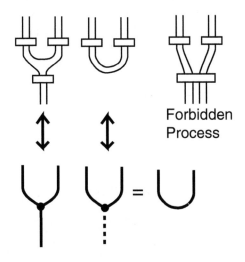

FIGURE 14.50: Fibonacci particle as 2-projector.

is a guided tour. Fig. 14.51 illustrates three basic nets in the case of two strands. These are the theta, delta and tetrahedron nets. In this figure we have shown the decomposition on the theta and delta nets in terms of 2-projectors. The tetrahedron net will be similarly decomposed in Figs. 14.55 and 14.56. The theta net is denoted Θ, the delta by Δ, and the tetrahedron by T. In Fig. 14.52 we illustrate how a pendant loop has a zero evaluation. In Fig. 14.53 we use the identity in Fig. 14.52 to show how an interior loop (formed by two trivalent vertices) can be removed and replaced by a factor of Θ/Δ. Note how, in this figure, line two proves that one network is a multiple of the other, while line three determines the value of the multiple by closing both nets.

Fig. 14.54 illustrates the explicit calculation of the delta and theta nets. The figure begins with a calculation of the result of closing a single strand of the 2-projector. The result is a single stand multiplied by $(\delta - 1/\delta)$ where $\delta = -A^2 - A^{-2}$, and A is the bracket polynomial parameter. We then find that

$$\Delta = \delta^2 - 1$$

and
$$\Theta = (\delta - 1/\delta)^2\delta - \Delta/\delta = (\delta - 1/\delta)(\delta^2 - 2).$$

Figs. 14.55 and 14.56 illustrate the calculation of the value of the tetrahedral network T. The reader should note the first line of Fig. 14.55 where the tetrahedral net is translated into a pattern of 2-projectors, and simplified. The rest of these two figures are diagrammatic calculations, using the expansion formula for the 2-projector. At the end of Fig. 14.56 we obtain the formula for the tetrahedron
$$T = (\delta - 1/\delta)^2(\delta^2 - 2) - 2\Theta/\delta.$$

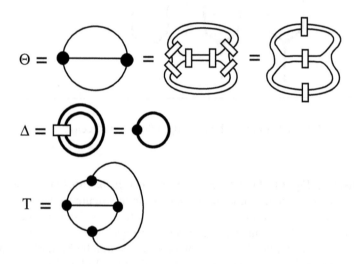

FIGURE 14.51: Theta, delta and tetrahedron.

Fig. 14.57 is the key calculation for this model. In this figure we assume that the recoupling formulas involve only 0 and 2 strands, with 0 corresponding to the null particle and 2 corresponding to the 2-projector. ($2+2=4$ is forbidden as in Fig. 14.50.) From this assumption we calculate that the recoupling matrix is given by

$$F = \begin{pmatrix} a & b \\ c & d \end{pmatrix} = \begin{pmatrix} 1/\Delta & \Delta/\Theta \\ \Theta/\Delta^2 & T\Delta/\Theta^2 \end{pmatrix}.$$

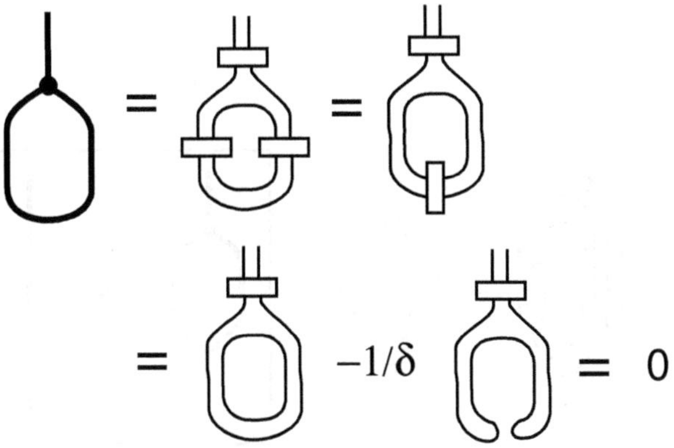

FIGURE 14.52: Loop evaluation (1).

Figs. 14.58 and 14.59 work out the exact formulas for the braiding at a three-vertex in this theory. When the three-vertex has three marked lines, then the braiding operator is multiplication by $-A^4$, as in Fig. 14.58. When the three-vertex has two marked lines, then the braiding operator is multiplied by A^8, as shown in Fig. 14.59.

Notice that it follows from the symmetry of the diagrammatic recoupling formulas of Fig. 14.57 that *the square of the recoupling matrix F is equal to the identity.* That is,

$$\begin{pmatrix} 1 & 0 \\ 0 & 1 \end{pmatrix} = F^2 = \begin{pmatrix} 1/\Delta & \Delta/\Theta \\ \Theta/\Delta^2 & T\Delta/\Theta^2 \end{pmatrix} \begin{pmatrix} 1/\Delta & \Delta/\Theta \\ \Theta/\Delta^2 & T\Delta/\Theta^2 \end{pmatrix}$$

$$= \begin{pmatrix} 1/\Delta^2 + 1/\Delta & 1/\Theta + T\Delta^2/\Theta^3 \\ \Theta/\Delta^3 + T/(\Delta\Theta) & 1/\Delta + \Delta^2 T^2/\Theta^4 \end{pmatrix}.$$

Thus we need the relation

$$1/\Delta + 1/\Delta^2 = 1.$$

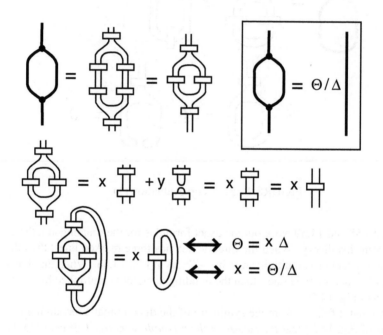

FIGURE 14.53: Loop evaluation (2).

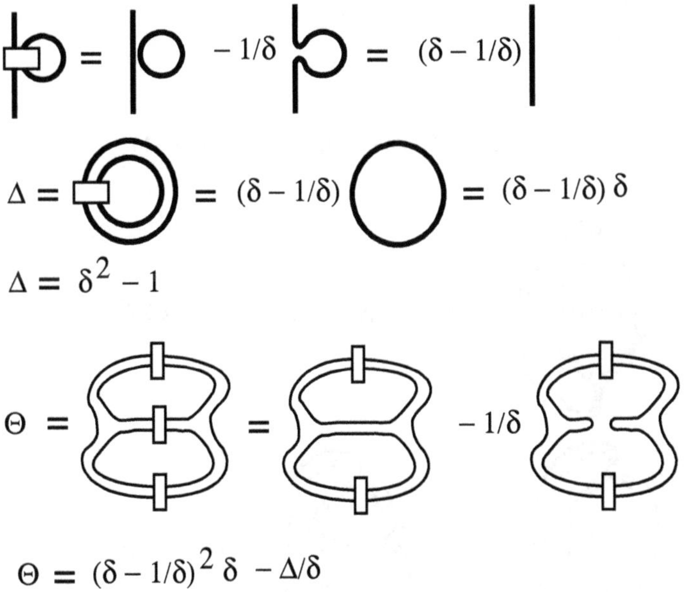

$$\Delta = \delta^2 - 1$$

$$\Theta = (\delta - 1/\delta)^2 \, \delta \; - \Delta/\delta$$

FIGURE 14.54: Calculations of theta and delta.

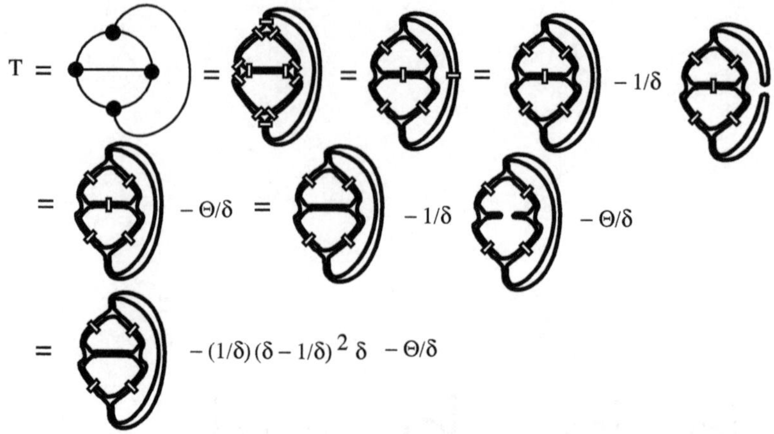

FIGURE 14.55: Calculation of tetrahedron (1).

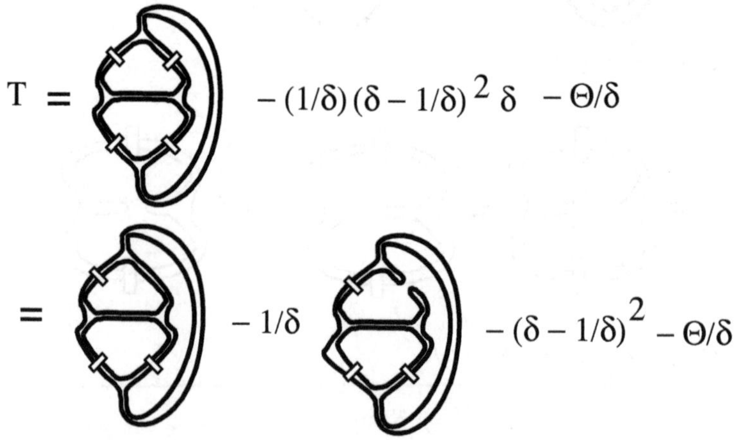

$$T = \quad -(1/\delta)(\delta - 1/\delta)^2 \, \delta \ - \Theta/\delta$$

$$= \quad -1/\delta \qquad\qquad -(\delta - 1/\delta)^2 - \Theta/\delta$$

$$= \quad (\delta - 1/\delta)^3 \, \delta \ - (1/\delta)\Theta \ -(\delta - 1/\delta)^2 - \Theta/\delta$$

$$= \quad (\delta - 1/\delta)^2 (\delta^2 - 2) \ - 2\Theta/\delta$$

FIGURE 14.56: Calculation of tetrahedron (2).

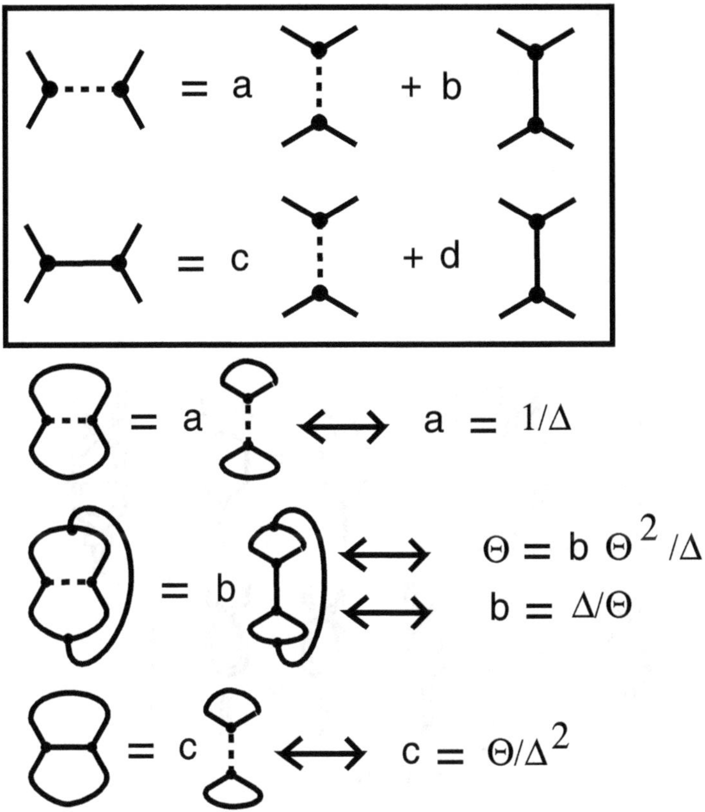

FIGURE 14.57: Recoupling for 2-projectors.

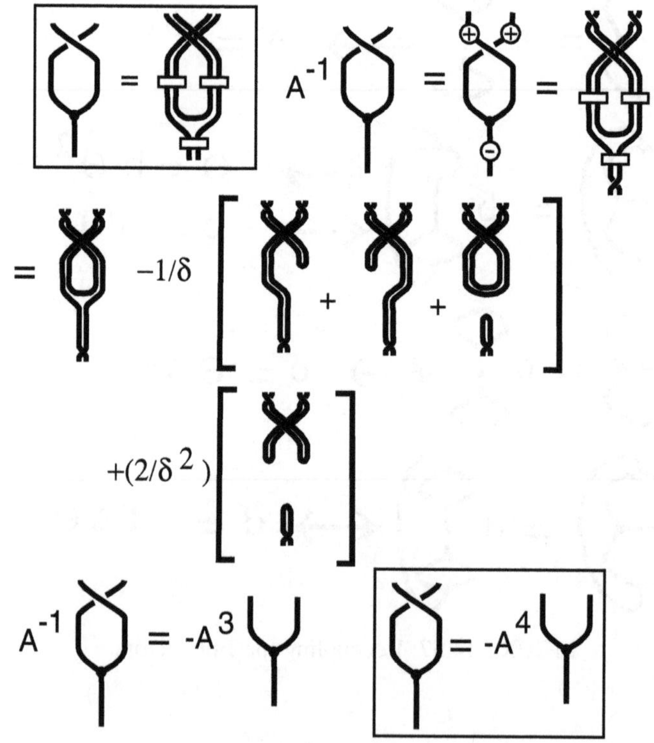

FIGURE 14.58: Braiding at the three-vertex.

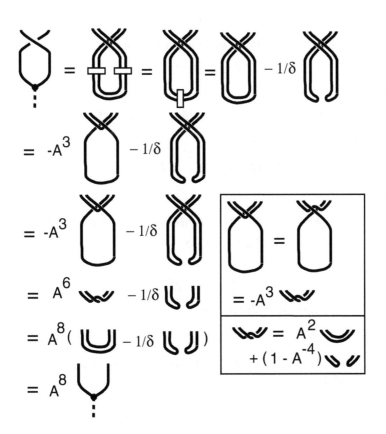

FIGURE 14.59: Braiding at the null three-vertex.

This is equivalent to saying that

$$\Delta^2 = 1 + \Delta,$$

a quadratic equation whose solutions are

$$\Delta = (1 \pm \sqrt{5})/2.$$

Furthermore, we know that

$$\Delta = \delta^2 - 1$$

from Fig. 14.54. Hence

$$\Delta^2 = \Delta + 1 = \delta^2.$$

We shall now specialize to the case where

$$\Delta = \delta = (1 + \sqrt{5})/2,$$

leaving the other cases for the exploration of the reader. We then take

$$A = e^{3\pi i/5}$$

so that

$$\delta = -A^2 - A^{-2} = -2cos(6\pi/5) = (1 + \sqrt{5})/2.$$

Note that $\delta - 1/\delta = 1$. Thus

$$\Theta = (\delta - 1/\delta)^2 \delta - \Delta/\delta = \delta - 1$$

and

$$\begin{aligned} T &= (\delta - 1/\delta)^2(\delta^2 - 2) - 2\Theta/\delta = (\delta^2 - 2) - 2(\delta - 1)/\delta \\ &= (\delta - 1)(\delta - 2)/\delta = 3\delta - 5. \end{aligned}$$

Note that

$$T = -\Theta^2/\Delta^2,$$

from which it follows immediately that

$$F^2 = I.$$

This proves that we can satisfy this model when $\Delta = \delta = (1 + \sqrt{5})/2$. For this specialization we see that the matrix F becomes

$$F = \begin{pmatrix} 1/\Delta & \Delta/\Theta \\ \Theta/\Delta^2 & T\Delta/\Theta^2 \end{pmatrix} = \begin{pmatrix} 1/\Delta & \Delta/\Theta \\ \Theta/\Delta^2 & (-\Theta^2/\Delta^2)\Delta/\Theta^2 \end{pmatrix} = \begin{pmatrix} 1/\Delta & \Delta/\Theta \\ \Theta/\Delta^2 & -1/\Delta \end{pmatrix}.$$

This version of F has square equal to the identity independent of the value of Θ, so long as $\Delta^2 = \Delta + 1$.

The Final Adjustment. Our last version of F suffers from a lack of symmetry. It is not a symmetric matrix, and hence not unitary. A final adjustment of the model gives this desired symmetry. *Consider the result of replacing each trivalent vertex (with three 2-projector strands) by a multiple of a given quantity α.* Since the Θ has two vertices, it will be multiplied by α^2. Similarly, the tetrahedron T will be multiplied by α^4. The Δ and the δ will be unchanged. Other properties of the model will remain unchanged. The new recoupling matrix, after such an adjustment is made, becomes

$$\begin{pmatrix} 1/\Delta & \Delta/\alpha^2\Theta \\ \alpha^2\Theta/\Delta^2 & -1/\Delta \end{pmatrix}.$$

For symmetry we require

$$\Delta/(\alpha^2\Theta) = \alpha^2\Theta/\Delta^2.$$

We take

$$\alpha^2 = \sqrt{\Delta^3}/\Theta.$$

With this choice of α we have

$$\Delta/(\alpha^2\Theta) = \Delta\Theta/(\Theta\sqrt{\Delta^3}) = 1/\sqrt{\Delta}.$$

Hence the new symmetric F is given by the equation

$$F = \begin{pmatrix} 1/\Delta & 1/\sqrt{\Delta} \\ 1/\sqrt{\Delta} & -1/\Delta \end{pmatrix} = \begin{pmatrix} \tau & \sqrt{\tau} \\ \sqrt{\tau} & -\tau \end{pmatrix}$$

where Δ is the golden ratio and $\tau = 1/\Delta$. This gives the Fibonacci model. Using Figs. 14.58 and 14.59, we have that the local braiding matrix for the model is given by the formula below with $A = e^{3\pi i/5}$.

$$R = \begin{pmatrix} A^8 & 0 \\ 0 & -A^4 \end{pmatrix} = \begin{pmatrix} e^{4\pi i/5} & 0 \\ 0 & -e^{2\pi i/5} \end{pmatrix}.$$

The simplest example of a braid group representation arising from this theory is the representation of the three-strand braid group generated by $S_1 = R$ and $S_2 = FRF$. (Remember that $F = F^T = F^{-1}$.) The matrices S_1 and S_2 are both unitary, and they generate a dense subset of the unitary group $U(2)$, supplying the first part of the transformations needed for quantum computing.

14.14 Quantum computation of colored Jones polynomials and the Witten–Reshetikhin–Turaev invariant

In this section we make some brief comments on the quantum computation of colored Jones polynomials. This material will be expanded in a subsequent publication.

First, consider Fig. 14.60. In that figure we illustrate the calculation of the evaluation of the *(a)—colored bracket polynomial* for the *plat closure* $P(B)$ of a braid B. The reader can infer the definition of the plat closure from Fig. 14.60. One takes a braid of an even number of strands and closes the top strands with each other in a row of maxima. Similarly, the bottom strands are closed with a row of minima. It is not hard to see that any knot or link can be represented as the plat closure of some braid. Note that in this figure we indicate the action of the braid group on the process spaces corresponding to the small trees attached below the braids.

The (a)—colored bracket polynomial of a link L, denoted $\langle L \rangle_a$, is the evaluation of that link where each single strand has been replaced by a parallel strands and the insertion of a Jones–Wenzl projector (as discussed in Section 14.7). We then see that we can use our discussion of the Temperley–Lieb recoupling theory as in Sections 14.7, 14.8 and 14.9 to compute the value of the colored bracket polynomial for the plat closure PB. As shown in Fig. 14.60, we regard the braid as acting on a process space $V_0^{a,a,\cdots,a}$ and take the case of the action on the vector v whose process space coordinates are all zero. Then the action of the braid takes the form

$$Bv(0,\cdots,0) = \Sigma_{x_1,\cdots,x_n} B(x_1,\cdots,x_n) v(x_1,\cdots,x_n)$$

where $B(x_1,\cdots,x_n)$ denotes the matrix entries for this recoupling transformation and $v(x_1,\cdots,x_n)$ runs over a basis for the space $V_0^{a,a,\cdots,a}$. Here n is even and equal to the number of braid strands. In the figure we illustrate with $n = 4$. Then, as the figure shows, when we close the top of the braid action to form PB, we cut the sum down to the evaluation of just one term. In the general case we will get

$$\langle PB \rangle_a = B(0,\cdots,0)\Delta_a^{n/2}.$$

The calculation simplifies to this degree because of the vanishing of loops in the recoupling graphs. The vanishing result is shown in Fig. 14.60, and it is proved in the case $a = 2$ in Fig. 14.52.

The *colored Jones polynomials* are normalized versions of the colored bracket polynomials, differing just by a normalization factor.

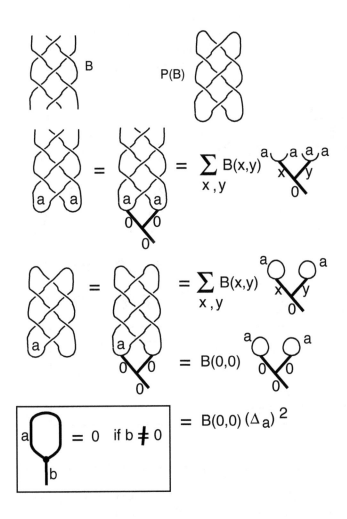

FIGURE 14.60: Evaluation of the plat closure of a braid.

In order to consider quantum computation of the colored bracket or colored Jones polynomials, we therefore can consider quantum computation of the matrix entries $B(0,\cdots,0)$. These matrix entries in the case of the roots of unity $A = e^{i\pi/2r}$ and for the $a = 2$ Fibonacci model with $A = e^{3i\pi/5}$ are parts of the diagonal entries of the unitary transformation that represents the braid group on the process space $V_0^{a,a,\cdots,a}$. *We can obtain these matrix entries by using the Hadamard test as described in Section 14.4.* As a result we get relatively efficient quantum algorithms for the colored Jones polynomials at these roots of unity, in essentially the same framework as we described in Section 14.4, but for braids of arbitrary size. The computational complexity of these models is essentially the same as the models for the Jones polynomial discussed in [1]. We reserve discussion of these issues to a subsequent publication.

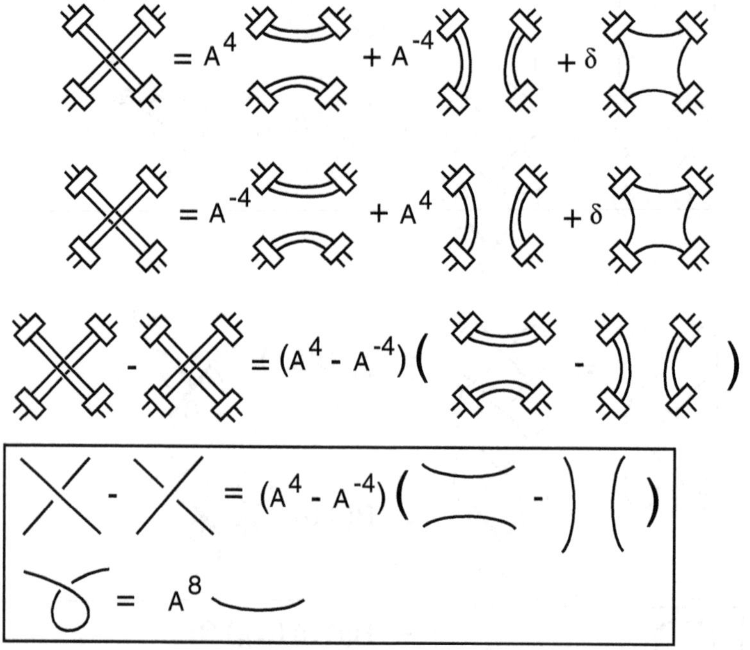

FIGURE 14.61: Dubrovnik polynomial specialization at two strands.

It is worth remarking here that these algorithms give not only quantum algorithms for computing the colored bracket and Jones polynomials, but also for computing the Witten–Reshetikhin–Turaev (*WRT*) invariants at the above roots of unity. The reason for this is that the *WRT* invariant in unnormalized form is given as a finite sum of colored bracket polynomials:

$$WRT(L) = \Sigma_{a=0}^{r-2} \Delta_a \langle L \rangle_a,$$

and so the same computation as shown in Fig. 14.60 applies to the WRT. This means that we have, in principle, a quantum algorithm for the computation of the Witten functional integral [85] via this knot-theoretic combinatorial topology. It would be very interesting to understand a more direct approach to such a computation via quantum field theory and functional integration.

Finally, we note that in the case of the Fibonacci model, the two-colored bracket polynomial is a special case of the Dubrovnik version of the Kauffman polynomial [40]. See Fig. 14.61 for diagammatics that resolve this fact. The skein relation for the Dubrovnik polynomial is boxed in this figure. Above the box, we show how the double strands with projectors reproduce this relation. This observation means that in the Fibonacci model, the natural underlying knot polynomial is a special evaluation of the Dubrovnik polynomial, and the Fibonacci model can be used to perform quantum computation for the values of this invariant.

Acknowledgement

The first author thanks the National Science Foundation for support of this research under NSF Grant DMS-0245588. Much of this effort was sponsored by the Defense Advanced Research Projects Agency (DARPA) and Air Force Research Laboratory, Air Force Materiel Command, USAF, under agreement F30602-01-2-05022. The U.S. government is authorized to reproduce and distribute reprints for government purposes notwithstanding any copyright annotations thereon. The views and conclusions contained herein are those of the authors and should not be interpreted as necessarily representing the official policies or endorsements, either expressed or implied, of the Defense Advanced Research Projects Agency, the Air Force Research Laboratory, or the U.S. government. (Copyright 2006.) It gives the authors pleasure to thank the Newton Institute in Cambridge England and ISI in Torino, Italy for their hospitality during the inception of this research and to thank Hilary Carteret for useful conversations.

References

[1] D. Aharonov, V. Jones, and Z. Landau, A polynomial quantum algorithm for approximating the Jones polynomial, quant-ph/0511096.

[2] D. Aharonov and I. Arad, The BQP-hardness of approximating the Jones polynomial, quant-ph/0605181.

[3] Y. Akutsu and M. Wadati, Knot invariants and critical statistical systems, *J. Phys. Soc. Japan* **56** (1987), 839–842.

[4] J. W. Alexander, Topological invariants of knots and links, *Trans. Amer. Math. Soc.* **20** (1923), 275–306.

[5] P. K. Aravind, Borromean entanglement of the GHZ state, in *Potentiality, Entanglement and Passion-at-a-Distance*, R. S. Cohen et al. (eds.), pp. 53-59, Kluwer, 1997.

[6] M. F. Atiyah, *The Geometry and Physics of Knots*, Cambridge University Press, 1990.

[7] A. Ashtekar, C. Rovelli, and L. Smolin, Weaving a classical geometry with quantum threads, *Phys. Rev. Lett.* **69** (1992), 237.

[8] A. Ashetekar and J. Lewandowski, Quantum theory of geometry I: Area operators, *Class. Quant. Grav.* **14** (1997), A55–A81.

[9] J. Baez and J. P. Muniain, *Gauge Fields, Knots and Gravity*, World Scientific Series on Knots and Everything, Vol. 4 (1994).

[10] R. J. Baxter, *Exactly Solved Models in Statistical Mechanics*, Academic Press (1982).

[11] D. Bar-Natan, On the Vassiliev knot invariants, *Topology* **34** (1995), 423–472.

[12] Dror Bar-Natan, *Perturbative Aspects of the Chern–Simons Topological Quantum field Theory*, Ph.D. Thesis, Princeton University, June 1991.

[13] G. Benkart, Commuting actions – a tale of two groups, in *Lie algebras and their representations (Seoul 1995)*, Contemp. Math. Series, Vol. 194, American Mathematical Society (1996), pp. 1–46.

[14] J. Birman and X. S. Lin, Knot polynomials and Vassilievs invariants, *Invent. Math.* **111** No. 2 (1993), 225–270.

[15] N. E. Bonesteel, L. Hormozi, G. Zikos, and S. H. Simon, Braid topologies for quantum computation, quant-ph/0505665.

[16] S. H. Simon, N. E. Bonesteel, M. H. Freedman, N. Petrovic, and L. Hormozi, Topological quantum computing with only one mobile quasiparticle, quant-ph/0509175.

[17] J. L. Brylinski and R. Brylinski, Universal quantum gates, in *Mathematics of Quantum Computation*, R. Brylinski and G. Chen (eds.) Chapman & Hall/CRC Press, Boca Raton, Florida, 2002.

[18] L. Crane, 2-d physics and 3-d topology, *Comm. Math. Phys.* **135** (1991), no. 3, 615–640.

[19] B. Coecke, The logic of entanglement, quant-phy/0402014.

[20] S. Abramsky and B. Coecke, A categorical semantics of quantum protocols, quant-ph/0402130.

[21] C. Dewitt-Morette, P. Cartier and A. Folacci, *Functional Integration: Basics and Applications*, NATO ASI Series, Series B: Physics Vol. 361 (1997).

[22] P.A.M. Dirac, *Principles of Quantum Mechanics*, Oxford University Press, 1958.

[23] V. G. Drinfeld, *Quantum Groups*, Proc. Intl. Congress Math., Berkeley, Calif., USA (1986), 789–820.

[24] H. Dye, Unitary solutions to the Yang–Baxter equation in dimension four, arXiv:quant-ph/0211050, v2, 22 January 2003.

[25] E. Fradkin and P. Fendley, Realizing non-abelian statistics in time-reversal invariant systems, Theory Seminar, Physics Department, UIUC, 4/25/2005.

[26] M. Freedman, A magnetic model with a possible Chern–Simons phase, quant-ph/0110060v1, 9 Oct 2001, (2001), preprint.

[27] M. Freedman, *Topological Views on Computational Complexity*, Documenta Mathematica - Extra Volume ICM, 1998, pp. 453–464.

[28] M. Freedman, M. Larsen, and Z. Wang, A modular functor which is universal for quantum computation, quant-ph/0001108v2, 1 Feb 2000.

[29] M. H. Freedman, A. Kitaev, and Z. Wang, Simulation of topological field theories by quantum computers, *Commun. Math. Phys.*, **227**, (2002), 587–603, quant-ph/0001071.

[30] M. Freedman, Quantum computation and the localization of modular functors, quant-ph/0003128.

[31] J. Fröhlich and C. King, The Chern–Simons theory and knot polynomials, *Commun. Math. Phys.* **126** (1989), 167–199.

[32] C. Frohman and J. Kania-Bartoszynska, $SO(3)$ topological quantum field theory, *Comm. Anal. Geom.* **4**, (1996), no. 4, 589–679.

[33] V. F. R. Jones, A polynomial invariant for links via von Neumann algebras, *Bull. Amer. Math. Soc.* **129** (1985), 103–112.

[34] V. F. R. Jones, Hecke algebra representations of braid groups and link polynomials, *Ann. Math.* **126** (1987), 335–388.

[35] V. F. R. Jones, On knot invariants related to some statistical mechanics models, *Pacific J. Math.* **137**, no. 2 (1989), 311–334.

[36] L. H. Kauffman, State models and the Jones polynomial, *Topology* **26** (1987), 395–407.

[37] L. H. Kauffman, Statistical mechanics and the Jones polynomial, *AMS Contemp. Math. Series* **78** (1989), 263–297.

[38] L. H. Kauffman, *Temperley–Lieb Recoupling Theory and Invariants of Three-Manifolds*, Princeton University Press, Annals Studies, Vol. 114, 1994.

[39] L. H. Kauffman, New invariants in the theory of knots, *Amer. Math. Monthly*, **95**, No. 3, March 1988, pp. 195–242.

[40] L. H. Kauffman, An invariant of regular isotopy, *Trans. Amer. Math. Soc.* **318** (1990), No. 2, 417–471.

[41] L. H. Kauffman and P. Vogel, Link polynomials and a graphical calculus, *Journal of Knot Theory and Its Ramifications*, **1**, No. 1, March 1992, pp. 59–104.

[42] L. H. Kauffman (ed.), *The Interface of Knots and Physics*, AMS PSAPM, Vol. 51, Providence, RI, 1996.

[43] L. H. Kauffman, *Knots and Physics*, World Scientific Publishers (1991), Second Edition (1993), Third Edition (2002).

[44] L. H. Kauffman and D. E. Radford, Invariants of 3-manifolds derived from finite dimensional Hopf algebras, *Journal of Knot Theory and its Ramifications* **4**, No. 1 (1995), 131–162.

[45] L. H. Kauffman, Functional Integration and the theory of knots, *J. Math. Physics*, **36(5)**, May 1995, pp. 2402–2429.

[46] L. H. Kauffman, Witten's Integral and the Kontsevich Integrals, in *Particles, Fields, and Gravitation*, Proceedings of the Lodz, Poland (April 1998) Conference on Mathematical Physics edited by Jakub Rembliens-ki, AIP Conference Proceedings **453** (1998), pp. 368–381.

[47] L. H. Kauffman, Knot theory and the heuristics of functional integration, *Physica A* **281** (2000), 173–200.

[48] L.H. Kauffman, Quantum computing and the Jones polynomial, math.QA/0105255, in *Quantum Computation and Information*, S. Lomonaco, Jr. (ed.), AMS CONM/305, 2002, pp. 101–137.

[49] L. H. Kauffman and S. J. Lomonaco Jr., Quantum entanglement and topological entanglement, *New Journal of Physics* **4** (2002), 73.1–73.18 (http://www.njp.org/).

[50] L. H. Kauffman, Teleportation Topology, quant-ph/0407224, in *The Proceedings of the 2004 Byelorus Conference on Quantum Optics*, Opt. Spectrosc. **9** (2005), 227–232.

[51] L. H. Kauffman, math.GN/0410329, Knot diagrammatics, *Handbook of Knot Theory*, Menasco and Thistlethwaite (eds.), 233–318, Elsevier B. V., Amsterdam, 2005.

[52] L. H. Kauffman and T. Liko, hep-th/0505069, Knot theory and a physical state of quantum gravity, *Classical and Quantum Gravity*, Vol. 23, ppR63 (2006).

[53] L.H. Kauffman and S. J. Lomonaco Jr., Entanglement Criteria: Quantum and Topological, in *Quantum Information and Computation: Spie Proceedings, 21-22 April, 2003, Orlando, FL*, Donkor, Pinch and Brandt (eds.), Vol. 5105, pp. 51–58.

[54] L. H. Kauffman and S. J. Lomonaco Jr., Quantum knots, in *Quantum Information and Computation II, Proceedings of Spie, 12-14 April 2004* (2004), Donkor Pirich and Brandt (eds.), pp. 268–284.

[55] L. H. Kauffman and S. J. Lomonaco, Braiding operators are universal quantum gates, *New Journal of Physics* **6** (2004) 134, 1–39.

[56] L. H. Kauffman and S. J. Lomonaco Jr., Spin Networks and Anyonic Topological Quantum Computing, quant-ph/0603131, v3, Apr 2006 in *Quantum Information and Computation IV*, edited by Donkor, Pirich and

Brandt, Vol. 6244 (2006), Intl. Soc. for Opt. Eng. (Spie), pp. 62440Y-1 to 62440Y-12.

[57] L. H. Kauffman and S. J. Lomonaco Jr., q - deformed spin networks, knot polynomials and anyonic topological quantum computation, quant-ph/0606114, J. Knot Theory and Its Ramifications, Vol. 16, No. 3, March 2007, pp. 267–332.

[58] L. H. Kauffman and S. J. Lomonaco Jr., Topological quantum computing and the Jones polynomial, quant-ph/0605004, in *Quantum Information and Computation IV*, edited by Donkor, Pirich and Brandt, Vol. 6244 (2006), Intl. Soc. for Opt. Eng. (Spie), pp. 62440Z-1 to 62440Z-18.

[59] L. H. Kauffman (ed.), *Knots and Applications*, World Scientific Pub. Co. (1996).

[60] A. Kitaev, Anyons in an exactly solved model and beyond, *arXiv:cond-mat/0506438 v1 17 June 2005*.

[61] H. Kleinert, *Path Integrals in Quantum Mechanics, Statistics and Polymer Physics*, 2nd edition, World Scientific, Singapore (1995).

[62] H. Kleinert, *Grand Treatise on Functional Integration*, World Scientific Pub. Co. (1999).

[63] T. Kohno, *Conformal Field Theory and Topology*, AMS Translations of Mathematical Monographs, Vol. 210 (1998).

[64] J. M. F. Labastida and E. Pérez, Kontsevich integral for Vassiliev invariants from Chern–Simons perturbation theory in the light-cone gauge, *J. Math. Phys.* **39** (1998), 5183–5198.

[65] A. Marzuoli and M. Rasetti, Spin network quantum simulator, *Physics Letters A* **306** (2002), 79–87.

[66] S. Garnerone, A. Marzuoli, and M. Rasetti, Quantum automata, braid group and link polynomials, quant-ph/0601169.

[67] S. A. Major, A spin network primer, arXiv:gr-qc/9905020.

[68] G. Moore and N. Seiberg, Classical and quantum conformal field theory, *Comm. Math. Phys.* **123** (1989), No. 2, 177–254.

[69] M. A. Nielsen and I. L. Chuang, *Quantum Computation and Quantum Information*, Cambridge University Press, Cambridge (2000).

[70] R. Penrose, Angular momentum: An approach to combinatorial space-time, in *Quantum Theory and Beyond*, T. Bastin (ed.), Cambridge University Press (1969).

[71] J. Preskill, Topological computing for beginners, (slide presentation), Lecture Notes for Chapter 9 - Physics 219 - Quantum Computation. *http://www.iqi.caltech.edu/ preskill/ph219.*

[72] P. Cotta-Ramusino, E. Guadagnini, M. Martellini, and M. Mintchev, Quantum field theory and link invariants, *Nucl. Phys. B* **330**, Nos. 2–3 (1990), 557–574.

[73] N. Y. Reshetikhin and V. Turaev, Ribbon graphs and their invariants derived from quantum groups, *Comm. Math. Phys.* **127** (1990), 1–26.

[74] N. Y. Reshetikhin and V. Turaev, Invariants of three manifolds via link polynomials and quantum groups, *Invent. Math.* **103** (1991), 547–597.

[75] M. Roetteles, (private conversation, fall 2003).

[76] C. Rovelli and L. Smolin, Spin networks and quantum gravity, *Phys Rev. D* **52** (1995), 5743–5759.

[77] B. Schumacher, Ph.D. Thesis.

[78] V. V. Shende, S. S. Bullock, and I. L. Markov, Recognizing small circuit structure in two-qubit operators, (arXiv:quant-ph/030845 v2 8 Aug 2003).

[79] C. Ernst and D. W. Sumners, A calculus for rational tangles: Applications to DNA recombination, *Math. Proc. Camb. Phil. Soc.*, **108** (1990), 489–515.

[80] G. Spencer–Brown, *Laws of Form*, George Allen and Unwin Ltd. London (1969).

[81] V. G. Turaev, The Yang–Baxter equations and invariants of links, LOMI preprint E-3-87, Steklov Institute, Leningrad, USSR, *Inventiones Math.* **92** Fasc. 3, 527–553.

[82] V. G. Turaev and O. Viro, State sum invariants of 3-manifolds and quantum 6j symbols, *Topology* **31** No. 4, (1992), 865–902.

[83] Lee Smolin, Link polynomials and critical points of the Chern–Simons path integrals, *Mod. Phys. Lett. A*, **4** No. 12, (1989), 1091–1112.

[84] F. Wilczek, *Fractional Statistics and Anyon Superconductivity,* World Scientific Publishing Company (1990).

[85] E. Witten, Quantum field theory and the Jones polynomial, *Commun. Math. Phys.* **121** (1989), 351–399.

[86] P. Wocjan and J. Yard The Jones polynomial: quantum algorithms and applications in quantum complexity theory, quant-ph/0603069.

[87] C. N. Yang, *Phys. Rev. Lett.* **19** (1967), 1312.

[88] Y. Zhang, L. H. Kauffman, and M. L. Ge, Yang–Baxterizations, universal quantum gates and Hamiltonians. *Quantum Inf. Process.* **4** (2005), No. 3, 159–197.

Chapter 15

Temperley–Lieb Algebra: From Knot Theory to Logic and Computation via Quantum Mechanics

Samson Abramsky

Abstract   We study the Temperley–Lieb algebra, central to the Jones polynomial invariant of knots and ensuing developments, from a novel point of view. We relate the Temperley–Lieb category to the categorical formulation of quantum mechanics introduced by Abramsky and Coecke as the basis for the development of high-level methods for quantum information and computation. We develop some structural properties of the Temperley–Lieb category, giving a simple diagrammatic description of epi-monic factorization, and hence of splitting idempotents. We then relate the Temperley–Lieb category to some topics in proof theory and computation. We give a direct, "fully abstract" description of the Temperley–Lieb category, in which arrows are just relations on discrete finite sets, with planarity being characterized by simple order-theoretic properties. The composition is described in terms of the "Geometry of Interaction" construction, originally introduced to analyze cut elimination in Linear Logic. Thus we obtain a planar version of Geometry of Interaction. Moreover, we get an explicit description of the free pivotal category on one self-dual object, which is easily generalized to an arbitrary generating category. Moreover, we show that the construction naturally lifts a dagger structure on the underlying category, thus exhibiting a key feature of the Abramsky–Coecke axiomatization. The dagger or "adjoint", and the "complex conjugate", acquire natural diagrammatic readings in this context. Finally, we interpret a non-commutative lambda calculus (a variant of the Lambek calculus, widely used in computational linguistics) in the Temperley–Lieb category, and thus show how diagrammatic simplification can be viewed as functional computation.

15.1   Introduction

Our aim in this paper is to trace some of the surprising and beautiful connections which are beginning to emerge among a number of apparently disparate topics.

15.1.1   Knot theory

Vaughan Jones' discovery of his new polynomial invariant of knots in 1984 [26] triggered a spate of mathematical developments relating knot theory, topological quantum field theory, and statistical physics *inter alia* [44, 30]. A central role, both in the initial work by Jones and in the subsequent developments, was played by what has come to be known as the *Temperley–Lieb algebra*.[1]

15.1.2   Categorical quantum mechanics

Recently, motivated by the needs of quantum information and computation, Abramsky and Coecke have recast *the foundations of quantum mechanics itself*, in the more abstract language of category theory. The key contribution is the paper [4], which develops an axiomatic presentation of quantum mechanics in the general setting of *strongly compact closed categories*, which is adequate for the needs of quantum information and computation. Moreover, these categorical axiomatics can be presented in terms of a *diagrammatic calculus* which is both intuitive and effective, and can replace low-level computation with matrices by much more conceptual reasoning. This diagrammatic calculus can be seen as a proof system for a logic [6], leading to a radically new perspective on what the right logical formulation for quantum mechanics should be.

---

[1]The original work of Temperley and Lieb [43] was in discrete lattice models of statistical physics. In finding exact solutions for a certain class of systems, they had identified the same relations which Jones, quite independently, found later in his work.

This line of work has a direct connection to the Temperley–Lieb algebra, which can be put in a categorical framework, in which it can be described essentially as *the free pivotal dagger category on one self-dual generator* [21].[2] Here *pivotal dagger category* is a non-symmetric ("planar") version of *(strongly or dagger) compact closed category*—the key notion in the Abramsky–Coecke axiomatics.

15.1.3 Logic and computation

The Temperley–Lieb algebra itself has some direct and striking connections to basic ideas in logic and computation, which offer an intriguing and promising bridge between these *prima facie* very different areas. We shall focus in particular on the following two topics:

- The Temperley–Lieb algebra has always hitherto been presented as a *quotient* of some sort: either algebraically by generators and relations as in Jones' original presentation [26], or as a diagram algebra modulo planar isotopy as in Kauffman's presentation [29]. We shall use tools from *Geometry of Interaction* [23], a dynamical interpretation of proofs under cut elimination developed as an off-shoot of Linear Logic [22], to give a *direct description* of the Temperley–Lieb category—a *fully abstract presentation*, in computer science terminology [37]. This also brings something new to the Geometry of Interaction, since we are led to develop a planar version of it, and to verify that the interpretation of cut elimination (the "execution formula" [23], or "composition by feedback" [8, 1]) preserves planarity.

- We shall also show how the Temperley–Lieb algebra provides a natural setting in which computation can be performed diagrammatically as *geometric simplification*—"yanking lines straight". We shall introduce a "planar λ-calculus" for this purpose, and show how it can be interpreted in the Temperley–Lieb category.

15.1.4 Outline of the paper

We briefly summarize the further contents of this paper. In Section 15.2 we introduce the Temperley–Lieb algebras, emphasizing Kauffman's diagrammatic formulation. We also briefly outline how the Temperley–Lieb algebra figures in the construction of the Jones polynomial. In Section 15.3 we describe

---

[2]Strictly speaking, the full Temperley–Lieb category over a ring R is the free R-linear enrichment of this free pivotal dagger category.

the Temperley–Lieb category, which provides a more structured perspective on the Temperley–Lieb algebras. In Section 15.4, we discuss some features of this category, which have apparently not been considered previously, namely a characterization of monics and epics, leading to results on image factorization and splitting of idempotents. In Section 15.5, we briefly discuss the connections with the Abramsky–Coecke categorical formulation of quantum mechanics, and raise some issues and questions about the possible relationship betwen planar, braided and symmetric settings for quantum information and computation. In Section 15.6 we develop a planar version of Geometry of Interaction, and the direct "fully abstract" presentation of the Temperley–Lieb category. In Section 15.7 we discuss the planar λ-calculus and its interpretation in the Temperley–Lieb category. We conclude in Section 15.8 with some further directions.

Note to the Reader Since this paper aims at indicating cross-currents between several fields, it has been written in a somewhat expansive style, and an attempt has been made to explain the context of the various ideas we will discuss. We hope it will be accessible to readers with a variety of backgrounds.

15.2   The Temperley–Lieb algebra

Our starting point is the Temperley–Lieb algebra, which has played a central role in the discovery by Vaughan Jones of his new polynomial invariant of knots and links [26], and in the subsequent developments over the past two decades relating knot theory, topological quantum field theory, and statistical physics [30].

Jones' approach was algebraic: in his work, the Temperley–Lieb algebra was originally presented, rather forbiddingly, in terms of abstract generators and relations. It was recast in beautifully elementary and conceptual terms by Louis Kauffman as a *planar diagram algebra* [29]. We begin with the algebraic presentation.

15.2.1   Temperley–Lieb algebra: generators and relations

We fix a ring R; in applications to knot polynomials, this is taken to be a ring of Laurent polynomials $\mathbb{C}[X,X^{-1}]$. Given a choice of *parameter* $\tau \in R$ and a *dimension* $n \in \mathbb{N}$, we define the Temperley–Lieb algebra $\mathscr{A}_n(\tau)$ to be the unital, associative R-linear algebra with generators

$$U_1, \ldots, U_{n-1}$$

and relations

$$U_i U_j U_i = U_i \qquad |i - j| = 1$$
$$U_i^2 = \tau \cdot U_i$$
$$U_i U_j = U_j U_i \quad |i - j| > 1.$$

Note that the only relations used in defining the algebra are multiplicative ones. This suggests that we can obtain the algebra $\mathscr{A}_n(\tau)$ by presenting the multiplicative monoid $\mathscr{M}_n$, and then obtaining $\mathscr{A}_n(\tau)$ as the *monoid algebra* of formal R-linear combinations $\sum_i r_i \cdot a_i$ over $\mathscr{M}_n$, with the multiplication in $\mathscr{A}_n(\tau)$ defined as the bilinear extension of the monoid multiplication in $\mathscr{M}_n$:

$$\left(\sum_i r_i \cdot a_i \right) \left(\sum_j s_j \cdot b_j \right) = \sum_{i,j} (r_i s_j) \cdot (a_i b_j).$$

We define $\mathscr{M}_n$ as the monoid with generators

$$\delta, U_1, \ldots, U_{n-1}$$

and relations

$$U_i U_j U_i = U_i \qquad |i - j| = 1$$
$$U_i^2 = \delta U_i$$
$$U_i U_j = U_j U_i \quad |i - j| > 1$$
$$\delta U_i = U_i \delta.$$

We can then obtain $\mathscr{A}_n(\tau)$ as the monoid algebra over $\mathscr{M}_n$, subject to the identification

$$\delta = \tau \cdot 1.$$

15.2.2 Diagram monoids

These formal algebraic ideas are brought to vivid geometric life by Kauffman's interpretation of the monoids $\mathscr{M}_n$ as *diagram monoids*.

We start with two parallel rows of n dots (geometrically, the dots are points in the plane). The general form of an element of the monoid is obtained by "joining up the dots" pairwise in a smooth, planar fashion, where the arc connecting each pair of dots must lie within the rectangle framing the two parallel rows of dots. Such diagrams are identified up to planar isotopy, i.e., continuous deformation within the portion of the plane bounded by the framing rectangle.

Thus the generators $U_1, \ldots, U_{n-1}$ can be drawn as follows:

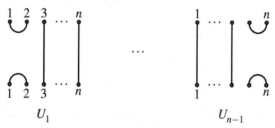

$$U_1 \qquad\qquad\qquad U_{n-1}$$

The generator δ corresponds to a loop ◯—all such loops are identified up to isotopy.

We refer to arcs connecting dots in the top row as *cups*, those connecting dots in the bottom row as *caps*, and those connecting a dot in the top row to a dot in the bottom row as *through lines*.

Multiplication xy is defined by identifying the bottom row of x with the top row of y, and composing paths. In general loops may be formed—these are "scalars", which can float freely across these figures. The relations can be illustrated as follows:

$$U_1 U_2 U_1 = U_1 \qquad\qquad\qquad U_1^2 = \delta U_1$$

$$U_1 U_3 = U_3 U_1$$

15.2.3 Expressiveness of the generators

The fact that all planar diagrams can be expressed as products of generators is not entirely obvious. For proofs, see [20, 29]. As an illustrative example, consider the planar diagrams in $\mathcal{M}_3$. Apart from the generators U_1, U_2, and ignoring loops, there are three:

The first is the identity for the monoid; we refer to the other two as the *left wave* and *right wave* respectively. The left wave can be expressed as the product U_2U_1:

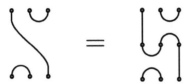

The right wave has a similar expression.

Once we are in dimension four or higher, we can have *nested cups and caps*. These can be built using waves, as illustrated by the following:

15.2.4 The trace

There is a natural *trace function* on the Temperley–Lieb algebra, which can be defined diagrammatically on $\mathcal{M}_n$ by connecting each dot in the top row to the corresponding dot in the bottom row, using auxiliary cups and cups. This always yields a diagram isotopic to a number of loops—hence to a *scalar*, as expected. This trace can then be extended linearly to $\mathcal{A}_n(\tau)$.

We illustrate this firstly by taking the trace of a wave—which is equal to a single loop:

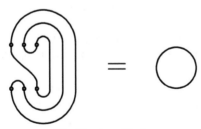

The Ear is a Circle

Our second example illustrates the important general point that *the trace of the identity in $\mathcal{M}_n$ is δ^n*:

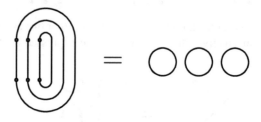

15.2.5   The connection to knots

How does this connect to knots? Again, a key conceptual insight is due to Kauffman, who saw how to recast the Jones polynomial in elementary combinatorial form in terms of his *bracket polynomial*. The basic idea of the bracket polynomial is expressed by the following equation:

$$\left\langle \times \right\rangle \;=\; A\left\langle \smile \atop \frown \right\rangle \;+\; B\left\langle\right)\,\left(\right\rangle$$

Each over-crossing in a knot or link is evaluated to a weighted sum of the two possible planar smoothings. With suitable choices for the coefficients A and B (as Laurent polynomials), this is invariant under the second and third Reidemeister moves. With an ingenious choice of normalizing factor, it becomes invariant under the first Reidemeister move—and yields the Jones polynomial! What this means algebraically is that the braid group $\mathcal{B}_n$ has a representation in the Temperley–Lieb algebra $\mathcal{A}_n(\tau)$—the above bracket evaluation shows how the generators β_i of the braid group are mapped into the Temperley–Lieb algebra:

$$\beta_i \;\mapsto\; A \cdot U_i + B \cdot 1.$$

Every knot arises as the closure (*i.e.* the diagrammatic trace) of a braid; the invariant arises by mapping the *open braid* into the Temperley–Lieb algebra, and taking the trace there.

This is just the beginning of a huge swathe of further developments, including: topological quantum field theories [44], quantum groups [28], quantum statistical mechanics [30], diagram algebras and representation theory [25], and more.

15.3   The Temperley–Lieb category

We can expose more structure by gathering all the Temperley–Lieb algebras into a single category. We begin with the category $\mathcal{D}$ which plays a similar role with respect to the diagram monoids $\mathcal{M}_n$.

The objects of $\mathcal{D}$ are the natural numbers. An arrow $\mathbf{n} \to \mathbf{m}$ is given by

- a number $k \in \mathbb{N}$ of loops.

- a diagram which joins the top row of n dots and the bottom row of m dots up pairwise, in the same smooth planar fashion as we have already specified for the diagram monoids. As before, diagrams are identified up to planar isotopy.

Composition of arrows $f : \mathbf{n} \to \mathbf{m}$ and $g : \mathbf{m} \to \mathbf{p}$ is defined by identifying the bottom row of m dots for f with the top row of m dots for g, and composing paths. The loops in the resulting arrow are those of f and of g, together with any formed by the process of composing paths.

Clearly we recover each $\mathcal{M}_n$ as the endomorphism monoid $\mathcal{D}(\mathbf{n}, \mathbf{n})$. Moreover, we can define the Temperley–Lieb category $\mathcal{T}$ over a ring R as the free R-linear category generated by $\mathcal{D}$, with a construction which generalizes that of the monoid algebra: the objects of $\mathcal{T}$ are the same as those of $\mathcal{D}$, and arrows are R-linear combinations of arrows of $\mathcal{D}$, with composition defined by bilinear extension from that in $\mathcal{D}$:

$$\left(\sum_i r_i \cdot g_i\right) \circ \left(\sum_j s_j \cdot f_j\right) = \sum_{i,j} (r_i s_j) \cdot (g_i \circ f_j).$$

If we fix a parameter $\tau \in R$, then we obtain the category $\mathcal{T}_\tau$ by the identification of the loop $\bigcirc$ in $\mathcal{D}$ with the scalar τ in $\mathcal{T}$.[3] We then recover the Temperley–Lieb algebras as

$$\mathcal{A}_n(\tau) = \mathcal{T}_\tau(\mathbf{n}, \mathbf{n}).$$

New possibilities also arise in $\mathcal{D}$. In particular, we get the *pure cap*

---

[3]The full justification of this step requires the identification of $\mathcal{D}$ as a free pivotal category, as discussed below.

as (the unique) arrow $\mathbf{0} \to \mathbf{2}$, and similarly the *pure cup* as the unique arrow $\mathbf{2} \to \mathbf{0}$. More generally, for each n we have arrows $\eta_{\mathbf{n}} : \mathbf{0} \to \mathbf{n} + \mathbf{n}$, and $\varepsilon_{\mathbf{n}} : \mathbf{n} + \mathbf{n} \to \mathbf{0}$:

We refer to the arrows $\eta_{\mathbf{n}}$ as *units*, and the arrows $\varepsilon_{\mathbf{n}}$ as *counits*.

The category $\mathscr{D}$ has a natural *strict monoidal structure*. On objects, we define $\mathbf{n} \otimes \mathbf{m} = \mathbf{n} + \mathbf{m}$, with unit given by $I = \mathbf{0}$. The tensor product of morphisms

$$\frac{f : \mathbf{n} \to \mathbf{m} \quad g : \mathbf{p} \to \mathbf{q}}{f \otimes g : \mathbf{n} + \mathbf{p} \to \mathbf{p} + \mathbf{q}}$$

is given by juxtaposition of diagrams in the evident fashion, with (multiset) union of loops. Thus we can write the units and counits as arrows

$$\eta_{\mathbf{n}} : I \to \mathbf{n} \otimes \mathbf{n}, \qquad \varepsilon_{\mathbf{n}} : \mathbf{n} \otimes \mathbf{n} \to I.$$

These units and counits satisfy important identities, which we illustrate diagrammatically

and write algebraically as

$$(\varepsilon_{\mathbf{n}} \otimes 1_{\mathbf{n}}) \circ (1_{\mathbf{n}} \otimes \eta_{\mathbf{n}}) = 1_{\mathbf{n}} = (1_{\mathbf{n}} \otimes \varepsilon_{\mathbf{n}}) \circ (\eta_{\mathbf{n}} \otimes 1_{\mathbf{n}}). \qquad (15.1)$$

15.3.1 Pivotal categories

From these observations, we see that $\mathscr{D}$ is a *strict pivotal category* [21], in which the duality on objects is trivial: $A = A^*$. We recall that a strict pivotal category is a strict monoidal category $(\mathscr{C}, \otimes, I)$ with an assignment $A \mapsto A^*$ on objects satisfying

$$A^{**} = A, \qquad (A \otimes B)^* = B^* \otimes A^*, \qquad I^* = I,$$

and for each object A, arrows

$$\eta_A : I \rightarrow A^* \otimes A, \qquad \varepsilon_A : A \otimes A^* \rightarrow I$$

satisfying the triangular identities:

$$(\varepsilon_A \otimes 1_A) \circ (1_A \otimes \eta_A) = 1_A, \qquad (1_{A^*} \otimes \varepsilon_A) \circ (\eta_A \otimes 1_{A^*}) = 1_{A^*}. \qquad (15.2)$$

In addition, the following coherence equations are required to hold:

$$\eta_I = 1_I, \qquad \eta_{A \otimes B} = (1_{B^*} \otimes \eta_A \otimes 1_B) \circ \eta_B,$$

and, for $f : A \rightarrow B$:

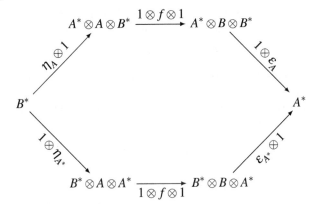

This last equation is illustrated diagrammatically by

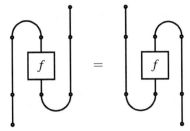

We extend $()^*$ to a contravariant involutive functor:

$$\frac{f : A \rightarrow B}{f^* : B^* \rightarrow A^*} \qquad f^* = (1 \otimes \varepsilon_A) \circ (1 \otimes f \otimes 1) \circ (\eta_A \otimes 1)$$

which indeed satisfies

$$1^* = 1, \qquad (g \circ f)^* = f^* \circ g^*, \qquad f^{**} = f,$$

the last equation being illustrated diagrammatically by

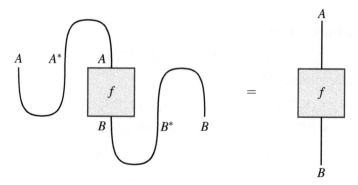

These axioms have powerful consequences. In particular, $\mathscr{C}$ is *monoidal closed*, with internal hom given by $A^* \otimes B$, and the adjunction:

$$\mathscr{C}(A \otimes B, C) \simeq \mathscr{C}(B, A^* \otimes C) :: f \mapsto (1 \otimes f) \circ (\eta_A \otimes 1).$$

This means that a restricted form of λ-calculus can interpreted in such categories—a point we shall return to in Section 15.7.

A *trace function* can be defined in pivotal categories, which takes an endomorphism $f : A \to A$ to a *scalar* in $\mathscr{C}(I,I)$:

$$\mathrm{Tr}(f) = \varepsilon_A \circ (f \otimes 1) \circ \eta_{A^*}.$$

It satisfies:

$$\mathrm{Tr}(g \circ f) = \mathrm{Tr}(f \circ g).$$

In $\mathscr{D}$, this definition yields exactly the diagrammatic trace we discussed previously.

We have the following important characterization of the diagrammatic category $\mathscr{D}$:

PROPOSITION 15.1

$\mathscr{D}$ is the free pivotal category over one self-dual generator; that is, freely generated over the one-object one-arrow category, with object A say, subject to the equation $A = A^*$.

This was mentioned (although not proved) in [21]; see also [20]. The methods in [3] can be adapted to prove this result, using the ideas we shall develop in Section 15.6.

The idea of "identifying the loop with the scalar τ" in passing from $\mathscr{D}$ to the full Temperley–Lieb category $\mathscr{T}_\tau$ can be made precise using the construction given in [3] of gluing a specified ring R of scalars onto a free compact closed category, along a given map from the loops in the generating category to R. In this case, there is a single loop in the generating category, and we send it to τ.

15.3.2 Pivotal dagger categories

We now mention a strengthening of the axioms for pivotal categories, corresponding to the notion of *strongly compact closed* or *dagger compact closed* category which has proved to be important in the categorical approach to quantum mechanics [4, 5]. Again we give the strict version for simplicity. We assume that the strict monoidal category $(\mathscr{C}, \otimes, I)$ comes equipped with an identity-on-objects, contravariant involutive functor $()^{\dagger}$ such that $\varepsilon_A = \eta_{A^*}^{\dagger}$. The idea is that $f^{\dagger}$ abstracts from the *adjoint* of a linear map, and allows the extra structure arising from the use of *complex* Hilbert spaces in quantum mechanics to be expressed in the abstract setting.

Note that there is a clear diagrammatic distinction between the dual f^* and the adjoint $f^{\dagger}$. The dual corresponds to $180°$ rotation in the plane:

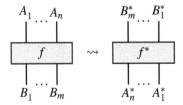

while the adjoint is reflection in the *x*-axis:

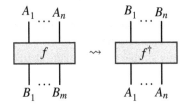

For example in $\mathscr{D}$, if we consider the left and right wave morphisms L and R:

then we have

$$L^* = L, \quad L^{\dagger} = R, \quad R^* = R, \quad R^{\dagger} = L.$$

Using the adjoint, we can define a *covariant* functor

$$\frac{f :\to B}{f_* : A^* \to B^*} \qquad f \mapsto f^{*\dagger}.$$

We have

$$(f^*)_* = f^\dagger = (f_*)^*.$$

In terms of complex matrices, f^* is transpose, while f_* is complex conjugation. Diagrammatically, f_* is "reflection in the y-axis".

$$f \qquad\qquad f^* \qquad\qquad f^\dagger \qquad\qquad f_*$$

We have the following refinement of Proposition 15.1, by similar methods to those used for free strongly compact closed categories in [3].

PROPOSITION 15.2

$\mathscr{D}$ *is the free pivotal dagger category over one self-dual generator.*

15.4   Factorization and idempotents

We now consider some structural properties of the category $\mathscr{D}$ which we have not found elsewhere in the literature.[4]

We begin with a pleasingly simple diagrammatic characterization of *monics* and *epics* in $\mathscr{D}$.

PROPOSITION 15.3

An arrow in $\mathscr{D}$ is monic iff it has no cups; it is epic iff it has no caps.

PROOF   Suppose that $f : \mathbf{n} \to \mathbf{m}$ has no cups. Thus all dots in $\mathbf{n}$ are connected by through lines to dots in $\mathbf{m}$. Now consider a composition $f \circ g$. No loops can be formed by this composition; hence we can recover g from $f \circ g$ by erasing the caps of f. Moreover, the number of loops in $f \circ g$ will simply be the sum of the loops in f and g, so we can recover

---

[4]The idea of considering these properties arose from a discussion with Louis Kauffman, who showed the author a direct diagrammatic characterization of idempotents in $\mathscr{D}$, which has subsequently appeared in [32].

the loops of g by subtracting the loops of f from the composition. It follows that

$$f \circ g = f \circ h \implies g = h,$$

i.e., that f is monic, as required.

For the converse, suppose that f has a cup, which we can assume to be connecting dots i and $i+1$ in the top row. (Note that if $i < j$ are connected by a cup, then by planarity, every k with $i < k < j$ must also be connected in a cup to some l with $i < l < j$.) Then $f \circ \delta \cdot 1 = f \circ (1 \otimes U_i \otimes 1)$, so f is not monic. Diagrammatically, this says that we can either form a loop using the cup of f, or simply add a loop which is attached to an identity morphism.

The characterization of epics is entirely similar. ∎

This immediately yields an "image factorization" structure for $\mathcal{D}$.

PROPOSITION 15.4

Every arrow in $\mathcal{D}$ has an epi-mono factorization.

PROOF   Given an arrow $f : \mathbf{n} \to \mathbf{m}$, suppose it has p cups and q caps. Then we obtain arrows $e : \mathbf{n} \to (\mathbf{m} - 2\mathbf{q})$ by erasing the caps, and $m : (\mathbf{n} - 2\mathbf{p}) \to \mathbf{m}$ by erasing the cups. By Proposition 15.3, e is epic and m monic. Moreover, the number of dots in the top and bottom rows connected by through lines must be the same. Hence

$$(\mathbf{m} - 2\mathbf{q}) = \mathbf{k} = (\mathbf{n} - 2\mathbf{p}),$$

and we can compose e and m to recover f. Note that by planarity, once we have assigned cups and caps, there is no choice about the correspondence between top and bottom row dots by through lines.

This factorization is "essentially" unique. However, we are free to split the l loops of f between e and m in any way we wish, so there is a distinct factorization $\delta^a \cdot m \circ \delta^b \cdot e$ for all $a, b \in \mathbb{N}$ with $a + b = l$. ∎

We illustrate the epi-mono factorization for the left wave:

We recall that an *idempotent* in a category is an arrow $i : A \to A$ such that $i^2 = i$. We say that an idempotent i *splits* if there are arrows $r : A \to B$ and $s : B \to A$ such that

$$i = s \circ r, \qquad r \circ s = 1_B.$$

PROPOSITION 15.5
All idempotents split in $\mathcal{D}$.

PROOF   Let $i : \mathbf{n} \to \mathbf{n}$ be an idempotent in $\mathcal{D}$. By Proposition 15.4, $i = m \circ e$, where $e : \mathbf{n} \to \mathbf{k}$ is epic and $m : \mathbf{k} \to \mathbf{n}$ is monic. Now

$$m \circ e \circ m \circ e = m \circ e.$$

Since m is monic, this implies that $e \circ m \circ e = e = 1 \circ e$. Since e is epic, this implies that $e \circ m = 1$. ∎

15.5   Categorical quantum mechanics

We now relate our discussion to the Abramsky–Coecke programme of categorical quantum mechanics.

This approach is very different from previous work on the computer science side of this interdisciplinary area, which has focused on quantum algorithms and complexity. The focus has rather been on developing *high-level methods* for quantum information and computation (QIC)—languages, logics, calculi, type systems etc.—analogous to those which have proved so essential in classical computing [2]. This has led to nothing less than a recasting of *the foundations of quantum mechanics itself*, in the more abstract language of category theory. The key contribution is the paper with Coecke [4], in which we develop an axiomatic presentation of quantum mechanics in the general setting of *strongly compact closed categories*, which is adequate for all the needs of QIC.

Specifically, we show that we can recover the key quantum mechanical notions of *inner product, unitarity, full and partial trace, Hilbert–Schmidt inner product and map-state duality, projection, positivity, measurement*, and *Born rule* (which provides the quantum *probabilities*), axiomatically at this high level of abstraction and generality. Moreover, we can derive the correctness of protocols such as quantum teleportation, entanglement swapping and logic-gate teleportation [10, 24, 45] in a transparent and very conceptual fashion.

Also, while at this level of abstraction there is no underlying field of complex numbers, there *is* still an intrinsic notion of 'scalar', and we can still make sense of *dual vs. adjoint* [4, 5], and *global phase and elimination thereof* [15]. Peter Selinger recovered *mixed state, complete positivity* and *Jamiolkowski map-state duality* [42]. Recently, in collaboration with Dusko Pavlovic and Eric Paquette, *decoherence, generalized measurements* and *Naimark's theorem* have been recovered [17, 16].

Moreover, this formalism has two important additional features. Firstly, it goes *beyond* the standard Hilbert space formalism, in that it is able to capture classical as well as quantum information flows, and the interaction between them, *within the formalism*. For example, we can capture the idea that the result of a measurement is used to determine a further stage of quantum evolution, as, e.g., in the teleportation protocol [10], where a unitary correction must be performed after a measurement; or also in measurement-based quantum computation [39, 40]. Secondly, this categorical axiomatics can be presented in terms of a *diagrammatic calculus* which is extremely intuitive, and potentially can replace low-level computation with matrices by much more conceptual— and automatable—reasoning. Moreover, this diagrammatic calculus can be seen as a proof system for a logic, leading to a radically new perspective on what the right logical formulation for quantum mechanics should be. This latter topic is initiated in [6], and developed further in the forthcoming thesis of Ross Duncan.

15.5.1 Outline of the approach

We now give some further details of the approach. The general setting is that of *strongly (or dagger) compact closed categories*, which are the symmetric version of the pivotal dagger categories we encountered in Section 15.3. Thus, in addition to the structure mentioned there, we have a symmetry natural isomorphism

$$\sigma_{A,B} : A \otimes B \simeq B \otimes A.$$

See [5] for an extended discussion. An important feature of the Abramsky–Coecke approach is the use of an intuitive *graphical calculus*, which is essentially the diagrammatic formalism we have seen in the Temperley–Lieb setting, extended with more general basic types and arrows. They key point is that this formalism admits a very direct *physical interpretation* in quantum mechanics.

In the graphical calculus we depict physical processes by boxes, and we label the inputs and outputs of these boxes by *types* which indicate the kind of system on which these boxes act, e.g., one qubit, several qubits, classical data, etc. Sequential composition (in time) is depicted by connecting matching outputs and inputs by wires, and parallel composition (tensor) by locating entities side by side, e.g.,

$$1_A : A \to A \quad f : A \to B \quad g \circ f \quad 1_A \otimes 1_B \quad f \otimes 1_C \quad f \otimes g \quad (f \otimes g) \circ h$$

for $g : B \to C$ and $h : E \to A \otimes B$ are respectively depicted as:

(The convention in these diagrams is that the 'upward' vertical direction represents progress of time.) A special role is played by boxes with either no input or no output, called *states* and *costates*, respectively (cf. Dirac's kets and bras [19]), which we depict by triangles. Finally, we also need to consider diamonds which arise by post-composing a state with a matching costate (cf. inner product or Dirac's bra-ket):

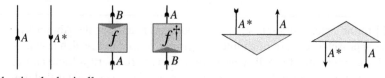

that is, algebraically,

$$\psi : I \to A \qquad \pi : A \to I \qquad \pi \circ \psi : I \to I$$

where I is the *tensor unit*: $A \otimes I \simeq A \simeq I \otimes A$. Extra structure is represented by (i) assigning a direction to the wires, where reversal of this direction is denoted by $A \mapsto A^*$, (ii) allowing reversal of boxes (cf. the *adjoint* for vector spaces), and, (iii) assuming that for each type A there exists a special bipartite *Bell state* and its adjoint *Bell costate*:

that is, algebraically,

$$A \quad A^* \quad f : A \to B \quad f^\dagger : B \to A \quad \eta_A : I \to A^* \otimes A \quad \eta_A^\dagger : A^* \otimes A \to I.$$

Hence, bras and kets are adjoint and the inner product has the form $(-)^\dagger \circ (-)$ on states. Essentially the sole *axiom* we impose is:

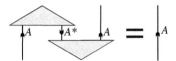

that is, algebraically,

$$(\eta_{A^*}^\dagger \otimes 1_A) \circ (1_A \otimes \eta_A) = 1_A.$$

If we extend the graphical notation of Bell-(co)states to:

we obtain a clear graphical interpretation for the axiom:

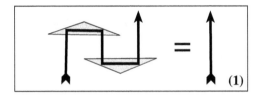

(1)

which now tells us that we are allowed to *yank the black line straight* :

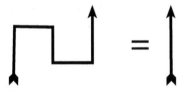

This equation and its diagrammatic counterpart should of course be compared to equation (15.2), and equation (15.1) and its accompanying diagram, in Section 15.3—they are one and the same, subject to minor differences in diagrammatic conventions.

This intuitive graphical calculus is an important benefit of the categorical axiomatics. Other advantages can be found in [4, 2].

15.5.2 Quantum non-logic vs. quantum hyper-logic

The term *quantum logic* is usually understood in connection with the 1936 Birkhoff–von Neumann proposal [11, 41] to consider the (closed) linear subspaces of a Hilbert space ordered by inclusion as the formal expression of the logical distinction between quantum and classical physics. While in classical logic we have deduction, the linear subspaces of a Hilbert space form a non-distributive lattice and hence there is no obvious notion of implication or deduction. Quantum logic was therefore always seen as logically very weak, or even as a non-logic. In addition, it has never given a satisfactory account of compound systems and entanglement.

On the other hand, *compact closed logic* in a sense goes beyond ordinary logic in the principles it admits. Indeed, while in ordinary categorical logic "logical deduction" implies that *morphisms internalize as elements* (which above we referred to above as *states*), i.e.,

$$B \xrightarrow{f} C \quad \overset{\sim}{\longleftrightarrow} \quad I \xrightarrow{\ulcorner f \urcorner} B \Rightarrow C$$

(where I is the tensor unit), in *compact closed logic* they internalize *both* as states *and* as costates, *i.e.*

$$A \otimes B^* \xrightarrow{\llcorner f \lrcorner} I \quad \overset{\sim}{\longleftrightarrow} \quad A \xrightarrow{f} B \quad \overset{\sim}{\longleftrightarrow} \quad I \xrightarrow{\ulcorner f \urcorner} A^* \otimes B$$

where we introduce the following notation:

$$\ulcorner f \urcorner = (1_{A^*} \otimes f) \circ \eta_A : I \to A^* \otimes B \quad \llcorner f \lrcorner = \varepsilon_B \circ (f \otimes 1_{B^*}) : A \otimes B^* \to I.$$

It is exactly this dual internalization which allows the *straightening axiom* in picture **(1)** to be expressed. In the graphical calculus this is witnessed by the fact that we can define both a state and a costate

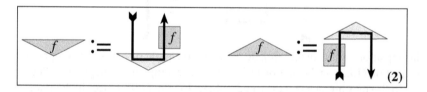

$$(2)$$

for each operation f. Physically, costates form the (destructive parts of) *projectors*, i.e. branches of projective measurements.

15.5.2.1   Compositionality

The semantics is obviously compositional, both with respect to sequential composition of operations and parallel composition of types and operations, allowing the description of systems to be built up from smaller components. But we also have something more specific in mind: a form of compositionality with direct applications to the analysis of compound entangled systems. Since we have:

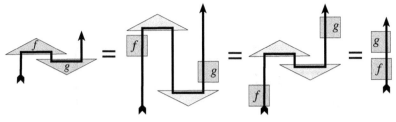

we obtain:

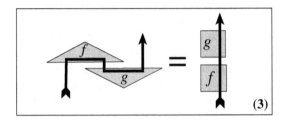

$$(3)$$

i.e., composition of operations can be *internalized* in the behavior of entangled states and costates. Note in particular the interesting phenomenon of "apparent reversal of the causal order" which is the source of many quite mystical interpretations of quantum teleportation in terms of "traveling backward in time"—cf. [35]. Indeed, while on the left, physically, we first prepare the state labeled g and then apply the costate labeled f, the global effect is *as if* we first applied f itself first, and only then g.

15.5.2.2   Derivation of quantum teleportation

This is the most basic application of compositionality in action. Immediately from picture (1) we can read the quantum mechanical potential for teleportation:

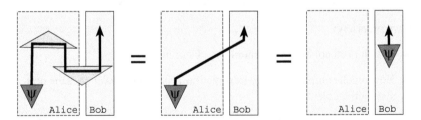

This is not quite the whole story, because of the non-deterministic nature of measurements. But it suffices to introduce a unitary correction. Using picture (3) the full description of teleportation becomes:

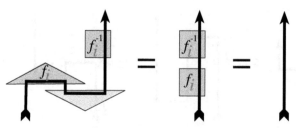

where the *classical communication* is now implicit in the fact that the index i is both present in the costate (= measurement branch) and the correction, and hence needs to be sent from Alice to Bob.

The classical communication can be made *explicit* as a fully fledged part of the formalism, using *additional types*: biproducts in [4], and "classical objects" in [17]. This allows entire protocols, including the interplay between quantum and classical information which is often their most subtle ingredient, to be captured and reasoned about rigorously in a single formal framework.

15.5.3   Remarks

We close this section with some remarks. We have seen that the categorical and diagrammatic setting for quantum mechanics developed by Abramsky and Coecke is strikingly close to that in which the Temperley–Lieb category lives. The main *difference* is the free recourse to symmetry allowed in the Abramsky–Coecke setting (and in the main intended models for that setting, namely finite-dimensional Hilbert spaces with linear or completely positive maps). However, it is interesting to note that in the various protocols and constructions in quantum information and computation which have been modelled in that setting to date [4], the symmetry has not played an essential role. The example of teleportation given above serves as an example.

This raises some natural questions:

How much of QM/QIC lives in the plane?

More precisely:

- Which protocols make essential use of symmetry?

- How much computational or information-processing power does the non-symmetric calculus have?

- Does *braiding* have some computational significance? Remember that between *pivotal* and *symmetric* we have *braided* strongly compact closed categories [21].

15.6 Planar geometry of interaction and the Temperley–Lieb algebra

We now address the issue of giving what, so far as I know, is the first *direct*—or "fully abstract"—description of the Temperley–Lieb category. Since the category $\mathscr{T}$ is directly and simply described as the free R-linear category generated by $\mathscr{D}$, we focus on the direct description of $\mathscr{D}$.

Previous descriptions:

- Algebraic, by generators and relations, whether "locally", of the Temperley–Lieb algebras $\mathscr{A}_n(\tau)$, as in Jones' presentation, or "globally", by a description of $\mathscr{D}$ as the free pivotal category, as in Proposition 15.1

- Kauffman's topological description: diagrams "up to planar isotopy"

In fact, it is well known (see [33]) that the diagrams are completely characterized by how the dots are joined up—i.e., by discrete relations on finite sets. This leaves us with the problem of how to capture

1. Planarity

2. The multiplication of diagrams—i.e., composition in $\mathscr{D}$

purely in terms of the data given by these relations.

The answers to these questions exhibit the connections that exist between the Temperley–Lieb category and what is commonly known as the *Geometry of Interaction*. This is a dynamical/geometrical interpretation of proofs and cut elimination initiated by Girard [23] as an off-shoot of Linear Logic [22]. The general setting for these notions is now known to be that of *traced monoidal* and *compact closed* categories—in particular, in the free construction of compact closed categories over traced monoidal categories [1, 7]. In fact, this general construction was first clearly described in [27], where one of the leading motivations was the knot-theoretic context.

Our results in this section establish a two-way connection. In one direction, we shall use ideas from Geometry of Interaction to answer Question 2 above: that is, to define path composition (including the formation of loops) purely in terms of the discrete relations tabulating how the dots are joined up. In the other direction, our answer to Question 1 will allow us to consider a natural *planar variant* of the Geometry of Interaction.

15.6.1   Some preliminary notions

15.6.1.1   Partial orders

We use the notation $P = (|P|, \leq_P)$ for partial orders. Thus $|P|$ is the underlying set, and $\leq_P$ is the order relation (reflexive, transitive and antisymmetric) on this set. An order relation is *linear* if for all $x, y \in |P|$, $x \leq_P y$ or $y \leq_P x$.

Given a natural number n, we define $[n] := \{1 < \cdots < n\}$, the linear order of length n. We define several constructions on partial orders. Given partial orders P, Q, we define:

- The disjoint sum $P \oplus Q$, where $|P \oplus Q| = |P| + |Q|$, the disjoint union of $|P|$ and $|Q|$, and

$$x \leq_{P \oplus Q} y \iff (x \leq_P y) \vee (x \leq_Q y).$$

- The concatenation $P \lhd Q$, where $|P \lhd Q| = |P| + |Q|$, with the following order:

$$x \leq_{P \lhd Q} y \iff (x \leq_P y) \vee (x \leq_Q y) \vee (x \in P \wedge y \in Q).$$

- $P^{\mathrm{op}} = (|P|, \geq_P)$.

Given elements x, y of a partial order P, we define:

$$x \uparrow y \iff (x \leq_P y) \vee (y \leq_P x)$$
$$x \# y \iff \neg(x \uparrow y).$$

15.6.1.2   Relations

A relation on a set X is a subset of the cartesian product: $R \subseteq X \times X$. Since relations are sets, they are closed under unions and intersections. We shall also use the following operations of relation algebra:

Identity relation:	$1_X := \{(x,x) \mid x \in X\}$
Relation composition:	$R; S := \{(x,z) \mid \exists y. (x,y) \in R$
	$\wedge (y,z) \in S\}$
Relational converse:	$R^c := \{(y,x) \mid (x,y) \in R\}$
Transitive closure:	$R^+ := \bigcup_{k \geq 1} R^k$
Reflexive transitive closure:	$R^* := \bigcup_{k \geq 0} R^k$

Here R^k is defined inductively: $R^0 := 1_X$, $R^1 := R$, $R^{k+1} := R; R^k$. A relation R is *single-valued* or a *partial function* if $R^c; R \subseteq 1_X$. It is *total* if $R; R^c \supseteq 1_X$. A *function* $f : X \to X$ is a single-valued, total relation.

These notions extend naturally to relations $R \subseteq X \times Y$.

15.6.1.3 Involutions

A *fixed-point free involution* on a set X is a function $f : X \to X$ such that

$$f^2 = 1_X, \qquad f \cap 1_X = \varnothing.$$

Thus for such a function $f(x) = y \Leftrightarrow x = f(y)$ and $f(x) \neq x$. We write $\mathsf{Inv}(X)$ for the set of fixed-point free involutions on a set X. Note that $\mathsf{Inv}(X)$ is *not* closed under function composition; nor does it contain the identity function. We must look elsewhere for suitable notions of composition and identity.

An involution is equivalently described as a partition of X into two-element subsets:

$$X = \bigcup E, \qquad \text{where } E = \{\{x,y\} \mid f(x) = y\}. \tag{15.3}$$

This defines an undirected graph $G_f = (X, E)$. Clearly G_f is one-regular [18]: each vertex has exactly one incident edge. Conversely, every graph $G = (X, E)$ with this property determines a unique $f \in \mathsf{Inv}(X)$ with $G_f = G$. Note that a finite set can only carry such a structure if its cardinality is even.

15.6.2 Formalizing diagrams

From our previous discussion, it is fairly clear how we will proceed to formalize morphisms $\mathbf{n} \to \mathbf{m}$ in $\mathscr{D}$. Given $n, m \in \mathbb{N}$, we define $\mathsf{N}(n,m) = [n] \oplus [m]$. We visualize this partial order as

We use the notation i' to distinguish the elements of $[m]$ in this disjoint union from those of $[n]$, which are unprimed. Note that the order on $\mathsf{N}(n,m)$ has an immediate spatial interpretation in the diagrammatic representation: $i <$

j just in case i lies to the left of j on either the top or bottom line of dots, corresponding to $[n]$ and $[m]$ respectively.

A diagram connecting up dots pairwise will be formalized as a map $f \in \mathsf{Inv}(|\mathsf{N}(n,m)|)$. Such a map can be visualized by drawing *undirected arcs* between the pairs of nodes i, j such that $f(i) = j$.

15.6.2.1 Example

The map $f \in \mathsf{Inv}(|\mathsf{N}(4,2)|)$ such that

$$f : 1 \leftrightarrow 2', \qquad 2 \leftrightarrow 4, \qquad 3 \leftrightarrow 1'$$

is depicted thus:

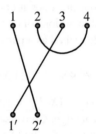

Our task is now is *characterize those involutions which are planar*. The key idea is that this can be done using just the order relations we have introduced.

15.6.3 Characterizing planarity

A map $f \in \mathsf{Inv}(|\mathsf{N}(n,m)|)$ will be called *planar* if it satisfies the following two conditions, for all $i, j \in \mathsf{N}(n,m)$:

$$\text{(PL1)} \quad i < j < f(i) \quad \Longrightarrow i < f(j) < f(i)$$

$$\text{(PL2)} \; f(i) \, \# \, i < j \, \# \, f(j) \; \Longrightarrow \quad f(i) < f(j).$$

It is instructive to see which possibilities are *excluded* by these conditions.

15.6.3.1 First condition

$$\text{(PL1)} \; i < j < f(i) \; \Longrightarrow i < f(j) < f(i)$$

(PL1) rules out

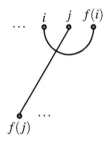

where $f(j) \# f(i)$, and also

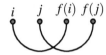

where $f(i) < f(j)$.

15.6.3.2   Second condition

$$(PL2)\ f(i) \# i < j \# f(j) \implies f(i) < f(j).$$

Similarly, (PL2) rules out

We write $\mathscr{P}(n,m)$ for the set of planar maps in $\mathsf{Inv}(|\mathsf{N}(n,m)|)$.

PROPOSITION 15.6

1. *Every planar diagram satisfies the two conditions.*

2. *Every involution satisfying the two conditions can be drawn as a planar diagram.*

Rather than proving this directly, it is simpler, and also instructive, to reduce it to a special case. We consider arrows in $\mathscr{D}$ of the special form $I \to \mathbf{n}$.

Such arrows consist only of caps. They correspond to *points*, or *states* in the terminology of Section 15.5.

Since the top row of dots is empty, in this case we have a linear order, and the premise of condition (PL2) can never arise. Hence planarity for such arrows is just the simple condition (PL1)—which can be seen to be equivalent to saying that, if we write a left parenthesis for each left end of a cap, and a right parenthesis for each right end, we get a well-formed string of parentheses. Thus

corresponds to

$$()(()).$$

(Of course, exactly similar comments apply to arrows of the form $\mathbf{n} \to I$, i.e., *costates*.) It is also clear[5] that Proposition 15.6 holds for such arrows.

Now we recall that quite generally, in any pivotal category we have the hom-tensor adjunction

$$A \otimes B^* \xrightarrow{\llcorner f \lrcorner} I \overset{\simeq}{\longleftrightarrow} A \xrightarrow{f} B \overset{\simeq}{\longleftrightarrow} I \xrightarrow{\ulcorner f \urcorner} A^* \otimes B$$

$$\ulcorner f \urcorner = (1_{A^*} \otimes f) \circ \eta_A : I \to A^* \otimes B \qquad \llcorner f \lrcorner = \varepsilon_B \circ (f \otimes 1_{B^*}) : A \otimes B^* \to I.$$

We call $\ulcorner f \urcorner$ the *name* of f, and $\llcorner f \lrcorner$ the *coname*. The inverse to the map $f \mapsto \ulcorner f \urcorner$ is defined by

$$g : I \to A^* \otimes B \quad \mapsto \quad (\varepsilon_A \otimes 1_B) \circ (1_A \otimes g) : A \to B.$$

For example, we compute the name of the left wave:

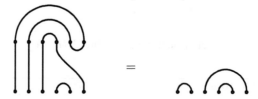

Applying the inverse transformation:

Note also that *the unit is the name of the identity*: $\eta_{\mathbf{n}} = \ulcorner 1_{\mathbf{n}} \urcorner$, and similarly $\varepsilon_{\mathbf{n}} = \llcorner 1_{\mathbf{n}} \lrcorner$.

Thus we see that diagrammatically the process of forming the name of an arrow involves reversing the left-right order of the top row of dots by rotating them concentrically, and sliding them down to lie parallel with, and to the left of, the bottom row. In this process, cups are turned into caps, while through lines are stretched out and turned to also form caps.

This transposition of the top row of dots can be described order theoretically, as replacing the partial order $[n] \oplus [m]$ by the linear order $[n]^{\mathrm{op}} \lhd [m]$. Note that the underlying sets of these two partial orders are the same: $\|[n] \oplus [m]\| = \|[n]^{\mathrm{op}} \lhd [m]\|$. Thus $\ulcorner f \urcorner$ is essentially *the same function* as f.

PROPOSITION 15.7

For $f \in \mathsf{Inv}(\|\mathsf{N}(n,m)\|)$, the following are equivalent:

1. *f satisfies (PL1) and (PL2) with respect to $[n] \oplus [m]$.*

2. *f satisfies (PL1) with respect to $[n]^{\mathrm{op}} \lhd [m]$.*

PROOF Firstly, assume (2), and suppose $f(i) \# i < j \# f(j)$. If $i < j$ in the bottom row, then $f(j) < i < j$ in $[n]^{\mathrm{op}} \lhd [m]$, so by (PL1), $f(j) < f(i) < j$, i.e. $f(i) < f(j)$ in $[n] \oplus [m]$, as required. Now suppose $i < j$ in the top row. Then $j < i < f(j)$ in $[n]^{\mathrm{op}} \lhd [m]$, so by (PL1), $j < f(i) < f(j)$, and in particular $f(i) < f(j)$.

Now assume (1), and suppose that $i < j < f(i)$ in $[n]^{\mathrm{op}} \lhd [m]$. The interesting case is where i is in the top row and $f(i)$ in the bottom row. We need to do some case analysis. Suppose firstly that j is in the top row. If $f(j)$ is in the bottom row, then $f(j) \# j < i \# f(i)$ in $[n] \oplus [m]$, and we can apply (PL2) to conclude that $f(j) < f(i)$, and hence $i < f(j) < f(i)$ in $[n]^{\mathrm{op}} \lhd [m]$. If $f(j)$ is in the top row, we must have $f(j) < i$ by (PL1) for $[n] \oplus [m]$, and hence $i < f(j) < f(i)$ in $[n]^{\mathrm{op}} \lhd [m]$.

Now suppose that j is in the bottom row. If $f(j)$ is in the bottom row, we must have $f(j) < f(i)$ by (PL1). If $f(j)$ is in the top row, then we have $f(j) \# j < f(i) \# i$ in $[n] \oplus [m]$, so by (PL2) we have $f(j) < i$, and hence $i < f(j) < f(i)$ in $[n]^{\mathrm{op}} \lhd [m]$. ∎

Since (PL1) characterizes planarity for $\ulcorner f \urcorner$, it follows that (PL1) and (PL2) characterize planarity for f.

15.6.4   The Temperley–Lieb category

Our aim is now to define a category **TL**, which will yield the desired description of the diagrammatic category $\mathscr{D}$. The *objects* of **TL** are the natural numbers. The homset $\mathbf{TL}(\mathbf{n},\mathbf{m})$ is defined to be the cartesian product $\mathbb{N} \times \mathscr{P}(n,m)$. Thus a morphism $\mathbf{n} \to \mathbf{m}$ in **TL** consists of a pair (k,f), where k is a natural number, and $f \in \mathscr{P}(n,m)$ is a planar map in $\mathsf{Inv}(|\mathsf{N}(n,m)|)$. The idea is that k is a counter for the number of loops, so such an arrow can be written $\delta^k \cdot f$ in the notation used previously.

It remains to define the composition and identities in this category. Clearly (even leaving aside the natural number components of morphisms) composition cannot be defined as ordinary function composition. This does not even make sense—the codomain of an involution $f \in \mathscr{P}(n,m)$ does not match the domain of an involution $g \in \mathscr{P}(m,p)$—let alone yield a function with the necessary properties to be a morphism in the category.

15.6.4.1   Composition: the "execution formula"

Consider a map $f : [n] + [m] \longrightarrow [n] + [m]$. Each input lies in *either* $[n]$ or $[m]$ (exclusive or), and similarly for the corresponding output. This leads to a decomposition of f into four *disjoint partial maps*:

$$f_{n,n} : [n] \longrightarrow [n] \qquad f_{n,m} : [n] \longrightarrow [m]$$
$$f_{m,n} : [m] \longrightarrow [n] \qquad f_{m,m} : [m] \longrightarrow [m]$$

so that f can be recovered as the disjoint union of these four maps. If f is an involution, then these maps will be partial involutions.

Note that these components have a natural diagrammatic reading: $f_{n,n}$ describes the *cups* of f, $f_{m,m}$ the *caps*, and $f_{n,m} = f^c_{m,n}$ the *through lines*.

Now suppose we have maps $f : [n] + [m] \to [n] + [m]$ and $g : [m] + [p] \to [m] + [p]$. We write the decompositions of f and g as above in matrix form:

$$f = \begin{pmatrix} f_{n,n} & f_{n,m} \\ f_{m,n} & f_{m,m} \end{pmatrix} \qquad g = \begin{pmatrix} g_{m,m} & g_{m,p} \\ g_{p,m} & g_{p,p} \end{pmatrix}.$$

We can view these maps as *binary relations* on $[n] + [m]$ and $[m] + [p]$ respectively, and use relational algebra (union $R \cup S$, relational composition $R; S$ and reflexive transitive closure R^*) to define a *new relation* θ on $[n] + [p]$. If we write

$$\theta = \begin{pmatrix} \theta_{n,n} & \theta_{n,p} \\ \theta_{p,n} & \theta_{p,p} \end{pmatrix}$$

so that θ is the disjoint union of these four components, then we can define it component-wise as follows:

$$\theta_{n,n} \;=\; f_{n,n} \,\cup\, f_{n,m}; g_{m,m}; (f_{m,m}; g_{m,m})^*; f_{m,n}$$
$$\theta_{n,p} \;=\; f_{n,m}; (g_{m,m}; f_{m,m})^*; g_{m,p}$$
$$\theta_{p,n} \;=\; g_{p,m}; (f_{m,m}; g_{m,m})^*; f_{m,n}$$
$$\theta_{p,p} \;=\; g_{p,p} \,\cup\, g_{p,m}; f_{m,m}; (g_{m,m}; f_{m,m})^*; g_{m,p}.$$

We can give clear intuitive readings for how these formulas express composition of paths in diagrams in terms of relational algebra:

- The component $\theta_{n,n}$ describes the *cups* of the diagram resulting from the composition. These are the union of the cups of f ($f_{n,n}$), together with paths that start from the top row with a through line of f, given by $f_{n,m}$, then go through an alternating odd-length sequence of cups of g ($g_{m,m}$) and caps of f ($f_{m,m}$), and finally return to the top row by a through line of f ($f_{m,n}$).

- Similarly, $\theta_{p,p}$ describes the caps of the composition.

- $\theta_{n,p} = \theta_{p,n}^c$ describes the through lines. Thus $\theta_{n,p}$ describes paths which start with a through line of f from n to m, continue with an alternating even length (and possibly empty) sequence of cups of g and caps of f, and finish with a through line of g from m to p.

All through lines from n to p must have this form.

This formula corresponds to the interpretation of cut elimination in the Geometry of Interaction interpretation of proofs in Linear Logic (and by extension in related logics and type theories) [23]. A more abstract and general perspective on how this construction arises can be given in the setting of traced monoidal categories [27, 1].

PROPOSITION 15.8

If f and g are planar, so is θ.

We write $\theta = g \odot f \in \mathscr{P}(n,p)$.

15.6.4.2   Cycles

Given $f \in \mathscr{P}(n,m)$, $g \in \mathscr{P}(m,p)$, we define $\chi(f,g) := f_{m,m}; g_{m,m}$. Note that $\chi(f,g)^c = (g_{m,m}; f_{m,m})$, and

$$\chi(f,g); \chi(f,g)^c \subseteq 1_{[m]}, \qquad\qquad \chi(f,g)^c; \chi(f,g) \subseteq 1_{[m]}.$$

Thus $\chi(f,g)$ is a *partial bijection*. However, in general it is neither an involution nor fixpoint-free. The *cyclic elements* of $\chi(f,g)$ are those elements of $[m]$ which lie in the intersection

$$\chi(f,g)^+ \cap 1_{[m]}.$$

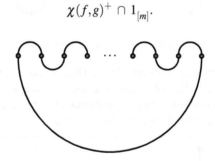

Thus if i is a cyclic element, there is a least $k > 0$ such that $\chi(f,g)^k(i) = i$. The corresponding *cycle* is

$$\{i, \chi(f,g)(i), \ldots, \chi(f,g)^{k-1}(i)\}.$$

Distinct cycles are disjoint. We write $Z(f,g)$ for the number of distinct cycles of $\chi(f,g)$.

15.6.4.3 Composition and identities

Finally, we define the composition of morphisms in **TL**. Given $(s, f) : \mathbf{n} \to \mathbf{m}$ and $(t, g) : \mathbf{m} \to \mathbf{p}$:

$$(t, g) \circ (s, f) = (s + t + Z(f, g), g \odot f).$$

The identity morphism $\mathrm{id}_{\mathbf{n}} : \mathbf{n} \to \mathbf{n}$ is defined to be the pair $(0, \tau_{n,n})$, where $\tau_{n,n}$ is the *twist map* on $[n] + [n]$; i.e., the involution $i \leftrightarrow i'$. Diagrammatically, this is just

Note that this is *not* the identity map on $[n] + [n]$—indeed it is (necessarily) fixpoint free!

PROPOSITION 15.9

TL *with composition and identities as defined above is a category.*

15.6.4.4 TL as a pivotal category

The monoidal structure of **TL** is straightforward. If $(k, f) : \mathbf{n} \to \mathbf{m}$ and $(l, g) : \mathbf{p} \to \mathbf{q}$, then $(k + l, f + g) : \mathbf{n} + \mathbf{p} \to \mathbf{m} + \mathbf{q}$, where $f + g \in \mathscr{P}(n + p, m + q)$ is the evident disjoint union of the involutions f and g.

The unit $\eta_{\mathbf{n}} : I \to \mathbf{n} + \mathbf{n}$ is given by

$$i \leftrightarrow i' \quad (1 \leq i \leq n),$$

and similarly for the counit. Note that identities, units and counits are all essentially the same maps, but with distinct *types*, which partition their arguments between inputs and outputs differently.

We describe the dual, adjoint and conjugate of an arrow $(k, f) : \mathbf{n} \to \mathbf{m}$. Let $\tau_{n,m} : [n] + [m] \xrightarrow{\cong} [m] + [n]$ be the symmetry isomorphism of the disjoint union, and

$$\rho_n : [n] \xrightarrow{\cong} [n] :: i \mapsto n - i + 1$$

be the order-reversal isomorphism. Note that

$$\tau_{n,m}^{-1} = \tau_{m,n}, \quad \rho_n^{-1} = \rho_n, \quad \tau_{n,m} \circ (\rho_n + \rho_m) = (\rho_m + \rho_n) \circ \tau_{n,m}.$$

Then we have $(k,f)^\bullet = (k, f^\bullet)$, where:

$$f^\dagger = \tau_{n.m} \circ f \circ \tau_{n,m}^{-1}$$
$$f_* = (\rho_n + \rho_m) \circ f \circ (\rho_n + \rho_m)^{-1}$$
$$f^* = (f^\dagger)_*.$$

15.6.4.5   The main result

THEOREM 15.1

TL *is isomorphic as a strict, pivotal dagger category to* $\mathscr{D}$.

As an immediate corollary of this result and Proposition 15.2, we have:

THEOREM 15.2

TL *is the free strict, pivotal dagger category on one self-dual generator.*

This is in the same spirit as the characterizations of free compact and dagger compact categories in [34, 3].

These results can easily be extended to descriptions of the free pivotal dagger category over an arbitrary generating category, leading to *oriented Temperley–Lieb algebras* with *primitive (physical) operations*. We refer to [3] for a more detailed presentation (in the symmetric case).

15.7   Planar λ-calculus

Our aim in this section is to show how a restricted form of λ-calculus can be interpreted in the Temperley–Lieb category, and how β-reduction of λ-terms, which is an important foundational paradigm for computation, is then reflected diagrammatically as geometric simplification, i.e., "yanking lines straight". We can give only a brief indication of what is in fact a rich topic in its own right. See [6, 7, 31] for discussions of related matters.

15.7.1 The λ-calculus

We begin with a (very) brief review of the λ-calculus [14, 9], which is an important foundational paradigm in logic and computation, and in particular forms the basis for all modern functional programming languages.

The syntax of the λ-calculus is beguilingly simple. Given a set of variables $x, y, z, \ldots$ we define the set of terms as follows:

$$ t ::= x \mid \underbrace{tu}_{\text{application}} \mid \underbrace{\lambda x.t}_{\text{abstraction}} $$

Notational Convention: We write

$$ t_1 t_2 \cdots t_k \equiv (\cdots (t_1 t_2) \cdots) t_k. $$

Some examples of terms:

$\lambda x.x$	identity function
$\lambda f. \lambda x. fx$	application
$\lambda f. \lambda x. f(fx)$	double application
$\lambda f. \lambda g. \lambda x. g(f(x))$	composition $g \circ f$

The basic equation governing this calculus is β-*conversion*:

$$ (\lambda x.t)u = t[u/x], $$

e.g. (assuming some arithmetic operations are given),

$$ (\lambda f. \lambda x. f(fx))(\lambda x.x + 1)0 = (0+1)+1 = 2. $$

By orienting this equation, we get a 'dynamics'—β-*reduction*

$$ (\lambda x.t)u \rightarrow t[u/x]. $$

Despite its sparse syntax, λ-calculus is very expressive—it is in fact a universal model of computation, equivalent to Turing machines.

15.7.2 Types

One important way of constraining the λ-calculus is to introduce types.

> **Types are there to stop you doing (bad) things**

Types are in fact one of the most fruitful *positive* ideas in computer science!

We shall introduce a (highly restrictive) type system, such that the typeable terms can be interpreted in the Temperley–Lieb category (in fact, in any pivotal category).

Firstly, assuming some set of *basic types B*, we define a syntax of general types:

$$T \ := \ B \mid T \to T.$$

Intuitively, $T \to U$ represents the type of functions which take inputs of type T to outputs of type U.

Notational Convention: We write

$$T_1 \to T_2 \to \cdots T_k \to T_{k+1} \quad \equiv \quad T_1 \to (T_2 \to \cdots (T_k \to T_{k+1}) \cdots).$$

Examples:

$$A \to A \to A \qquad \text{first-order function type}$$

$$(A \to A) \to A \qquad \text{second-order function type}$$

We now introduce a formal system for deriving *typing judgments*, of the form:

$$x_1 : T_1, \ldots x_k : T_k \vdash t : T.$$

Such a judgment asserts that the term t has type T *under the assumption* (or *in the context*) that the variable x_1 has type T_1, ..., x_k has type T_k. All the variables x_i appearing in the context must be distinct—and in our setting, the order in which the variables appear in the list is significant.

There is one basic form of axiom for typing variables:

Variable

$$\overline{x : T \vdash x : T}$$

and two inference rules for typing abstractions and applications, respectively:

Function

$$\frac{\Gamma, x : U \vdash t : T}{\Gamma \vdash \lambda x. t : U \to T} \qquad \frac{\Gamma \vdash t : U \to T \qquad \Delta \vdash u : U}{\Gamma, \Delta \vdash tu : T}.$$

Note that Γ, Δ represents the concatenation of the lists Γ, Δ. This implies that the variables appearing in Γ and Δ are distinct—an important *linearity constraint* in the sense of Linear Logic [22].

15.7.3 Interpretation in pivotal categories

We now show how terms typeable in our system can be interpreted in a pivotal category $\mathscr{C}$. We assume firstly that the basic types B have been interpreted as objects $[\![B]\!]$ of $\mathscr{C}$. We then extend this to general types by:

$$[\![T \to U]\!] = [\![U]\!] \otimes [\![T]\!]^*.$$

Now we show how, for each typing judgment $\Gamma \vdash t : T$, to assign an arrow

$$[\![\Gamma]\!] \longrightarrow [\![T]\!],$$

where if $\Gamma = x_1 : T_1, \ldots x_k : T_k$,

$$[\![\Gamma]\!] = [\![T_1]\!] \otimes \cdots \otimes [\![T_k]\!].$$

This assignment is defined by induction on the derivation of the typing judgment in the formal system.

Variable

$$\frac{}{x : T \vdash x : T} \qquad\qquad \frac{}{1_{[\![T]\!]} : [\![T]\!] \longrightarrow [\![T]\!]}.$$

Abstraction

To interpret λ-abstraction, we use the adjunction

$$\Lambda_r : \mathscr{C}(A \otimes B, C) \simeq \mathscr{C}(A, C \otimes B^*)$$

$$\Lambda_r(f) = A \xrightarrow{1_A \otimes \eta_{B^*}} A \otimes B \otimes B^* \xrightarrow{f \otimes 1_{B^*}} C \otimes B^*.$$

We can then define:

$$\frac{\Gamma, x : U \vdash t : T}{\Gamma \vdash \lambda x : U.t : U \to T} \qquad\qquad \frac{[\![t]\!] : [\![\Gamma]\!] \otimes [\![U]\!] \longrightarrow [\![T]\!]}{\Lambda_r([\![t]\!]) : [\![\Gamma]\!] \longrightarrow [\![T]\!] \otimes [\![U]\!]^*}.$$

Application

We use the following operation of *right application*:

$$\mathsf{RApp} : \mathscr{C}(C, B \otimes A^*) \times \mathscr{C}(D, A) \longrightarrow \mathscr{C}(C \otimes D, B)$$

$$\mathsf{RApp}(f, g) = C \otimes D \xrightarrow{f \otimes g} B \otimes A^* \otimes A \xrightarrow{1_B \otimes \varepsilon_{B^*}} B.$$

We can then define:

$$\frac{\Gamma \vdash t : U \to T \qquad \Delta \vdash u : U}{\Gamma, \Delta \vdash tu : T}$$

$$\frac{[\![t]\!] : [\![\Gamma]\!] \longrightarrow [\![T]\!] \otimes [\![U]\!]^* \qquad [\![u]\!] : [\![\Delta]\!] \longrightarrow [\![U]\!]}{\mathsf{RApp}([\![t]\!], [\![u]\!]) : [\![\Gamma]\!] \otimes [\![\Delta]\!] \longrightarrow [\![T]\!]}.$$

It can be proved that this interpretation *is sound for β conversion*, i.e.,

$$\llbracket (\lambda x.t)u \rrbracket = \llbracket t[u/x] \rrbracket$$

in any pivotal category.

15.7.4 An example

We now discuss an example to show how all this works diagrammatically in **TL**. We shall consider the *bracketing combinator*

$$\mathbf{B} \equiv \lambda x.\lambda y.\lambda z.x(yz).$$

This is characterized by the equation

$$\mathbf{B}abc = a(bc).$$

Firstly, we derive a typing judgment for this term:

$$
\cfrac{
\cfrac{
x:B \to C \vdash x:B \to C \qquad
\cfrac{
y:A \to B \vdash y:A \to B \qquad z:A \vdash z:A
}{
y:A \to B, z:A \vdash yz:B
}
}{
\cfrac{
x:B \to C, y:A \to B, z:A \vdash x(yz):C
}{
\cfrac{
x:B \to C, y:A \to B \vdash \lambda z.x(yz):A \to C
}{
\cfrac{
x:B \to C \vdash \lambda y.\lambda z.x(yz):(A \to B) \to (A \to C)
}{
\vdash \lambda x.\lambda y.\lambda z.x(yz):(B \to C) \to (A \to B) \to (A \to C)
}
}
}
}
{}
$$

Now we take $A = B = C = \mathbf{1}$ in **TL**. The interpretation of the open term

$$x:B \to C, y:A \to B, z:A \vdash x(yz):C$$

is as follows:

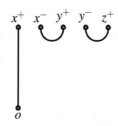

Here x^+ is the output of x, and x^- the input, and similarly for y. The output of the whole expression is o. When we abstract the variables, we obtain the following caps-only diagram:

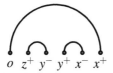

Now we consider an application **B**abc:

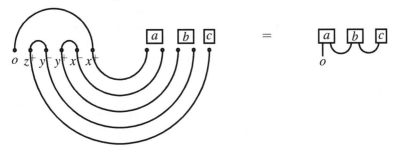

15.7.5 Discussion

The typed λ-calculus we have used here is in fact a fragment of the *Lambek calculus* [36], a basic non-commutative logic and λ-calculus, which has found extensive applications in computational linguistics [13, 38]. The Lambek calculus can be interpreted in any monoidal biclosed category, and has notions of *left abstraction and application*, as well as the right-handed versions we have described here. Pivotal categories have stronger properties than monoidal biclosure; for example, duality and adjoints allow the left- and right-handed versions of abstraction and application to be defined in terms of each other. Moreover, the duality means that the corresponding logic has a *classical* format, with an involutive negation. Thus there is much more to this topic than we have had the time to discuss here. We merely hope to have given an impression of how the geometric ideas expressed in the Temperley–Lieb category have natural connections to a central topic in logic and computation.

15.8 Further directions

We hope to have given an indication of the rich and suggestive connections which exist between ideas stemming from knot theory, topology and mathematical physics, on the one hand, and logic and computation on the other, with the Temperley–Lieb category serving as an intuitive and compelling meeting point. We hope that further investigation will uncover deeper links and interplays, leading to new insights in both directions.

We conclude with a few specific directions for future work:

- The symmetric case, where we drop the planarity constraint, is also interesting. The algebraic object corresponding to the Temperley–Lieb algebra in this case is the *Brauer algebra* [12], important in the representation theory of the orthogonal group (Schur–Weyl duality). Indeed, there are now a family of various kinds of diagram algebras: partition algebras, rook algebras etc., arising in quantum statistical mechanics, and studied in representation theory [25].

- The categorical perspective suggests *oriented* versions of the Temperley–Lieb algebra and related structures, where we no longer have $A = A^*$. This is also natural from the point of view of quantum mechanics, where this non-trivial duality on objects distinguishes complex from real Hilbert spaces.

- We can ask how expressive planar Geometry of Interaction is; and what rôle may be played by braiding or other geometric information.

- Again, it would be interesting to understand the scope and limits of planar quantum mechanics and quantum information processing.

References

[1] S. Abramsky. Retracing some paths in process algebra. In U. Montanari and V. Sassone, editors, *Proceedings of CONCUR '96*, volume 1119 of *Springer Lecture Notes in Computer Science*, pages 1–17. Springer-Verlag, 1996.

[2] S. Abramsky. High-level methods for quantum computation and information. In *Proceedings of the 19th Annual IEEE Symposium on Logic in Computer Science*, pages 410–414. IEEE Computer Science Press, 2004.

[3] S. Abramsky. Abstract scalars, loops, and free traced and strongly compact closed categories. In J. Fiadeiro, editor, *Proceedings of CALCO*

2005, volume 3629 of *Springer Lecture Notes in Computer Science*, pages 1–31. Springer-Verlag, 2005.

[4]   S. Abramsky and B. Coecke. A categorical semantics of quantum protocols. In *Proceedings of the 19th Annual IEEE Symposium on Logic in Computer Science*, pages 415–425. IEEE Computer Science Press. quant-ph/0402130, 2004.

[5]   S. Abramsky and B. Coecke. Abstract physical traces. *Theory and Applications of Categories*, 14:111–124, 2005.

[6]   S. Abramsky and R. W. Duncan. A categorical quantum logic. *Mathematical Structures in Computer Science*, 16:469–489, 2006.

[7]   S. Abramsky, E. Haghverdi, and P. J. Scott. Geometry of interaction and linear combinatory algebras. *Mathematical Structures in Computer Science*, 12:625–665, 2002.

[8]   S. Abramsky and R. Jagadeesan. New foundations for the geometry of interaction. *Information and Computation*, 111:53–119, 1994.

[9]   H. P. Barendregt. *The Lambda Calculus*, volume 103 of *Studies in Logic*. North-Holland, 1984.

[10] C. H. Bennet, G. Brassard, C. Crépeau, R. Jozsa, A. Peres, and W. K. Wooters. Teleporting an unknown quantum state via dual classical and Einstein-Podolsky-Rosen channels. *Physical Review Letters*, 70:1895–1899, 1993.

[11] G. Birkhoff and J. von Neumann. The logic of quantum mechanics. *Annals of Mathematics*, 37:823–843, 1936.

[12] R. Brauer. On algebras which are connected with the semisimple continuous groups. *Ann. Math.*, 38:854–872, 1937.

[13] W. Buszkowski. Mathematical linguistics and proof theory. In J. van Benthem and A. ter Meulen, editors, *Handbook of Logic and Language*, chapter 12, pages 683–736. Elsevier, 1997.

[14] A. Church. *The Calculi of Lambda Conversion*. Princeton University Press, 1941.

[15] B. Coecke. De-linearizing linearity: projective quantum axiomatics from strong compact closure. *Electronic Notes in Theoretical Computer Science*, 2006. To appear.

[16] B. Coecke and E. O. Paquette. Generalized measurements and Naimark's theorem without sums. To appear in Fourth Workshop on Quantum Programming Languages, 2006.

[17] B. Coecke and D. Pavlovic. Quantum measurements without sums. In G. Chen, L. Kauffman, and S. Lamonaco, editors, *Mathematics of Quantum Computing and Technology*. Taylor and Francis, 2006. To appear.

[18] J. Diestel. *Graph Theory*. Springer-Verlag, 1997.

[19] P. A. M. Dirac. *The Principles of Quantum Mechanics (third edition)*. Oxford University Press, 1947.

[20] K. Dosen and Z. Petric. Self-adjunctions and matrices. *Journal of Pure and Applied Algebra*, 184:7–39, 2003.

[21] P. Freyd and D. Yetter. Braided compact closed categories with applications to low-dimensional topology. *Advances in Mathematics*, 77:156–182, 1989.

[22] J.-Y. Girard. Linear Logic. *Theoretical Computer Science*, 50(1):1–102, 1987.

[23] J.-Y. Girard. Geometry of Interaction I: Interpretation of System F. In R. Ferro, editor, *Logic Colloquium '88*, pages 221–260. North-Holland, 1989.

[24] D. Gottesman and I. L. Chuang. Quantum teleportation is a universal computational primitive. *Nature*, 402:390–393, 1999.

[25] T. Halvorson and A. Ram. Partition algebras. *European J. of Combinatorics*, 26(1):869–921, 2005.

[26] V. F. R. Jones. A polynomial invariant for links via von Neumann algebras. *Bulletin of the Amer. Math. Soc.*, 129:103–112, 1985.

[27] A. Joyal, R. Street, and D. Verity. Traced monoidal categories. *Mathematical Proceedings of the Cambridge Philosophical Society*, 119:447–468, 1996.

[28] C. Kassel. *Quantum Groups*. Springer, 1995.

[29] L. H. Kauffman. An invariant of regular isotopy. *Trans. Amer. Math. Soc.*, 318(2):417–471, 1990.

[30] L. H. Kauffman. *Knots in Physics*. World Scientific Press, 1994.

[31] L. H. Kauffman. Knot Logic. In L. H. Kauffman, editor, *Knots and Applications*, pages 1–110. World Scientific Press, 1995.

[32] L. H. Kauffman. Biologic II. In N. Tongring and R. C. Penner, editors, *Woods Hole Mathematics*, pages 94–132. World Scientific Press, 2004.

[33] L. H. Kauffman. Knot diagrammatics. In W. W. Menasco and M. Thistlethwaite, editors, *Handbook of Knot Theory*. Elsevier, 2005.

[34] G. M. Kelly and M. L. Laplaza. Coherence for compact closed categories. *Journal of Pure and Applied Algebra*, 19:193–213, 1980.

[35] M. Laforest, R. Laflamme, and J. Baugh. Time-reversal formalism applied to maximal bipartite entanglement: Theoretical and experimental exploration. quant-ph/0510048. Unpublished.

[36] J. Lambek. The mathematics of sentence structure. *Amer. Math. Monthly*, 65:154–170, 1958.

[37] R. Milner. Fully abstract models of typed lambda-calculus. *Theoretical Computer Science*, 4:1–22, 1977.

[38] M. Moortgat. Categorial type logic. In J. van Benthem and A. ter Meulen, editors, *Handbook of Logic and Language*, chapter 2, pages 93–177. Elsevier, 1997.

[39] R. Raussendorf and H.-J. Briegel. A one-way quantum computer. *Physical Review Letters*, 86:5188, 2001.

[40] R. Raussendorf, D. Browne, and H.-J. Briegel. Measurement-based quantum computation on cluster states. *Physical Review A*, 68:022312, 2003.

[41] M. Rédei. Why John von Neumann did not like the Hilbert space formalism of quantum mechanics (and what he liked instead). *Studies in History and Philosophy of Modern Physics*, 27:493–510, 1997.

[42] P. Selinger. Dagger compact closed categories and completely positive maps. *Electronic Notes in Theoretical Computer Science*, 2006. To appear.

[43] H. N. V. Temperley and E. H. Lieb. Relations between the 'percolation' and 'coloring' problem and other graph-theoretical problems associated with regular planar lattices: some exact results for the 'percolation' problem. *Proc. Roy. Soc. Lond. A*, 322:251–280, 1971.

[44] E. Witten. Topological quantum field theory. *Communications in Mathematical Physics*, 117:353—386, 1988.

[45] M. Zukowski, A. Zeilinger, M. A. Horne, and A. K. Ekert. 'Event-ready-detectors' Bell experiment via entanglement swapping. *Physical Review Letters*, 71:4287–4290, 1993.

Chapter 16

Quantum measurements without sums

Bob Coecke and Dusko Pavlovic

Abstract   Sums play a prominent role in the formalisms of quantum mechanics, whether for mixing and superposing states, or for composing state
spaces. Surprisingly, a conceptual analysis of quantum measurement seems to
suggest that quantum mechanics can be done without direct sums, *expressed
entirely in terms of the tensor product*. The corresponding axioms define classical spaces as objects that allow copying and deleting data. Indeed, the information exchange between the quantum and the classical worlds is essentially
determined by their *distinct capabilities to copy and delete data*. The sums turn
out to be an implicit implementation of this capability. Realizing it through explicit axioms not only dispenses with the unnecessary structural baggage, but
also allows a simple and intuitive *graphical calculus*. In category-theoretic
terms, classical data types are †-*compact Frobenius algebras*, and quantum
spectra underlying quantum measurements are *Eilenberg–Moore coalgebras*
induced by these Frobenius algebras.

16.1   Introduction

Ever since John von Neumann denounced, back in 1935 [34], his own foundation of quantum mechanics in terms of Hilbert spaces, there has been an
ongoing search for a high-level, fully abstract formalism of quantum mechan-

ics. With the emergence of quantum information technology, this quest became more important than ever. The low-level matrix manipulations in quantum informatics are akin to machine programming with bit strings from the early days of computing, which are of course inadequate.[1]

A recent research thread, initiated by Abramsky and the first author [2], aims at recasting the quantum mechanical formalism in *categorical* terms. The upshot of categorical semantics is that it displays concepts in a *compositional* and *typed* framework. In the case of quantum mechanics, it uncovers the *quantum information flows* [7] which are hidden in the usual formalism. Moreover, while the investigations of quantum structures have so far been predominantly academic, categorical semantics open an alley towards a practical, low-overhead tool for the design and analysis of quantum informatic protocols, versatile enough to capture both quantitative and qualitative aspects of quantum information [2, 8, 11, 12, 15, 35]. In fact, some otherwise complicated quantum informatic protocols become trivial exercises in this framework [9]. On the other hand, compared with the order-theoretic framework for quantum mechanics in terms of Birkhoff–von Neumann's quantum logic [33], this categorical setting comes with logical derivations, topologically embodied into something as simple as "yanking a rope".[2] Moreover, in terms of deductive mechanism, it turns out to be some kind of "hyper-logic" [15], as compared to Birkhoff–von Neumann logic which as a consequence of being non-distributive fails to admit a deduction mechanism.

The core of categorical semantics are †-*compact categories*, originally proposed in [2, 3] under the name *strongly compact closed categories*, extending the structure of *compact closed categories*, which have been familiar in various communities since the 1970s [25]. A salient feature of *categorical tensor calculi* of this kind is that they admit sound and complete graphical representations, in the sense that *a well-typed equation in such a tensor calculus is provable from its axioms if and only if the graphical interpretation of that equation is valid in the graphical language*. Various graphical calculi have been an important vehicle of computation in physics [32, and subsequent work], and a prominent research topic of category theory e.g. [24, 25, 21]. Soundness and completeness of the graphical language of †-compact categories, which can be viewed as a two-dimensional formalization and extension of Dirac's bra-ket notation [9], has been demonstrated by Selinger in [35]. Besides this refer-

---

[1] But while computing devices do manipulate strings of 0s and 1s, and high-level modern programming is a matter of providing a convenient interface with that process, the language for quantum information and computation we seek is not a convenient superstructure, but the meaningful infrastructure.

[2] A closely related knot-theoretical scheme has been put forward by Kauffman in [23].

ence, the interested reader may wish to consult [1, 21, 35] for methods and proofs, and [9, 10] and also Baez [4] for a more leisurely introduction into †-categories.

An important aspect of the †-compact semantics of quantum protocols proposed in [2, 8, 35] was the interplay of the multiplicative and additive structures of tensor products and direct sums, respectively. The direct sums (in fact biproducts, since all compact categories are self-dual) seemed essential for specifying classical data types, families of mutually orthogonal projectors, and ultimately for defining measurements. The drawback of this was that the additive types do not yield to a simple graphical calculus; in fact, they make it unusable for many practical purposes.

The main contribution of the present paper is a description of quantum measurement entirely in terms of tensor products, with no recourse to additive structure. The conceptual substance of this description is expressed in the framework of †-compact categories through a simple, operationally motivated definition of *classical objects*, introduced in our work in 2005, and first presented in print here. A classical object, as a †-*compact Frobenius algebra*, equipped with copying and deleting operations, also provides an abstract counterpart to *GHZ states* [18]. We moreover expose an intriguing conceptual and structural connection between the classical capabilities to copy and delete data, as compared to quantum [30, 38], and the mechanism of quantum measurement: the classical interactions emerge as comonoid homomorphisms, i.e., those morphisms that commute with copying and deleting. While each classical object canonically induces a non-degenerate quantum measurement, we show that general quantum measurements arise as *coalgebras* for the comonads induced by classical objects. Quite remarkably, this coalgebra structure exactly captures von Neumann's projection postulate *in a resource-sensitive fashion*. Furthermore, the probabilistic content of quantum measurements is then captured using the abstract version of completely positive maps, due to Selinger [35]. Using these conceptual components, captured in a succinct categorical signature, we provide a purely graphical derivation of teleportation and dense coding.

As a first application of the introduced classical structure, we spell out a purely multiplicative form of *projective quantum measurements*. In subsequent work [11], Paquette and the first author extend this treatment to POVMs, and prove *Naimark's theorem* entirely within our graphical calculus. Extended abstract [12] surveys several important directions and results of further work. The fact that quantum theory can be developed without the additive type constructors suggests a new angle on the question of *parallelism vs. entanglement*. In the final sections of the present paper, we show that superposition too can be described entirely in terms of the monoidal structure, in contrast with the usual Hilbert space view, where entanglement is described as a special case of a superposition.

16.2 Categorical semantics

In this section, we present both the simple categorical algebra of †-compact categories, and the corresponding graphical calculus. In a formal sense, they capture exactly the same structure, and the reader is welcome to pick her favorite flavor (and sort of ignore the other one).

16.2.1 †-compact categories

In a *symmetric monoidal category* [29] the objects form a monoid with the tensor $\otimes$ as multiplication and an object I as the multiplicative unit, up to the coherent[3] natural isomorphisms

$$\lambda_A : A \simeq I \otimes A \qquad \rho_A : A \simeq A \otimes I \qquad \alpha_{A,B,C} : A \otimes (B \otimes C) \simeq (A \otimes B) \otimes C.$$

The fact that a monoidal category is symmetric means that this monoid is commutative, up to the natural transformation

$$\sigma_{A,B} : A \otimes B \simeq B \otimes A$$

coherent with the previous ones. We shall assume that α is strict, i.e., realized by identity, but it will be convenient to carry λ and ρ as explicit structures. Physically, we interpret the objects of a symmetric monoidal category as system types, e.g. qubit, two qubits, classical data, qubit + classical data etc. A morphism should be viewed as a *physical operation*, e.g., unitary, or a measurement, classical communication etc. The tensor captures *compoundness* i.e., conceiving two systems or two operations as one. Morphisms of type

---

[3]*Coherence* here means that all diagrams composed of these natural transformations commute. In particular, there is at most one natural isomorphism between any two functors composed from $\otimes$ and I [28]. As a consequence, some functors can be transferred along these canonical isomorphisms, which then become identities. Without loss of generality, one can thus assume that α, λ and ρ are identities, and that the objects form an actual monoid with $\otimes$ as multiplication and *I* as unit. Such monoidal categories are called *strict*. For every monoidal category, there is an equivalent strict one.

$I \rightarrow A$ represent *states* conceived through their respective preparations, whereas morphisms of the type $I \rightarrow I$ capture *scalars* e.g., *probabilistic weights*— cf. complex numbers $c \in \mathbb{C}$ are in bijective correspondence with linear maps $\mathbb{C} \rightarrow \mathbb{C} :: 1 \mapsto c$. Details of this interpretation are in [10].

A symmetric monoidal category is *compact* [24, 25] if each of its objects has a *dual*. An object B is dual to A when it is given with a pair of morphisms $\eta : I \rightarrow B \otimes A$ and $\varepsilon : A \otimes B \rightarrow I$ often called *unit* and *counit*, satisfying

$$(\varepsilon \otimes 1_A) \circ (1_A \otimes \eta) = 1_A \quad \text{and} \quad (1_B \otimes \varepsilon) \circ (\eta \otimes 1_B) = 1_B. \tag{16.1}$$

It follows that any two duals of A must be isomorphic.[4] A representative of the isomorphism class of the duals of A is usually denoted by A^*. The corresponding unit and counit are then denoted η_A and ε_A.

A *symmetric monoidal †-category* **C** comes with a contravariant functor $(-)^\dagger : \mathbf{C}^{op} \rightarrow \mathbf{C}$, which is identity on the objects, involutive on the morphisms, and preserves the tensor structure [35]. The image $f^\dagger$ of a morphism f is called its (abstract) *adjoint*.

Finally, †-compact categories [2, 3] sum up all of the above structure, subject to the additional coherence requirements that

- every natural isomorphism χ, derived from the symmetric monoidal structure, must be *unitary*, i.e. satisfies $\chi^\dagger \circ \chi = 1$ and $\chi \circ \chi^\dagger = 1$, and

- $\eta_{A^*} = \varepsilon_A^\dagger = \sigma_{A^*A} \circ \eta_A$.

Since in a †-compact category $\varepsilon_A = \eta_{A^*}^\dagger$ some of the structure of the duals becomes redundant. In particular, it is sufficient to stipulate the units $\eta : I \rightarrow A^* \otimes A$, which we call *Bell states*, in reference to their physical meaning. In fact, one can skip the above stepwise introduction, and define †-compact categories [3, 8] simply as a symmetric monoidal category with

- an involution $A \mapsto A^*$,

- a contravariant, identity-on-objects, monoidal involution $f \mapsto f^\dagger$,

- for each object A a distinguished morphism $\eta_A : I \rightarrow A^* \otimes A$,

---

[4]If η, ε make B dual to A, while $\tilde{\eta}, \tilde{\varepsilon}$ make $\tilde{B}$ dual to A, then $(1_{\tilde{B}} \otimes \varepsilon) \circ (\tilde{\eta} \otimes 1_B) : B \rightarrow \tilde{B}$ and $(1_B \otimes \tilde{\varepsilon}) \circ (\eta \otimes 1_{\tilde{B}}) : \tilde{B} \rightarrow B$ make B and $\tilde{B}$ isomorphic.

which make the diagram

$$
\begin{array}{ccccc}
A & \xleftarrow{\;\simeq\;} & I \otimes A & \xleftarrow{\;\eta^{\dagger}_{A^*} \otimes 1_A\;} & A \otimes A^* \otimes A \\[2mm]
\Big\uparrow{\scriptstyle 1_A} & & & & \Big\uparrow{\scriptstyle 1_{A \otimes A^* \otimes A}} \\[2mm]
A & \xrightarrow[\;\simeq\;]{} & A \otimes I & \xrightarrow[\;1_A \otimes \eta_A\;]{} & A \otimes A^* \otimes A
\end{array}
\qquad (16.2)
$$

commute. In a sense, †-compact categories can thus be construed as an abstract axiomatization of the Bell states, familiar in the Hilbert space formalism

$$
\eta_{\mathscr{H}} : \mathbb{C} \to \mathscr{H}^* \otimes \mathscr{H} :: 1 \mapsto \sum_{i \in I} |ii\rangle,
$$

where $\mathscr{H}^*$ is the *conjugate space* to $\mathscr{H}$. This apparently simple axiomatic turns out to generate an amazing amount of the Hilbert space machinery, including the Hilbert–Schmidt inner product, completely positive map and POVMs [3, 8, 11, 35], to mention just a few.

16.2.2   Graphical calculus

The algebraic structure of †-compact categories satisfies exactly those equations that can be proven in its *graphic language*, which we shall now describe. In other words, the morphisms of the free †-compact category can be presented as the well-formed diagrams of this graphic language. Proving such statements, and extracting sound and complete graphic languages for particular categorical varieties, has a long tradition in categorical algebra [28, 24, 25, 21, 22]. Using the deep results about coherent categories in particular [25], Selinger has elegantly derived a succinct coherence argument for the graphic language of †-compact categories in [35, **Thm**. 3.9]. We briefly summarize a version of this graphic language.

The objects of a †-compact category are represented by tuples of wires, whereas the morphisms are the I/O-boxes. Sequential composition connects the output wires of one box with the input wires of the other one. The tensor product is the union of the wires, and it places the boxes next to each other. A physicist-friendly introduction to this graphical language for symmetric monoidal categories is in [10]. The main power of the graphical language lies in its representation of duality. The Bell state (unit) and its adjoint (counit) correspond to a wire from A returning into A^*, with the directions reversed:

Graphically, the composition of η_A and $\varepsilon_A = \eta_{A^*}^\dagger$ as expressed in commutative diagram (16.2) boils down to

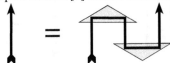

Note that in related papers such as [9] a more involved notation

appears. The triangles witness the fact that in physical terms

respectively stand for a preparations procedure, or state, or *ket*, and for the corresponding *bra*, with an inner-product or *bra-ket*

then yielding a *diamond shaped scalar* (cf. [9]), while the wire itself is now a *loop*. In this paper we will omit these special bipartite triangles.

Given a choice of the duals $A \mapsto A^*$, one can follow the same pattern to define the arrow part $f \mapsto f^*$ of the duality functor $(-)^* : \mathbf{C}^{op} \to \mathbf{C}$ by the commutativity of the following diagram:

$$
\begin{array}{ccccc}
A^* & \xleftarrow{\;\simeq\;} & A^* \otimes I & \xleftarrow{\;1_{A^*} \otimes \eta_{B^*}^\dagger\;} & A^* \otimes B \otimes B^* \\
\big\uparrow{\scriptstyle f^*} & & & & \big\uparrow{\scriptstyle 1_{A^*} \otimes f \otimes 1_{B^*}} \\
B^* & \xrightarrow[\simeq]{} & I \otimes B^* & \xrightarrow[\eta_A \otimes 1_{B^*}]{} & A^* \otimes A \otimes B^*
\end{array}
$$

Replacing $f : A \to B$ by $f^\dagger : B \to A$, we can similarly define $f_* : A^* \to B^*$, and thus extend the duality assignment $A \mapsto A^*$ by the morphism assignment

$f \mapsto f_*$ to the covariant functor $(-)_* : \mathbf{C} \to \mathbf{C}$. It can be shown [2] that the adjoint decomposes in every †-compact category as

$$f^\dagger = (f^*)_* = (f_*)^*$$

with both $(-)^*$ and $(-)_*$ involutive. In finite-dimensional Hilbert spaces and linear maps **FdHilb**, these two functors respectively correspond to *transposition* and *complex conjugation*. The functor $(-)_* : \mathbf{C} \to \mathbf{C}$ will thus be called *conjugation*; the image f_* is a conjugate of f. Graphically, the above diagram defining f^*, and the similar one for f_*, respectively become

The direction of the arrows is, of course, just relative, and we have chosen to direct the arrows down in order to indicate that both f^* and f_* have the duals as their domain and codomain types. We will use horizontal reflection to depict $(-)^\dagger$ and Selinger's 180° rotation [35] to depict $(-)_*$, resulting in:

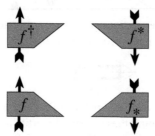

16.2.3   Scalars, trace, and partial transpose

One can prove that the monoid $\mathbf{C}(I,I)$ is always commutative [25] and induces a *scalar multiplication*

$$s \bullet f := \lambda_B^{-1} \circ (s \otimes f) \circ \lambda_A : A \to B$$

which by naturality satisfies

$$(s \bullet f) \circ (t \bullet g) = (s \circ t) \bullet (f \circ g) \quad (s \bullet f) \otimes (t \bullet g) = (s \circ t) \bullet (f \otimes g). \quad (16.3)$$

As already indicated above, we will depict these scalars by diamonds, and such scalars can arise as loops. The equations (16.3) show that these diamonds capturing probabilistic weights can be 'freely moved in the pictures'.

The compact structure of †-compact categories induces the familiar *trace* operation [22],

$$\mathrm{tr}^C_{A,B} : \mathbf{C}(C \otimes A, C \otimes B) \to \mathbf{C}(A,B)$$

which maps $f : C \otimes A \to C \otimes B$ to

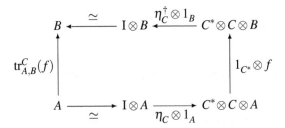

The graphic form of $\mathrm{tr}^C_{A,B}(f)$ is:

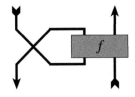

A less familiar operation is *partial transpose*

$$\mathrm{pt}^{C,D}_{A,B} : \mathbf{C}(C \otimes A, D \otimes B) \to \mathbf{C}(D^* \otimes A, C^* \otimes B)$$

which maps $f : C \otimes A \to D \otimes B$

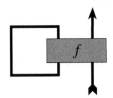

In a picture, $\mathrm{pt}^{C,D}_{A,B}(f)$ is:

Partial transpose can in fact be viewed as an internalisation of the swap actions $\sigma_{C,D} \circ -$ and $- \circ \sigma_{C,D}$, combined with a transposition of the dual space, so that it does not swap two inputs, or two outputs, but an input and an output.

16.3   Sums and bases in Hilbert spaces

To motivate the algebraic and diagrammatic analysis of quantum measurement in the next section, we first discuss some particular aspects of the Hilbert space model of quantum mechanics.

16.3.1   Sums in quantum mechanics

Sums occur in the Hilbert space formalism both as a part of the linear structure of states, as well as a part of their projective (convex) structure, through the fundamental theorem of projective geometry and Gleason's theorem [33]. Viewed categorically, these structures lift, respectively, to a vector space enrichment and a projective space enrichment of operators, typically yielding a C^*-algebra. They appear to be necessary because of the specific nature of quantum measurement, and the resulting quantum probabilistic structure. The additive structure permeates not only states, but also state spaces; it is crucial not only for adding vectors, but also for composing and decomposing spaces. In fact, one verifies that operator sums arise from the direct sum:

$$
\begin{array}{ccc}
\mathbb{C}^{\oplus n} & \xrightarrow{\;\; f+g \;\;} & \mathbb{C}^{\oplus m} \\[2pt]
\Big\downarrow{\scriptstyle d} & & \Big\uparrow{\scriptstyle d^\dagger} \\[2pt]
\mathbb{C}^{\oplus n} \oplus \mathbb{C}^{\oplus n} & \xrightarrow[\;\; f \oplus g \;\;]{} & \mathbb{C}^{\oplus m} \oplus \mathbb{C}^{\oplus m}
\end{array}
$$

where $d :: |i\rangle \mapsto |i\rangle \oplus |i\rangle$ is the additive diagonal. As a particular case we have that the vector sums arise from

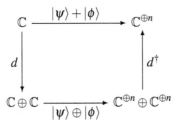

where $|\psi\rangle, |\phi\rangle : \mathbb{C} \to \mathbb{C}^{\oplus n}$, recalling that vectors $|\psi\rangle \in \mathbb{C}^{\oplus n}$ are indeed, by linearity, in bijective correspondence with the linear maps

$$\mathbb{C} \to \mathbb{C}^{\oplus n} :: 1 \mapsto |\psi\rangle .$$

In addition to this, the direct sum canonically also defines bases (cf. the computational base) in terms of the n canonical injections

$$\mathbb{C} \hookrightarrow \mathbb{C}^{\oplus n} :: 1 \mapsto (0, \ldots, 0, 1, 0, \ldots, 0) .$$

16.3.2 No-cloning and existence of a natural diagonal

The classic No-Cloning Theorem [38] states that there exists no unitary operation

$$Clone : \mathcal{H} \otimes \mathcal{H} \to \mathcal{H} \otimes \mathcal{H} :: |\psi\rangle \otimes |0\rangle \mapsto |\psi\rangle \otimes |\psi\rangle . \qquad (16.4)$$

In categorical terms, this means that there is no *natural* diagonal for the Hilbert space tensor product. Formally, a diagonal is a family of linear maps

$$\Delta_{\mathcal{H}} : \mathcal{H} \to \mathcal{H} \otimes \mathcal{H} ,$$

one for each of Hilbert space $\mathcal{H}$. Such a family is said to be natural if for every linear map $f : \mathcal{H} \to \mathcal{H}'$ the diagram

$$
\begin{array}{ccc}
\mathcal{H} & \xrightarrow{\;\;f\;\;} & \mathcal{H}' \\
\Delta_{\mathcal{H}} \downarrow & & \downarrow \Delta_{\mathcal{H}'} \\
\mathcal{H} \otimes \mathcal{H} & \xrightarrow[f \otimes f]{} & \mathcal{H}' \otimes \mathcal{H}'
\end{array}
\qquad (16.5)
$$

commutes. For instance, the family

$$\delta : \mathcal{H} \to \mathcal{H} \otimes \mathcal{H} :: |i\rangle \mapsto |ii\rangle$$

is a diagonal, but the No-Cloning Theorem implies that it cannot be natural. Indeed, it is not hard to see that the above diagram fails to commute, say, for $\mathcal{H} := \mathbb{C}$, $\mathcal{H}' := \mathbb{C} \oplus \mathbb{C}$ and $f : 1 \mapsto |0\rangle + |1\rangle$. In general, given a cloning machine (16.4), one can define a natural diagonal

$$\Delta_{\mathcal{H}} := Clone \circ (- \otimes |0\rangle) : \mathcal{H} \to \mathcal{H} \otimes \mathcal{H} :: |\psi\rangle \mapsto |\psi\rangle \otimes |\psi\rangle .$$

To prove its naturality, note that (16.4) holds for every $|\psi\rangle$, including $|\psi\rangle = |f(\varphi)\rangle$, which gives

$$(\Delta_{\mathcal{H}} \circ f)(|\varphi\rangle) = \Delta_{\mathcal{H}}(|f(\varphi)\rangle) = |f(\varphi)\rangle \otimes |f(\varphi)\rangle = (f \otimes f)(|\varphi\rangle \otimes |\varphi\rangle)$$
$$= (f \otimes f)(\Delta_{\mathcal{H}}(|\varphi\rangle)) = ((f \otimes f) \circ \Delta_{\mathcal{H}})(|\varphi\rangle)$$

which shows that diagram (16.5) commutes.

A diligent reader may have noticed that commutativity of (16.5) actually implies that a diagonal Δ must be independent on the bases, because a change of base can be viewed as just another linear map f (e.g., [10]). In fact, invariance under the base change was one of the original motivations behind the categorical concept of naturality, viz. of natural transformations [29].

16.3.3   Measurement and bases

When diagonalized, self-adjoint operators, which represent measurements in quantum mechanics boil down, modulo a change of base, to two families of data: eigenvalues and eigenvectors. Viewed quantum informatically, eigenvalues are merely token witnesses which discriminate outcomes. A nondegenerate measurement thus essentially corresponds to a base, and a degenerate one can also be captured by a base, and an equivalence relation over it. Taking another look at the map $|i\rangle \overset{\delta}{\mapsto} |ii\rangle$,[5] we see that it *does copy* the base vectors, but *not* other states:

$$|\psi\rangle = \sum_i c_i |i\rangle \overset{\delta}{\mapsto} \sum_i c_i |ii\rangle \neq |\psi\rangle \otimes |\psi\rangle .$$

---

[5]This map, when assigning agents i.e. $|i\rangle_A \overset{\delta}{\mapsto} |i\rangle_A \otimes |i\rangle_B$, has appeared in the literature under the name *coherent bit*, as a 'between classical and quantum'-channel [13, 20]. A more detailed study of this connection can be found in [12].

This map, in fact, exactly captures the base $\{|i\rangle\}_i$, because

$$\delta :: \sum_{i \in I} c_i |i\rangle \mapsto \sum_{i \in I} c_i |ii\rangle,$$

yields a disentangled state if and only if the index set I is a singleton, i.e., if and only if the linear combination $\sum_{i \in I} c_i |i\rangle$ boils down to a base vector. Going in the opposite direction, we can also recover the base as the image of pure tensors under the map[6]

$$\delta^\dagger :: \begin{cases} i \neq j : |ij\rangle \mapsto \vec{o} \\ \\ else \ : |ii\rangle \mapsto |i\rangle \end{cases}.$$

Since the linear diagonal δ thus captures the base, it is of course not independent of the base, and cannot be a natural transformation, in the categorical sense. We shall see that its importance essentially arises from this "unnaturality". Restricted to the base vectors, δ is a 'classical' *copying* operation par excellence; viewed as a linear operation on all of the Hilbert space, it drastically fails naturality tests.

The upshot is that this operation allows us to characterize *classical measurement context* as the domain where it faithfully copies data, with no recourse to an explicit base. If needed, however, the base can be extracted from among the quantum states as consisting of just those vectors that can be copied.

16.3.4 Vanishing of non-diagonal elements and deletion

The map δ also allows capturing the 'formal decohering' in quantum measurement, i.e. the vanishing of the non-diagonal elements in the passage of the initial state represented as a density matrix within the measurement base to the density matrix describing the resulting ensemble of possible outcome states.[7] Indeed, non-diagonal elements get erased setting

$$\delta \circ \delta^\dagger :: \begin{cases} i \neq j : |ij\rangle \mapsto \vec{o} \mapsto \vec{o} \\ \\ else \ : |ii\rangle \mapsto |i\rangle \mapsto |ii\rangle \end{cases}.$$

Note also that δ's adjoint $\delta^\dagger$ doesn't delete classical data, but *compares* its two inputs and only passes on data if they coincide. *Deletion* is

$$\varepsilon :: |i\rangle \mapsto 1 \qquad \text{that is} \qquad 1 \otimes \varepsilon :: |ij\rangle \mapsto |i\rangle.$$

---

[6]This operation $\delta^\dagger : \mathscr{H} \otimes \mathscr{H} \to \mathscr{H}$ has appeared in the quantum informatics literature under the name *fusion*, providing a means for constructing cluster states [5, 37].

[7]See [17] for a discussion why we call this 'formal decohering'.

What ε and $\delta^\dagger$ do have in common is the fact that

$$\delta^\dagger \circ \delta = (1 \otimes \varepsilon) \circ \delta :: |i\rangle \mapsto |ii\rangle \mapsto |i\rangle.$$

Also, since in Dirac notation we have $\delta = \sum_i |ii\rangle\langle i|$, the (base-dependent) isomorphism $\theta :: |i\rangle \mapsto \langle i|$ applied to the *bra* turns δ into the generalized GHZ state $\sum_i |iii\rangle$ [19] exposing that δ is 'up to θ' symmetric in all variables.

16.3.5 Canonical bases

While all Hilbert spaces of the same dimension are obviously isomorphic, they are not all equivalent. Indeed, above we already mentioned that the direct sum structure provides the Hilbert space $\mathbb{C}^{\oplus n}$ with a canonical base, from which it also follows that it is canonically isomorphic to its conjugate space $(\mathbb{C}^{\oplus n})^* = (\mathbb{C}^*)^{\oplus n}$, namely for the isomorphism

$$\mathbb{C}^{\oplus n} \to (\mathbb{C}^*)^{\oplus n} :: (c_1, \ldots, c_n) \mapsto (\bar{c}_1, \ldots, \bar{c}_n).$$

In fact, one should not think of $\mathbb{C}^{\oplus n}$ as just being a Hilbert space, but as the pair consisting of a Hilbert space $\mathscr{H}$ and a base $\{|i\rangle\}_{i=1}^{i=n}$, which by the above discussion boils down to the pair consisting of a Hilbert space $\mathscr{H}$ and a linear map $\delta : \mathscr{H} \to \mathscr{H} \otimes \mathscr{H}$ satisfying certain properties, in particular, its matrix being self-transposed in the canonical base. Below, we will assume the correspondence between $\mathbb{C}^{\oplus n}$ and its dual to be strict, something which can always be established by standard methods. The special status of the objects $\mathbb{C}^{\oplus n}$ in **FdHilb**, in category-theoretic terms, is due to the fact the direct sum is both a product and a coproduct and $\mathbb{C}$ the tensor unit [2].

16.4 Classical objects

Consider a quantum measurement. It takes a quantum state as its input and produces a measurement outcome together with a quantum state, which is typically different from the input state due to the *collapse*. Hence the type of a quantum measurement should be

$$\mathscr{M} : A \to X \otimes A$$

where A is of the type *quantum state* while X is of the type *classical data*. But how do we distinguish between classical and quantum data types?

We will take a very operational view on this matter, and define classical data types as objects which come together with a copying operation

$$\delta_{(X)} : X \to X \otimes X$$

and a deleting operation

$$\varepsilon_{(X)} : X \to I,$$

counterfactually exploiting the fact that such operations do not exist for quantum data. We will refer to these structured objects (X, δ, ε) as *classical objects*. The axioms which we require the morphisms δ and ε to satisfy are motivated by the operational interpretation of δ and ε as copying and deleting operations of classical data. This leads us to introducing the notion of a *special †-compact Frobenius algebra*, which refines the usual topological quantum field theoretic notion of a normalized special Frobenius algebra [26]. The defining equality is due to Carboni and Walters [6].[8]

16.4.1   Special †-compact Frobenius algebras

An *internal monoid* (X, μ, ν) in a monoidal category $(\mathbf{C}, \otimes, I)$ is a pair of morphisms

$$X \otimes X \xrightarrow{\ \mu\ } X \xleftarrow{\ \nu\ } I,$$

called the *multiplication* and the *multiplicative unit*, such that

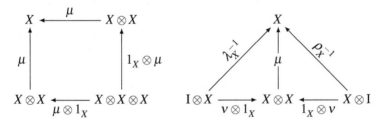

commute. Dually, an *internal comonoid* (X, δ, ε) is a pair of morphisms

$$X \otimes X \xleftarrow{\ \delta\ } X \xrightarrow{\ \varepsilon\ } I,$$

---

[8]They introduced it as a characteristic categorical property of relations. The connection between the work presented in this paper and Carboni and Walters' categories of relations is in [12].

the *comultiplication* and the *comultiplicative unit*, such that

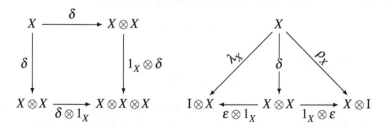

commute. Graphically these conditions are:

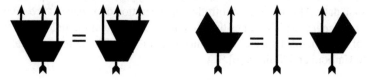

When $(\mathbf{C}, \otimes, I)$ is *symmetric*, the monoid is commutative iff $\mu \circ \sigma_{X,X} = \mu$, and the comonoid is commutative iff $\sigma_{X,X} \circ \delta = \delta$, in a picture:

Note that the conditions defining an internal commutative comonoid are indeed what we expect a copying and deleting operation to satisfy.

A *symmetric Frobenius algebra* is an internal commutative monoid (X, μ, ν) together with an internal commutative comonoid (X, δ, ε) which satisfies

$$\delta \circ \mu = (\mu \otimes 1_X) \circ (1_X \otimes \delta),\tag{16.6}$$

that is, in a picture:

It is moreover *special* iff $\mu \circ \delta = 1_X$, in a picture:

In a symmetric monoidal †-category every internal commutative comonoid (X, δ, ε) also defines an internal commutative monoid $(X, \delta^\dagger, \varepsilon^\dagger)$, yielding a notion of †-*Frobenius algebra* (X, δ, ε) in the obvious manner. In such a †-Frobenius algebra we have:

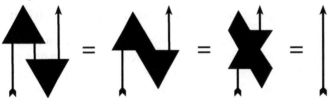

that is, $\delta \circ \varepsilon^\dagger : I \to X \otimes X$ and $\varepsilon \circ \delta^\dagger : X \otimes X \to I$ satisfy equations (16.1) of Section 16.2 and hence canonically provide a unit $\eta = \delta \circ \varepsilon^\dagger$ and counit $\varepsilon = \varepsilon \circ \delta^\dagger$ which realizes $X^* = X$ (cf. Section 16.2). In a picture this choice stands for:

$$\cup = \blacktriangledown$$

One easily verifies that the linear maps δ and ε as defined in the previous section indeed yield an internal comonoid structure on the Hilbert space $\mathbb{C}^{\oplus n}$ which satisfies the *Frobenius identity* (16.6), and that $\delta \circ \varepsilon^\dagger$ is the Bell state.

DEFINITION 16.1   *A* classical object *in a* †-*compact category is a special* †-*compact Frobenius algebra* (X, δ, ε), *i.e., a special* †-*Frobenius algebra for which we choose* $\eta_X = \delta_X \circ \varepsilon_X^\dagger$ *in order to realize* $X^* = X$.

So typical examples of classical objects are the ones existing in **FdHilb** which were implicitly discussed in Section 16.3, namely

$$(\mathbb{C}^{\oplus n}, \delta^{(n)} : \mathbb{C}^{\oplus n} \to \mathbb{C}^{\oplus n} \otimes \mathbb{C}^{\oplus n} :: |i\rangle \mapsto |ii\rangle, \varepsilon^{(n)} : \mathbb{C}^{\oplus n} \to \mathbb{C} :: |i\rangle \mapsto 1).$$

Since the Frobenius identity (16.6) allows us to set $X^* = X$ we can now compare $\delta_*, \delta : X \to X \otimes X$, and also, $\varepsilon_*, \varepsilon : X \to I$, them having the same type. Recalling that in **FdHilb** the covariant functor $(-)_*$ stands for complex conjugation, the structure of a †-compact Frobenius algebra guarantees the highly significant and crucial property that the operations of copying and deleting classical data carry no phase information:

THEOREM 16.1
For a classical object we have $\delta_* = \delta$ *and* $\varepsilon_* = \varepsilon$.

Before we prove this fact we introduce some additional concepts.

16.4.2   Self-adjointness relative to a classical object

From now on we will denote classical objects as X whenever it is clear from the context that we are considering the structured classical data type (X, δ, ε) and not the unstructured quantum data type X. Given a classical object X we call a morphism $\mathscr{F} : A \to X \otimes A$ *self-adjoint relative to X* if the diagram

$$
\begin{array}{ccc}
A & \xrightarrow{\ \ \mathscr{F}\ \ } & X \otimes A \\
{\scriptstyle \lambda_A}\downarrow & & \uparrow{\scriptstyle 1_X \otimes \mathscr{F}^{\dagger}} \\
I \otimes A & \xrightarrow[\ \eta_X \otimes 1_A\ ]{} & X \otimes X \otimes A
\end{array}
\qquad (16.7)
$$

commutes. In a picture, this is:

A morphism $\mathscr{F} : X \otimes A \to A$ is self-adjoint relative to X whenever $\mathscr{F}^{\dagger}$ is. Note furthermore that in every monoidal category, the unit I carries a canonical comonoid structure, with $\delta = \lambda_{\mathrm{I}} = \rho_{\mathrm{I}} : \mathrm{I} \to \mathrm{I} \otimes \mathrm{I}$ and $\varepsilon = 1_{\mathrm{I}} : \mathrm{I} \to \mathrm{I}$. In every †-compact category, this comonoid is in fact a degenerate classical object. Self-adjointness in the usual sense of $f^{\dagger} = f : A \to A$ corresponds to self-adjointness relative to I. For a general classical object X, a morphism $\mathscr{F} : A \to X \otimes A$ can be thought of as an X-indexed family of morphisms of type $A \to A$. Self-adjointness relative to X then means that each of the elements of this indexed family are required to be self-adjoint in the ordinary sense. We abbreviate 'self-adjoint relative to X' to 'X-self-adjoint'. There are several analogous generalizations of standard notions e.g. X-scalar, X-inverse, X-unitarity, X-

idempotence, X-positivity etc. In Section 16.5.1 we discuss a systemic way of defining these 'relative to X'-concepts.

PROPOSITION 16.1

Both the comultiplication δ and the unit ε of a classical object X are always X-self-adjoint, that is, in a picture:

PROOF

Note that X-self-adjointness of ε is exactly $\varepsilon_* = \varepsilon$, already providing part of the proof of Theorem 16.1. In fact, given an internal commutative comonoid (X, δ, ε) diagram (16.7) implicitly stipulates that, of course $X^* = X$, but also that this self-duality of X is realized through $\eta = \delta \circ \varepsilon^\dagger$ since we have

Hence it makes sense to speak of an *X-self-adjoint internal comonoid* in a †-compact category. From X-self-adjointness we can straightforwardly derive many other useful properties, including the Frobenius identity itself, hence providing an alternative characterization of classical objects, and also $\delta_* = \delta$, providing the remainder of the proof of Theorem 16.1.

LEMMA 16.1

The comultiplication of an X-self-adjoint commutative internal monoid satisfies the Frobenius identity (16.6), is partial-transpose-invariant $\mathrm{pt}_{1,X}^{X,X}(\delta) = \delta$, and is self-dual $\delta_* = \delta$ (or $\delta^* = \delta^\dagger$). The latter two depict as:

PROOF   For the Frobenius identity, apply X-self-adjointness to the left hand side, use associativity of the comultiplication, and apply X-self-adjointness again, for partial-transpose-invariance apply X-self-adjointness twice, and for self-duality apply X-self-adjointness three times.   ∎

THEOREM 16.2

A classical object can equivalently be defined as a special X-self-adjoint internal commutative comonoid (X, δ, ε).

16.4.3   GHZ states as classical objects

Analogously to the Hilbert space case (cf. Section 16.3), each classical object X induces an abstract counterpart to generalized GHZ states, namely

$$\mathsf{GHZ}_X := (1_X \otimes \delta) \circ \eta : I \to X \otimes X \otimes X.$$

In a picture that is:

The unit property of the comonoid structure, together with the particular choice for the unit of compact closure $\varepsilon = \varepsilon \circ \delta^\dagger$ becomes pleasingly symmetric:

The same is the case for commutativity of the comonoid structure, together with partial-transpose-invariance:

16.4.4 Extracting the classical world

If **C** comes with a †-structure then any internal comonoid yields an internal monoid. But there is a clear conceptual distinction between the two structures, in the sense that the comultiplication and its unit admit interpretation in terms of copying and deleting. We will be able to extract the classical world by defining *morphisms of classical objects* to be those which preserve the copying and deleting operations of these classical objects, or, in other words, by restricting to those morphisms with respect to which the copying and deleting operations *become natural* (cf. Section 16.3).

Given a †-compact category **C**, we define a new category $\mathbf{C}_\times$ of which the objects are the classical objects and with the morphisms restricted to those which preserve both δ and ε. So there is a forgetful functor

$$\mathbf{C}_\times \to \mathbf{C}.$$

In **FdHilb**, a linear map $f : \mathbb{C}^{\oplus m} \to \mathbb{C}^{\oplus n}$ preserves $\varepsilon^{(n)}$ if it is a 'pseudo stochastic operator' i.e. $\sum_{j=1}^{j=n} f_{ij} = 1$ for all i (note that f_{ij} can still be properly complex), and it preserves $\delta^{(n)}$ if $f_{ij} f_{ij} = f_{ij}$ and $f_{ij} f_{ik} = 0$ for $j \neq k$, hence, there is a function $\varphi : m \to n$ such that

$$f(|i\rangle) = |\varphi(i)\rangle.$$

So **FdHilb**$_\times$ = **FSet**, the latter being the category of finite sets and functions. Hence morphisms in $\mathbf{C}_\times$ are to be conceived as deterministic manipulations of classical data, i.e., while **C** represents the quantum world, $\mathbf{C}_\times$ represents the classical world. The canonical status of $\mathbf{C}_\times$ is exposed by the following result due to Fox [16].

THEOREM 16.3

Let **C** *be a symmetric monoidal category. The category* $\mathbf{C}_\times$ *of its commutative comonoids and corresponding morphisms, with the forgetful functor* $\mathbf{C}_\times \to \mathbf{C}$, *is final among all cartesian categories with a monoidal functor to* **C**, *mapping the cartesian product* $\times$ *to the tensor* $\otimes$.

In fact, there are many other categories of classical operations which can be extracted from **C** using classical object structure, including Carboni and Walters' categories of relations, and categories of (doubly) stochastic maps which, in turns, induce information ordering. For this we refer the reader to [12] and other forthcoming papers.

16.5   Quantum spectra

Given a classical object X, a morphism $\mathscr{F} : A \to X \otimes A$ is *idempotent relative to X*, or shorter, *X-idempotent*, if

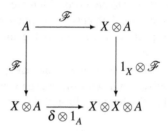

commutes. In a picture that is:

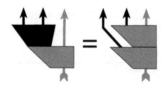

Continuing in the same vein, an *X-projector* is a morphism $\mathscr{P} : A \to X \otimes A$ which is both *X*-self-adjoint and *X*-idempotent. The following proposition shows that an *X*-projector is not just an indexed family of projectors.

PROPOSITION 16.2

A $\mathbb{C}^{\oplus k}$-projector in **FdHilb** *of type $\mathscr{H} \to \mathbb{C}^{\oplus k} \otimes \mathscr{H}$ with $\mathscr{H} \simeq \mathbb{C}^{\oplus n}$ exactly corresponds to a family of k mutually orthogonal projectors $\{P_i\}_{i=1}^{i=k}$, hence we have $\sum_{i=1}^{i=k} P_i \leq 1_{\mathscr{H}}$.*

PROOF   One verifies that from $\mathbb{C}^{\oplus k}$-idempotence follows idempotence $P_i^2 = P_i$ and mutual orthogonality $P_i \circ P_{j \neq i} = \mathbf{0}$, and that from $\mathbb{C}^{\oplus k}$-self-adjointness follows orthogonality of projectors $P_i^\dagger = P_i$.   ∎

DEFINITION 16.2   *A morphism $\mathscr{P} : A \to X \otimes A$ is said to be X-complete if*

$$\lambda_A^\dagger \circ (\varepsilon \otimes 1_A) \circ \mathscr{P} = 1_A .$$

In a picture that is:

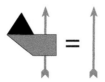

A morphism $\mathscr{P} : A \to X \otimes A$ is a projector-valued spectrum *if it is an X-projector for some classical object X, and if it is moreover X-complete.*

THEOREM 16.4
Projector-valued spectra in **FdHilb** *exactly correspond to complete families of mutually orthogonal projectors* $\{P_i\}_i$, *i.e.* $\sum_{i=1}^{i=k} P_i = 1_{\mathscr{H}}$.

Each classical object (X, δ, ε) canonically induces a projector-valued spectrum $\delta : X \to X \otimes X$ since associativity of the comultiplication coincides with X-idempotence and the defining property of the comultiplicative unit coincides with completeness—the reader should not be confused by the fact that the quantum data type X is now also the carrier of the classical data type (X, δ, ε). Having in mind the characterization of classical objects of Theorem 16.2, mathematically, projector-valued spectra consitute a generalization of classical objects by admitting *degeneracy*.

16.5.1 Coalgebraic characterization of spectra

Recall from [31] that the internal commutative (co)monoid structures over an object X in a monoidal category $\mathbf{C}$ are in one-to-one correspondence with commutative (co)monad structures on the functor

$$X \otimes - : \mathbf{C} \to \mathbf{C}.$$

Hence we can attribute a notion of (co)algebra to internal commutative (co)monoids.

THEOREM 16.5
Let $\mathbf{C}$ *be a* †-*compact category. Its projector-valued spectra are exactly the X-self-adjoint Eilenberg–Moore coalgebras for the comonads* $X \otimes - : \mathbf{C} \to \mathbf{C}$ *canonically induced by some classical object X.*

PROOF The requirements for Eilenberg–Moore coalgebras with respect to the comonad $(X \otimes -)$ are exactly X-idempotence and X-completeness. ∎

We can now rephrase all the above as follows.

THEOREM 16.6

X-self-adjoint coalgebras in **FdHilb** *exactly correspond to complete families of mutually orthogonal projectors* $\{P_i\}_i$.

PROOF   We also rephrase the proof. From the Eilenberg–Moore commuting square we obtain idempotence $P_i^2 = P_i$ and mutual orthogonality $P_i \circ P_{j \neq i} = \mathbf{0}$, from the Eilenberg–Moore commuting triangle we obtain completeness and from X-self-adjointness follows orthogonality of projectors $P_i^\dagger = P_i$.   ∎

16.5.2   Characterization of X-concepts

All 'relative to X'-concepts can now be defined as the corresponding standard concept in the Kleisli category for the comonad $(X \otimes -)$. For example, X-unitarity of a morphism $\mathscr{U} : X \otimes A \to A$ simply means that $\mathscr{U}$ is unitary in the Kleisli category for the comonad $(X \otimes -)$. This approach immediately provides all coherence conditions which are required for these X-concepts to be sound with respect to the categorical structure with which one works. Below we also define X-unitarity in an ad hoc manner for those readers who are not very familiar yet with categorical language.

16.6   Quantum measurements

Given projector-valued spectra we are very close to having an abstract notion of quantum measurement. In fact, the type $A \to X \otimes A$ which we attributed to the spectra is indeed the compositional type of a (*non-demolition*) measurement. But what is even more compelling is the following. The fact that a spectrum is X-idempotent, or equivalently, that it satisfies the coalgebraic

Eilenberg–Moore commuting square, i.e.

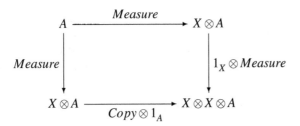

exactly captures *von Neumann's projection postulate*, stating that repeating a measurement is equivalent to copying the data obtained in its first execution. Note here in particular the manifest *resource sensitivity* of this statement, accounting for the fact that two measurements provide two sets of data, even if this data turns out to be identical.

However, what we get in **FdHilb** is not (yet) a quantum measurement. For $A = X := \mathbb{C}^{\oplus n}$ the canonical projector-valued spectrum $\delta^{(n)} : A \to X \otimes A$ expressed in the computational base yields

$$|\psi\rangle = \sum_i \langle i | \psi \rangle | i \rangle_A \longmapsto \sum_i \langle i | \psi \rangle \left(| i \rangle_X \otimes | i \rangle_A \right)$$

where $|i\rangle_X \in X$ is the measurement outcome, $|i\rangle_A \in A$ is the resulting state of the system for that outcome, and the coefficients $\langle i | \psi \rangle$ in the sum capture the respective probabilities for these outcomes i.e. $|\langle i | \psi \rangle|^2$. This however does not reflect the fact that we cannot retain the relative phase factors present in the probability amplitudes $\langle i | \psi \rangle$. In other words, the passage from physics to the semantics is not *fully abstract*. It is moreover well-known that the operation which erases these relative phases does not live in **FdHilb**, but is *quadratic* in the state, hence lives in **CPM(FdHilb)**, the category of Hilbert spaces and completely positive maps.

Fortunately, for many practical purposes (such as those outlined in Sections 16.7 and 16.8 of this paper) this 'approximate' notion of measurement suffices,[9] and in all other cases it turns out that we can rely on Selinger's abstract counterpart for the passage from **FdHilb** to **CPM(FdHilb)**, a construction which applies to any †-compact category [35], to turn those approximate quantum measurements into true quantum measurements.

---

[9] This approximate notion of quantum measurement is also the one considered in [2].

16.6.1 The CPM-construction

This construction takes a †-compact category $\mathbf{C}$ as its input and produces an 'almost inclusion' (it in fact kills redundant global phases) of $\mathbf{C}$ into a bigger one $\mathbf{CPM}(\mathbf{C})$. While $\mathbf{C}$ is to be conceived as containing pure operations with those of type $I \to A$ being the pure states, $\mathbf{CPM}(\mathbf{FdHilb})$ consists of mixed operations with those of type $I \to A$ being the mixed states. Explicitly we have the †-compact functor

$$\text{Pure} : \mathbf{C} \to \mathbf{CPM}(\mathbf{C}) :: f \to f \otimes f_*$$

where

$$\mathbf{CPM}(\mathbf{C})(A,B) := \left\{ (1_B \otimes \eta_{C^*}^\dagger \otimes 1_{B^*}) \circ (f \otimes f_*) \,\Big|\, f : A \to B \otimes C \right\}$$

and the †-compact structure on $\mathbf{CPM}(\mathbf{C})$ covariantly inherits its composition, its tensor, its adjoints and its Bell states from $\mathbf{C}$. In a picture the morphisms of $\mathbf{CPM}(\mathbf{C})$ are:

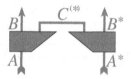

Note in particular that the two copies of each $\mathbf{C}$-morphism in these $\mathbf{CPM}(\mathbf{C})$-morphisms is also present in Dirac's notation when working with density matrices. However, in Dirac notation one considers the pair of a *ket*-vector $|\psi\rangle$ and its adjoint $\langle\psi|$ resulting in the action of an operation being

$$|\psi\rangle\langle\psi| \mapsto f|\psi\rangle\langle\psi|f^\dagger$$

for an ordinary operation, while it becomes

$$|\psi\rangle\langle\psi| \mapsto f(1_C \otimes |\psi\rangle\langle\psi|)f^\dagger$$

for a completely positive map. What we do here is quite similar but now we consider pairs $|\psi\rangle \otimes |\psi\rangle_*$ allowing for more intuitive covariant composition

$$|\psi\rangle \otimes |\psi\rangle_* \mapsto (f \otimes f_*)(|\psi\rangle \otimes |\psi\rangle_*)$$

for an ordinary operation, while it becomes

$$|\psi\rangle \otimes |\psi\rangle_* \mapsto (1_B \otimes \eta_C^\dagger \otimes 1_{B^*})(f \otimes f_*)(|\psi\rangle \otimes |\psi\rangle_*)$$

for a completely positive map. The most important benefit of this covariance is two-dimensional display-ability i.e., it enables graphical calculus.

16.6.2   Formal decoherence

Given a classical object X in a $\dagger$-compact category $\mathbf{C}$ we define the following morphism

$$\Gamma_X := (1_X \otimes \eta^\dagger \otimes 1_X) \circ (\delta \otimes \delta) \in \mathbf{CPM}(\mathbf{C})(X,X).$$

In a picture that is:

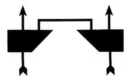

PROPOSITION 16.3

In $\dagger$-compact category with X a classical object we have

$$\Gamma_X = \delta \circ \delta^\dagger : X \otimes X \to X \otimes X$$

so in particular is Γ_X idempotent.

PROOF   Using the Frobenius identity we have

where the highlighted part expresses the use of X-self-adjointness.   ∎

In particular in **FdHilb** we have

$$\sum_{ij} \alpha_i \bar{\alpha}_j |i\rangle \otimes |j\rangle_*$$

$$\delta^{(k)} \overset{\top}{\otimes} \delta^{(k)} \Bigg\downarrow$$

$$\sum_{ij} \alpha_i \bar{\alpha}_j |ii\rangle \otimes |jj\rangle_*$$

$$1_{\mathbb{C}^{\oplus n}} \otimes \eta^{\dagger}_{\mathbb{C}^{\oplus n}} \otimes 1_{\mathbb{C}^{\oplus n}} \Bigg\downarrow$$

$$\Gamma_{\mathbb{C}^{\oplus n}}$$

$$\sum_{ij} \delta_{ij} \alpha_i \bar{\alpha}_j |i\rangle \otimes |j\rangle_* =\!=\!=\!=\!=\!=\!=\!=\!= \sum_i \alpha_i \bar{\alpha}_i |i\rangle \otimes |i\rangle_*$$

i.e. we obtain the desired effect of elimination of the relative phases. Hence, given a projector-valued spectrum now represented in **CPM(C)** through the functor Pure, which depicts as

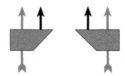

we obtain a genuine quantum measurement by adjoining Γ_X as in

$$Meas := (1_B \otimes \Gamma_X \otimes 1_B) \circ (\mathcal{M} \otimes \mathcal{M}_*),$$

which in a picture becomes:

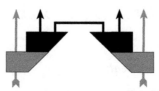

16.6.3   Demolition measurements

As compared to the type $A \to X \otimes A$ of a non-demolition measurement, a *demolition measurement* has type $A \to X$. We claim that the demolition analogue to a projector-valued spectrum $\mathcal{M} : A \to X \otimes A$ is the adjoint to an *isometry* $m^{\dagger} : X \to A$, i.e. $m \circ m^{\dagger} = 1_X$—or equivalently put in our X-jargon, a *normalized X-bra*. Indeed, setting

$$\mathcal{M}_m := (1_X \otimes m^{\dagger}) \circ \delta \circ m : A \to X \otimes A$$

we exactly obtain a projector-valued spectrum since $\mathcal{M}_m$ is trivially X-self-adjoint, and $m \circ m^\dagger = 1_X$ yields X-idempotence. In a picture $\mathcal{M}_m$ is:

The corresponding demolition measurement arises by adjoining Γ_X i.e.

$$DeMeas := \Gamma_X \circ (m \otimes m_*),$$

that is, in a picture:

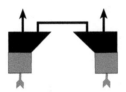

Such a demolition measurement is *non-degenerate* iff m is unitary.

16.7 Quantum teleportation

The notion of measurements proposed in this paper abstracts over the structure of classical data, and we will show that we can describe and prove correctness of the teleportation protocol without making the classical data structure explicit, nor by relying on the cartesian structure of $\mathbf{C}_\times$.

DEFINITION 16.3 *Given a classical object X a morphism $\mathcal{U}$: $X \otimes A \to B$, and at the same time $\mathcal{U}^\dagger$ and $\mathcal{U} \circ \sigma_{X,A}$, are* unitary relative to X *or* X-unitary *iff*

$$(1_X \otimes \mathcal{U}) \circ (\delta \otimes 1_A) : X \otimes A \to X \otimes B$$

is unitary in the usual sense i.e., its adjoint is its inverse. In a picture:

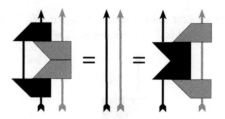

A trivial example of such a unitary morphism is $\varepsilon \otimes 1_A : X \otimes A \to A$.

PROPOSITION 16.4

In **FdHilb** *morphisms that are* $(\mathbb{C}^{\oplus n}, \delta^{(n)})$*-unitary are in bijective correspondence with n-tuples unitary operators of the same type.*

Let the *size of a classical object* be the scalar

$$s_X := \eta_X^\dagger \circ \eta_X = \varepsilon_X \circ \varepsilon_X^\dagger : I \to I$$

i.e., in a picture:

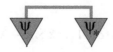

using in the last two steps respectively $\delta^\dagger \circ \delta = 1_X$ and $\eta = \delta \circ \varepsilon^\dagger$.

PROPOSITION 16.5

The positive scalars in the scalar monoid $\mathbf{C}(I, I)$*, i.e. those scalars* $s : I \to I$ *that can be written as* $s = \psi^\dagger \circ \psi$ *for some* $\psi : I \to A$*, have self-adjoint square-roots when embedded in* **CPM(C)** *via* Pure.

PROOF   The image of a positive scalar s under Pure is $s \otimes s_*$. For $t = \eta_{A^*}^\dagger \circ (\psi \otimes \psi_*) \in \mathbf{CPM(C)}(I, I)$ which we depict in a picture as:

we have $t \circ t = s \otimes s_*$ since

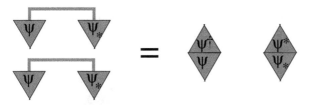

follows from $(f^* \otimes 1_B) \circ \eta_B = (1_A \otimes f) \circ \eta_A$ [2]. Self-adjointness follows from:

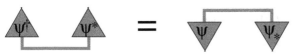

Hence $\mathrm{Pure}(s)$ indeed has a scalar in $\mathbf{CPM}(\mathbf{C})(\mathrm{I},\mathrm{I})$ as a square root. ∎

This implies that square root $\sqrt{s_A} : \mathrm{I} \to \mathrm{I}$ of the *dimension* $s_A := \eta_A^\dagger \circ \eta_A$ of an object A always exists whenever we are within $\mathbf{CPM}(\mathbf{C})$. It can be shown that each †-compact category also admits a canonical embedding in another †-compact category in which all scalars have inverses. For scalars

$$s_A \qquad \sqrt{s_A} \qquad \frac{1}{s_A} \qquad \frac{1}{\sqrt{s_A}},$$

respectively, we introduce the following graphical notations:

—the reversed symbols representing inverses needn't be confused with the adjoint since these scalar dimensions are always self-adjoint.

DEFINITION 16.4 *Let X be a classical object in a †-compact category. A (non-degenerate) demolition Bell measurement is a unitary morphism*

$$DeMeas_{Bell} := \frac{1}{\sqrt{s_A}} \bullet \rho_A^\dagger \circ (1_X \otimes \eta_A^\dagger) \circ (\mathscr{U}^\dagger \otimes 1_A) : A \otimes A^* \to X$$

which is such that $\mathscr{U} : X \otimes A \to A$ is X-unitary.

The corresponding projector-valued spectrum is

$$\mathscr{M}_{Bell} := (DeMeas_{Bell}^\dagger \otimes 1_X) \circ \delta \circ DeMeas_{Bell} : A \otimes A^* \to X \otimes A \otimes A^*,$$

from which the corresponding non-demolition Bell measurement arises by ad-joining Γ_X. In a picture the demolition Bell measurement and corresponding projector-valued spectrum are:

and unitarity of $DeMeas_{Bell}$ is:

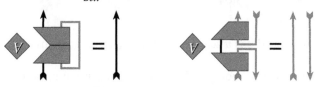

that is, in formulae, respectively,

$$\frac{1}{\sqrt{s_A}} \bullet \operatorname{tr}^A_{X,X}(\mathscr{U}^\dagger \circ \mathscr{U}) = DeMeas_{Bell} \circ DeMeas^\dagger_{Bell} = 1_X \qquad (16.8)$$

and of course

$$DeMeas^\dagger_{Bell} \circ DeMeas_{Bell} = 1_{A \otimes A^*}. \qquad (16.9)$$

Let us verify that $\mathscr{M}_{Bell}$ is indeed a projector-valued spectrum. Using Eqn. (16.8) we obtain X-idempotence:

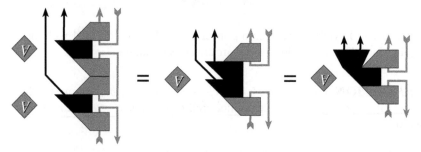

and Eqn. (16.9) assures X-completeness:

Finally, unitarity of $DeMeas_{Bell}$ yields $X \simeq A \otimes A^*$, so $DeMeas_{Bell}$ can be conceived as *non-degenerate*. We normalize the *Bell states* of type A, i.e.,

$$\frac{1}{\sqrt{s_A}} \bullet \eta_A : I \to A^* \otimes A.$$

Now we will describe the teleportation protocol and prove its correctness. For simplicity we will not explicitly depict Γ_X since it doesn't play an essential role in the topological manipulations of the picture.[10] Here it is:

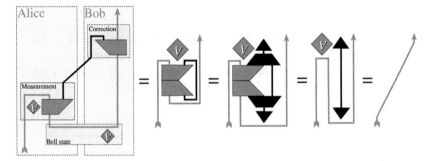

The red box in `Measurement` is a unitary morphism

$$\sigma_{X,A} \circ \mathcal{U}_* : A^* \to A^* \otimes X,$$

which defines a demolition Bell-base measurement, the red box in `Correction` is the unitary morphism

$$\mathcal{U} : A \to X \otimes A,$$

and the bottom red box in the second picture obtained by 'sliding' $\mathcal{U}$ along the red line is

$$\mathcal{U}^* \circ \sigma_{A,X} : A^* \otimes X \to A^*,$$

the adjoint to $\sigma_{X,A} \circ \mathcal{U}_*$—note that the σ-isomorphisms are introduced to avoid crossing of lines. Then we apply the decomposition $\eta := \delta \circ \varepsilon^\dagger$ which enables us to use X-unitarity of $\mathcal{U}$. The reason why the black and the red scalars cancel out requires considering Γ_X as part of the measurement—we refer the reader to [12] for details. We could *copy* the measurement outcome before *consuming* it:

---

[10]A case where Γ_X does play a crucial role is the proof of Naimark's theorem in [11].

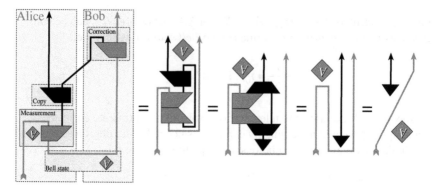

Note that now we explicitly used X-self-adjointness. We can of course still choose to delete this data at a later stage using ε, nicely illustrating *resource sensitivity*. If we wish to use the resulting available classical data for other purposes we possibly now might have to introduce Γ_X explicitly.

Categorically we can fully specify this protocol as[11]

$$
\begin{array}{ccccc}
A & \xrightarrow{\;(1_A \otimes \eta_A)\circ\rho_A\;} & A\otimes A^*\otimes A & \xrightarrow{\;DeMeas_{Bell}\otimes 1_A\;} & X\otimes A \\[2pt]
{\scriptstyle \varepsilon^\dagger \otimes 1_A}\Big\downarrow & & & & \Big\| \\[2pt]
X\otimes A & \xleftarrow[\;1_X \otimes ((\eta_X \otimes 1_A)\circ(1_X \otimes \mathscr{U}^\dagger))\;]{} & X\otimes X\otimes A & \xleftarrow{\;\delta\otimes 1_A\;} & X\otimes A
\end{array}
$$

The morphism $\varepsilon^\dagger \otimes 1_A$ together with commutation of this diagram specifies the intended behavior, i.e., teleporting a state of type A with the creation of classical data as a biproduct, while the other morphisms respectively are: (i) creation of a Bell state η_A; (ii) a demolition Bell-base measurement $DeMeas_{Bell}$; (iii) copying of classical data using δ; (iv) unitary correction using the X-adjoint to $\mathscr{U}$. The above depicted graphical proof can be converted into an explicit category-theoretic one.

16.8   Dense coding

We can also give a similar description and proof of dense coding.

---

[11]The first specification of quantum teleportation as a commutative diagram, together with a purely categorical correctness proof, is due to Abramsky and one of the authors [2]. However, their work relied heavily on the 'unphysical' assumption of biproducts to establish this—see [8] for a discussion of this issue.

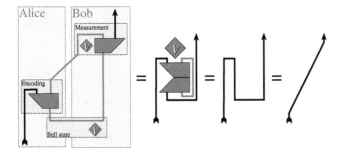

The remarks made above concerning Γ_X apply again here. Note in particular that we rely on a very different property in this derivation than in the derivation of teleportation: here we use (one-sided) unitarity of $DeMeas_{Bell}$ while for teleportation we use X-unitarity of $\mathcal{U}$. Hence it follows that teleportation and dense coding are not as closely related as one usually thinks: they are in fact *axiomatically independent*.

For a more systematic and more elaborate presentation of classical objects as the structure of classical data, together with several more quantum protocols, we refer the reader to [12].

Note and acknowledgements

An earlier version of this chapter has been in circulation since November 2005 with the title *Quantum measurements as coalgebras*. This work was supported from EPSRC grant EP/C500032/1 entitled *High-level methods in quantum computation and quantum information* and NSF project 0209004 entitled *Coalgebraic methods for embedded and hybrid systems*. Samson Abramsky, Dan Browne, Bill Edwards, Eric Oliver Paquette, Gordon Plotkin, Peter Selinger and Frank Valckenborgh provided useful comments. We used Paul Taylor's 2006 package to generate commutative diagrams.

References

[1] Abramsky, S. (2005) *Abstract scalars, loops, free traced and strongly compact closed categories.* In: *Proceedings of CALCO 2005*, pp. 1–31, Springer Lecture Notes in Computer Science **3629**.

[2] Abramsky, S. and Coecke, B. (2004) *A categorical semantics of quantum protocols.* Proceedings of the 19th Annual IEEE Symposium on Logic in Computer Science, pp. 415–425, IEEE Computer Science Press. arXiv:quant-ph/0402130

[3] Abramsky, S. and Coecke, B. (2005) *Abstract physical traces.* Theory and Applications of Categories **14**, 111–124. www.tac.mta.ca/tac/volumes/14/6/14-06abs.html

[4] Baez, J. (2004) *Quantum quandaries: a category-theoretic perspective.* In: *Structural Foundations of Quantum Gravity*, Oxford University Press. arXiv:quant-ph/0404040

[5] Browne, D. E. and Rudolph, T. (2005) *Resource-efficient linear optical quantum computation.* Physical Review Letters **95**, 010501.

[6] Carboni, A. and Walters, R. F. C. (1987) *Cartesian bicategories* I. Journal of Pure and Applied Algebra **49**, 11–32.

[7] Coecke, B. (2005) *Quantum information-flow, concretely, and axiomatically.* In: *Proceedings of Quantum Informatics 2004*, pp. 15–29, Proceedings of SPIE Vol. 5833. arXiv:quant-ph/0506132

[8] Coecke, B. (2005) *De-linearizing linearity: projective quantum axiomatics from strong compact closure.* Electronic Notes in Theoretical Computer Science, to appear. arXiv:quant-ph/0506134

[9] Coecke, B. (2005) *Kindergarten quantum mechanics—lecture notes.* In: *Quantum Theory: Reconsiderations of the Foundations* III, pp. 81–98, AIP Press. arXiv:quant-ph/0510032

[10] Coecke, B. (2006) *Introducing categories to the practicing physicist.* In: *What is Category Theory?* Advanced Studies in Mathematics and Logic **30**, pp.45–74, Polimetrica Publishing.

[11] Coecke, B. and Paquette, E. O. (2006) *POVMs and Naimark's theorem without sums*. Electronic Notes in Theoretical Computer Science, to appear.

[12] Coecke, B., Paquette, E. O. and Pavlovic, D. (2007) *Classical and quantum structures*. Preprint.

[13] Devetak, I., Harrow, A. W. and Winter, A. (2004) *A family of quantum protocols*. Physical Review Letters **93**, 230504. arXiv:quant-ph/0307091

[14] Dirac, P. A. M. (1947) *The Principles of Quantum Mechanics*, 3rd edition. Oxford University Press.

[15] Duncan, R. (2007) *Types for Quantum Computing*. D.Phil. thesis, Oxford University Computing Laboratory.

[16] Fox, T. (1976) *Coalgebras and cartesian categories*. Communications in Algebra **4**, 665–667.

[17] Gisin, N. and Piron, C. (1981) *Collapse of the wave packet without mixture*. Letters on Mathematical Physics **5**, 379–385.

[18] Greenberger, D. M., Horne, M. A. and Zeilinger, A. (1989) *Going beyond Bell's theorem*. In: Bell's theorem, quantum theory, and conceptions (ed. Kafatos, M.).

[19] Greenberger, D. M., Horne, M. A., Shimony, A. and Zeilinger, A. (1990) *Bell's theorem without inequalities*. American Journal of Physics **58**, 1131–1143.

[20] Harrow, A. (2004) *Coherent communication of classical messages*. Physical Review Letters **92**, 097902. arXiv:quant-ph/0307091

[21] Joyal, A. and Street, R. (1991) *The Geometry of tensor calculus* I. Advances in Mathematics **88**, 55–112.

[22] Joyal, A., Street, R. and Verity, D. (1996) *Traced monoidal categories*. Proceedings of the Cambridge Philosophical Society **119**, 447–468.

[23] Kauffman, L. H. (2005) *Teleportation topology*. Optics and Spectroscopy **99**, 227–232. arXiv:quant-ph/0407224

[24] Kelly, G. M. (1972) *Many-variable functorial calculus* I. In: *Coherence in Categories*, pp.66–105, G. M. Kelly, M. Laplaza, G. Lewis and S. Mac Lane, Eds., Lecture Notes in Mathematics **281**, Springer.

[25] Kelly, G. M. and Laplaza, M. L. (1980) *Coherence for compact closed categories.* Journal of Pure and Applied Algebra **19**, 193–213.

[26] Kock, J. (2003) *Frobenius Algebras and 2D Topological Quantum Field Theories.* Cambridge University Press.

[27] Lauda, A. T. (2005) *Frobenius algebras and planar open string topological field theories.* arXiv:math.QA/0508349

[28] Mac Lane, S. (1963) *Natural associativity and commutativity.* Rice Univ. Studies **49**, 28–46.

[29] Mac Lane, S. (1971) *Categories for the Working Mathematician.* Springer–Verlag.

[30] Pati, A. K. and Braunstein, S. L. (2000) *Impossibility of deleting an unknown quantum state.* Nature **404**, 164–165.

[31] Pavlovic, D. (1997) *Categorical logic of names and abstraction in action calculus.* Mathematical Structures in Computer Science **7**, 619–637.

[32] Penrose, R. (1971) *Applications of negative dimensional tensors.* In: *Combinatorial Mathematics and its Applications*, pp. 221–244, Academic Press.

[33] Piron, C. (1976) *Foundations of Quantum Physics.* W. A. Benjamin.

[34] Rédei, M. (1997) *Why John von Neumann did not like the Hilbert space formalism of quantum mechanics (and what he liked instead).* Studies in History and Philosophy of Modern Physics **27**, 493–510.

[35] Selinger, P. (2005) *Dagger compact closed categories and completely positive maps.* Electronic Notes in Theoretical Computer Science, to appear.

[36] Street, R. (2004) *Frobenius monads and pseudomonads.* Journal of Mathematical Physics **45**, 3930–3948.

[37] Verstraete, F. and Cirac, J. I. (2004) *Valence-bond states for quantum computation.* Physical Review A **70**, 060302(R).

[38] Wootters, W. and Zurek, W. (1982) *A single quantum cannot be cloned.* Nature **299**, 802–803.

Appendix

Panel Report on the Forward Looking Discussion that took place during the NSF Conference on the Mathematics of Quantum Computation and Quantum Technology held at Texas A&M University, November 13–16, 2005, written by Samuel Lomonaco with cooperation of and input from the panelists and the panel discussion participants.

On Tuesday, November 15, 2006 at the NSF Conference on the Mathematics of Quantum Computation and Quantum Technology held at Texas A&M University, a two-hour panel discussion was held, chaired by Professor Samuel Lomonaco of the Department of Computer Science and Electrical Engineering at the University of Maryland Baltimore County (UMBC). The three other panelists were Professor Goong Chen of Texas A&M University (TAMU), Professor Louis Kauffman of the University of Illinois at Chicago (UIC), and Dr. Howard Brandt of Army Research Laboratory (ARL). Approximately thirty-five conference participants attended and participated in the panel discussion. The lively and enthusiastic discussion that transpired focused on the topic of the future research directions and opportunities to be found in quantum computing.

The panel discussion essentially centered on the four basic questions given below. Accordingly, this report is organized as a summary of the participants' responses to these questions.

Question 1. Why quantum Computing? Why bother?

Within the next 10 to 15 years, the US computer industry will be facing one of its greatest challenges, the end of Moore's law. Suddenly and abruptly, as the size of the components of small scale integration approaches the quantum level, the computer industry will find that the requirements of the laws of physics will have dramatically changed to those of the microscopic quantum world. No longer will the industry be able to double the computing power of computing devices every 1.5 years at half the price. This is an inevitable and looming future economic downturn now facing the computer industry. It can only be averted if there is enough foresight to substantially invest in the basic research necessary to develop new science and technology for overcoming this fundamental technological barrier.

This is one among many reasons for the fundamental and central importance of quantum computing. Quantum computing shows the great promise of making it possible for the computer industry to leap far beyond the demise of Moore's law to a domain of exponentially faster computing. This potential for quantum computing is emphatically confirmed by Simon's quantum algorithm, which Simon has rigorously proven to be exponentially faster than any possible classical algorithms. This is reconfirmed by Shor's quantum factoring algorithm which is exponentially faster than any known classical factoring algorithm.

One more reason for the central importance of quantum computing is that this field is using the theory of information as a fundamental tool for probing and exploring the boundaries of quantum mechanics. With quantum computing, researchers are seeking to find answers to fundamental and central questions about the microscopic quantum world.

Even if we assume, just for the sake of argument, that future research in quantum computing will fail to deliver a practical scalable quantum computing device, the inevitable scientific and technological spin-offs from this research will no doubt lead to new technologies in computing and in areas other than computing. For, by learning how to control systems at the quantum level, we will most certainly produce fundamentally new useful and practical technologies for industry, such as for example, nano-technology.

Question 2. What are the obstacles to achieving the promise of quantum computing? Can these obstacles be overcome so that this promise will be fulfilled?

But will the potential of quantum computing actually be realized? Will future quantum computers be general purpose or special purpose computing devices?

Much research must be done before these questions can be answered. The obstacles to scalable quantum computing are monumental, and a rewarding challenge for the best scientific minds.

The major and fundamental obstacle to quantum computing is quantum decoherence. Quantum systems simply do not want to be isolated. But instead, they seek to quickly become quantum entangled with their environment. The more a quantum system entangles with its environment, the more it appears to one observing ONLY the quantum system, that the quantum system becomes noisy and classically random (i.e., loses quantum coherence), and hence, becomes uncontrollable. By this process, the qubits appear to the observer to be degenerating into random classical bits or decaying. This phenomenon is called quantum decoherence.

Much research needs to be done to overcome the obstacle of decoherence. Some of the possible approaches to overcoming decoherence are:

- Quantum error correcting codes, i.e., adding additional redundant qubits for correcting the effects of decoherence.

- Decoherence free subspaces, i.e., creating for the system an environment with symmetries that can be exploited to preserve quantum coherence, such as for example those created in NMR systems.

- Topological quantum computing, i.e., creating quantum systems (e.g., fractional quantum Hall effect systems) with natural global barriers to decoherence (called topological obstructions). The barrier to decoherence in those systems is, for example, very much analogous to the topological obstruction in a doughnut which prevents one from shrinking a longitudinal loop to a point without letting the loop leave the doughnut. The doughnut hole (which is the topological obstruction) makes that impossible.

- Bang-bang control of quantum systems, i.e., zapping a quantum system with a strong calculated laser pulse that reverses the decoherence dynamics.

- Distributed quantum computing, i.e., distributing the computational task to small computing devices interconnected by EPR channels. This is a divide-and-conquer approach that protects against the most probable effects of decoherence.

- Quantum Zeno effect, i.e., preserving quantum states by making appropriate frequent measurements.

Another major obstacle to fulfilling the promise of quantum computing is the current scarcity of quantum algorithms. We simply have not yet found enough quantum algorithms to determine whether or not future quantum computers will be general purpose or special purpose computing devices. More research is crucially needed to determine the algorithmic limits of quantum computing.

Question 3. Why should mathematicians work on quantum computing? Why are they needed? Isn't this simply a research field for physicists and engineers?

Quantum computing is a multidisciplinary research field requiring the team efforts of the best minds in mathematics, as well as in computer science, physics, electrical engineering, chemistry, and in many other fields. It is indeed shortsighted and unwise to call this a research field only for physicists and engineers.

Mathematicians are critically needed to solve fundamental research problems arising in quantum computing in, e.g., the following mathematical fields:

- Group representation theory

- Lie theory

- Invariant theory

- Differential geometry

- Mathematical physics

- Mathematical modeling of quantum devices

- Control theory

- Partial differential equations

- Quantum topology

- Knot theory

- Algebraic topology

- Algebraic geometry

- Quantum information theory

Question 4. What kind of support is needed to help mathematicians to continue working in this field? What kind of support is needed to encourage mathematicians to start working in this field?

More NSF financial support is needed for mathematicians to work in this field. Currently, because of a lack of sufficient support in the United States, many US researchers (especially young researchers doing more mathematically oriented work) seeking to continue their work in quantum computing are forced to look for job opportunities outside the United States.

Given below is a list of suggestions and proposals made during the panel discussion for the consideration of the National Science Foundation:

- We suggest that NSF create a grant program to fund mathematicians to team with physicists, electrical engineers, computer scientists, and other scientists for the purpose of collectively and synergistically carrying out research in quantum computing.

- We propose that NSF consider creating more funding for interdisciplinary NSF workshops focused on the mathematical problems arising in quantum computing. Such workshops should consist of an appropriate mix of tutorial and research talks, and of domestic and international quantum researchers from many disciplines.

- To encourage interdisciplinary research, we propose that the NSF consider creating an NSF postdoctoral program for young mathematicians (who are just beginning their careers) to team and to work with senior quantum computing researchers in other fields such as, e.g., physics, computer science, electrical engineering.

- And vice versa, we propose that the NSF also consider creating a postdoctoral program for young researchers in disciplines other than mathematics to team with established senior mathematicians working in quantum computing.

- We also propose that the NSF consider creating a program to fund multidisciplinary team-building workshops in quantum computing.

- We also suggest increased NSF funding for the creation of new instructional materials and courses in quantum computing for the training of future mathematicians

- For the purpose of encouraging international cooperation, we propose that the NSF consider increasing NSF funding for US and international quantum computing researchers to work together.

Index

Milton Keynes UK
Ingram Content Group UK Ltd.
UKHW021932071024
449327UK00022B/1780